High Performance Computing in Science and Engineering ’21

Wolfgang E. Nagel • Dietmar H. Kröner
Michael M. Resch
Editors

High Performance Computing in Science and Engineering '21

Transactions of the High Performance Computing Center, Stuttgart (HLRS) 2021

Editors
Wolfgang E. Nagel
Zentrum für Informationsdienste
und Hochleistungsrechnen (ZIH)
Technische Universität Dresden
Dresden, Germany

Dietmar H. Kröner
Abteilung für Angewandte
Mathematik
Universität Freiburg
Freiburg, Germany

Michael M. Resch
Höchstleistungsrechenzentrum
Stuttgart (HLRS)
Universität Stuttgart
Stuttgart, Germany

ISBN 978-3-031-17939-6 ISBN 978-3-031-17937-2 (eBook)
https://doi.org/10.1007/978-3-031-17937-2

Mathematics Subject Classification (2020): 65Cxx, 65C99, 68U20, 90-08, 97M50

Cover illustration: The image shows the resulting polycrystalline microstructure of a martensitic phase transformation predicted by a multiphase-field simulation performed with PACE3D. The regions shown in shades of gray represent retained austenite, while the regions depicted in color are martensitic variants. Details can be found in "High-performance multiphase-field simulations of solid-state phase transformations using PACE3D" by E. Schoof, T. Mittnacht, M. Seiz, P. Hoffrogge, H. Hierl, and B. Nestler, Institute of Applied Materials (IAM), Karlsruhe Institute of Technology (KIT), Straße am Forum 7, 76131 Karlsruhe, Germany, on pages 167ff.

This Springer imprint is published by the registered company Springer Nature Switzerland AG
The registered company address is: Gewerbestrasse 11, 6330 Cham, Switzerland

Contents

Physics

In this part, three physics projects are presented, which achieved important scientific results in 2020/21 by using Hawk/Hazel Hen at the HLRS and ForHLR II of the Steinbuch Center.

Fascinating new results are being presented in the following pages on soft matter/biochemical systems (ligand-induced protein stabilization) and on quantum systems (anomalous magnetic moment of the muon, ultracold-boson quantum simulators, phase transitions, resonant tunneling, and variances).

Studies of the soft matter/biochemical systems have focused on ligand-induced protein stabilization.

T. Schäfer, A.C. Joerger, J. Spencer, F. Schmid, and G. Settanni from Mainz (T.S., F.S., G.S.), Frankfurt (A.C.J.), and Sussex (J.S.) present interesting new results on ligand-induced protein stabilization in their project *Flexadfg*. The authors show how Molecular Dynamics simulations of several cancer mutants of the DNA-binding domain of the tumor suppressor protein p53 allowed to establish the destabilizing effect of the mutations as well as the stabilizing effects of bound ligands. In addition, the authors report on the development of a new reweighting technique for metadynamics simulations that speeds up convergence and may provide an advantage in the case of simulations of large systems.

Studies of the quantum systems have focused on the anomalous magnetic moment of the muon, and on ultracold-boson quantum simulators, phase transitions, resonant tunneling, and variances.

M. Cè, E. Chao, A. Gérardin, J.R. Green, G. von Hippel, B. Hörz, R.J. Hudspith, H.B. Meyer, K. Miura, D. Mohler, K. Ottnad, S. Paul, A. Risch, T. San José, and H. Wittig from Mainz (E.C., G.v.H., R.J.H., H.B.M., K.M., K.O., S.P., T.S.J., H.W.), Darmstadt (K.M., D.M., T.S.J.), Zeuthen (A.R.), Geneva (M.C., J.R.G.), Marseille (A.G.), and Berkeley (B.H.) present interesting results obtained by their lattice QCD Monte Carlo simulations on Hawk/Hazel Hen in their project *GCS-HQCD* on leading hadronic contributions to the anomalous magnetic moment of the muon, on the energy dependence of the electromagnetic coupling, on the electroweak mixing angle, and on the hadronic vacuum polarisation and light-by-light scattering contributions.

The authors focus will turn to increasing the overall precision of their determination of the hadronic vacuum polarization contribution to the muon anomalous magnetic moment to the sub-percent level.

A.U.J. Lode, O.E. Alon, J. Arnold, A. Bhowmik, M. Büttner, L.S. Cederbaum, B. Chatterjee, R. Chitra, S. Dutta, C. Georges, A. Hemmerich, H. Keßler, J. Klinder, C. Lévêque, R. Lin, P. Molignini, F. Schäfer, J. Schmiedmayer, and M. Žonda from Freiburg (A.U.J.L., M.B.), Haifa (O.E.A., A.B., S.D.), Basel (J.A., F.S.), Heidelberg (L.S.C.), Kanpur (B.C.), Zürich (R.C., R.L.), Hamburg (C.G., A.H., H.K., J.K.), Wien (C.L., J.S.), Oxford (P.M.), and Prague (M.Z.) present interesting results obtained in their project *MCTDHB* with their multiconfigurational time-dependent Hartree method for indistinguishable particles (MCTDH-X) on Hazel Hen and Hawk. In the past the authors have implemented their method to solve the many-particle Schrödinger equation for time-dependent and time-independent systems in various software packages. The authors present interesting new results of their investigations on ultracold boson quantum simulators for crystallization and superconductors in a magnetic field, on phase transitions of ultracold bosons interacting with a cavity, and of charged fermions in lattices described by the Falicov–Kimball model. In addition, the authors report on new results on the many-body dynamics of tunneling and variances, in two- and three-dimensional ultracold-boson systems.

Fachbereich Physik,
Universität Konstanz,
78457 Konstanz,
Germany,
e-mail: peter.nielaba@uni-konstanz.de

Peter Nielaba

Ligand-induced protein stabilization and enhanced molecular dynamics sampling techniques

Timo Schäfer, Andreas C. Joerger, John Spencer, Friederike Schmid and Giovanni Settanni

Abstract Molecular dynamics (MD) simulations provide an increasingly important instrument to study protein-materials interaction phenomena, thanks to both the constant improvement of the available computational resources and the refinement of the modeling methods. Here, we summarize the results obtained along two different research directions within our project. First, we show how MD simulations of several cancer mutants of the DNA-binding domain of the tumor suppressor protein p53 allowed to establish the destabilizing effect of the mutations as well as the stabilizing effects of bound ligands. Second, we report on the development of a new reweighting technique for metadynamics simulations that speeds up convergence and may provide an advantage in the case of simulation of large systems.

1 Introduction

Molecular dynamics (MD) simulations provide a way to observe the motions of molecular objects at the atomic scale. In classical MD simulations, each system is composed of a set of particles and is represented by a Hamiltonian energy function of the coordinates and momenta of the particles. By numerical integration of the Hamiltonian equations of motion, MD allows to follow the trajectories of all the particles as a function of time and, in this way, to extrapolate its static and dynamic properties. A crucial step to guarantee the accuracy of the observations is the

Timo Schäfer, Friederike Schmid and Giovanni Settanni
Department of Physics, Johannes Gutenberg University, Mainz, Germany,
e-mail: settanni@uni-mainz.de

Andreas C. Joerger
Institute of Pharmaceutical Chemistry, Johann Wolfgang Goethe University, Frankfurt am Main, Germany

John Spencer
Department of Chemistry, School of Life Sciences, University of Sussex, United Kingdom

W. E. Nagel et al. (eds.), *High Performance Computing in Science and Engineering '21*,
https://doi.org/10.1007/978-3-031-17937-2_1

availability of highly optimized force fields, i.e., the classical Hamiltonian function, which approximates the underlying quantum chemical nature of the system. During the course of the last 40 years, MD force fields for the simulations of biomolecules have been dramatically improved in terms of accuracy to the point that it is now possible to simulate phenomena like protein folding, at least for some small proteins [1], as well as protein-ligand and protein-materials interactions [2–11]. In this report, we will show how we have been able to use classical MD simulations to characterize the stability of several cancer mutants of p53, a protein that is found mutated in about 50% of the cancer cases diagnosed every year [12]. In that study simulations were used also to asses the effect of small ligands which, by binding to a mutation-induced pocket on the protein surface, are capable of stabilizing the protein, thus representing possible lead compounds for the development of cancer therapeutics. In addition to protein folding, many biological phenomena, however, occur on time scales that are still inaccessible to standard classical MD. For this reason, a wide range of enhanced sampling techniques have been proposed. Metadynamics [13–15] is one of the most popular techniques. In metadynamics, a time-dependent energy term is added to the Hamiltonian, to drive the system away from regions of conformational space already sampled. We started to use this method in the past to characterize the conformational properties of a large protein complex, fibrinogen [9]. We soon discovered that in this case, the available methods to unbias the metadynamics sampling and obtain an equilibrium distribution of the most significant observables of the system were inadequate. We have then developed a new method, which is more accurate than those previously available in the limit of short trajectories [16]. In what follows, we also review these findings.

2 Methods

GROMACS[17] was used to perform the MD simulations. GROMACS exploits intra-node OpenMP parallelization and inter-node MPI parallelization and was optimally compiled to run efficiently both on Hazelhen and on Hawk HPC infrastructures at HLRS. Unless stated differently, in what follows, the simulations were performed with the CHARMM36m force field [18], while the ligands were modeled using the Charmm Generalized force field (CGenFF)[19]. The time step for the simulations was set to 2 fs. The LINCS [20] algorithm was used to constrain the length of bonds involving hydrogen atoms. A cut-off of 1.2 nm with a switch function starting at 1.0 nm was used for direct non-bonded interactions. A cell-list like algorithm [21] was used. Periodic boundary conditions were adopted along all directions and a smooth particle-mesh Ewald (sPME) approach [22] was used for long-range electrostatics. The water was modeled explicitly using a modified TIP3P model[23]. Pressure and temperature were regulated at 1 atm and 300 K, using the Parrinello–Rahman [24] and Nose–Hoover [25, 26] algorithms, respectively. Few water molecules were replaced by sodium and chlorine ions during simulation setup to neutralize the charge of the simulation box and to achieve the physiological ion concentration (0.15 M).

In the case of the p53 DNA-binding domain, the simulations were based on the structure of a stabilized pseudo-wild-type (pdb id 1UOL) [27] and mutant structures determined by X-ray crystallography or modeled in ref. [28]. The systems were minimized for a max 50'000 steps, then equilibrated in the NVT ensemble for 1 ns with positional restraints on the heavy atoms of the protein. Then, they were further equilibrated for 1 ns in the NPT ensemble with no restraints. Four production runs were started for each mutant. Each run was 200 ns long. Further methodological details are provided in the original publication ref. [28]. System sizes ranged from 37000 to 43000 atoms, which are sufficient to ensure good scaling on Hazelhen and Hawk running on 100 and 64 nodes, respectively. Simulations were set up as job chains with each job not exceeding 3 hours length, writing a single restart file at the end. The trajectories were analyzed using the program VMD [29] and WORDOM [30].

3 Simulations of p53 cancer mutants and stabilization by ligand binding

The tumor suppressor protein p53 is involved in several processes protecting the human genome, including activation of DNA repair mechanisms or induction of apoptosis (cell death) in case of extensive DNA damage. Mutations of this protein can often result in cancer and, indeed, mutations in this protein occur in half of the diagnosed cancers [12]. Among the most frequently found cancer mutations are those of residue 220 in the DNA-binding domain of the protein, with the most abundant being Y220C, found in about 100'000 new cancer cases each year. The wild-type protein (WT) (i.e., the one without mutations) is only marginally stable, and many cancer mutations induce a loss of stability, which reduces the folding transition temperature of the protein, leading to unfolding and, consequently, to a loss of function at body temperature. In some cases, such as Y220C, the mutation creates a crevice on the protein surface. This offers a possible strategy to reactivate the mutant protein: a drug that binds to the mutation-induced pocket may actually stabilize the protein and thereby rescue its tumor suppressor function. This strategy has already been successfully used to rescue the Y220C mutant [2,31]. In collaboration with our experimental partners, we investigated other frequent cancer mutants with a mutation of Y220.

In ref. [28], we analyzed the cancer associated mutants Y220H, Y220N, Y220S, and Y220C using experimental biophysical techniques including, among others, X-ray crystallography as well as MD simulations. The latter have been used to estimate the effects of mutations on protein stability by monitoring the root-mean-square fluctuations (RMSF) [32], that is the average amplitude of the fluctuations of the atomic positions of the protein around the average structure. We used the same approach in the context of the p53 DNA-binding domain mutants, and we verified that the RMSF measured on the simulations correlated with experimentally determined differences in melting temperature of the mutant proteins (Tab. 1). Our simulations also provided additional information complementing the structural data obtained by X-ray crystallography, for example, by sampling the dynamics of different regions of the protein. We monitored several atomic distances characterizing the mutation-induced crevice on the surface of the protein, which can be split into a central crevice and different subsites (Fig. 1). These distances did not show significant fluctuations in the WT, whereas in the cancer mutants, they fluctuated between two states which can be associated with open and collapsed conformations of the mutation-induced crevice (Fig. 1c).

Several carbazole-based compounds were shown to bind to the mutation-induced crevice in Y220C and stabilize the protein structure, thus increasing the folding transition temperature [2,31]. Given the similarity between the crevices generated by the various mutations at site 220, in ref. [28] the effects of some of these carbazole compounds on the Y220S and Y220C mutants were analyzed. It was shown that several compounds that bind Y220C in the mutation-induced crevice can also bind to the equivalent pocket in Y220S, to a lesser extent also in the Y220N pocket. The binding of the compounds resulted in a considerable increase in the melting temperature of the mutants. MD simulations of the mutants in the presence of the carbazole compounds in the crevice revealed that the ligands dramatically reduced the RMSF of the protein, which correlated with the binding constant measured experimentally and the ligand-induced increase of the folding transition temperature (Tab. 1). In particular, the simulations showed that the collapsed state of the crevice is completely absent in the presence of the ligands (Fig. 2).

Summarizing this part of the report, the simulations of p53 DNA-binding domain mutants, with and without ligands, provided important insights into the molecular basis of the experimentally observed stabilizing effect upon binding. This information will aid the development of more potent small-molecule stabilizers and molecules targeting more than one mutant.

Codon 220 mutation	Average RMSF (Å)	Tm (°C)
Native structure		
WT	1.46	51.5
Y220H	1.56	45.1
Y220C	1.61	43.8
Y220N	1.64	39.9
Y220S	1.67	39.4
Ligand complexes		
Y220C-PK9323	1.35	
Y220S-PK9301	1.42	
Y220S-PK9323	1.58	

Table 1: Root-mean-square fluctuations (RMSF) of the simulated p53 DNA-binding domain constructs averaged over residues 97-289 and the experimentally determined folding transition temperature. Adapted from ref. [28], Bauer et al. ©2020 licensed under CC-BY.

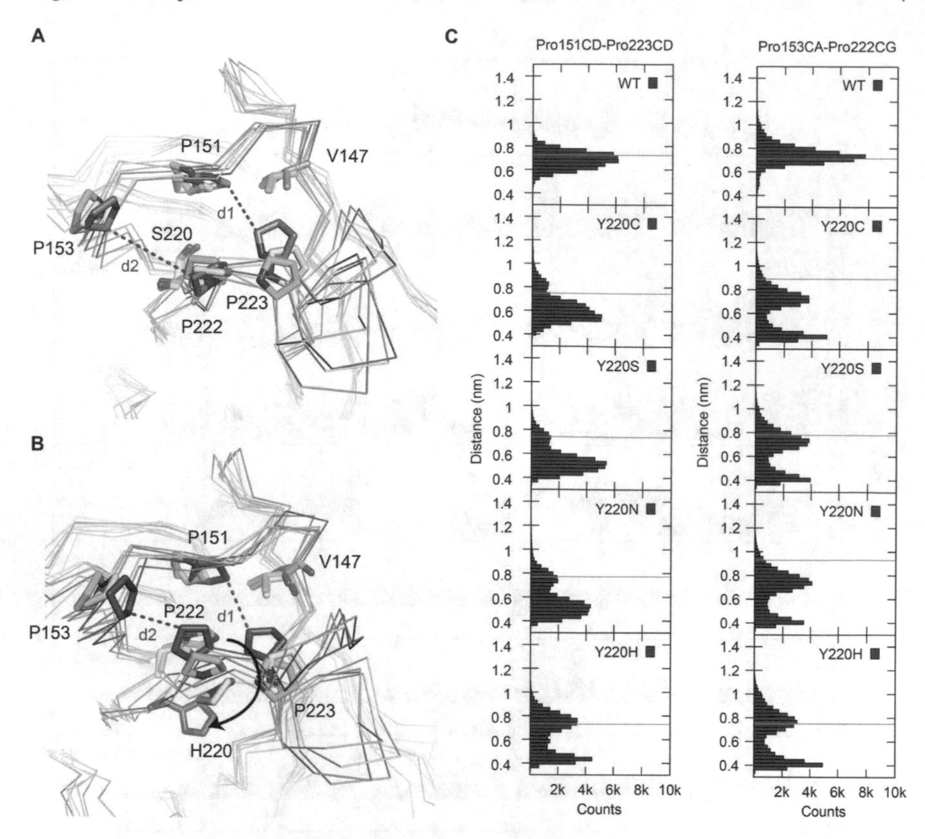

Fig. 1: MD simulation of p53 cancer mutants. (A) Representative C_{α}-trace structures of the open and closed states from the simulations of the p53 mutant Y220S superimposed onto the crystal structure (green). Highlighted residues are represented as sticks. Distances defining the size of the central cavity (d1) and the subsite 2 (d2) are represented with magenta dashed lines. (B) Same as (A) but for the Y220H mutant. The closed state is determined by the H220 side chain swinging out of the pocket. (C) Distribution of the d1 and d2 distances in the simulations of the p53 variants under consideration. The green line represent the distance in the crystal structure. Only WT shows an unimodal distribution. All mutants show the presence of a collapsed state of the crevice beside the open state. Adapted from ref. [28], Bauer et al. ©2020 licensed under CC-BY.

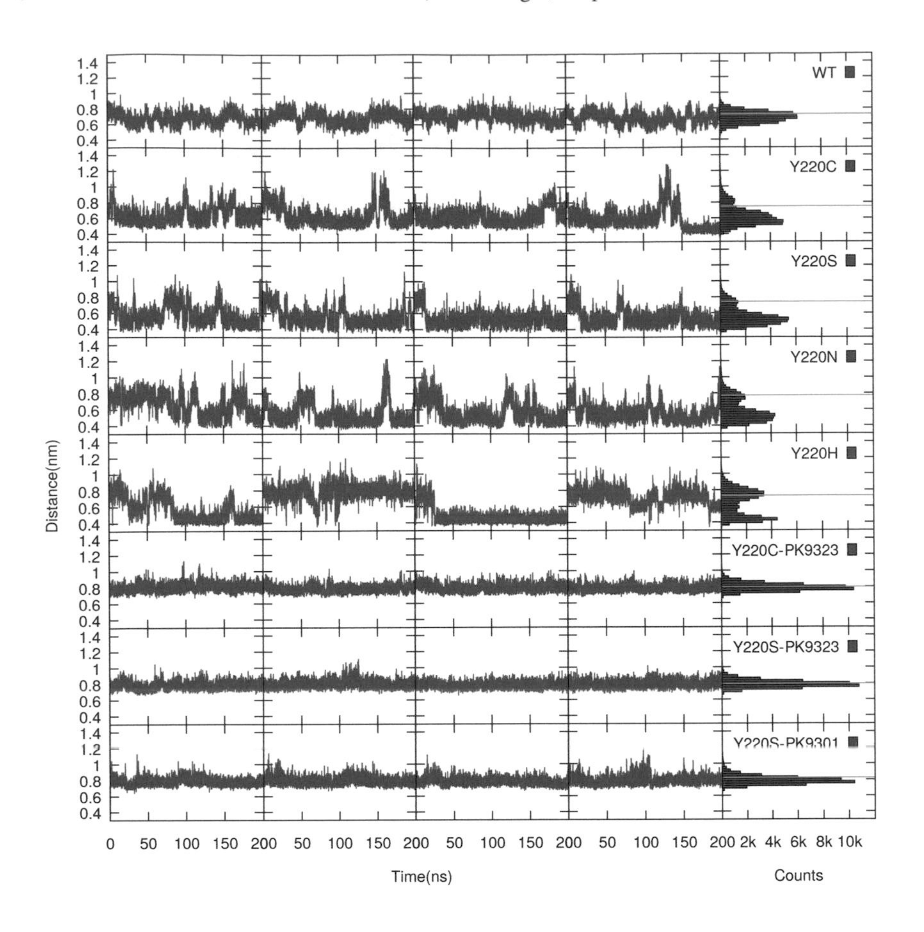

Fig. 2: Time series and resulting distributions of the d1 distance (see fig. 1) in all the simulated constructs. The presence of the ligands PK9323 and PK9301 in the crevice dramatically reduces structural fluctuations and prevents the population of collapsed states of the mutation-induced pocket. Adapted from ref. [28], Bauer et al. ©2020 licensed under CC-BY.

3.1 Data reweighting in metadynamics simulations

In this section, we review the results presented in ref. [16]. Metadynamics introduces a time-dependent bias potential in the classical energy of the system. The bias is dependent on selected collective variables $\boldsymbol{s}(\boldsymbol{r})$ of the system's coordinates, and it is built as a sum of Gaussian functions that are deposited at regular time intervals on the points reached by the trajectory:

$$V(\boldsymbol{s}(\boldsymbol{r}),t) = \sum_{t'=\Delta t,2\Delta t,\ldots}^{t} W\mathrm{e}^{-\sum_i^d (s_i(\boldsymbol{r})-s_i(\boldsymbol{r}(t')))^2/2\sigma_i^2} = \sum_{t'=\Delta t,2\Delta t,\ldots}^{t} g_{t'}(\boldsymbol{r},\boldsymbol{s}(t')) \quad (1)$$

where the Gaussian hills are $g_{t'}(\boldsymbol{s}(\boldsymbol{r})) = W\mathrm{e}^{-\sum_i^d (s_i(\boldsymbol{r})-s_i(\boldsymbol{r}(t')))^2/2\sigma_i^2}$. This results in pushing the system away from those values of the collective variables that have been already sampled.

In order to obtain equilibrium properties from the trajectories sampled using metadynamics, a reweighting procedure is necessary that takes into account the time-dependent influence of the bias. In other words, a weight $w(s,t)$ has to be assigned to each sampled conformation of the system, which is dependent on the time evolution of the bias. Once we have that, we can measure equilibrium properties from the biased simulations:

$$\langle A \rangle_0 = \langle Aw \rangle_\mathrm{b} / \langle w \rangle_\mathrm{b} \quad (2)$$

where A is any observables defined on the system, the subscript 0 indicates the unbiased (i.e. equilibrium) average and the subscript b indicates the average done over the biased simulations.

Several reweighting techniques have been proposed for metadynamics simulation [33–37], which come with some limitations, for example, some of them are specifically suited for well-tempered metadynamics, which is a modified version of the original algorithm where the height of the Gaussians decreases during the simulation. The simulations we ran on Hazelhen to determine the conformational states of the fibrinogen complex [9] showed however that the popular reweighting scheme from Tiwary and Parrinello [37], when used to build the free energy landscape as a function of the collective variables, did not match exactly the negative bias potential, which also represents an estimate of the free energy of the system. Tiwary's method is based on the assumption that in between two Gaussian depositions, the system samples a biased energy function $U + V$ where U is the unbiased energy function of the system and V is the metadynamics bias. Thus the conformations of the system follow a canonical distribution with biased energy.

$$p_b(\boldsymbol{r},t) = \frac{\mathrm{e}^{-\beta(U(\boldsymbol{r})+V(\boldsymbol{s}(\boldsymbol{r}),t))}}{\int \mathrm{d}\boldsymbol{r}\mathrm{e}^{-\beta(U(\boldsymbol{r})+V(\boldsymbol{s}(\boldsymbol{r}),t))}} \quad (3)$$

For the unbiased system, however, we would have the following distribution:

$$p_0(\boldsymbol{r}) = \frac{\mathrm{e}^{-\beta U(\boldsymbol{r})}}{\int \mathrm{d}\boldsymbol{r}\mathrm{e}^{-\beta U(\boldsymbol{r})}} \quad (4)$$

We can then obtain the equilibrium distribution from the biased distribution by means of the following reweighting factor [34, 37]:

$$w(\boldsymbol{r},t) = \frac{p_0(\boldsymbol{r})}{p_b(\boldsymbol{r},t)} \tag{5}$$

$$= e^{\beta V(s(\boldsymbol{r}),t))} \frac{\int d\boldsymbol{r} e^{-\beta(U(\boldsymbol{r},t)+V(s(\boldsymbol{r}),t))}}{\int d\boldsymbol{r} e^{-\beta U(\boldsymbol{r},t)}} \tag{6}$$

which can be further rewritten as [34, 37]:

$$w(\boldsymbol{r},t) = e^{\beta V(s(\boldsymbol{r}),t))} \frac{\int d\boldsymbol{r} \int ds\, \delta(s - s(\boldsymbol{r}))e^{-\beta(U(\boldsymbol{r},t)+V(s(\boldsymbol{r}),t))}}{\int d\boldsymbol{r} \int ds\, \delta(s - s(\boldsymbol{r}))e^{-\beta U(\boldsymbol{r},t)}} \tag{7}$$

$$= e^{\beta V(s(\boldsymbol{r}),t))} \frac{\int ds\, p_0(s)e^{-\beta V(s,t)}}{\int ds\, p_0(s)} \tag{8}$$

$$= e^{\beta V(s(\boldsymbol{r}),t))} \frac{\int ds\, e^{-\beta F(s)} e^{-\beta V(s,t)}}{\int ds\, e^{-\beta F(s)}} \tag{9}$$

where

$$p_0(s) = \int d\boldsymbol{r}\delta(s - s(\boldsymbol{r}))p_0(\boldsymbol{r}) = \frac{e^{-\beta F(s)}}{\int ds\, e^{-\beta F(s)}} \tag{10}$$

is the unbiased distribution projected on the low-dimensional CV space Ω and $F(s)$ is the unbiased free energy.

Eq. 9, although formally exact, contains the term $F(s)$, which is not known a priori (it is actually often the main aim of the simulation). In metadynamics the negative bias potential asymptotically approximates the free energy of the system [14, 15, 36, 38]:

$$F(s) \approx -\frac{\gamma}{\gamma - 1} V(s,t) + c(t) \tag{11}$$

where γ is the so-called bias factor of well-tempered metadynamics, which goes to infinity in the case of standard metadynamics, and $c(t)$ is a time-dependent offset of the free energy profile. In the Tiwary and Parrinello method [37], $F(s)$ is approximated using eq. 11, leading to the following expression for the weight (for standard metadynamics):

$$w^{tw}(\boldsymbol{r},t) = e^{\beta V(s(\boldsymbol{r}),t)} \frac{\int ds}{\int ds e^{\beta V(s,t)}} = e^{\beta V(s(\boldsymbol{r}),t)} \frac{1}{\langle e^{\beta V(s,t)} \rangle_s} \tag{12}$$

The approximation eq. 11, however, is only valid asymptotically, thus at short time scales the reweighting scheme may not provide accurate results.

Alternatively, we suggested [16] to approximate $F(s)$ with the value of the negative potential at the end of the simulation, which is supposedly more accurate:

$$w(\boldsymbol{r},t) = e^{\beta V(s(\boldsymbol{r}),t))} \frac{\int ds\, e^{\beta(V(s,t_f)-V(s,t)}}{\int ds\, e^{\beta V(s,t_f)}} \tag{13}$$

where t_f corresponds to the time at the end of the simulation. Eq. 13 would converge to eq. 12 at large t, but it will behave differently at small t. At short simulation times, a simple exponential reweighting would be more accurate than the Tiwary's eq. 12. A simple exponential reweighting, however, in the standard metadynamics setting where the bias increases with time would result in an underweighting of the initial part of the simulation. In ref. [16] we propose a simple way to correct it by subtracting the average value of the bias at every time step. This results in the balanced exponential reweighting:

$$w^{bex}(\boldsymbol{r},t) \propto e^{\beta V'(s(\boldsymbol{r}),t)} = e^{\beta\left(V(s(\boldsymbol{r}),t)-V_a(t)\right)} = e^{\beta V(s(\boldsymbol{r}),t)} \frac{1}{e^{\beta \langle V(s,t)\rangle_s}}. \tag{14}$$

The scheme proposed above differs from the previously proposed scheme of eq. 12 by the normalization factor of the exponential weight: in the case of eq. 12 the average value over Ω of the exponential of the bias potential is used, whereas in the new scheme eq. 14 we propose to use the exponential of the average bias. The average of the exponential (eq. 12) is very sensitive to small changes in the upper tail of the distribution of the bias potential and is, therefore, less robust in the initial part of the trajectory where the global free energy minimum of the system may not have been reached, yet.

The newly proposed scheme can be implemented without changes to the output of the popular metadynamics software PLUMED [39]. To demonstrate its advantages, we have tested it in several different scenarios and compared it with existing schemes. The standard mean of comparison that we have adopted consists in recovering the free energy landscape of a given system as a function of the collective variables by reweighting the conformations of the system sampled along the metadynamics trajectory. We did that for a series of systems for which an accurate free energy landscape is accessible and can be used as reference.

The first system we studied is a particle in a uni-dimensional double-well potential of the form $U(x) = (x^2 - 1)^2$. Simulations are performed at a temperature such that k_bT is 1/10 of the barrier separating the two energy wells. The system is discretized along the x direction into equally sized bins, and pseudo standard metadynamics simulation are performed by moving the particle to the left or right bin using a Metropolis criterion for accepting the move. The energy function for the Metropolis criterion is $U(x) + V(x,t)$, $V(x,t)$, where:

$$V(x,t) = \sum_{t'=\Delta t, 2\Delta t, \ldots}^{t} \frac{v}{\sqrt{2\pi}\sigma} e^{\left(-(x-x(t'))^2/2\sigma\right)} \tag{15}$$

In the above expression, the metadynamics Gaussian hills have volume v (that is height $v/\sqrt{2\pi}\sigma$) and width σ. The deposition period is Δt. Several simulations were run using different metadynamics parameters but keeping the length and step size (bin width) fixed. We then estimated the free energy landscape of the system using eq. 12 and 14:

$$F_{est}(x,t) = -\beta^{-1} \log \left(\frac{\sum_{t'=0}^{t} \delta(x - x(t'))w(x(t'),t')}{\sum_{t'=0}^{t} w(x(t'),t')} \right) \tag{16}$$

where the δ functions are the characteristic functions of the discrete bins along the x axis.

In Fig. 3a we report the estimated and reference free energy landscape of the system as a function of the simulated time. We repeated the simulations 72 times with different initial conditions and measured the error of the estimated free energy with respect to the reference in each of the simulations (the error is computed as the root-mean-square deviation of the two free energy profiles limited to the interval (-2,2) after subtracting the average). The data (Fig. 3b) show that the estimate obtained with the newly proposed eq. 14 converges faster than the other tested methods to the reference landscape. Also other estimates of the quality of the free energy landscape, like the estimate of the free energy difference between the minima (Fig. 3c) and the estimate of the height of the barrier (Fig. 3d) reveal a similar picture. Another advantage of the balanced exponential reweighting is the low run-to-run variability reported by the error bar in Fig.3b-d. A detailed look at the weights of the sampled conformations (Fig. 3) shows that while the balanced exponential weights are generally constant along the simulation, they are smaller than average in the very initial part of the simulation when the system has not yet explored both minima. On the other hand, Tiwary's method produces weights that are larger than average in the initial part of the simulation. This overestimate reduces the quality of the free energy profile for the early part of the run.

Although the uni-dimensional system offers already a good overview of the advantages of the newly proposed reweighting scheme, a test with a more realistic system is necessary to assess the performance in a normal-use scenario. For that, in the same work [16], we used an alanine dipeptide, which represents a standard benchmark of enhanced sampling techniques. The alanine dipeptide can be considered as the smallest protein-like unit as its structure can be characterized by the two protein backbone dihedral angles ϕ and ψ. We performed the metadynamics simulations using GROMACS [17] with the PLUMED [39] plugin. The system was simulated for short trajectories (8 ns) in vacuum using the standard force field AMBER03 [40] and standard values for time step and non-bonded interaction cutoff. The backbone dihedral angles ϕ and ψ were biased during the metadynamics simulations (for all the details of the simulations and the set of metadynamics parameters used please refer to our original work [16]). Adopting a strategy similar to the uni-dimensional case, we estimated the free energy landscape of the system as a function of ϕ and ψ using several different reweighting schemes including the newly proposed balanced exponential. We did that at several different time points along the simulation. We then compared the free energy estimates with the reference obtained by running an extremely long well-tempered metadynamics simulations (Fig. 4).

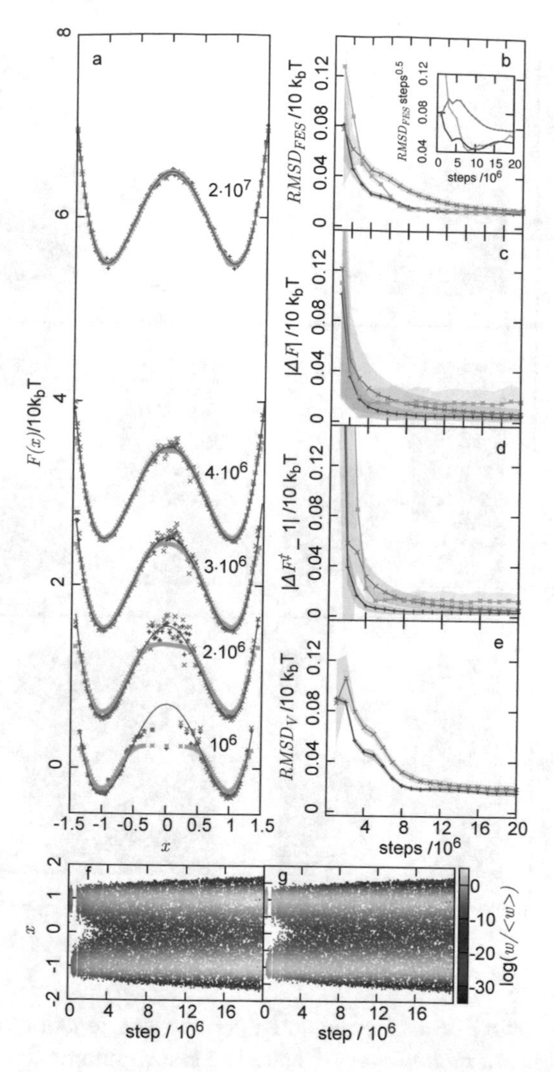

Fig. 3: (a) FES obtained along a single run at different trajectory lengths using balanced exponential reweighting, Tiwary reweighting and negative bias potential (red, blue and green points, respectively). The reference FES is plotted in purple. (b-e) Time series of (b) the RMSD (in inset is the RMSD$\cdot\sqrt{(t)}$), (c) estimated free energy difference between the two minima (absolute value), (d) estimated error on height of free energy barrier, and (e) the RMSD between reweighted FES and negative bias. Same color scheme as in part a. The solid lines represent the average values of the quantities across the 72 independent runs. Shaded bands indicate the standard deviations. (f-g) Time series of the position of the particle along one run where balanced exponential (f) and Tiwary (g) weights are reported according to a color scale. Adapted with permission from ref. [16], ©2020 American Chemical Society.

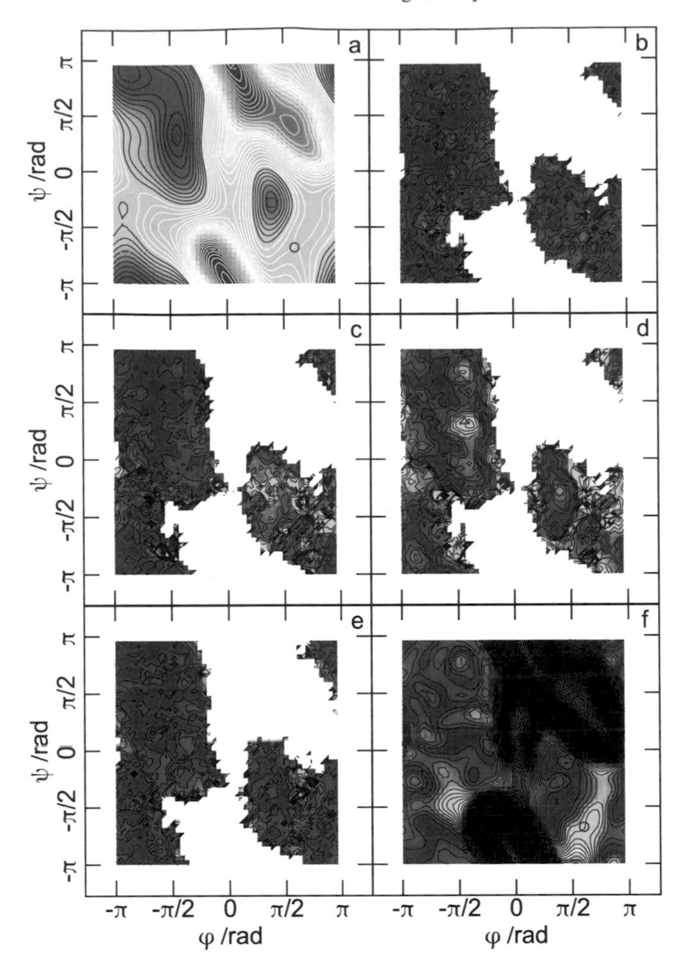

Fig. 4: (a) Reference FES of the alanine dipeptide as a function of the CV ϕ and ψ. Contour levels are plotted every 1 kbT. The black contours indicate the region within 10 kbT from the global minimum, which is used for the RMSD calculations. The red triangle, square and circle indicate the position of the minima $C7_{eq}$ and C_{ax} and the transition state ‡, respectively. The color shades are guides for the eye. (b-f) Difference between the reference FES and those estimated using the balanced exponential, Tiwary's, Branduardi's, and Bonomi's reweighting, and the negative bias, respectively, after 1.4 ns in one of the runs. Contour levels are plotted every 0.25 kbT, and the range of color shades is 4 times smaller than that in part a. Adapted with permission from ref. [16], ©2020 American Chemical Society.

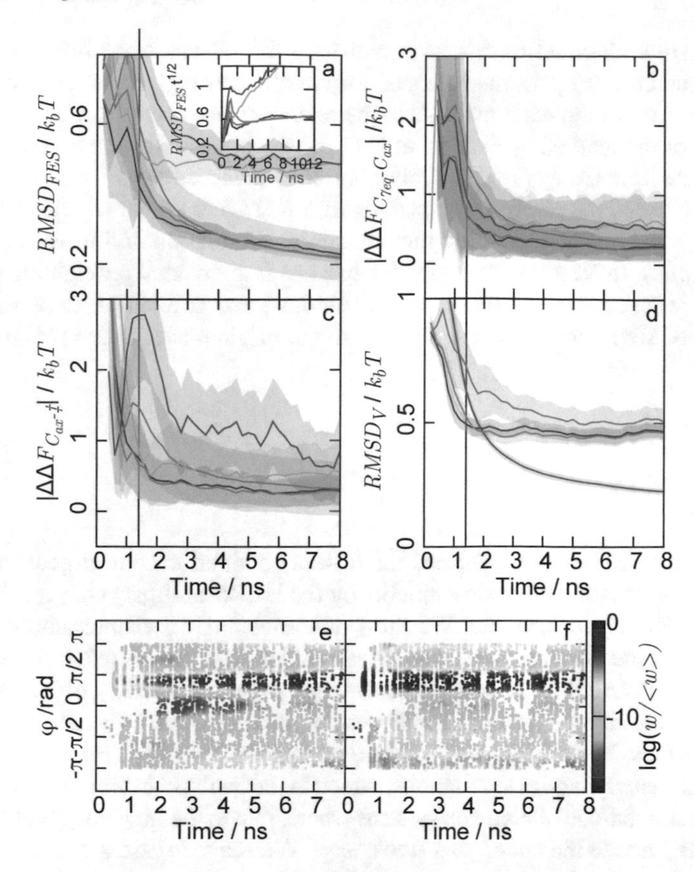

Fig. 5: (a), (b) and (c) provide the time series of the $RMSD$ from reference FES, and the error on the $\Delta F_{C_{7eq}-C_{ax}}$ and $\Delta F_{C_{ax}-\ddagger}$, respectively, for the balanced exponential (red), Tiwary's (blue), Branduardi's (purple), Bonomi's (orange) reweighting and negative bias (green) estimate of the FES. The data are averaged over 96 runs. Shaded bands indicate standard deviations. The black line at 1.4 ns indicates the time point where the FES in Fig. 4 have been extracted. (a inset) the $RMSD_{FES}\sqrt{t}$ shows to approximately reach a plateau for balanced exponential, Tiwary's and Bonomi's reweighting schemes. (d) $RMSD_V$ between the different reweighting schemes and negative bias (same color scheme as above). (e), (f) Time series of ψ when ϕ=-1.88 rad along one selected run. Balanced exponential (e) and Tiwary (f) weights are reported according to the color scale. $\langle w \rangle$ is the average of the weights along the run. Adapted with permission from ref. [16], ©2020 American Chemical Society.

The results obtained in this test confirmed the observations made in the uni-dimensional case (Fig. 5): the balanced exponential reweighting scheme converges faster than most of the other methods to the reference free energy landscape, with the exception of the method by Bonomi et.al [34], which, as we demonstrate in ref. [16], within some limits, may provide similar but not better results.

We also tested the newly developed algorithm in other scenarios: the reweighting of observables not biased in the metadynamics simulation and in well-tempered metadynamics. In all tested scenarios the balanced exponential reweighting provided similar or faster convergence than the other methods and, in addition, lower run-to-run fluctuations. We refer the interested reader to our original publication [16] for further details.

4 Conclusions

In this report, we have summarized the results obtained for our projects using the computational resources made available by the HLRS Stuttgart with the Hazelhen and Hawk HPC infrastructure. We show how molecular dynamics simulations of the tumor suppressor protein p53, an important target in cancer research, have been used to understand the effect of cancer mutations on the stability of the protein as well as the stabilizing effect of ligands binding to a mutation-induced pocket on the protein surface. In addition, we proposed an improved reweighting method for the analysis of metadynamics simulations, which is particularly useful in the context of large MD simulations of complex systems where, due to the high computational cost, fast convergence to the underlying free energy landscape is essential.

Acknowledgements This work was partly funded by the German Research Foundation (DFG) grant SFB1066 project Q1 and SFB TRR 146 as well as DFG grant JO 1473/1-1 and Worldwide Cancer Research (grants 14-1002, 18-0043). We gratefully acknowledge support with computing time from the HPC facility Hazelhen and Hawk at the High Performance Computing Center Stuttgart (HLRS) project Flexadfg and the HPC facility Mogon at the University of Mainz.

References

1. K. Lindorff-Larsen, S. Piana, R.O. Dror, D.E. Shaw, Science **334**(6055), 517 (2011)
2. N. Basse, J. Kaar, G. Settanni, A. Joerger, T. Rutherford, A. Fersht, Chemistry and Biology **17**(1) (2010)
3. S. Köhler, F. Schmid, G. Settanni, PLoS Computational Biology **11**(9) (2015)
4. S. Köhler, F. Schmid, G. Settanni, Langmuir **31**(48), 13180 (2015)
5. G. Settanni, J. Zhou, T. Suo, S. Schöttler, K. Landfester, F. Schmid, V. Mailänder, Nanoscale **9**(6) (2017)
6. G. Settanni, J. Zhou, F. Schmid, Journal of Physics: Conference Series **921**(1), 012002 (2017)
7. G. Settanni, T. Schäfer, C. Muhl, M. Barz, F. Schmid, Computational and Structural Biotechnology Journal **16**, 543 (2018)

8. S. Köhler, F. Schmid, G. Settanni, in *High Performance Computing in Science and Engineering '16: Transactions of the High Performance Computing Center Stuttgart (HLRS) 2016* (Springer International Publishing, 2017), pp. 61–78
9. T. Schäfer, J. Zhou, F. Schmid, G. Settanni, *Blood proteins and their interactions with nanoparticles investigated using molecular dynamics simulations* (2018)
10. T. Schafer, C. Muhl, M. Barz, F. Schmid, G. Settanni, in *High Performance Computing in Science and Engineering ' 18* (Springer International Publishing, 2019), pp. 63–74
11. T. Schäfer, C. Muhl, M. Barz, F. Schmid, G. Settanni, in *High Performance Computing in Science and Engineering '19* (Springer International Publishing, Cham, 2021), pp. 65–76
12. A.C. Joerger, A.R. Fersht, Annual Review of Biochemistry **85**, 375 (2016)
13. A. Laio, M. Parrinello, Proceedings of the National Academy of Sciences of the United States of America **99**(20), 12562 (2002)
14. G. Bussi, A. Laio, M. Parrinello, Physical review letters **96**(9), 090601 (2006)
15. A. Barducci, G. Bussi, M. Parrinello, Physical Review Letters (2008)
16. T.M. Schäfer, G. Settanni, Journal of Chemical Theory and Computation **16**(4), 2042 (2020)
17. H.J.C. Berendsen, D. van der Spoel, R. van Drunen, Computer Physics Communications **91**(1–3), 43 (1995)
18. J. Huang, S. Rauscher, G. Nawrocki, T. Ran, M. Feig, B.L. De Groot, H. Grubmüller, A.D. MacKerell, Nature Methods **14**(1), 71 (2016)
19. K. Vanommeslaeghe, E. Hatcher, C. Acharya, S. Kundu, S. Zhong, J. Shim, E. Darian, O. Guvench, P. Lopes, I. Vorobyov, A.D. Mackerell, J. Comput. Chem. **31**(4), 671 (2010)
20. B. Hess, H. Bekker, H.J.C. Berendsen, J.G.E.M. Fraaije, LINCS: A Linear Constraint Solver for Molecular Simulations. Tech. rep. (1997)
21. M.P. Allen, D.J. Tildesley, *Computer Simulation of Liquids* (Clarendon Press., 1987)
22. U. Essmann, L. Perera, M.L. Berkowitz, T. Darden, H. Lee, L.G. Pedersen, The Journal of Chemical Physics **103**(19), 8577 (1995)
23. W.L. Jorgensen, J. Chandrasekhar, J.D. Madura, R.W. Impey, M.L. Klein, Journal of Chemical Physics **79**(2), 926 (1983)
24. M. Parrinello, A. Rahman, Physical Review Letters **45**(14), 1196 (1980)
25. S. Nosé, The Journal of Chemical Physics **81**(1), 511 (1984)
26. W.G. Hoover, Physical Review A **31**(3), 1695 (1985)
27. A.C. Joerger, M.D. Allen, A.R. Fersht, Journal of Biological Chemistry **279**(2), 1291 (2004)
28. M.R. Bauer, A. Krämer, G. Settanni, R.N. Jones, X. Ni, R. Khan Tareque, A.R. Fersht, J. Spencer, A.C. Joerger, ACS Chemical Biology **15**(3), 657 (2020)
29. W. Humphrey, A. Dalke, K. Schulten, Journal of Molecular Graphics **14**, 33 (1996)
30. M. Seeber, M. Cecchini, F. Rao, G. Settanni, A. Caflisch, Bioinformatics **23**(19), 2625 (2007)
31. M.R. Bauer, R.N. Jones, R.K. Tareque, B. Springett, F.A. Dingler, L. Verduci, K.J. Patel, A.R. Fersht, A.C. Joerger, A.C. Joerger, Future Medicinal Chemistry **11**(19), 2491 (2019)
32. G. Interlandi, S.K. Wetzel, G. Settanni, A. Plückthun, A. Caflisch, Journal of Molecular Biology **375**(3), 837 (2008)
33. G. Tiana, European Physical Journal B **63**(2), 235 (2008)
34. M. Bonomi, A. Barducci, M. Parrinello, Journal of Computational Chemistry **30**(11), 1615 (2009)
35. J. Smiatek, A. Heuer, Journal of Computational Chemistry **32**(10), 2084 (2011)
36. D. Branduardi, G. Bussi, M. Parrinello, Journal of Chemical Theory and Computation (2012)
37. P. Tiwary, M. Parrinello, The Journal of Physical Chemistry B **119**(3), 736 (2015)
38. J.F. Dama, M. Parrinello, G.A. Voth, Physical Review Letters **112**(24), 240602 (2014)
39. G.A. Tribello, M. Bonomi, D. Branduardi, C. Camilloni, G. Bussi, Computer Physics Communications **185**(2), 604 (2014)
40. D.A. Case, T.E. Cheatham, T. Darden, H. Gohlke, R. Luo, K.M. Merz, A. Onufriev, C. Simmerling, B. Wang, R.J. Woods, Journal of Computational Chemistry **26**(16), 1668 (2005)

Hadronic contributions to the anomalous magnetic moment of the muon from Lattice QCD

M. Cè, E.-H. Chao, A. Gérardin, J.R. Green, G. von Hippel, B. Hörz, R.J. Hudspith, H.B. Meyer, K. Miura, D. Mohler, K. Ottnad, S. Paul, A. Risch, T. San José and H. Wittig

Abstract The recently reported new measurement of the anomalous magnetic moment of the muon, a_μ, by the E989 collaboration at Fermilab has increased the tension with the Standard Model (SM) prediction to 4.2 standard deviations. In order to increase the sensitivity of SM tests, the precision of the theoretical prediction, which is limited by the strong interaction, must be further improved. In our project we employ lattice QCD to compute the leading hadronic contributions to a_μ and various other precision observables, such as the energy dependence ("running") of the electromagnetic coupling, α, and the electroweak mixing angle, $\sin^2\theta_W$. Here we report on the performance of our simulation codes used for the generation of gauge ensembles at (near-)physical pion masses and fine lattice spacings. Furthermore, we present results for the hadronic running of α, the electroweak mixing angle, as well

H.B. Meyer, K. Miura, T. San José and H. Wittig
Helmholtz Institut Mainz, Johannes Gutenberg Universität, 55099 Mainz, Germany

K. Miura, D. Mohler and T. San José
GSI Helmholtzzentrum für Schwerionenforschung, Darmstadt, Germany

M. Cè and J. R. Green
Theoretical Physics Department, CERN, Geneva, Switzerland

E.-H. Chao, G. von Hippel, R.J. Hudspith, H.B. Meyer, K. Ottnad, S. Paul, T. San José and H. Wittig
Institut für Kernphysik and PRISMA$^+$ Cluster of Excellence, Universität Mainz, Johann-Joachim-Becher-Weg 45, D-55099 Mainz, Germany

A. Risch
John von Neumann Institute for Computing NIC, Deutsches Elektronen-Synchrotron DESY, Platanenallee 6, 15738 Zeuthen, Germany

A. Gérardin
Aix Marseille Univ, Université de Toulon, CNRS, CPT, Marseille, France

B. Hörz
Nuclear Science Division, Lawrence Berkeley National Laboratory, Berkeley, CA 94720, USA

K. Miura
Kobayashi-Maskawa Institute for the Origin of Particles and the Universe, Nagoya University, Nagoya 464-8602, Japan

W. E. Nagel et al. (eds.), *High Performance Computing in Science and Engineering '21*,
https://doi.org/10.1007/978-3-031-17937-2_2

as the hadronic vacuum polarisation and light-by-light scattering contributions to a_μ. Results from an ancillary calculation of the spectrum in the isovector channel are crucial in order to further increase the precision of our determination of the hadronic vacuum polarisation contribution.

1 Introduction

The Standard Model of Particle Physics provides a quantitative and precise description of the properties of the known constituents of matter in terms of a uniform theoretical formalism. However, despite its enormous success, the Standard Model (SM) does not explain some of the most pressing problems in particle physics, such as the nature of dark matter or the asymmetry between matter and antimatter. The world-wide quest for discovering physics beyond the SM involves several different strategies, namely (1) the search for new particles and interactions that are not described by the SM, (2) the search for the enhancement of rare processes by new interactions, and (3) the comparison of precision measurements with theoretical, SM-based predictions of the same quantity. These complementary activities form an integral part of the future European strategy for particle physics [1].

Precision observables, such as the anomalous magnetic moment of the muon, a_μ, have provided intriguing hints for the possible existence of "new physics". The longstanding tension between the direct measurement of a_μ and its theoretical prediction has recently increased to 4.2 standard deviations, following the publication of the first result from the E989 experiment at Fermilab [2]. As E989 prepares to improve the experimental precision further, it is clear that the theoretical prediction must be pushed to a higher level of accuracy as well, in order to increase the sensitivity of the SM test. Since the main uncertainties of the SM prediction arise from strong interaction effects, current efforts are focussed on quantifying the contributions from hadronic vacuum polarisation (HVP) and hadronic light-by-light scattering (HLbL). This has also been emphasised in the 2020 White Paper [3] in which the status of the theoretical prediction is reviewed.

Our project is focussed on calculations of the hadronic contributions to the muon anomalous magnetic moment from first principles, using the methodology of Lattice QCD. To this end, we perform calculations of the HVP contribution at the physical value of the pion mass, in order to reduce systematic errors. Another highly important ingredient of our calculation is the determination of the spectrum in the isovector channel of the electromagnetic current correlator, which constrains the long-distance contribution to the HVP. Our group has also developed a new formalism for the direct calculation of the HLbL contribution, which has produced the most precise estimate from first principles so far [4].

The HVP contribution to the muon anomalous magnetic moment is closely linked to the hadronic effects that modify the value of the electromagnetic coupling, $\Delta\alpha$. Since $\Delta\alpha$ depends on other SM parameters such as the mass of the W-boson, a precise determination provides important information for precision tests of the SM. Finally, we also compute the hadronic contributions to the "running" of the electroweak mixing angle, a precision observable which is particularly sensitive to the effects of physics beyond the SM in the regime of low energies.

2 Computational setup

One of the major computational tasks of our project is the generation of gauge-field ensembles at (close to) physical light-quark masses. For a particular challenge encountered in these simulations please refer to [5]. The generation of a gauge field ensemble dubbed E250 at physical pion and kaon masses has been a long standing goal of our programme on Hazel Hen and Hawk, and has been finalized since the last report. Furthermore we recently produced two somewhat coarser lattices named D452 and D152[1] at light pion masses. For both ensembles, the generation of 500 gauge field configurations, corresponding to 2000 molecular dynamics units (MDU) had been proposed. As it turned out, the rather coarse lattice spacing of D152 lead to sustained algorithmic problems, hence the run was stopped after 275 gauge configurations (1100 MDU). For D452 no such issues were observed, and we were able to produce 1000 gauge configurations (4000 MDU) due to better than expected performance for this run. Preliminary results for observables suggest that ensemble D452 will play a vital role for obtaining more precise results for the observables in our project.

Figure 1 shows the Hamiltonian deficits ΔH as well as the Monte Carlo history of the topological charge for ensembles E250 and D452. The acceptance rate resulting from the history of ΔH is $(87.1 \pm 1.0)\%$ for E250 and $(91.5 \pm 0.7)\%$ for D452. The generation of these chains is now complete and the calculation of physics observables has been started on compute clusters operated by JGU Mainz.

The `openQCD` code used in our calculations exhibits excellent scaling properties over a wide range of problem sizes. Figure 3 shows the strong-scaling behaviour for the system size corresponding to ensemble D452, i.e. for a 128×64^3 lattice, as measured on Hawk (left pane). The timings refer to the application of the even-odd preconditioned $O(a)$ improved Wilson–Dirac operator $\hat{D}_{\mathrm{w}}$ to a spinor field which accounts for the largest fraction of the total computing time for several of our projects.

For the gauge field generation runs on Hawk and for the HLbL runs on lattices of size 128×64^3 we used the following setup:

A Local lattice volume of size 8^4 per MPI rank with 8192 MPI ranks on 64 nodes.

In addition to this setup, we performed spectroscopy runs on J303 (192×64^3) with setup B and on E250 192×96^3 with setup C:

B Local lattice of size 12×8^3 per MPI rank with 8192 MPI ranks on 64 nodes.
C Local lattice volume of size $12 \times 6^2 \times 12$ per MPI rank with 32768 MPI ranks on 256 nodes.

[1] This ensemble was called D151 in the proposal.

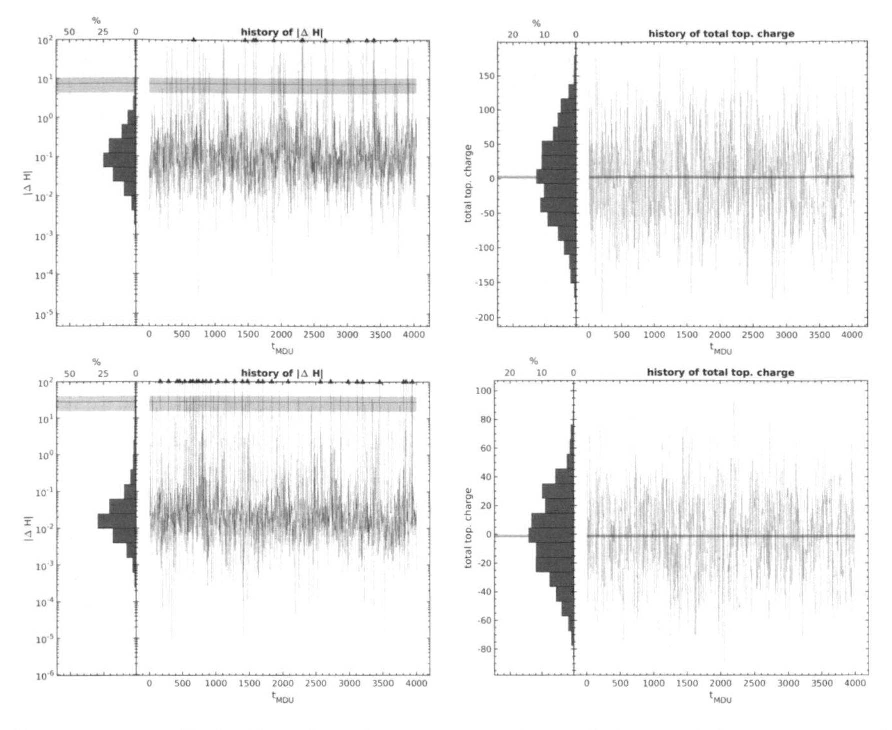

Fig. 1: Monte-Carlo histories of the Hamiltonian deficit ΔH (left) and the total topological charge (right) for E250 (top) and D452 (bottom).

When possible a hypercube of size $2^2 \times 4^2$ processes was grouped onto a single CPU, to minimise off-CPU communication.

For the current allocation year we received a total of 100 MCore hours. It was foreseen that 32% / 40% / 27% would be spent on the gauge field generation / HVP spectroscopy / HLbL respectively, while the actual percentages of computing time spent by the time of writing this report are given by 30% / 44% / 26%.

3 The hadronic running of the electroweak couplings

We start our discussion of precision observables with the hadronic contributions to the energy dependence ("running") of the electromagnetic coupling, α, since its calculation shares many features and definitions with the determination of the HVP contribution to the muon anomalous moment.

The electromagnetic coupling in the Thomson limit is one of the most precisely known quantities, $\alpha = 1/137.035999084(21)$ [6]. However, for scales above a few hundred MeV, the low-energy hadronic contribution to the vacuum polarization

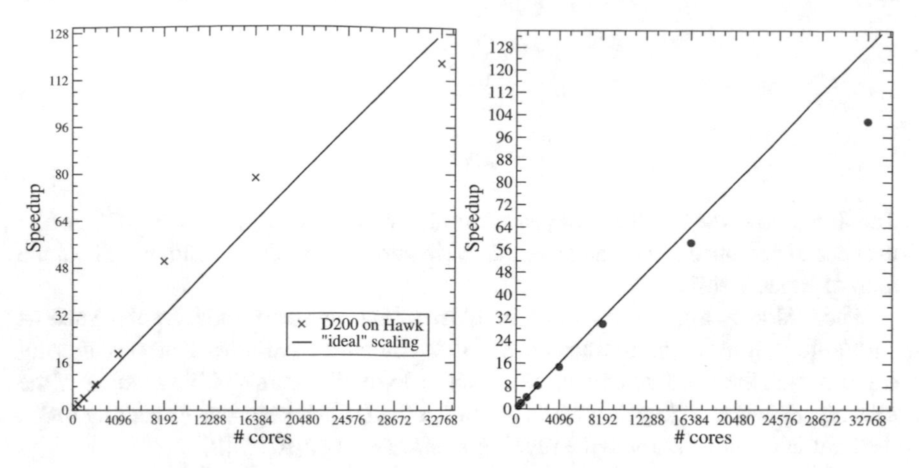

Fig. 2: Left: Strong-scaling behaviour of the openQCD code on Hawk. The plot shows the application of the even-odd preconditioned Wilson–Dirac operator on a 128×64^3 lattice. Speedup factors are defined relative to 256 cores. A clear indication of hyperscaling is seen in this regime. Right: Weak scaling with a local volume of 8^4 on Hawk.

induces a theoretical uncertainty. If one parameterizes the energy dependence, or "running", of the coupling in the so-called on-shell scheme,

$$\alpha(q^2) = \frac{\alpha}{1 - \Delta\alpha(q^2)}, \tag{1}$$

the five flavour quark contribution at the Z-pole is found to be $\Delta\alpha^{(5)}_{\text{had}}(M_Z^2) = 0.02766(7)$ [6]. Its poorer precision is a limiting factor for the global fit of the electroweak sector of the Standard Model (SM). Thus, the goal is to improve the precision of this estimate, by reducing the uncertainties associated with the strong interaction. We concentrate on the quark contribution at low energies, which is given by the one-subtracted dispersion relation

$$\Delta\alpha_{\text{had}}(q^2) = 4\pi\alpha\bar{\Pi}^{\gamma\gamma}(q^2), \qquad \bar{\Pi}^{\gamma\gamma}(q^2) = \text{Re}\left[\Pi^{\gamma\gamma}(q^2) - \Pi^{\gamma\gamma}(0)\right]. \tag{2}$$

The standard method to obtain eq. (2) employs the optical theorem, which relates the HVP function with the so-called R-ratio, i.e. the total hadronic cross section $\sigma(e^+e^- \to \text{hadrons})$ normalized by $\sigma(e^+e^- \to \mu^+\mu^-)$, via a dispersion integral. While the integral can be evaluated using experimental data for the R-ratio in the low-energy domain, this procedure introduces experimental uncertainties into a theoretical prediction. Therefore, lattice computations in the space-like region $Q^2 = -q^2$ provide a valuable *ab initio* crosscheck. In order to estimate $\Delta\alpha^{(5)}_{\text{had}}(M_Z^2)$, we use the so-called Euclidean split technique [7, 8]

$$\Delta\alpha^{(5)}_{\rm had}(M_Z^2) = \Delta\alpha^{(5)}_{\rm had}(-Q_0^2) + \left[\Delta\alpha^{(5)}_{\rm had}(-M_Z^2) - \Delta\alpha^{(5)}_{\rm had}(-Q_0^2)\right] + \left[\Delta\alpha^{(5)}_{\rm had}(M_Z^2) - \Delta\alpha^{(5)}_{\rm had}(-M_Z^2)\right]. \quad (3)$$

The first term is the result of this project at the threshold energy $Q_0^2 = 5\ \mathrm{GeV}^2$, while the second and third terms can be evaluated in perturbative QCD, with or without the help of experimental data.

There is growing interest to probe electroweak precision observables such as $\sin^2\theta_{\rm W}(q^2)$ at momentum transfers $q^2 \ll M_Z^2$ in parity-violating lepton scattering experiments. Such measurements are sensitive to modifications of the running of the mixing angle due to beyond the Standard Model (BSM) physics. At leading order, the hadronic contribution to the running of $\sin^2\theta_{\rm W}$ is given by [9]

$$\Delta_{\rm had}\sin^2\theta_{\rm W}(q^2) = -\frac{4\pi\alpha}{\sin^2\theta_{\rm W}}\bar{\Pi}^{Z\gamma}(q^2). \quad (4)$$

We employ the time-momentum representation (TMR) [10, 11] to compute the vacuum polarization functions $\bar{\Pi}^{\gamma\gamma}$ and $\bar{\Pi}^{Z\gamma}$. This allows us to represent our observables in terms of integrals over two-point functions with a known Q^2-dependent kernel $K(t, Q^2) = t^2 - 4/Q^2\sin^2(Qt/2)$,

$$\bar{\Pi}(\ Q^2) = \int_0^\infty dt\, G(t)K(t, Q^2), \quad (5)$$

where $G(t)$ is the correlator, projected to zero momentum and averaged over the three spatial directions to improve the signal,

$$G(t) = -\frac{1}{3}\sum_{k=1}^{3}\sum_{\mathbf{x}}\left\langle j_k(x)j_k(0)\right\rangle. \quad (6)$$

The two currents j_k are either the electromagnetic j_μ^γ or the vector part of the weak neutral current j_μ^Z,

$$j_\mu^\gamma = j_\mu^3 + \frac{1}{\sqrt{3}}j_\mu^8 + \frac{2}{3}j_\mu^c, \quad (7a)$$

$$j_\mu^Z|_{\rm vector} = \left(\frac{1}{2} - \sin^2\theta_{\rm W}\right)j_\mu^\gamma - \frac{1}{6}j_\mu^0 - \frac{1}{12}j_\mu^c. \quad (7b)$$

Figure 3 shows the hadronic contribution to the running of α and $\sin^2(\theta_W)$ for a domain of low Q^2. The error bands represent the total uncertainties affecting these observables: statistical, scale setting, extrapolation to the physical point and isospin breaking corrections.

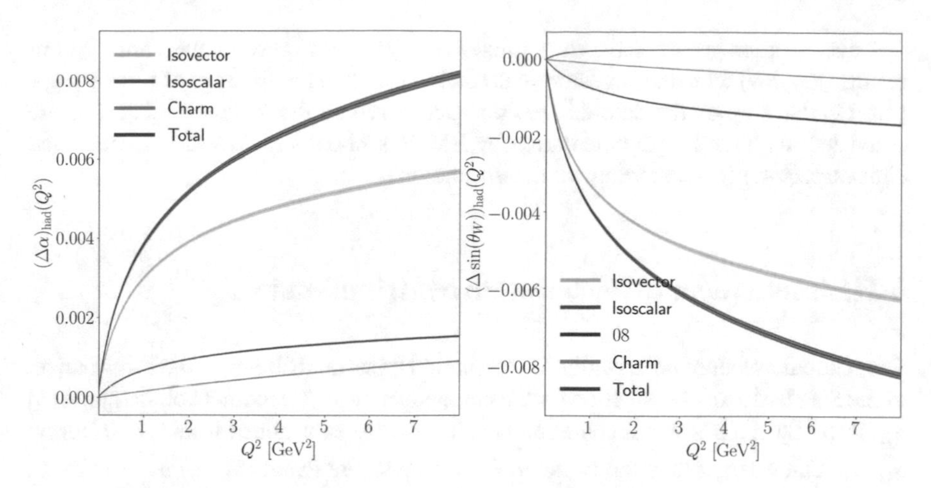

Fig. 3: Hadronic contributions to the running of α (left) and $\sin^2\theta_{\mathrm{W}}$ (right) with the energy. We depict both the total running as well as the different components according to the SU(3)-isospin decomposition and the charm quark contribution.

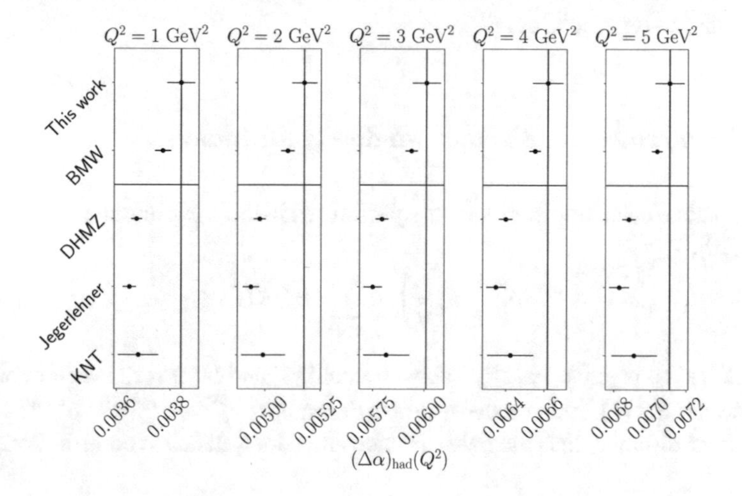

Fig. 4: Comparison of the hadronic contribution to the running of α at several energies. The top row and band show our result. The horizontal line separates lattice from phenomenological results. We find agreement with BMW [12], slight tensions with KNT [13] and bigger discrepancy with DHMZ [14] and Jegerlehner [15].

Several other groups have determined $\Delta\alpha(Q^2)$. In Figure 4, we compare our results (top row) with another lattice calculation [12] and the findings of three groups [13–15] that employ the data-driven approach based on the R-ratio. Our results are consistent with the lattice calculation by BMW, while a slight tension is observed with respect to phenomenological determinations.

4 Hadronic vacuum polarization contribution to a_μ

The hadronic vacuum polarization is the single largest contributor to the error budget of the Standard Model prediction for the anomalous magnetic moment of the muon [3] a_μ. Recently, the BMW collaboration [16] has published a value for this contribution $a_\mu^{\rm hvp}$, which is larger than the value quoted in the White Paper [3]; the latter is based on a dispersive relation with input from experimentally measured $e^+e^- \to$ hadrons cross-sections. Taken at face value, the BMW results would strongly reduce the current tension with the experimental world average for a_μ [17]. Therefore, it is vital for this important test of the Standard Model to resolve the tension between the BMW and the dispersive result for $a_\mu^{\rm hvp}$.

4.1 The intermediate-distance window contribution

Current lattice calculations of $a_\mu^{\rm hvp}$ are performed in the 'time-momentum' representation [10],

$$a_\mu^{\rm hvp} = \left(\frac{\alpha}{\pi}\right)^2 a \sum_{t\geq 0} G(t)\widetilde{K}(t), \tag{8}$$

where $\widetilde{K}(t)$ is a positive, exactly known kernel [18] and $G(t)$ is the vector correlator introduced in Eq. (6). It has proved useful to decompose $a_\mu^{\rm hvp} = a_\mu^{\rm winSD} + a_\mu^{\rm winID} + a_\mu^{\rm winLD}$ into a short-distance, intermediate-distance and a long-distance contribution,

$$a_\mu^{\rm winSD} = \left(\frac{\alpha}{\pi}\right)^2 a \sum_{t\geq 0} G(t)\widetilde{K}(t)\ [1 - \Theta(t, t_0, \Delta)], \tag{9}$$

$$a_\mu^{\rm winID} = \left(\frac{\alpha}{\pi}\right)^2 a \sum_{t>0} G(t)\widetilde{K}(t)\ [\Theta(t, t_0, \Delta) - \Theta(t, t_1, \Delta)], \tag{10}$$

$$a_\mu^{\rm winLD} = \left(\frac{\alpha}{\pi}\right)^2 a \sum_{t\geq 0} G(t)\widetilde{K}(t)\ \Theta(t, t_1, \Delta). \tag{11}$$

Here $\Theta(t, \bar{t}, \Delta) = (1 + \tanh[(t - \bar{t})/\Delta])/2$ is a smoothened step function, and the standard choice of parameters is $t_0 = 0.4$ fm, $t_1 = 1.0$ fm and $\Delta = 0.15$ fm. Particularly the intermediate-distance contribution $a_\mu^{\rm winID}$ can be computed on the lattice with smaller (relative) statistical and systematic uncertainty than the total $a_\mu^{\rm hvp}$. Therefore $a_\mu^{\rm winID}$ has emerged as an excellent benchmark quantity to first test the consistency of different lattice QCD calculations, and in a second step their consistency with the data-driven evaluation based on the R-ratio. Furthermore, the contributions of different quark-flavour combinations can be compared separately between lattice calculations.

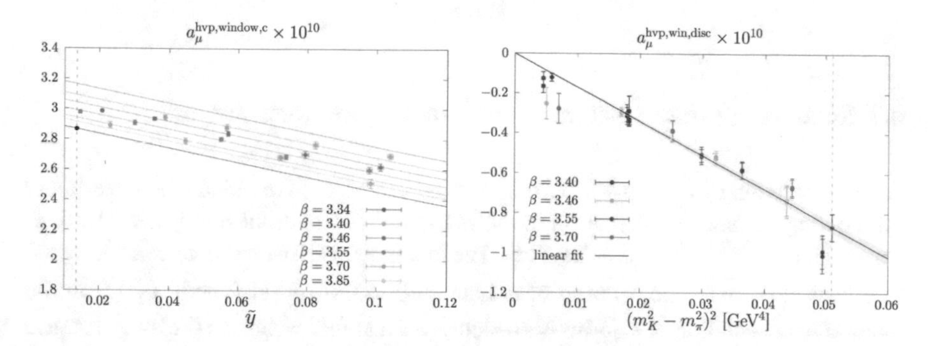

Fig. 5: Left: chiral and continuum extrapolation of the connected charm contribution to $a_\mu^{\rm winID}$, with $\tilde{y} = (m_\pi/(4\pi f_\pi))^2$, f_π being the pion decay constant. Right: Extrapolation in the SU(3)$_{\rm f}$-breaking difference of squared kaon and pion masses of the u, d, s quark-disconnected contribution to $a_\mu^{\rm winID}$. In both panels, the target extrapolation point is marked by a dashed vertical line. The β values are in one-to-one correspondence with the lattice spacing a.

As sample results, we present the charm-quark as well as the quark-disconnected contributions to $a_\mu^{\rm winID}$ in Figure 5. The results are

$$a_\mu^{\rm winID,charm} = (+2.85 \pm 0.12) \times 10^{-10}, \tag{12}$$

$$a_\mu^{\rm winID,disc} = (-0.87 \pm 0.03_{\rm stat} \pm 0.07_{\rm syst}) \times 10^{-10}. \tag{13}$$

In the case of the disconnected contribution, the systematic error is estimated as the difference between the linear fit in $(m_K^2 - m_\pi^2)^2$ of all data and our most chiral ensemble E250. We also point out that the physical values of the lattice spacings a were determined using the results of Ref. [19]. The analysis of the light-quark contributions are in the final stage, and we provide the relative-error budget for the isovector contribution, which represents about 80% of the total $a_\mu^{\rm winID}$ (Table 1).

The dominance of the chiral and continuum extrapolation error highlights the importance of the physical-quark-mass ensemble E250.

Type	%
Extrapolation	0.6
Scale setting	0.4
Statistical error of $G(t)$	0.2
Finite-size effect	≤ 0.2
Renormalization	< 0.01
Total	0.77

Table 1

4.2 Scattering phase shift and the timelike pion form factor

The computation of $I = 1$ p-wave $\pi\pi$ scattering can be used to reduce the uncertainty on the long-distance contribution to the integrand for the light-quark HVP [20], in particular on a_μ^{winLD} defined in Eq. (11). The fact that this issue has not been addressed in our discussion of the running of α is related to the kernel function $\widetilde{K}(t)$ in the integral representation of a_μ^{hvp} in (8), which gives a higher weight to the long-distance regime.

In addition, our related determination of the timelike pion form factor [21, 22], can be used to access infinite-volume physics from the lattice calculation in a finite volume, providing a valuable cross-check of finite-volume effects encountered in lattice calculations of the HVP [23]. To this end we proposed the determination of the $l = 1$ pion-pion scattering phase shift δ_1 and of the pion timelike form factor on ensembles J303 and E250. During the current reporting period we finished the full statistics for J303 and obtained first results for E250. Figure 6 shows δ_1 for ensemble J303.

The next steps will include the reconstruction of the isovector part of the correlator $G(t)$ for this ensemble, from which an accurate determination of the long-distance regime is possible. The corresponding analysis for ensemble E250 is in progress. The precise determination of $G(t)$ at large distances will be crucial for reducing the statistical error on a_μ^{hvp} below the percent level.

5 Hadronic light-by-light scattering contribution to a_μ

Although sub-leading in the electromagnetic coupling (i.e. $O(\alpha^3)$ compared to $O(\alpha^2)$) and therefore a smaller contribution to a_μ than the HVP, the hadronic light-by-light contribution has a large associated theoretical error. This is mostly due to the difficulty in computing such a complicated contribution. To match the upcoming expected experimental precision from E989 and J-PARC, it has been suggested that the theory error should be around 10×10^{-11} [3], whereas the currently best estimate included in the White Paper estimate is $92(19) \times 10^{-11}$ [3]. Following our infinite-volume QED, position-space methodology outlined in [24], we have

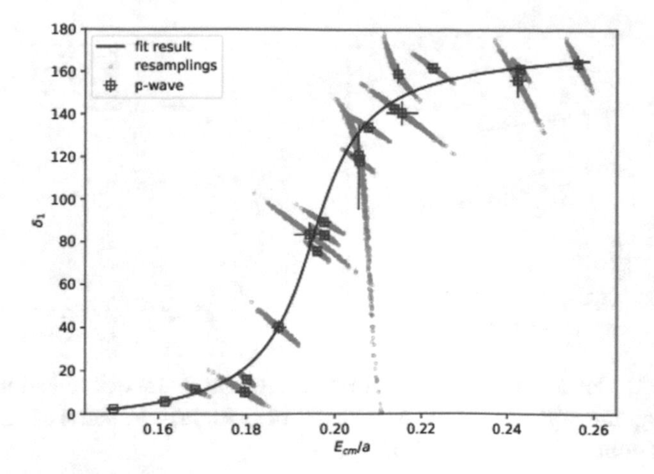

Fig. 6: Isospin 1, pion-pion p-wave phase shift δ_1 on ensemble J303 as a function of the center of momentum energy E_{cm} in lattice units. The lattice data is shown in blue, the red curve shows the results of a parameterization with a "pole plus constant" form for the scattering K-matrix.

determined the HLbL contribution at the physical pion mass [4]. The crucial results on the two largest-volume, lowest-pion mass results (i.e. ensembles D200 and D450, the left-most blue and green points in the left plot of Fig. 8) were computed on Hawk; without the resources granted to us, two of our near-physical determinations would have been absent and likely our approach to the physical pion mass would be less well-constrained.

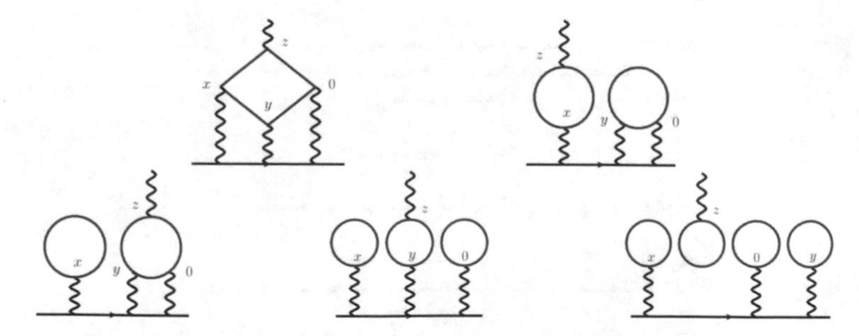

Fig. 7: Different quark Wick-contraction classes appearing in the computation of the QCD four-point correlation function. The straight horizontal lines represent muon propagators, wavy lines represent photon propagators. From left to right, top to bottom, they are the fully-connected, $(2+2)$, $(3+1)$, $(2+1+1)$ and $(1+1+1+1)$. Each class contains the diagrams obtained from all the possible permutations of the four points attached to photons.

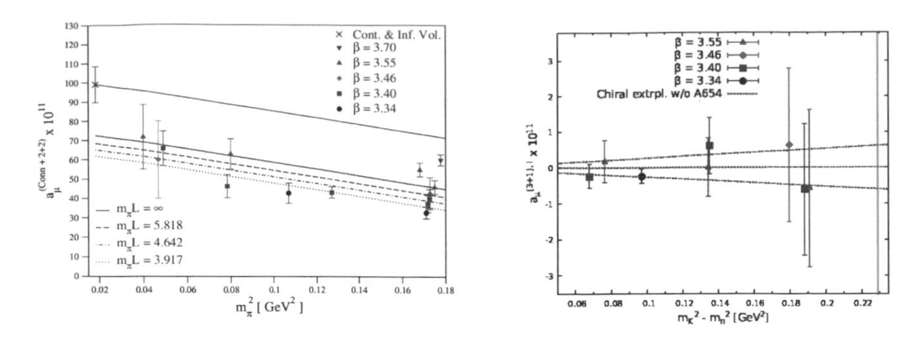

Fig. 8: (Left) The combined chiral/finite-volume/continuum extrapolation of the two leading diagrammatic contributions to the HLbL. (Right) The chiral extrapolation for the $(3 + 1)$ contribution to the HLbL.

As is expected by the charge factors and large-N_c arguments, only the leading light-quark connected and $(2 + 2)$-disconnected diagrams contribute significantly to hadronic light-by-light scattering. However, there are three additional topologies, whose magnitude is argued to be small but have not been directly computed so far, those are the $(3 + 1)$, the $(2 + 1 + 1)$, and the $(1 + 1 + 1 + 1)$ contributions, shown in Fig. 7. Our work [4] was the first to determine all of these contributions (which all turn out to be consistent with zero within our statistical precision), and the data for D200 and D450 generated on Hawk were used in the extrapolation of the $(3 + 1)$ light-quark topology (the right-most blue and green points of Fig. 8).

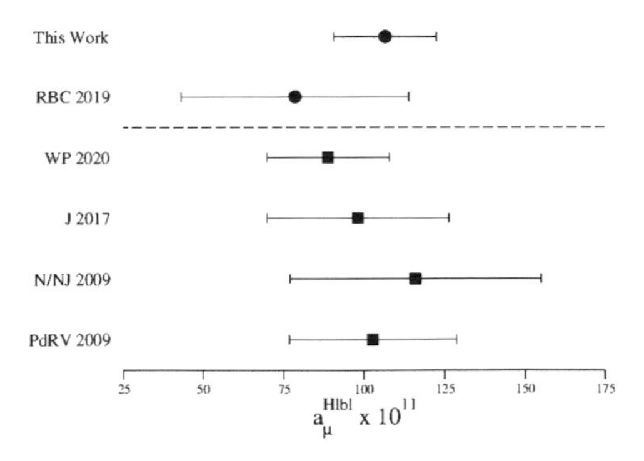

Fig. 9: A comparison of our HLbL result with the literature. The results in circles are the two available lattice determinations (this work and [25], above the horizontal dashed line). The results in squares are phenomenological predictions from [3],[26], [27, 28], and [29].

Following our precise determination of all relevant parts to this contribution our final result (including systematic effects) compares favourably to a previous lattice determination and to phenomenological predictions in general, as is illustrated in Figure 9. Our result is the most precise determination thus far, and including it in a global average with other independent determinations gives $a_\mu^{\mathrm{HLbL}} = 97.5(11.6) \times 10^{-11}$. This number is precise enough to meet the expected precision of the experiments and brings the focus back to the precision determination of the HVP.

6 Summary and outlook

We have reported on the current status of our determination of a set of precision observables that are crucial for exploring the limits of the Standard Model, using the *ab initio* formalism of lattice QCD. The continued production of gauge ensembles close to the physical pion mass is crucial for achieving full control over systematic effects at the desired overall accuracy.

Our results for the hadronic running of the electromagnetic coupling and weak mixing angle are already at the same level of precision as reached in conventional data-driven approaches and can be readily applied in phenomenological studies. A full account will be published soon. Regarding the HVP contribution a_μ^{hvp}, we have focussed mostly on the intermediate-window quantity which is emerging as a benchmark quantity that facilitates the comparison between different calculations. For the other most important hadronic contribution, arising from light-by-light scattering, we have published the most accurate result so far.

Our focus will turn to increasing the overall precision of our determination of a_μ^{hvp} to the sub-percent level.

Acknowledgements This work is partly supported by the Deutsche Forschungsgemeinschaft (DFG, German Research Foundation) grant HI 2048/1-2 (Project ID 399400745) and by the DFG-funded Collaborative Research Centre SFB 1044 *The low-energy frontier of the Standard Model*. The Mainz $(g-2)_\mu$ project is also supported by the Cluster of Excellence *Precision Physics, Fundamental Interactions, and Structure of Matter* (PRISMA+ EXC 2118/1) funded by the DFG within the German Excellence Strategy (Project ID 39083149). This work is supported by the European Research Council (ERC) under the European Union's Horizon 2020 research and innovation programme through grant agreement 771971-SIMDAMA. The work of M.C. is supported by the European Union's Horizon 2020 research and innovation program under the Marie Skłodowska-Curie Grant Agreement No. 843134. The authors gratefully acknowledge the Gauss Centre for Supercomputing e.V. (www.gauss-centre.eu) for supporting this project by providing computing time on the GCS Supercomputer Hazel Hen and Hawk at Höchstleistungsrechenzentrum Stuttgart (www.hlrs.de) under project GCS-HQCD. Additional calculations were performed on the HPC clusters "Clover" and "HIMster II" at the Helmholtz-Institut Mainz and "MOGON II" at JGU Mainz. We are grateful to our colleagues in the CLS initiative for sharing ensembles.

References

1. R.K. Ellis, et al., Physics Briefing Book: Input for the European Strategy for Particle Physics Update 2020 (2019)
2. B. Abi, et al., Measurement of the Positive Muon Anomalous Magnetic Moment to 0.46 ppm, Phys. Rev. Lett. **126**(14), 141801 (2021). DOI 10.1103/PhysRevLett.126.141801
3. T. Aoyama, et al., The anomalous magnetic moment of the muon in the Standard Model, Phys. Rept. **887**, 1 (2020). DOI 10.1016/j.physrep.2020.07.006
4. E.H. Chao, R.J. Hudspith, A. Gérardin, J.R. Green, H.B. Meyer, K. Ottnad, Hadronic light-by-light contribution to $(g-2)_\mu$ from lattice QCD: a complete calculation, Eur. Phys. J. C **81**(7), 651 (2021). DOI 10.1140/epjc/s10052-021-09455-4
5. D. Mohler, S. Schaefer, Remarks on strange-quark simulations with Wilson fermions, Phys. Rev. D **102**(7), 074506 (2020). DOI 10.1103/PhysRevD.102.074506
6. P. Zyla, et al., Review of Particle Physics, PTEP **2020**(8), 083C01 (2020). DOI 10.1093/ptep/ptaa104
7. S. Eidelman, F. Jegerlehner, A.L. Kataev, O. Veretin, Testing nonperturbative strong interaction effects via the Adler function, Phys. Lett. B **454**, 369 (1999). DOI 10.1016/S0370-2693(99)00389-5
8. F. Jegerlehner, The Running fine structure constant alpha(E) via the Adler function, Nucl. Phys. B Proc. Suppl. **181-182**, 135 (2008). DOI 10.1016/j.nuclphysbps.2008.09.010
9. F. Burger, K. Jansen, M. Petschlies, G. Pientka, Leading hadronic contributions to the running of the electroweak coupling constants from lattice QCD, JHEP **11**, 215 (2015). DOI 10.1007/JHEP11(2015)215
10. D. Bernecker, H.B. Meyer, Vector Correlators in Lattice QCD: Methods and applications, Eur. Phys. J. A **47**, 148 (2011). DOI 10.1140/epja/i2011-11148-6
11. A. Francis, B. Jaeger, H.B. Meyer, H. Wittig, A new representation of the Adler function for lattice QCD, Phys. Rev. D **88**, 054502 (2013). DOI 10.1103/PhysRevD.88.054502
12. S. Borsanyi, et al., Hadronic vacuum polarization contribution to the anomalous magnetic moments of leptons from first principles, Phys. Rev. Lett. **121**(2), 022002 (2018). DOI 10.1103/PhysRevLett.121.022002
13. A. Keshavarzi, D. Nomura, T. Teubner, $g-2$ of charged leptons, $\alpha(M_Z^2)$, and the hyperfine splitting of muonium, Phys. Rev. D **101**(1), 014029 (2020). DOI 10.1103/PhysRevD.101.014029
14. M. Davier, A. Hoecker, B. Malaescu, Z. Zhang, A new evaluation of the hadronic vacuum polarisation contributions to the muon anomalous magnetic moment and to $\alpha(\mathbf{m_Z^2})$, Eur. Phys. J. C **80**(3), 241 (2020). DOI 10.1140/epjc/s10052-020-7792-2. [Erratum: Eur.Phys.J.C 80, 410 (2020)]
15. F. Jegerlehner, $\alpha_{QED,eff}$(s) for precision physics at the FCC-ee/ILC, CERN Yellow Reports: Monographs **3**, 9 (2020). DOI 10.23731/CYRM-2020-003.9
16. S. Borsanyi, et al., Leading hadronic contribution to the muon magnetic moment from lattice QCD, Nature **593**(7857), 51 (2021). DOI 10.1038/s41586-021-03418-1
17. B. Abi, et al., Measurement of the Positive Muon Anomalous Magnetic Moment to 0.46 ppm, Phys. Rev. Lett. **126**(14), 141801 (2021). DOI 10.1103/PhysRevLett.126.141801
18. M. Della Morte, A. Francis, V. Gülpers, G. Herdoíza, G. von Hippel, H. Horch, B. Jäger, H.B. Meyer, A. Nyffeler, H. Wittig, The hadronic vacuum polarization contribution to the muon $g-2$ from lattice QCD, JHEP **10**, 020 (2017). DOI 10.1007/JHEP10(2017)020
19. M. Bruno, T. Korzec, S. Schaefer, Setting the scale for the CLS 2 + 1 flavor ensembles, Phys. Rev. D **95**(7), 074504 (2017). DOI 10.1103/PhysRevD.95.074504
20. A. Gérardin, M. Cè, G. von Hippel, B. Hörz, H.B. Meyer, D. Mohler, K. Ottnad, J. Wilhelm, H. Wittig, The leading hadronic contribution to $(g-2)_\mu$ from lattice QCD with $N_f = 2+1$ flavours of O(a) improved Wilson quarks, Phys. Rev. D **100**(1), 014510 (2019). DOI 10.1103/PhysRevD.100.014510
21. H.B. Meyer, Lattice QCD and the Timelike Pion Form Factor, Phys. Rev. Lett. **107**, 072002 (2011). DOI 10.1103/PhysRevLett.107.072002

22. C. Andersen, J. Bulava, B. Hörz, C. Morningstar, The $I = 1$ pion-pion scattering amplitude and timelike pion form factor from $N_f = 2 + 1$ lattice QCD, Nucl. Phys. B **939**, 145 (2019). DOI 10.1016/j.nuclphysb.2018.12.018
23. F. Erben, J.R. Green, D. Mohler, H. Wittig, Rho resonance, timelike pion form factor, and implications for lattice studies of the hadronic vacuum polarization, Phys. Rev. D **101**(5), 054504 (2020). DOI 10.1103/PhysRevD.101.054504
24. E.H. Chao, A. Gérardin, J.R. Green, R.J. Hudspith, H.B. Meyer, Hadronic light-by-light contribution to $(g-2)_\mu$ from lattice QCD with SU(3) flavor symmetry, Eur. Phys. J. C **80**(9), 869 (2020). DOI 10.1140/epjc/s10052-020-08444-3
25. T. Blum, N. Christ, M. Hayakawa, T. Izubuchi, L. Jin, C. Jung, C. Lehner, Hadronic Light-by-Light Scattering Contribution to the Muon Anomalous Magnetic Moment from Lattice QCD, Phys. Rev. Lett. **124**(13), 132002 (2020). DOI 10.1103/PhysRevLett.124.132002
26. F. Jegerlehner, *The Anomalous Magnetic Moment of the Muon*, vol. 274 (Springer, Cham, 2017). DOI 10.1007/978-3-319-63577-4
27. A. Nyffeler, Hadronic light-by-light scattering in the muon g-2: A New short-distance constraint on pion-exchange, Phys. Rev. D **79**, 073012 (2009). DOI 10.1103/PhysRevD.79.073012
28. F. Jegerlehner, A. Nyffeler, The Muon g-2, Phys. Rept. **477**, 1 (2009). DOI 10.1016/j.physrep.2009.04.003
29. J. Prades, E. de Rafael, A. Vainshtein, The Hadronic Light-by-Light Scattering Contribution to the Muon and Electron Anomalous Magnetic Moments, Adv. Ser. Direct. High Energy Phys. **20**, 303 (2009). DOI 10.1142/9789814271844_0009

Quantum simulators, phase transitions, resonant tunneling, and variances: A many-body perspective

A.U.J. Lode, O.E. Alon, J. Arnold, A. Bhowmik, M. Büttner, L.S. Cederbaum, B. Chatterjee, R. Chitra, S. Dutta, C. Georges, A. Hemmerich, H. Keßler, J. Klinder, C. Lévêque, R. Lin, P. Molignini, F. Schäfer, J. Schmiedmayer and M. Žonda

A.U.J. Lode
Institute of Physics, Albert-Ludwigs-Universität Freiburg, Hermann-Herder-Str. 3, 79104 Freiburg, Germany
e-mail: auj.lode@gmail.com

O.E. Alon
Department of Mathematics, University of Haifa, Haifa 3498838, Israel
Haifa Research Center for Theoretical Physics and Astrophysics, University of Haifa, Haifa 3498838, Israel
e-mail: ofir@research.haifa.ac.il

J. Arnold
Department of Physics, University of Basel, Klingelbergstrasse 82, CH-4056 Basel, Switzerland

A. Bhowmik
Department of Mathematics, University of Haifa, Haifa 3498838, Israel
Haifa Research Center for Theoretical Physics and Astrophysics, University of Haifa, Haifa 3498838, Israel

M. Büttner
Institute of Physics, Albert-Ludwigs-Universität Freiburg, Hermann-Herder-Str. 3, 79104 Freiburg, Germany

L.S. Cederbaum
Theoretische Chemie, Physikalisch-Chemisches Institut, Universität Heidelberg, Im Neuenheimer Feld 229, D-69120 Heidelberg, Germany
e-mail: lorenz.cederbaum@pci.uni-heidelberg.de

B. Chatterjee
Department of Physics, Indian Institute of Technology-Kanpur, Kanpur 208016, India

R. Chitra
Institute for Theoretical Physics, ETH Zürich, 8093 Zürich, Switzerland

S. Dutta
Department of Mathematics, University of Haifa, Haifa 3498838, Israel
Haifa Research Center for Theoretical Physics and Astrophysics, University of Haifa, Haifa 3498838, Israel

C. Georges
The Hamburg Center for Ultrafast Imaging, Luruper Chaussee 149, 22761 Hamburg, Germany

A. Hemmerich
The Hamburg Center for Ultrafast Imaging, Luruper Chaussee 149, 22761 Hamburg, Germany
Zentrum für Optische Quantentechnologien and Institut für Laser-Physik, Universität Hamburg, 22761 Hamburg, Germany

W. E. Nagel et al. (eds.), *High Performance Computing in Science and Engineering '21*,
https://doi.org/10.1007/978-3-031-17937-2_3

Abstract This 2021 report summarizes our activities at the HLRS facilities Hawk and Hazel Hen in the framework of the multiconfigurational time-dependent Hartree for indistinguishable particles (MCTDH-X) high-performance computation project. Our results are a bottom-up investigation into exciting and intriguing many-body physics and phase diagrams obtained via the direct solution of the Schrödinger equation and its comparison to experiments, and via machine learning approaches. We investigated ultracold-boson quantum simulators for crystallization and superconductors in a magnetic field, the phase transitions of ultracold bosons interacting with a cavity and of charged fermions in lattices described by the Falicov–Kimball model. Moreover, we report exciting findings on the many-body dynamics of tunneling and variances, in two and three-dimensional ultracold-boson systems, respectively.

1 Introduction

Quantum many-body physics represents a demanding field of extraordinary depth featuring an intriguing and physically rich set of phenomena. Here, we report our activities exploring these phenomena within the framework of our multiconfigurational time-dependent Hartree for indistinguishable particles high-performance computation project at the HLRS. In our quest to explore many-body phenomena and in continuation of our fruitful research in the past years [1–8] we developed and applied computational methods to solve the Schrödinger equation numerically: we investigated ultracold-

H. Keßler
The Hamburg Center for Ultrafast Imaging, Luruper Chaussee 149, 22761 Hamburg, Germany

J. Klinder
The Hamburg Center for Ultrafast Imaging, Luruper Chaussee 149, 22761 Hamburg, Germany

C. Lévêque
Vienna Center for Quantum Science and Technology, Atominstitut, TU Wien, Stadionallee 2, 1020 Vienna, Austria
Wolfgang Pauli Institute c/o Faculty of Mathematics, University of Vienna, Oskar-Morgenstern Platz 1, 1090 Vienna, Austria

R. Lin
Institute for Theoretical Physics, ETH Zürich, 8093 Zürich, Switzerland

P. Molignini
Clarendon Laboratory, Department of Physics, University of Oxford, OX1 3PU, United Kingdom
Institute for Theoretical Physics, ETH Zürich, 8093 Zürich, Switzerland

F. Schäfer
Department of Physics, University of Basel, Klingelbergstrasse 82, CH-4056 Basel, Switzerland

J. Schmiedmayer
Vienna Center for Quantum Science and Technology, Atominstitut, TU Wien, Stadionallee 2, 1020 Vienna, Austria

M. Žonda
Department of Condensed Matter Physics, Charles University in Prague, Ke Karlovu 5, 121 16 Prague 2, Czech Republic

boson quantum simulators for **i)** crystallization [9] and **ii)** superconductors in a magnetic field [10], phase transitions of **iii)** bosons in a cavity [11] and **iv)** mixtures of heavy and light fermions in a lattice [12], **v)** many-boson dynamics of **vi)** two-dimensional resonant tunneling [13] and **vii)** three-dimensional variance of position, momentum, and angular momentum [14]. The works **i) – vii)** report new fundamental insights into many-body-land and were obtained via software implementations [15–17] of the MCTDH-X family of methods [18–25] as well as novel computational methods for the simulation of the single-shot detection of quantum many-body systems [26–28] and phase diagrams [12, 29].

We base this report on the findings in Refs. [9–14] and structure it as follows. In Sec. 2, we provide an introduction to the MCTDH-X theory, in Sec. 3, we discuss the observables we used in this work, and in Sections 4,5,6,7,8, and 9 we – respectively – discuss the findings in Refs. [9–14]. We conclude with an outlook in Sec. 10.

2 Theory

Here we describe the theoretical method we chiefly used to solve the many-body Schrödinger equation, MCTDH-X, the multiconfigurational time-dependent Hartree method for indistinguishable particles [23–25].

MCTDH-X can be applied to a variety of systems – bosons [24, 25], fermions [19, 20], particles with internal degrees of freedom [30] or in a cavity [27]. The MCTDH-X comes from the MCTDH [31] family of methods for indistinguishable particles which also includes multilayering MCTDH [32, 33], restricted active space truncation [34–37], and MCTDH-X approaches [38, 39]. MCTDH-X has been reviewed [8, 18] and compared directly to experimental observations [11, 28].

The MCTDH-X method uses a time-dependent variational principle [40] applied to the Schrödinger equation in combination with the following ansatz that represents the state $|\Psi\rangle$ as a time-dependent superposition of time-dependent symmetric (bosons) or anti-symmetric (fermions) configurations:

$$|\Psi\rangle = \sum_{\mathbf{n}} C_{\mathbf{n}}(t)|\mathbf{n};t\rangle; \quad |\mathbf{n};t\rangle = \prod_{j=1}^{M} \frac{\left[\hat{b}_j^\dagger(t)\right]^{n_j}}{\sqrt{n_j!}}|vac\rangle = |n_1, ..., n_M; t\rangle. \quad (1)$$

Here, $\mathbf{n} = (n_1, ..., n_M)$ is an occupation-number state with fixed particle number $N = \sum_j n_j$ and $\{\mathbf{n}\}$ is a shorthand for the set of all possible such occupation-number vectors. The coefficients $C_{\mathbf{n}}(t)$ are the complex-valued weights of the time-dependent configurations $|\mathbf{n};t\rangle$ at time t. These time-dependent configurations are built by (anti-)symmetrizing a product of M orthonormal time-dependent single-particle states $\{\phi_1(\mathbf{r};t), ..., \phi_M(\mathbf{r};t)\}$.

In the variational derivation of the MCTDH-X working equations the coefficients $|\mathbf{n};t\rangle$ and the orbitals $\{\phi_1(\mathbf{r};t),...,\phi_M(\mathbf{r};t)\}$ are considered as variational parameters. The result of this variational derivation is a set of coupled non-linear integro-differential MCTDH-X working equations [23–25], see the review [18] and References therein for more details.

3 Quantities of interest

In this and the following sections, we will omit the dependency of quantities on time where we consider it notationally convenient.

3.1 One- and two-body densities

The one-body density matrix $\rho^{(1)}$ (the 1-RDM) is a standard probe for analyzing N-particle quantum states; it is defined as follows:

$$\rho^{(1)}(\mathbf{r},\mathbf{r}') = \mathbf{N}\langle\mathbf{\Psi}|\hat{\mathbf{\Psi}}^{\dagger}(\mathbf{r}')\hat{\mathbf{\Psi}}(\mathbf{r})|\mathbf{\Psi}\rangle. \tag{2}$$

Here, $\hat{\Psi}^{(\dagger)}$ represents the field annihilation (creation) operator.

3.2 Fragmentation and condensation

The eigenvalues λ_j and eigenfunctions $\phi_j^{(NO)}$ of the 1-RDM [Eq. (2)], "(NO)" being a shorthand for natural orbitals] determine the degree of fragmentation or condensation of the state $|\Psi\rangle$; we may rewrite

$$\rho^{(1)}(\mathbf{r},\mathbf{r}') = \sum_{\mathbf{j}} \lambda_{\mathbf{j}}\phi_{\mathbf{j}}^{(\mathbf{NO}),*}(\mathbf{r}')\phi_{\mathbf{j}}^{(\mathbf{NO})}(\mathbf{r}). \tag{3}$$

If only a single eigenvalue and eigenfunction are significant, the state $|\Psi\rangle$ represents a coherent condensate [41–43]. If, in contrast to the condensed case, two or more eigenvalues and eigenfunctions contribute to the 1-RDM, the state $|\Psi\rangle$ represents a fragmented condensate [44–46] that has lost some of its coherence [47].

3.3 Single-shot measurements

A single-shot measurement **s** is a projective measurement of a quantum state $|\Psi\rangle$. Single-shot measurements are random samples drawn from the probability $|\Psi|^2$. For ultracold atoms, single-shot measurements correspond to absorption images in real or momentum space. Ideally, such a single-shot image,

$$\mathbf{s} = (\mathbf{s_1}, ..., \mathbf{s_N}), \tag{4}$$

contains the positions or momenta $\mathbf{s_1}, ..., \mathbf{s_N}$ of all N particles in the state $|\Psi\rangle$.

3.3.1 Full distribution functions

Single-shot images allow for the extraction of the full distribution function $P_n(\mathbf{r})$ of the particle number at position **r**. From a set of N_s single-shot images, $\{\mathbf{s^1}, ..., \mathbf{s^{N_s}}\}$ one can find $P_n(\mathbf{r})$ by determining the relative abundance of the detection of n particles at position **r** by counting how many of the N_s single-shot images contain exactly n particles at position **r**. Doing this analysis for $n = 1, n = 2, ..., n = N$ and for all positions **r** yields the full distribution function $P_n(\mathbf{r})$.

4 Quantum simulation of crystallization

The quantum properties underlying crystal formation can be replicated and investigated with the help of ultracold atoms. In this work [9], how the use of dipolar atoms enables even the realization and precise measurement of structures that have not yet been observed in any material.

Crystals are ubiquitous in nature. They are formed by many different materials—from mineral salts to heavy metals like bismuth. Their structures emerge because a particular regular ordering of atoms or molecules is favorable, because it requires the smallest amount of energy. A cube with one constituent on each of its eight corners, for instance, is a crystal structure that is very common in nature. A crystal's structure determines many of its physical properties, such as how well it conducts a current or heat or how it cracks and behaves when it is illuminated by light. But what determines these crystal structures? They emerge as a consequence of the quantum properties of and the interactions between their constituents, which, however, are often scientifically hard to understand and also hard measure.

To nevertheless get to the bottom of the quantum properties of the formation of crystal structures, scientists can simulate the process using Bose–Einstein condensates—trapped ultracold atoms cooled down to temperatures close to absolute zero or minus 273.15 degrees Celsius. The atoms in these highly artificial and highly fragile systems are extremely well under control.

With careful tuning, the ultracold atoms behave exactly as if they were the constituents forming a crystal. Although building and running such a quantum simulator is a more demanding task than just growing a crystal from a certain material, the method offers two main advantages: First, scientists can tune the properties for the quantum simulator almost at will, which is not possible for conventional crystals. Second, the standard readout of cold-atom quantum simulators are images containing information about all crystal particles. For a conventional crystal, by contrast, only the exterior is visible, while the interior—and in particular its quantum properties—is difficult to observe.

In our work [9], we demonstrate that a flexible quantum simulator for crystal formation can be built using ultracold dipolar quantum particles. Dipolar quantum particles make it possible to realize and investigate not just conventional crystal structures, but also arrangements and many-body physics that were hitherto not seen for any material. The study explains how these crystal orders emerge from an intriguing competition between kinetic, potential, and interaction energy and how the structures and properties of the resulting crystals can be gauged in unprecedented detail, see Fig. 1.

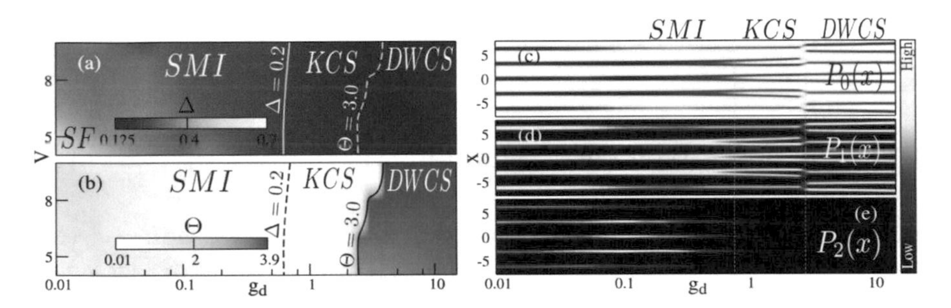

Fig. 1: Quantum simulation of crystallization with ultracold dipolar atoms: imbalance, order parameter, and full distribution functions. Panels (a) and (b) demonstrate that the superfluid/Mott-insulating (SMI), kinetic crystal (KCS) and density-wave crystal (DWCS) states can be mapped out with ultracold dipolar bosons in lattices and their crystal order parameter $\Delta = \sum_k \left[\frac{\lambda_k}{N}\right]^2$ and the imbalance of particles in even and odd sites, $\Theta = \sum_{e,o} \langle n_o \rangle - \langle n_e \rangle$. Alternatively, (c),(d),(e) show that the full distribution functions $P_n(x)$ with $n = 0, 1, 2$ [cf. Sec. 3.3.1] can be used to identify the SMI, KCS, and DWCS. Since Θ and $P_n(x)$ are extracted from single-shot images, our results demonstrate a viable experimental realization and readout of a tunable dipolar-ultracold-atom quantum simulator of crystallization. Figure adapted from [9].

5 Dynamics of ultracold bosons in artificial gauge fields: Angular momentum, fragmentation, and the variance of entropy

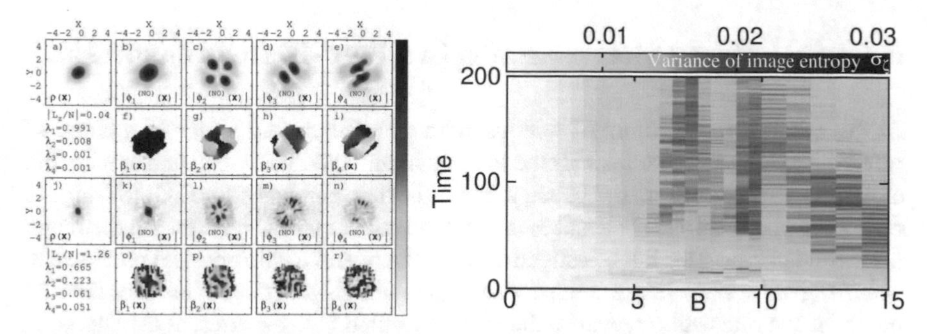

Fig. 2: Left: MCTDH-X simulations of one-body density [diagonal of Eq. (2)], natural orbitals [cf. Eq. (3)], and natural orbital phases, β_j, for two regimes of artificial gauge field strengths; weak in a)–i) and strong in j)–r) at time $t = 50.0$. Right: We show the variance of image entropy $\sigma_\zeta = \frac{1}{N_s}\left[\zeta - \zeta^j\right]^2$, where $\zeta^j = \int d\mathbf{r} s^j(\mathbf{r}) \ln s^j(\mathbf{r})$ and $\zeta = -\frac{1}{N_s}\sum_{j=1}^{N_s} \zeta^j$ as a function of propagation time and strength of the artificial gauge field. The variance of the image entropy is a sensitive probe of the evolution of fragmentation and angular momentum, see Ref. [10] for details and explanations on σ_ζ. Figure adapted from Ref. [10].

The charge neutrality inherent to ultracold trapped atoms introduces a constraint to explore many-body physics in presence of an external magnetic field. Magnetism in ultracold atoms can, however, effectively be realized via artificial gauge fields [48]. These artificial gauge fields can make the neutral ultracold atoms behave as if they were charged particles experiencing a magnetic field. In recent times, the successful realizations of artificial gauge fields in, both, continuum [49–51] and lattice potentials [52–54] have spearheaded novel ways for investigating the physics of charged particles with neutral atoms.

In this work [10], we analyze the quantum dynamics of two-dimensional weakly interacting ultracold bosons trapped in a harmonic potential triggered by an abrupt quench of an artificial gauge field. For a quench to a sufficiently strong artificial gauge field ghost vortices [55] at the edge of the density and phantom vortices [56] in the natural orbitals, accompanied by topological defects in their phase do emerge. These vortices and phase defects are hallmarks of the deposition of angular momentum in the quantum state as time proceeds, see Fig. 2(Left). We find that the quench in the artificial gauge field also triggers fragmentation and the emergence of non-trivial many-body correlations together with the dynamics of angular momentum. To test the possibility of the experimental detection of the emergent angular momentum and fragmentation in the BEC, we simulated single-shot images. We demonstrate that it is feasible to detect, angular momentum, fragmentation, and correlations via statistically analyzing the variance of the entropy of single-shot images, Fig. 2(Right).

6 Cavity-induced Mott transition in a Bose–Einstein condensate

At the absolute temperature of a thousandth of a Kelvin, all atoms in a cloud of rubidium gas condense into exactly the same quantum state. This collection of ultracold atoms is a new state of matter called a Bose–Einstein condensate (BEC) [57,58]. In recent years, technological advances in the preparation of BEC have turned them into a versatile platform for the investigation of different and exotic physical phenomena occurring at very low temperatures, where quantum physics dominates over thermal noise. In one interesting scenario, the BEC is excited with an external driving laser, and coupled to an optical cavity. This cavity can be realized as two mirrors facing each other, where photons from the laser can bounce back and forth and interact with the atoms. As the driving laser becomes stronger, the whole cavity-BEC system undergoes two consecutive phase transitions [59, 60]. In the first one, the atoms spontaneously self-organize into a lattice compatible with the profile generated by the cavity light-field [59]. Immediately after the self-organization the atoms are in coherence, or in other words, are still sharing one single quantum state as they do in the BEC. As a result, all atoms can flow freely in the whole lattice system, showing *superfluidity*. By pushing the laser intensity further, the atoms undergo the second phase transition, where they localize even more into the lattice structure and lose coherence [60], turning into a so-called *Mott insulator* state. Mott insulators are of great interest to physicists because it is proposed to underpin many condensed matter phenomena like high-temperature superconductivity [61,62], and could be applied for innovative electronics [63] like Mott transition field effect transistor (Mott-FET) [64]. The realization with a BEC in a cavity offers an extremely clean and controlled environment for the study of these phenomena. Although the Mott transition of an optical lattice system has been studied extensively over the past decades [27,60,65–69], it is still a difficult task to precisely determine its phase boundary. In the present work [11], we propose a new scheme to predict this phase boundary using MCTDH-X in a numerically exact fashion, and compare it quantitatively to experimental results. This is mainly performed by tracking the experimental and simulated momentum space density distributions $\rho(\boldsymbol{k})$ [diagonal of the momentum-space version of Eq. (2)] at various driving intensities, as presented in Fig. 3(a). Our main results are then presented in Fig. 3(b), where we compare the experimental and simulated phase diagrams consisting of the normal BEC phase (NP), the self-organized superfluid phase (SSF) and the self-organized Mott insulating phase (SMI). For a large range of parameter regimes, we see a remarkably good agreement in the SSF–SMI phase boundary between the experiments and the MCTDH-X simulations. Our study thus brings experimental realizations and theoretical predictions of Mott insulating physics one step closer to each other.

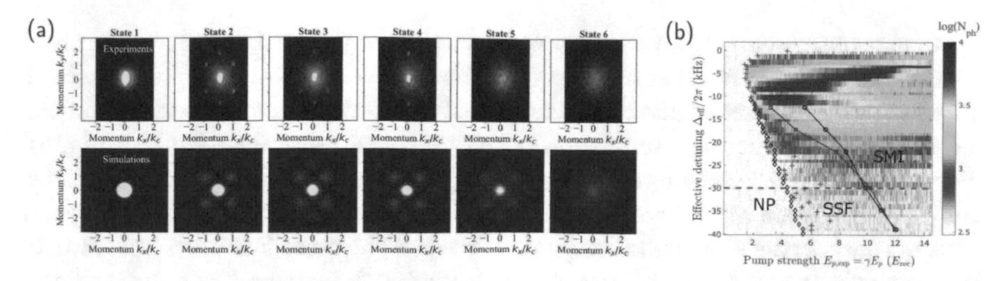

Fig. 3: (a) The momentum density distributions $\rho(k) = \langle \Psi_k^\dagger \Psi_k \rangle / N$ from (first row) experiments and (second row) MCTDH-X simulations for six different states, which are measured or simulated at various driving intensities. (b) The phase diagram delineating the normal BEC phase (NP), self-organized superfluid phase (SSF), and self-organized Mott insulator phase (SMI) from experiments and simulations. The brown crosses and the black circles are respectively the experimental NP–SSF and SSF–SMI boundaries; whereas the black diamonds and the blue squares are the experimental NP–SSF and SSF–SMI boundaries.

7 Interpretable and unsupervised phase classification

Phase diagrams and phase transitions are of paramount importance to physics [70–72]. While typical many-body systems have a large number of degrees of freedom, their phases are usually characterized by a small set of physical quantities like response functions or order parameters. For instance, the thermal phase transition in the celebrated two-dimensional classical Ising model [73] is revealed by the magnetization. However, in general the identification of phases and their order parameters is a complex problem involving a large state space [74, 75]. Machine learning methods are apt for this task [72, 76–85] as they can deal with large data sets and efficiently extract information from them. Ideally, such machine learning methods should not require prior knowledge about the phases, e.g., in the form of samples that are labelled by their correct phase, or even the number of distinct phases. That is, the methods should be unsupervised [77, 79, 86–103].

Yet, they should also allow for a straightforward physical insight into the character of phases. That is, we desire *interpretable* methods [104, 105] for which we understand why they yield a given phase classification, i.e., whose decision making is fully explainable. Significant progress in this direction has been made recently [98–103], but some open issues remain regarding the interpretability of phase classification methods. Many state-of-the-art phase classification methods rely on highly expressive machine learning models, such as deep neural networks [106] (DNNs), for which it is difficult to interpret the underlying functional dependence between their output and the input data [100, 101]. That is, these models are black-boxes which allow for a given phase classification task to be solved, but whose internal workings remain a mystery to the user.

In our contribution [12], we have proposed a novel interpretable and unsupervised phase classification method referred to as the *mean-based method*. The method requires input data that characterizes the state of a physical system for various choices of system parameters in a suitable symmetry-adapted representation (see Fig. 4). For each sampled point in parameter space, one calculates the magnitude of the difference between mean input features of neighbouring points in parameter space. This quantity serves as an indicator (label I in Fig. 4) of phase transitions and reveals the underlying phase diagram, because it is largest at phase boundaries. As such, the mean-based method is conceptually simple, computationally cheap, and generic in nature. In particular, the computation of its indicator does not rely on a black-box predictive model and allows for this data-driven method to be directly explainable.

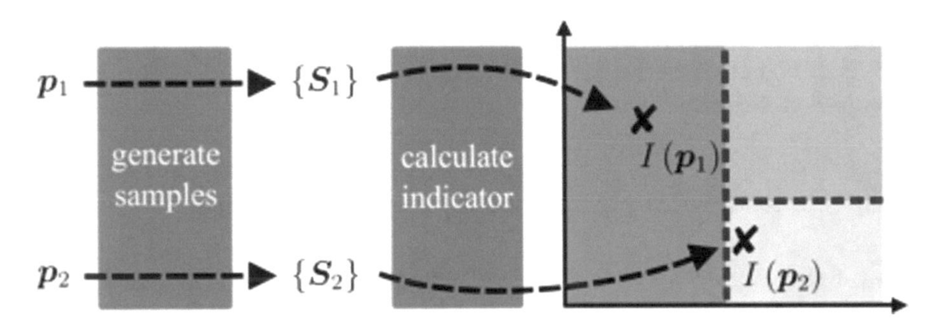

Fig. 4: Our workflow to predict a phase diagram with indicators I for phase transitions. Here, we illustrate the procedure for a two-dimensional parameter space: The parameter space is sampled on a grid which yields a set of points $\{\boldsymbol{p}_i\}$ of fixed system parameters. At each such point $\boldsymbol{p}_i$ a set of samples $\{\boldsymbol{S}_i\}$ is generated. Based on these samples, a scalar indicator for phase transitions, $I(\boldsymbol{p}_i)$, is calculated. This indicator highlights the boundaries (red) between phases (grey). Different unsupervised phase classification schemes are established via different indicators. Figure reprinted from [12].

As a physical system, we considered the two-dimensional spinless Falicov–Kimball model (FKM) [107–109]. This simple model of heavy and light fermions on a lattice is used to address a broad range of contemporary physical problems [110–115] and also serves as a standard test bed for the development of methods in the context of strongly correlated systems [109, 116–121]. The FKM ground-state phase diagram features a large number of different phases [122–124]. Figure 5(a) shows a sketch of the expected phase diagram in two-dimensional parameter space. It highlights the regions of stability of three main types of orderings, namely, (1) segregated, (2) diagonal, and (3) axial orderings. Hitherto, the classification of ground-state phases in the FKM was a manual and – due to the richness of the phase diagram [122–124] – lengthy and cumbersome task. The complexity of the FKM phase diagram makes it a challenging example for phase classification methods.

The probing of correlations is a standard procedure for studying phase transitions. This motivates the analysis of the FKM phase diagram with the set of correlation functions that measure square, axial, and diagonal order [Fig. 5(c)] as a suitable choice of representation for the input data. The resulting phase diagram is shown in Fig. 5(b). The mean-based method reproduces the main characteristics of the FKM phase diagram [compare Fig. 5(a),(b)]. Moreover, we obtain a detailed, physical subdivision of the phase diagram – see the identified orderings in Fig. 5 (top) and labels (1)–(9) in Fig. 5(b). In particular, the method is able to distinguish between dimer structures (7), different stable tile patterns (8),(9), or orderings exhibiting a complicated phase separation (4).

These results demonstrate that the mean-based method with correlation functions as input is an excellent tool to detect phase transitions. Moreover, the method can straightforwardly be extended to other types of inputs, such as the Fourier-transformed heavy-particle configurations, and higher-dimensional parameter spaces. This establishes the mean-based method as a general phase classification method: we infer that applications to arbitrary phase diagrams featuring, e.g., quantum or topological phase transitions are feasible. Specifically, applications to quantum-classical systems such as the FKM and its numerous generalizations [111, 116, 124–126] are now straightforward.

8 Longitudinal and transversal many-boson resonant tunneling in two-dimensional Josephson junctions

Going beyond our previous work on bosonic Josephson junctions [127–130], we investigate the out-of-equilibrium quantum dynamics of a few interacting bosonic clouds in a two-dimensional asymmetric double-well potential at the resonant-tunneling scenarios [13]. Recall that at the single-particle level of resonant tunneling, particles tunnel under the potential barrier from, typically, the ground-state in one well to an excited state in the other well, that is, states of distinct shapes and properties are coupled when their one-particle energies coincide. In two spatial dimensions, two kinds of resonant tunneling processes become possible, to which we call longitudinal and transversal resonant tunneling. Longitudinal resonant tunneling implies that the state in, say, the right well is longitudinally excited with respect to the state in the left well, whereas transversal resonant tunneling means that the state in the right well is transversely excited with respect to the state in the left well. We demonstrate that the inter-boson interaction makes resonant tunneling phenomena in two spatial dimensions profoundly rich. To this end, we analyze resonant-tunneling phenomena in terms of the loss of coherence of the junction and development of fragmentation, and the coupling between transversal and longitudinal degrees-of-freedom and excitations.

In [13], a comprehensive analysis of the tunneling dynamics is performed by exploring the time evolution of a few physical quantities, namely, the survival probability, occupation numbers of the reduced one-particle density matrix, and the many-particle position, momentum, and angular-momentum variances. To accurately

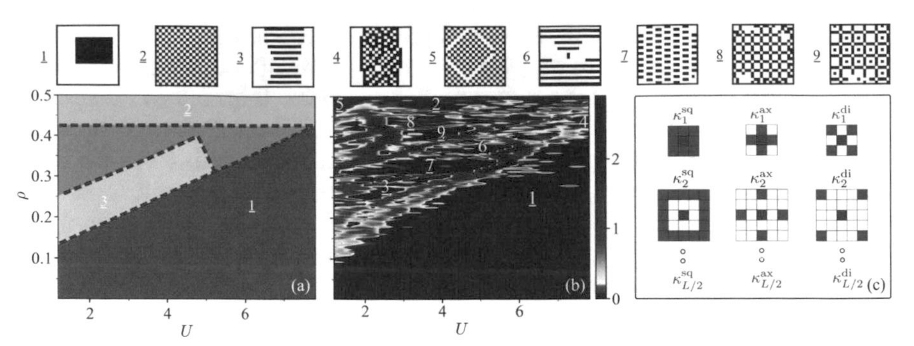

Fig. 5: (a) Sketch of the ground-state phase diagram of the spinless FKM. A multitude of other phases with smaller stability regions are expected to be present in the full diagram [122, 123]. Red-dashed lines highlight the boundaries of the phases with (1) segregated, (2) diagonal, and (3) axial orderings. For each ordering (1)–(3), an example of a typical ground-state heavy-particle configuration $\boldsymbol{w}_0$ on a square lattice with linear size $L = 20$ is shown on top. Here, the absence ($w_{0,i} = 0$) and presence ($w_{0,i} = 1$) of a heavy particle at lattice site i is denoted by a white or black square, respectively. (b) Indicator for phase transitions obtained with the mean-based method and correlation functions as input. Representative configurations (1)–(9) for some of the largest inferred regions of stability (connected regions marked in blue by the indicator), i.e., phases, are shown on top: These regions connect configurations of the same character. (c) Illustration of the correlation functions that measure square (κ_n^{sq}), axial (κ_n^{ax}), and diagonal (κ_n^{di}) correlation at a distance n from the origin. Blue squares denote the lattice sites marked by the corresponding stencil, where red denotes the origin. Figure adapted from [12].

calculate these physical quantities from the time-dependent many-boson wavefunction, we employ the MCTDHB method [18, 23–25, 30, 131, 132]. We use the numerical implementation in [15, 16]. By comparing the survival probabilities and variances at the many-body and mean-field levels of theory, and by following the development of fragmentation, we unravel the detailed mechanisms of many-body longitudinal as well as transversal resonant-tunneling processes in two-dimensional asymmetric double-well potentials. In particular, we show that the position and momentum variances along the transversal direction are almost negligible at the longitudinal resonant-tunneling conditions, whereas they are considerable at the transversal resonant-tunneling conditions which is administrated by the combination of density- and breathing-mode oscillations. We demonstrate explicitly that the range of the inter-boson interaction (width σ of the Gaussian interaction potential) does qualitatively not affect neither the mean-field nor the many-body physics of two-dimensional resonant-tunneling dynamics, see Fig. 6. All in all, we characterize the impact of the transversal and longitudinal degrees-of-freedom in the many-boson tunneling dynamics at the resonant-tunneling scenarios. The results are computed highly accurately in two spatial dimensions at the many-body level of theory using MCTDHB, see Fig. 7 for a selected few converged sensitive quantities taken from [13].

9 Morphology of a three-dimensional trapped Bose–Einstein condensate from position, momentum, and angular-momentum many-particle variances

The theory and properties of trapped BECs at the limit of an infinite number of particles have attracted much interest [133–144]. In [14], we analyze the many-particle position, momentum, and angular-momentum variances of a three-dimensional anisotropic trapped BEC at the limit of an infinite number of particles, addressing three-dimensional scenarios that have neither one-dimensional nor two-dimensional analogs [138, 140, 145–150]. The variance of the position operator is associated with the width of a wave-packet in position space, the variance of the momentum operator is similarly related to the width of wave-packet in momentum space, and the variance of the angular-momentum operator is correlated with how much a wave-packet is non-spherically symmetric.

To this end, we compute the variances of the three Cartesian components of the position, momentum, and angular-momentum operators of an interacting three-dimensional trapped BEC at the limit of an infinite-particle-number limit, and investigate their respective anisotropies [13]. We examine simple scenarios and show that the anisotropy of a BEC can be different at the many-body and mean-field levels of theory, although the BEC has identical many-body and mean-field densities per particle. The analysis offers a geometrical-based picture to classify correlations via the morphology of 100% condensed bosons in a three-dimensional trap at the limit of an infinite number of particles. Fig. 8 presents results for the out-of-equilibrium quench dynamics of the position ($\hat{X}$, $\hat{Y}$, and $\hat{Z}$) and momentum ($\hat{P}_X$, $\hat{P}_Y$, and $\hat{P}_Z$) variances per particle, and Fig. 9 depicts results for the ground-state angular-momentum ($\hat{L}_X$, $\hat{L}_Y$, and $\hat{L}_Z$) variances per particle with and without spatial translations. The position and momentum variances are given analytically within many-body theory and computed numerically within mean-field theory, whereas the angular-momentum variances are computed analytically both at the many-body and mean-field levels of theory using the anisotropic three-dimensional harmonic-interaction model, see in this context [151–166].

10 Conclusions and outlook

This 2021 report documents the substantial scientific activity in the framework of the MCTDHB project spurred by the computational resources at the HLRS: our high-performance computations resulted in significant contributions to the literature

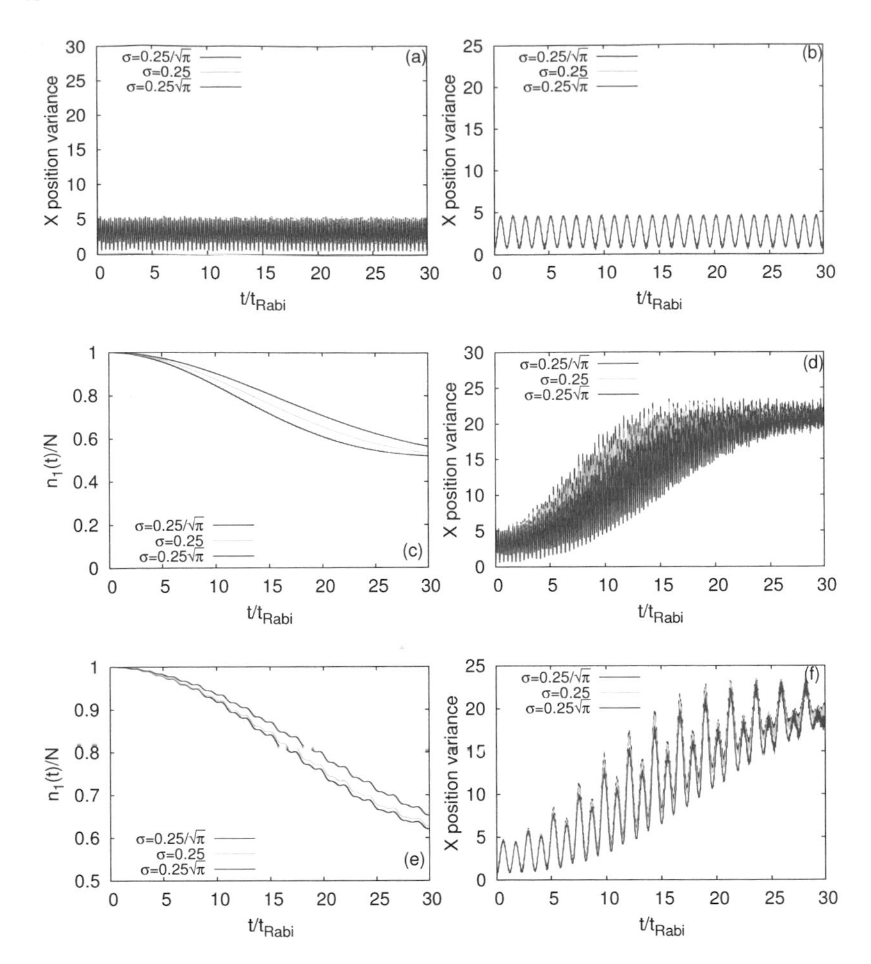

Fig. 6: Time-dependent mean-field variance per particle of the $\hat{X}$ component of the position operator for the ground state at the (a) longitudinal ($c = 0.25$) and (b) transversal ($\omega_n = 0.19$) resonant-tunneling scenarios. The three different widths of the inter-boson interaction potential are $\sigma = 0.25\sqrt{\pi}$, $\sigma = 0.25$, and $\sigma = 0.25/\sqrt{\pi}$. The interaction parameter is $\Lambda = 0.01\pi$. The mean-field results are found to be visibly independent of σ. Many-body results for the longitudinal [panels (c) and (d)] and transversal [panels (e) and (f)] dynamics. The dynamics are shown for the first occupation number per particle $\frac{n_1(t)}{N}$ [cf. λ_1 in Eq. (3)], see panels (c) and (e), and the $\hat{X}$ position variance per particle, see panels (d) and (f), respectively. The many-body dynamics are computed with $N = 10$ bosons, $M = 6$ time-adaptive orbitals, and the same interaction parameter. The quantitative many-body results are found to be weakly dependent on the inter-boson interaction width σ. All in all, we find that neither the mean-field nor the many-body physics of the two-dimensional resonant-tunneling dynamics qualitatively depend on σ. The color codes are explained in each panel. See [13] for more details and the definition of the variances shown in (b),(d),(f). The quantities shown are dimensionless. Figure panels reprinted from [13].

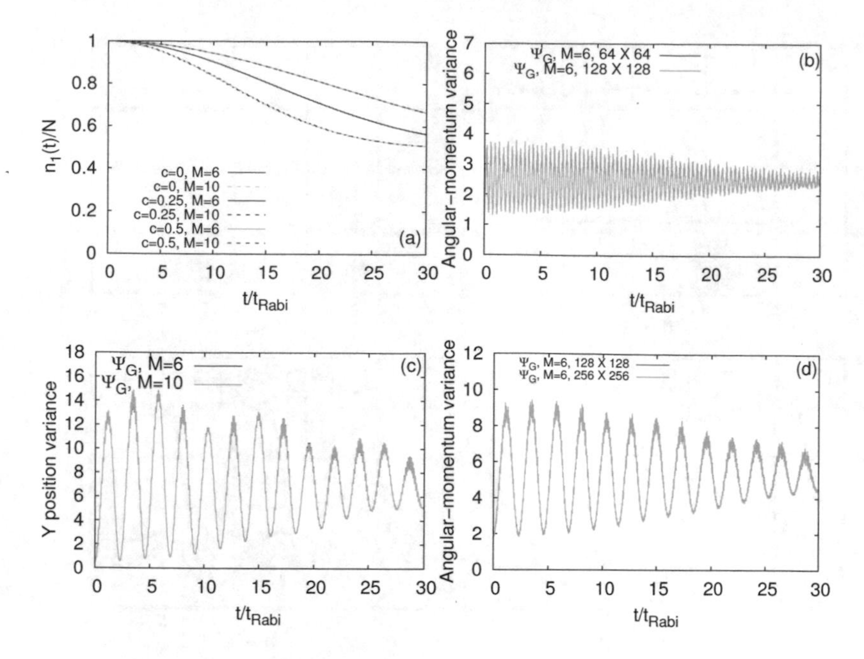

Fig. 7: Convergence of time-dependent many-body quantities with computational resources for the initial state Ψ_G for the longitudinally-asymmetric [panels (a) and (b)] and transversely-asymmetric [panels (c) and (d)] two-dimensional double-well potentials at the many-particle resonant-tunneling conditions. The bosonic cloud consists of $N = 10$ bosons with the interaction parameter $\Lambda = 0.01\pi$. (a) Convergence of the first occupation number per particle $\frac{n_1(t)}{N}$ [cf. λ_1 in Eq. (3)] with the number of time-dependent orbitals. The asymmetry parameters are $c = 0$, $c = 0.25$, and $c = 0.5$. The convergence is verified with $M = 6$ and $M = 10$ time-adaptive orbitals. (b) The variance per particle of the $\hat{L}_Z$ component of the angular-momentum operator with the number of grid points. The asymmetry parameter is $c = 0.25$. The convergence is verified with 64×64 and 128×128 grid points. (c) The variance per particle of the $\hat{Y}$ component of the position operator with the number of time-adaptive orbitals. The frequency is $\omega_n = 0.19$. (d) The variance per particle of the $\hat{L}_Z$ component of the angular-momentum operator with the number of grid points. The frequency is $\omega_n = 0.19$. The convergence is verified with 128×128 and 256×256 grid points. The color codes are explained in each panel. See [13] for more details and the definition of the variances shown in (b)–(d). The quantities shown are dimensionless. Figure panels reprinted from [13].

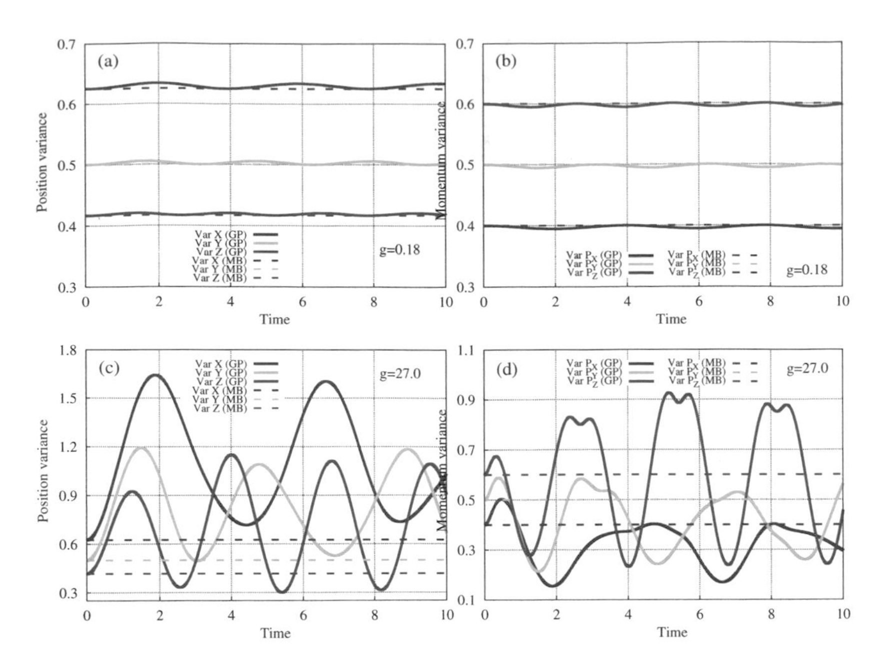

Fig. 8: Many-particle position [$\hat{X}$, $\hat{Y}$, and $\hat{Z}$; panels (a) and (c) in red, green, and blue] and momentum [$\hat{P}_X$, $\hat{P}_Y$, and $\hat{P}_Z$; panels (b) and (d) in red, green, and blue] variances per particle as a function of time computed at the limit of an infinite number of particles within many-body (dashed lines) and mean-field (solid lines) levels of theory in an interaction-quench scenario. The harmonic trap is 20% anisotropic. The coupling constant g in the Gross–Pitaevskii equation is indicated in each panel. Different anisotropy classes of the position variances and of the momentum variances emerge with time. See [14] for more details and the definition of the variances shown in (a)–(d). Figure panels reprinted from [14].

on many-body physics [9–14] as well as the implementation and distribution of an increasing amount of computational software [15–17, 29, 167] and datasets [168, 169]. We unveiled exciting fundamental many-body physics and novel computational tools and thus underpin the versatility and prosperity of the MCTDHB project at the HLRS. We look forward to continue our quest to compute the quantum many-body properties for ever larger systems and envision large-scale computations with, both, large configuration spaces and large spatial two- and three-dimensional spatial grids.

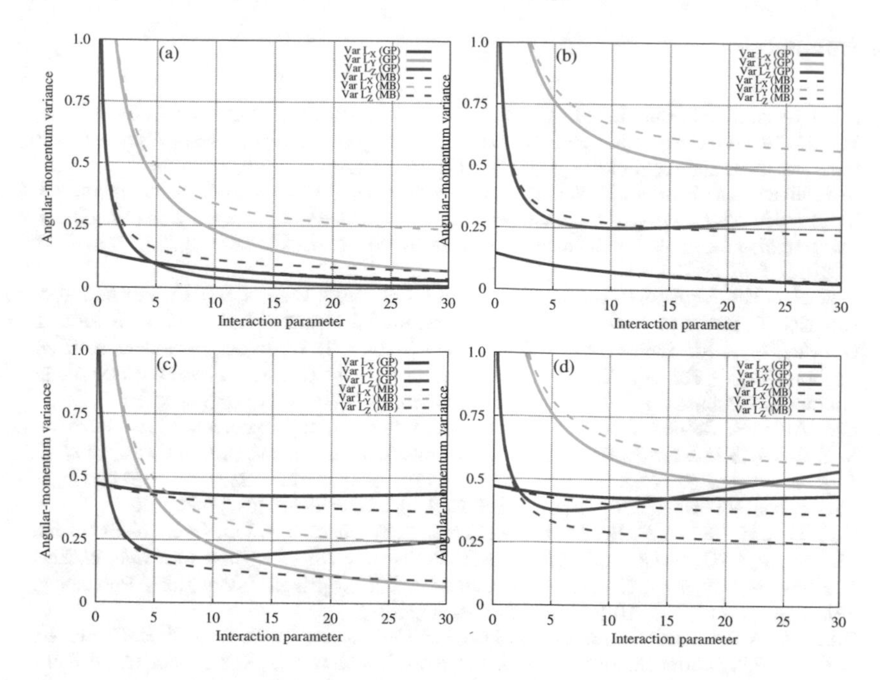

Fig. 9: Many-particle angular-momentum ($\hat{L}_X$, $\hat{L}_Y$, and $\hat{L}_Z$; in red, green, and blue) variances per particle as a function of the interaction parameter Λ computed at the limit of an infinite number of particles within many-body (dashed lines) and mean-field (solid lines) levels of theory for the ground state of the three-dimensional anisotropic harmonic-interaction model. The frequencies of the trap are $\omega_x = 0.7$, $\omega_y = 5.0$, and $\omega_z = 10.5$. Results for several translations $\mathbf{r}_0$ of the center of the trap are shown in the panels: (a) $\mathbf{r}_0 = (0, 0, 0)$; (b) $\mathbf{r}_0 = (0.25, 0, 0)$; (c) $\mathbf{r}_0 = (0, 0.25, 0)$; (d) $\mathbf{r}_0 = (0.25, 0.25, 0)$. Different anisotropy classes of the angular-momentum variances emerge with the interaction parameter. See [14] for more details and the definition of the variances shown in (a)–(d). The quantities shown are dimensionless. Figure panels reprinted from [14].

Acknowledgements Financial support by the Deutsche Forschungsgemeinschaft (DFG) is gratefully acknowledged. OEA acknowledges funding by the Israel Science Foundation (Grants No. 600/15 and No. 1516/19). We acknowledge financial support by the Austrian Science Foundation (FWF) under grants P32033 and M2653. We acknowledge financial support by the Swiss National Science Foundation (SNSF), the NCCR Quantum Science and Technology, and Mr. G. Anderheggen. Computation time on the HPC2013 cluster of the IIT Kanpur and on the HazelHen and Hawk clusters at the HLRS Stuttgart is gratefully acknowledged.

References

1. A.U.J. Lode, K. Sakmann, R.A. Doganov, J. Grond, O.E. Alon, A.I. Streltsov, L.S. Cederbaum, in *High Perform. Comput. Sci. Eng. '13 Trans. High Perform. Comput. Center, Stuttgart 2013* (Springer International Publishing, 2013), pp. 81–92. DOI 10.1007/978-3-319-02165-2_7
2. S. Klaiman, A.U.J. Lode, K. Sakmann, O.I. Streltsova, O.E. Alon, L.S. Cederbaum, A.I. Streltsov, in *High Perform. Comput. Sci. Eng. '14 Trans. High Perform. Comput. Center, Stuttgart 2014* (Springer International Publishing, 2015), pp. 63–86. DOI 10.1007/978-3-319-10810-0_5
3. O.E. Alon, V.S. Bagnato, R. Beinke, I. Brouzos, T. Calarco, T. Caneva, L.S. Cederbaum, M.A. Kasevich, S. Klaiman, A.U.J. Lode, S. Montangero, A. Negretti, R.S. Said, K. Sakmann, O.I. Streltsova, M. Theisen, M.C. Tsatsos, S.E. Weiner, T. Wells, A.I. Streltsov, in *High Perform. Comput. Sci. Eng. '15 Trans. High Perform. Comput. Center, Stuttgart 2015* (Springer International Publishing, 2016), pp. 23–49. DOI 10.1007/978-3-319-24633-8_3
4. O.E. Alon, R. Beinke, L.S. Cederbaum, M.J. Edmonds, E. Fasshauer, M.A. Kasevich, S. Klaiman, A.U.J. Lode, N.G. Parker, K. Sakmann, M.C. Tsatsos, A.I. Streltsov, in *High Perform. Comput. Sci. Eng. '16 Trans. High Perform. Comput. Cent. Stuttgart 2016* (Springer International Publishing, 2017), pp. 79–96. DOI 10.1007/978-3-319-47066-5_6
5. O.E. Alon, R. Beinke, C. Bruder, L.S. Cederbaum, S. Klaiman, A.U.J. Lode, K. Sakmann, M. Theisen, M.C. Tsatsos, S.E. Weiner, A.I. Streltsov, in *High Perform. Comput. Sci. Eng. 17 Trans. High Perform. Comput. Center, Stuttgart 2017* (Springer International Publishing, 2018), pp. 93–115. DOI 10.1007/978-3-319-68394-2_6
6. O.E. Alon, V.S. Bagnato, R. Beinke, S. Basu, L.S. Cederbaum, B. Chakrabarti, B. Chatterjee, R. Chitra, F.S. Diorico, S. Dutta, L. Exl, A. Gammal, S.K. Haldar, S. Klaiman, C. Lévêque, R. Lin, N.J. Mauser, P. Molignini, L. Papariello, R. Roy, K. Sakmann, A.I. Streltsov, G.D. Telles, M.C. Tsatsos, R. Wu, A.U.J. Lode, in *High Perform. Comput. Sci. Eng. ' 18* (Springer International Publishing, 2019), pp. 89–110. DOI 10.1007/978-3-030-13325-2_6
7. A.U.J. Lode, O.E. Alon, L.S. Cederbaum, B. Chakrabarti, B. Chatterjee, R. Chitra, A. Gammal, S.K. Haldar, M.L. Lekala, C. Lévêque, R. Lin, P. Molignini, L. Papariello, M.C. Tsatsos, in *High Perform. Comput. Sci. Eng. ' 19*, ed. by W.E. Nagel, D.H. Kröner, M.M. Resch (Springer International Publishing, Cham, 2021), pp. 77–87. DOI 10.1007/978-3-030-66792-4. URL https://link.springer.com/10.1007/978-3-030-66792-4
8. A. Lode, A. U. J. , Alon, O. E., Bastarrachea-Magnani, M. A. , Bhowmik, d.F. de Buchleitner, A., Cederbaum, L. S., Chitra, R., Fasshauer, E., S.E. Parny, L., Haldar, S. K., Lévêque, C., Lin, R., Madsen, L. B., Molignini, P., Papariello, L., Schäfer, F., Strelstov, A. I., Tsatsos, M. C., and Weiner, in *High Perform. Comput. Sci. Eng. '20 Trans. High Perform. Comput. Center, Stuttgart 2020* (2022), p. (in press)
9. B. Chatterjee, C. Lévêque, J. Schmiedmayer, A.U.J. Lode, Phys. Rev. Lett. **125**(9), 093602 (2020). DOI 10.1103/PhysRevLett.125.093602. URL https://link.aps.org/doi/10.1103/PhysRevLett.125.093602
10. A.U.J. Lode, S. Dutta, C. Lévêque, Entropy **23**(4), 392 (2021). DOI 10.3390/e23040392. URL https://www.mdpi.com/1099-4300/23/4/392
11. R. Lin, C. Georges, J. Klinder, P. Molignini, M. Büttner, A.U.J. Lode, R. Chitra, A. Hemmerich, H. Keßler, SciPost Phys. **11**, 30 (2021). DOI 10.21468/SciPostPhys.11.2.030. URL https://scipost.org/10.21468/SciPostPhys.11.2.030
12. J. Arnold, F. Schäfer, M. Žonda, A.U.J. Lode, Phys. Rev. Research **3**, 033052 (2021). DOI 10.1103/PhysRevResearch.3.033052. URL https://link.aps.org/doi/10.1103/PhysRevResearch.3.033052
13. A. Bhowmik, O.E. Alon, (2021). URL http://arxiv.org/abs/2101.04959
14. O.E. Alon, Symmetry **13**(7) (2021). DOI 10.3390/sym13071237. URL https://www.mdpi.com/2073-8994/13/7/1237
15. R. Streltsov, A. I., Cederbaum, L. S., Alon, O. E., Sakmann, K., Lode, A. U. J., Grond, J., Streltsova, O. I., Klaiman, S., Beinke. The Multiconfigurational Time-Dependent Hartree for Bosons Package, version 3.x, http://mctdhb.org, Heidelberg/Kassel (2006-Present). URL http://mctdhb.org

16. O.I. Streltsov, A. I., Streltsova. The multiconfigurational time-dependent Hartree for bosons laboratory, version 1.5, http://MCTDHB-lab.com. URL http://mctdhb-lab.com
17. A.U.J. Lode, M.C. Tsatsos, E. Fasshauer, R. Lin, L. Papariello, P. Molignini, S.E. Weiner, C. Lévêque. MCTDH-X: The multiconfigurational time-dependent Hartree for indistinguishable particles software, http://ultracold.org (2020). URL http://ultracold.org
18. A.U.J. Lode, C. Lévêque, L.B. Madsen, A.I. Streltsov, O.E. Alon, Rev. Mod. Phys. **92**, 011001 (2020). DOI 10.1103/RevModPhys.92.011001. URL https://link.aps.org/doi/10.1103/RevModPhys.92.011001
19. R. Lin, P. Molignini, L. Papariello, M.C. Tsatsos, C. Lévêque, S.E. Weiner, E. Fasshauer, R. Chitra, A.U.J. Lode, Quantum Sci. Technol. **5**, 024004 (2020). DOI 10.1088/2058-9565/ab788b. URL https://iopscience.iop.org/article/10.1088/2058-9565/ab788b
20. E. Fasshauer, A.U.J. Lode, Phys. Rev. A **93**(3), 033635 (2016). DOI 10.1103/PhysRevA.93.033635. URL https://link.aps.org/doi/10.1103/PhysRevA.93.033635
21. A.U.J. Lode, C. Bruder, Phys. Rev. A **94**(1), 013616 (2016). DOI 10.1103/PhysRevA.94.013616
22. A.U.J. Lode, Phys. Rev. A **93**(6), 063601 (2016). DOI 10.1103/PhysRevA.93.063601. URL https://link.aps.org/doi/10.1103/PhysRevA.93.063601
23. O.E. Alon, A.I. Streltsov, L.S. Cederbaum, J. Chem. Phys. **127**(15), 154103 (2007). DOI 10.1063/1.2771159
24. A.I. Streltsov, O.E. Alon, L.S. Cederbaum, Phys. Rev. Lett. **99**(3), 030402 (2007). DOI 10.1103/PhysRevLett.99.030402
25. O.E. Alon, A.I. Streltsov, L.S. Cederbaum, Phys. Rev. A **77**(3), 033613 (2008). DOI 10.1103/PhysRevA.77.033613
26. K. Sakmann, M. Kasevich, Nat. Phys. **12**(5), 451 (2016). DOI 10.1038/nphys3631. URL http://www.nature.com/articles/nphys3631
27. A.U.J. Lode, C. Bruder, Phys. Rev. Lett. **118**(1), 013603 (2017). DOI 10.1103/PhysRevLett.118.013603
28. J.H.V. Nguyen, M.C. Tsatsos, D. Luo, A.U.J. Lode, G.D. Telles, V.S. Bagnato, R.G. Hulet, Phys. Rev. X **9**(1), 011052 (2019). DOI 10.1103/PhysRevX.9.011052. URL https://link.aps.org/doi/10.1103/PhysRevX.9.011052
29. Arnold, Julian, Schäfer, Frank, Zonda, Martin, Lode, A. U. J. Interpretable-and-unsupervised-phase-classification. URL Interpretable-and-unsupervised-phase-classification
30. A.U.J. Lode, Phys. Rev. A **93**(6), 063601 (2016). DOI 10.1103/PhysRevA.93.063601. URL https://link.aps.org/doi/10.1103/PhysRevA.93.063601
31. M.H. Beck, A. Jäckle, G.A. Worth, H.D. Meyer, Phys. Rep. **324**(1), 1 (2000). DOI 10.1016/S0370-1573(99)00047-2
32. U. Manthe, J. Phys.: Condens. Matter **29**(25), 253001 (2017). DOI 10.1088/1361-648X/aa6e96. URL https://iopscience.iop.org/article/10.1088/1361-648X/aa6e96
33. H. Wang, M. Thoss, J. Chem. Phys. **119**, 1289 (2003). DOI 10.1063/1.1580111. URL http://aip.scitation.org/doi/10.1063/1.1580111
34. H. Miyagi, L.B. Madsen, Phys. Rev. A **87**(6), 062511 (2013). DOI 10.1103/PhysRevA.87.062511. URL https://link.aps.org/doi/10.1103/PhysRevA.87.062511
35. H. Miyagi, L.B. Madsen, Phys. Rev. A **89**(6), 063416 (2014). DOI 10.1103/PhysRevA.89.063416. URL https://link.aps.org/doi/10.1103/PhysRevA.89.063416
36. C. Lévêque, L.B. Madsen, New J. Phys. **19**, 043007 (2017). DOI 10.1088/1367-2630/aa6319. URL http://stacks.iop.org/1367-2630/19/i=4/a=043007
37. C. Lévêque, L.B. Madsen, J. Phys. B: At., Mol. Opt. Phys. **51**, 155302 (2018). DOI 10.1088/1361-6455/aacac6. URL https://iopscience.iop.org/article/10.1088/1361-6455/aacac6/pdf
38. L. Cao, S. Krönke, O. Vendrell, P. Schmelcher, J. Chem. Phys. **139**(13), 134103 (2013). DOI 10.1063/1.4821350
39. L. Cao, V. Bolsinger, S.I. Mistakidis, G.M. Koutentakis, S. Krönke, J.M. Schurer, P. Schmelcher, J. Chem. Phys. **147**(4), 044106 (2017). DOI 10.1063/1.4993512. URL http://aip.scitation.org/doi/10.1063/1.4993512

40. P. Kramer, M. Saraceno, *Geometry of the time-dependent variational principle in quantum mechanics* (Springer, Lecture Notes in Physics, 2007). DOI 10.1007/3-540-10271-x_317
41. O. Penrose, L. Onsager, Phys. Rev. **104**(3), 576 (1956). DOI 10.1103/PhysRev.104.576
42. E.P. Gross, Nuovo Cim. **20**(3), 454 (1961). DOI 10.1007/BF02731494. URL http://link.springer.com/10.1007/BF02731494
43. L.P. Pitaevskii, Sov. Phys. JETP **13**(2), 451 (1961). URL http://jetp.ac.ru/cgi-bin/dn/e{_}013{_}02{_}0451.pdf
44. E.J. Mueller, T.L. Ho, M. Ueda, G. Baym, Phys. Rev. A **74**, 033612 (2006)
45. P. Nozieres, D. St. James, J. Phys. Paris **43**(7), 1133 (1982). DOI 10.1051/jphys:019820043070113300. URL http://www.edpsciences.org/10.1051/jphys:019820043070113300
46. R.W. Spekkens, J.E. Sipe, Phys. Rev. A **59**(5), 3868 (1999). DOI 10.1103/PhysRevA.59.3868
47. K. Sakmann, A.I. Streltsov, O.E. Alon, L.S. Cederbaum, Phys. Rev. A **78**(2), 023615 (2008). DOI 10.1103/PhysRevA.78.023615
48. J. Dalibard, F. Gerbier, G. Juzeliūnas, P. Öhberg, Rev. Mod. Phys. **83**, 1523 (2011). DOI 10.1103/RevModPhys.83.1523. URL https://link.aps.org/doi/10.1103/RevModPhys.83.1523
49. I.B. Spielman, Phys. Rev. A **79**(6), 063613 (2009). DOI 10.1103/PhysRevA.79.063613. URL https://link.aps.org/doi/10.1103/PhysRevA.79.063613
50. Y.J. Lin, R.L. Compton, K. Jiménez-García, J.V. Porto, I.B. Spielman, Nature **462**(7273), 628 (2009). DOI 10.1038/nature08609. URL http://www.nature.com/articles/nature08609
51. Y.J. Lin, R. Compton, A. Perry, W. Phillips, J. Porto, I. Spielman, Phys. Rev. Lett. **102**(13), 130401 (2009). DOI 10.1103/PhysRevLett.102.130401. URL https://link.aps.org/doi/10.1103/PhysRevLett.102.130401
52. M. Aidelsburger, *Artificial Gauge Fields with Ultracold Atoms in Optical Lattices*. Springer Theses (Springer International Publishing, Cham, 2016). DOI 10.1007/978-3-319-25829-4. URL http://link.springer.com/10.1007/978-3-319-25829-4
53. M. Aidelsburger, M. Atala, S. Nascimbène, S. Trotzky, Y.A. Chen, I. Bloch, Phys. Rev. Lett. **107**(25), 255301 (2011). DOI 10.1103/PhysRevLett.107.255301. URL https://link.aps.org/doi/10.1103/PhysRevLett.107.255301
54. L.K. Lim, C.M. Smith, A. Hemmerich, Phys. Rev. Lett. **100**(13), 130402 (2008). DOI 10.1103/PhysRevLett.100.130402. URL https://link.aps.org/doi/10.1103/PhysRevLett.100.130402
55. M. Tsubota, K. Kasamatsu, M. Ueda, Phys. Rev. A **65**, 023603 (2002). DOI 10.1103/PhysRevA.65.023603. URL https://link.aps.org/doi/10.1103/PhysRevA.65.023603
56. S.E. Weiner, M.C. Tsatsos, L.S. Cederbaum, A.U.J. Lode, Sci. Reports **7**, 40122 (2017). DOI 10.1038/srep40122
57. M.H. Anderson, J.R. Ensher, M.R. Matthews, C.E. Wieman, E.A. Cornell, Collect. Pap. Carl Wieman **269**, 453 (2008). DOI 10.1142/9789812813787_0062
58. K.B. Davis, M.O. Mewes, M.R. Andrews, N.J. Van Druten, D.S. Durfee, D.M. Kurn, W. Ketterle, Phys. Rev. Lett. **75**(22), 3969 (1995). DOI 10.1103/PhysRevLett.75.3969
59. K. Baumann, C. Guerlin, F. Brennecke, T. Esslinger, Nature **464**, 1301 (2010). DOI 10.1038/nature09009. URL http://www.nature.com/articles/nature09009
60. J. Klinder, H. Keßler, M.R. Bakhtiari, M. Thorwart, A. Hemmerich, Phys. Rev. Lett. **115**(23), 230403 (2015). DOI 10.1103/PhysRevLett.115.230403. URL https://link.aps.org/doi/10.1103/PhysRevLett.115.230403
61. P.A. Lee, N. Nagaosa, X.G. Wen, Rev. Mod. Phys. **78**, 17 (2006)
62. T. Yanagisawa, M. Miyazaki, EPL (Europhysics Lett. **107**, 27004 (2014). DOI 10.1209/0295-5075/107/27004. URL https://iopscience.iop.org/article/10.1209/0295-5075/107/27004
63. You Zhou, S. Ramanathan, Proc. IEEE **103**, 1289 (2015). DOI 10.1109/JPROC.2015.2431914. URL http://ieeexplore.ieee.org/document/7137616/
64. D. Newns, T. Doderer, C. Tsuei, W. Donath, J. Misewich, A. Gupta, B. Grossman, A. Schrott, B. Scott, P. Pattnaik, R. von Gutfeld, J. Sun, J. Electroceramics **4**, 339 (2000). DOI 10.1023/A:1009914609532. URL http://link.springer.com/10.1023/A:1009914609532

65. M.R. Bakhtiari, A. Hemmerich, H. Ritsch, M. Thorwart, Phys. Rev. Lett. **114**, 123601 (2015). DOI 10.1103/PhysRevLett.114.123601. URL https://link.aps.org/doi/10.1103/PhysRevLett.114.123601
66. P. Molignini, L. Papariello, A.U.J. Lode, R. Chitra, Phys. Rev. A **98**, 053620 (2018). DOI 10.1103/PhysRevA.98.053620. URL https://link.aps.org/doi/10.1103/PhysRevA.98.053620
67. A.U.J. Lode, F.S. Diorico, R. Wu, P. Molignini, L. Papariello, R. Lin, C. Lévêque, L. Exl, M.C. Tsatsos, R. Chitra, N.J. Mauser, New J. Phys. **20**(5), 055006 (2018). DOI 10.1088/1367-2630/aabc3a. URL https://doi.org/10.1088/1367-2630/aabc3a
68. R. Lin, L. Papariello, P. Molignini, R. Chitra, A.U.J. Lode, Phys. Rev. A **100**(1), 013611 (2019). DOI 10.1103/PhysRevA.100.013611. URL https://link.aps.org/doi/10.1103/PhysRevA.100.013611
69. R. Lin, P. Molignini, A.U.J. Lode, R. Chitra, Phys. Rev. A **101**, 061602 (2020). DOI 10.1103/PhysRevA.101.061602. URL https://link.aps.org/doi/10.1103/PhysRevA.101.061602
70. S. Sachdev, *Quantum Phase Transitions* (Cambridge University Press, Cambridge, 2011). DOI 10.1017/CBO9780511973765. URL https://doi.org/10.1017{%}2Fcbo9780511973765
71. N. Goldenfeld, *Lectures on Phase Transitions and the Renormalization Group* (CRC Press, 2018). DOI 10.1201/9780429493492. URL https://www.taylorfrancis.com/books/9780429962042
72. G. Carleo, I. Cirac, K. Cranmer, L. Daudet, M. Schuld, N. Tishby, L. Vogt-Maranto, L. Zdeborová, Rev. Mod. Phys. **91**(4), 045002 (2019). DOI 10.1103/RevModPhys.91.045002. URL https://link.aps.org/doi/10.1103/RevModPhys.91.045002
73. L. Onsager, Phys. Rev. **65**(3-4), 117 (1944). DOI 10.1103/PhysRev.65.117. URL https://link.aps.org/doi/10.1103/PhysRev.65.117
74. J.P. Sethna, *Statistical Mechanics: Entropy, Order Parameters, and Complexity* (Oxford University Press, 2021). DOI 10.1093/oso/9780198865247.001.0001. URL https://oxford.universitypressscholarship.com/view/10.1093/oso/9780198865247.001.0001/oso-9780198865247
75. P.M. Chaikin, T.C. Lubensky, *Principles of Condensed Matter Physics* (Cambridge University Press, 1995). DOI 10.1017/CBO9780511813467. URL https://www.cambridge.org/core/product/identifier/9780511813467/type/book
76. J. Carrasquilla, R.G. Melko, Nat. Phys. **13**(5), 431 (2017). DOI 10.1038/nphys4035. URL http://www.nature.com/articles/nphys4035
77. E.P.L. van Nieuwenburg, Y.H. Liu, S.D. Huber, Nat. Phys. **13**(5), 435 (2017). DOI 10.1038/nphys4037. URL http://www.nature.com/articles/nphys4037
78. K. Ch'ng, J. Carrasquilla, R.G. Melko, E. Khatami, Phys. Rev. X **7**(3), 031038 (2017). DOI 10.1103/PhysRevX.7.031038. URL https://link.aps.org/doi/10.1103/PhysRevX.7.031038
79. L. Wang, Phys. Rev. B **94**(19), 195105 (2016). DOI 10.1103/PhysRevB.94.195105. URL https://link.aps.org/doi/10.1103/PhysRevB.94.195105
80. B.S. Rem, N. Käming, M. Tarnowski, L. Asteria, N. Fläschner, C. Becker, K. Sengstock, C. Weitenberg, Nat. Phys. **15**(9), 917 (2019). DOI 10.1038/s41567-019-0554-0. URL http://www.nature.com/articles/s41567-019-0554-0
81. A. Bohrdt, C.S. Chiu, G. Ji, M. Xu, D. Greif, M. Greiner, E. Demler, F. Grusdt, M. Knap, Nat. Phys. **15**(9), 921 (2019). DOI 10.1038/s41567-019-0565-x. URL http://www.nature.com/articles/s41567-019-0565-x
82. V. Dunjko, H.J. Briegel, Rep. Prog. Phys. **81**(7), 074001 (2018). DOI 10.1088/1361-6633/aab406. URL https://iopscience.iop.org/article/10.1088/1361-6633/aab406
83. T. Ohtsuki, T. Ohtsuki, J. Phys. Soc. Jpn. **86**(4), 044708 (2017). DOI 10.7566/JPSJ.86.044708. URL https://journals.jps.jp/doi/10.7566/JPSJ.86.044708
84. J. Carrasquilla, Adv. Phys. X **5**(1), 1797528 (2020). DOI 10.1080/23746149.2020.1797528. URL https://www.tandfonline.com/doi/full/10.1080/23746149.2020.1797528

85. A. Bohrdt, S. Kim, A. Lukin, M. Rispoli, R. Schittko, M. Knap, M. Greiner, J. Léonard, (2020). URL http://arxiv.org/abs/2012.11586
86. S.J. Wetzel, Phys. Rev. E **96**(2), 022140 (2017). DOI 10.1103/PhysRevE.96.022140. URL https://link.aps.org/doi/10.1103/PhysRevE.96.022140
87. Y.H. Liu, E.P.L. van Nieuwenburg, Phys. Rev. Lett. **120**(17), 176401 (2018). DOI 10.1103/PhysRevLett.120.176401. URL https://link.aps.org/doi/10.1103/PhysRevLett.120.176401
88. P. Huembeli, A. Dauphin, P. Wittek, Phys. Rev. B **97**(13), 134109 (2018). DOI 10.1103/PhysRevB.97.134109. URL https://link.aps.org/doi/10.1103/PhysRevB.97.134109
89. J.F. Rodriguez-Nieva, M.S. Scheurer, Nat. Phys. **15**(8), 790 (2019). DOI 10.1038/s41567-019-0512-x. URL http://www.nature.com/articles/s41567-019-0512-x
90. K. Liu, J. Greitemann, L. Pollet, Phys. Rev. B **99**(10), 104410 (2019). DOI 10.1103/PhysRevB.99.104410. URL https://link.aps.org/doi/10.1103/PhysRevB.99.104410
91. F. Schäfer, N. Lörch, Phys. Rev. E **99**(6), 062107 (2019). DOI 10.1103/PhysRevE.99.062107. URL https://link.aps.org/doi/10.1103/PhysRevE.99.062107
92. E. Greplova, A. Valenti, G. Boschung, F. Schäfer, N. Lörch, S.D. Huber, New J. Phys. **22**(4), 045003 (2020). DOI 10.1088/1367-2630/ab7771. URL https://iopscience.iop.org/article/10.1088/1367-2630/ab7771
93. Y. Che, C. Gneiting, T. Liu, F. Nori, Phys. Rev. B **102**(13), 134213 (2020). DOI 10.1103/PhysRevB.102.134213. URL https://link.aps.org/doi/10.1103/PhysRevB.102.134213
94. M.S. Scheurer, R.J. Slager, Phys. Rev. Lett. **124**(22), 226401 (2020). DOI 10.1103/PhysRevLett.124.226401. URL https://link.aps.org/doi/10.1103/PhysRevLett.124.226401
95. O. Balabanov, M. Granath, Phys. Rev. Res. **2**(1), 013354 (2020). DOI 10.1103/PhysRevResearch.2.013354. URL https://link.aps.org/doi/10.1103/PhysRevResearch.2.013354
96. Y. Long, J. Ren, H. Chen, Phys. Rev. Lett. **124**(18), 185501 (2020). DOI 10.1103/PhysRevLett.124.185501. URL https://link.aps.org/doi/10.1103/PhysRevLett.124.185501
97. N. Käming, A. Dawid, K. Kottmann, M. Lewenstein, K. Sengstock, A. Dauphin, C. Weitenberg, Mach. Learn. Sci. Technol. (2021). DOI 10.1088/2632-2153/abffe7. URL https://iopscience.iop.org/article/10.1088/2632-2153/abffe7
98. C. Casert, T. Vieijra, J. Nys, J. Ryckebusch, Phys. Rev. E **99**(2), 023304 (2019). DOI 10.1103/PhysRevE.99.023304. URL https://link.aps.org/doi/10.1103/PhysRevE.99.023304
99. S. Blücher, L. Kades, J.M. Pawlowski, N. Strodthoff, J.M. Urban, Phys. Rev. D **101**(9), 094507 (2020). DOI 10.1103/PhysRevD.101.094507. URL https://link.aps.org/doi/10.1103/PhysRevD.101.094507
100. Y. Zhang, P. Ginsparg, E.A. Kim, Phys. Rev. Res. **2**(2), 023283 (2020). DOI 10.1103/PhysRevResearch.2.023283. URL https://link.aps.org/doi/10.1103/PhysRevResearch.2.023283
101. A. Dawid, P. Huembeli, M. Tomza, M. Lewenstein, A. Dauphin, New J. Phys. **22**(11), 115001 (2020). DOI 10.1088/1367-2630/abc463. URL https://iopscience.iop.org/article/10.1088/1367-2630/abc463
102. A. Cole, G.J. Loges, G. Shiu, (2020). DOI https://arxiv.org/abs/2009.14231. URL http://arxiv.org/abs/2009.14231
103. N. Rao, K. Liu, M. Machaczek, L. Pollet, (2021). URL https://arxiv.org/abs/2102.01103http://arxiv.org/abs/2102.01103
104. R. Guidotti, A. Monreale, S. Ruggieri, F. Turini, F. Giannotti, D. Pedreschi, ACM Comput. Surv. **51**(5), 1 (2019). DOI 10.1145/3236009. URL https://dl.acm.org/doi/10.1145/3236009
105. C. Molnar, *Interpretable Machine Learning*. URL https://christophm.github.io/interpretable-ml-book/
106. I. Goodfellow, Y. Bengio, A. Courville, *Deep Learning* (MIT Press, 2016). URL http://www.deeplearningbook.org

107. L.M. Falicov, J.C. Kimball, Phys. Rev. Lett. **22**(19), 997 (1969). DOI 10.1103/PhysRevLett.22.997. URL https://journals.aps.org/prl/abstract/10.1103/PhysRevLett.22.997
108. J. Hubbard, Proc. R. Soc. London. Ser. A. Math. Phys. Sci. **276**(1365), 238 (1963). DOI 10.1098/rspa.1963.0204. URL https://royalsocietypublishing.org/doi/10.1098/rspa.1963.0204
109. J.K. Freericks, V. Zlatić, Rev. Mod. Phys. **75**(4), 1333 (2003). DOI 10.1103/RevModPhys.75.1333. URL https://link.aps.org/doi/10.1103/RevModPhys.75.1333
110. M. Hohenadler, F.F. Assaad, Phys. Rev. Lett. **121**(8), 086601 (2018). DOI 10.1103/PhysRevLett.121.086601. URL https://link.aps.org/doi/10.1103/PhysRevLett.121.086601
111. M. Gonçalves, P. Ribeiro, R. Mondaini, E.V. Castro, Phys. Rev. Lett. **122**(12), 126601 (2019). DOI 10.1103/PhysRevLett.122.126601. URL https://link.aps.org/doi/10.1103/PhysRevLett.122.126601
112. M. Eckstein, M. Kollar, Phys. Rev. Lett. **100**(12), 120404 (2008). DOI 10.1103/PhysRevLett.100.120404. URL https://link.aps.org/doi/10.1103/PhysRevLett.100.120404
113. M.M. Oliveira, P. Ribeiro, S. Kirchner, Phys. Rev. Lett. **122**(19), 197601 (2019). DOI 10.1103/PhysRevLett.122.197601. URL https://link.aps.org/doi/10.1103/PhysRevLett.122.197601
114. C. Prosko, S.P. Lee, J. Maciejko, Phys. Rev. B **96**(20), 205104 (2017). DOI 10.1103/PhysRevB.96.205104. URL https://link.aps.org/doi/10.1103/PhysRevB.96.205104
115. A. Kauch, P. Pudleiner, K. Astleithner, P. Thunström, T. Ribic, K. Held, Phys. Rev. Lett. **124**(4), 047401 (2020). DOI 10.1103/PhysRevLett.124.047401. URL https://doi.org/10.1103/PhysRevLett.124.047401https://link.aps.org/doi/10.1103/PhysRevLett.124.047401
116. J.K. Freericks, V.M. Turkowski, V. Zlatić, Phys. Rev. Lett. **97**(26), 266408 (2006). DOI 10.1103/PhysRevLett.97.266408. URL https://link.aps.org/doi/10.1103/PhysRevLett.97.266408
117. H. Aoki, N. Tsuji, M. Eckstein, M. Kollar, T. Oka, P. Werner, Rev. Mod. Phys. **86**(2), 779 (2014). DOI 10.1103/RevModPhys.86.779. URL https://link.aps.org/doi/10.1103/RevModPhys.86.779
118. T. Maier, M. Jarrell, T. Pruschke, M.H. Hettler, Rev. Mod. Phys. **77**(3), 1027 (2005). DOI 10.1103/RevModPhys.77.1027. URL https://link.aps.org/doi/10.1103/RevModPhys.77.1027
119. V. Turkowski, J.K. Freericks, Phys. Rev. B **75**(12), 125110 (2007). DOI 10.1103/PhysRevB.75.125110. URL https://link.aps.org/doi/10.1103/PhysRevB.75.125110
120. J. Kaye, D. Golez, SciPost Phys. **10**(4), 091 (2021). DOI 10.21468/SciPostPhys.10.4.091. URL https://scipost.org/10.21468/SciPostPhys.10.4.091
121. L. Huang, L. Wang, Phys. Rev. B **95**(3), 035105 (2017). DOI 10.1103/PhysRevB.95.035105. URL https://link.aps.org/doi/10.1103/PhysRevB.95.035105
122. R. Lemański, J.K. Freericks, G. Banach, Phys. Rev. Lett. **89**(19), 196403 (2002). DOI 10.1103/PhysRevLett.89.196403. URL https://link.aps.org/doi/10.1103/PhysRevLett.89.196403
123. R. Lemański, J.K. Freericks, G. Banach, J. Stat. Phys. **116**(1-4), 699 (2004). DOI 10.1023/B:JOSS.0000037213.25834.33. URL http://link.springer.com/10.1023/B:JOSS.0000037213.25834.33
124. Čenčariková, Farkašovský, Condens. Matter Phys. **14**(4), 42701 (2011). DOI 10.5488/CMP.14.42701. URL http://www.icmp.lviv.ua/journal/zbirnyk.68/42701/abstract.html
125. M.D. Petrović, B.S. Popescu, U. Bajpai, P. Plecháč, B.K. Nikolić, Phys. Rev. Appl. **10**(5), 054038 (2018). DOI 10.1103/PhysRevApplied.10.054038. URL https://link.aps.org/doi/10.1103/PhysRevApplied.10.054038
126. X.H. Li, Z. Chen, T.K. Ng, Phys. Rev. B **100**(9), 094519 (2019). DOI 10.1103/PhysRevB.100.094519. URL https://link.aps.org/doi/10.1103/PhysRevB.100.094519

127. K. Sakmann, A.I. Streltsov, O.E. Alon, L.S. Cederbaum, Phys. Rev. Lett. **103**(22), 220601 (2009). DOI 10.1103/PhysRevLett.103.220601. URL https://link.aps.org/doi/10.1103/PhysRevLett.103.220601
128. K. Sakmann, A.I. Streltsov, O.E. Alon, L.S. Cederbaum, Phys. Rev. A **89**(2), 023602 (2014). DOI 10.1103/PhysRevA.89.023602. URL https://link.aps.org/doi/10.1103/PhysRevA.89.023602
129. S.K. Haldar, O.E. Alon, New J. Phys. **21**(10), 103037 (2019). DOI 10.1088/1367-2630/ab4315. URL https://iopscience.iop.org/article/10.1088/1367-2630/ab4315
130. A. Bhowmik, S.K. Haldar, O.E. Alon, Sci. Reports **10**, 21476 (2020). DOI 10.1038/s41598-020-78173-w. URL http://www.nature.com/articles/s41598-020-78173-w
131. A.I. Streltsov, O.E. Alon, L.S. Cederbaum, Phys. Rev. A **73**(6), 063626 (2006). DOI 10.1103/PhysRevA.73.063626
132. A.U.J. Lode, K. Sakmann, O.E. Alon, L.S. Cederbaum, A.I. Streltsov, Phys. Rev. A **86**(6), 063606 (2012). DOI 10.1103/PhysRevA.86.063606
133. Y. Castin, R. Dum, *Low-temperature Bose-Einstein condensates in time-dependent traps: Beyond the U(1) symmetry-breaking approach*, vol. 57 (American Physical Society, 1998). DOI 10.1103/PhysRevA.57.3008. URL https://link.aps.org/doi/10.1103/PhysRevA.57.3008
134. E.H. Lieb, R. Seiringer, J. Yngvason, Phys. Rev. A **61**(4), 043602 (2000). DOI 10.1103/PhysRevA.61.043602. URL https://link.aps.org/doi/10.1103/PhysRevA.61.043602
135. E.H. Lieb, R. Seiringer, Phys. Rev. Lett. **88**(17), 170409 (2002). DOI 10.1103/PhysRevLett.88.170409. URL https://link.aps.org/doi/10.1103/PhysRevLett.88.170409
136. L. Erdős, B. Schlein, H.T. Yau, Invent. Math. **167**(3), 515 (2007). DOI 10.1007/s00222-006-0022-1. URL http://link.springer.com/10.1007/s00222-006-0022-1
137. L. Erdős, B. Schlein, H.T. Yau, Phys. Rev. Lett. **98**(4), 359 (2007). DOI 10.1103/PhysRevLett.98.040404. URL https://link.aps.org/doi/10.1103/PhysRevLett.98.040404
138. S. Klaiman, O.E. Alon, Phys. Rev. A **91**(6), 063613 (2015). DOI 10.1103/PhysRevA.91.063613. URL https://link.aps.org/doi/10.1103/PhysRevA.91.063613
139. S. Klaiman, L.S. Cederbaum, Phys. Rev. A **94**(6), 063648 (2016). DOI 10.1103/PhysRevA.94.063648. URL https://link.aps.org/doi/10.1103/PhysRevA.94.063648
140. S. Klaiman, A.I. Streltsov, O.E. Alon, *Uncertainty product of an out-of-equilibrium many-particle system*, vol. 93 (American Physical Society, 2016). DOI 10.1103/PhysRevA.93.023605. URL https://link.aps.org/doi/10.1103/PhysRevA.93.023605
141. I. Anapolitanos, M. Hott, D. Hundertmark, Rev. Math. Phys. **29**, 1750022 (2017). DOI 10.1142/S0129055X17500222. URL https://www.worldscientific.com/doi/abs/10.1142/S0129055X17500222
142. A. Michelangeli, A. Olgiati, Anal. Math. Phys. **7**, 377 (2017). DOI 10.1007/s13324-016-0147-3. URL http://link.springer.com/10.1007/s13324-016-0147-3
143. O.E. Alon, J. Phys. A Math. Theor. **50**, 295002 (2017). DOI 10.1088/1751-8121/aa78ad. URL https://iopscience.iop.org/article/10.1088/1751-8121/aa78ad
144. L.S. Cederbaum, Phys. Rev. A **96**, 013615 (2017). DOI 10.1103/PhysRevA.96.013615. URL http://link.aps.org/doi/10.1103/PhysRevA.96.013615
145. S. Klaiman, R. Beinke, L.S. Cederbaum, A.I. Streltsov, O.E. Alon, Chem. Phys. **509**, 45 (2018). DOI 10.1016/j.chemphys.2018.02.016. URL https://www.sciencedirect.com/science/article/abs/pii/S0301010417307668?via{%}3Dihub
146. K. Sakmann, J. Schmiedmayer, (2018). URL http://arxiv.org/abs/1802.03746
147. O.E. Alon, L.S. Cederbaum, Chem. Phys. **515**, 287 (2018). DOI 10.1016/j.chemphys.2018.09.029. URL https://www.sciencedirect.com/science/article/pii/S0301010418307183?via{%}3Dihub{#}b0210
148. O.E. Alon, Mol. Phys. **117**(15-16), 2108 (2019). DOI 10.1080/00268976.2019.1587533. URL https://www.tandfonline.com/doi/full/10.1080/00268976.2019.1587533
149. O.E. Alon, J. Phys. Conf. Ser. **1206**, 012009 (2019). DOI 10.1088/1742-6596/1206/1/012009. URL https://iopscience.iop.org/article/10.1088/1742-6596/1206/1/012009
150. O.E. Alon, Symmetry **11**, 1344 (2019). DOI 10.3390/sym11111344. URL https://www.mdpi.com/2073-8994/11/11/1344

151. P.D. Robinson, J. Chem. Phys. **66**, 3307 (1977). DOI 10.1063/1.434310. URL http://aip.scitation.org/doi/10.1063/1.434310
152. R.L. Hall, J. Phys. A: Math. Gen. **11**, 1235 (1978). DOI 10.1088/0305-4470/11/7/011. URL https://iopscience.iop.org/article/10.1088/0305-4470/11/7/011/meta
153. L. Cohen, C. Lee, J. Math. Phys. **26**(12), 3105 (1985). DOI 10.1063/1.526688
154. M.S. Osadchii, V.V. Murakhtanov, Int. J. Quantum Chem. **39**, 173 (1991). DOI 10.1002/qua.560390207. URL http://doi.wiley.com/10.1002/qua.560390207
155. M.A. Załuska-Kotur, M. Gajda, A. Orłowski, J. Mostowski, Phys. Rev. A **61**, 8 (2000). DOI 10.1103/PhysRevA.61.033613. URL https://journals.aps.org/pra/abstract/10.1103/PhysRevA.61.033613
156. J. Yan, J. Stat. Phys. **113**, 623 (2003). DOI 10.1023/A:1026029104217
157. M. Gajda, *Criterion for Bose-Einstein condensation in a harmonic trap in the case with attractive interactions*, vol. 73 (American Physical Society, 2006). DOI 10.1103/PhysRevA.73.023603. URL https://link.aps.org/doi/10.1103/PhysRevA.73.023603
158. J.R. Armstrong, N.T. Zinner, D.V. Fedorov, A.S. Jensen, J. Phys. B: At., Mol. Opt. Phys. **44**(5), 055303 (2011). DOI 10.1088/0953-4075/44/5/055303. URL http://stacks.iop.org/0953-4075/44/i=5/a=055303?key=crossref.aa534c8a7543acdd895681648ff1992e
159. J.R. Armstrong, N.T. Zinner, D.V. Fedorov, A.S. Jensen, Phys. Rev. E **86**, 021115 (2012). DOI 10.1103/PhysRevE.86.021115. URL https://journals.aps.org/pre/abstract/10.1103/PhysRevE.86.021115
160. C. Schilling, Phys. Rev. A **88**, 042105 (2013). DOI 10.1103/PhysRevA.88.042105. URL https://journals.aps.org/pra/abstract/10.1103/PhysRevA.88.042105
161. C.L. Benavides-Riveros, I.V. Toranzo, J.S. Dehesa, J. Phys. B: At., Mol. Opt. Phys. **47**, 195503 (2014). DOI 10.1088/0953-4075/47/19/195503. URL https://iopscience.iop.org/article/10.1088/0953-4075/47/19/195503
162. P.A. Bouvrie, A.P. Majtey, M.C. Tichy, J.S. Dehesa, A.R. Plastino, Eur. Phys. J. D **68**, 1 (2014). DOI 10.1140/epjd/e2014-50349-2. URL https://link.springer.com/article/10.1140/epjd/e2014-50349-2
163. J.R. Armstrong, A.G. Volosniev, D.V. Fedorov, A.S. Jensen, N.T. Zinner, J. Phys. A Math. Theor. **48**(8), 085301 (2015). DOI 10.1088/1751-8113/48/8/085301. URL https://iopscience.iop.org/article/10.1088/1751-8113/48/8/085301
164. C. Schilling, R. Schilling, Phys. Rev. A **93**, 021601 (2016). DOI 10.1103/PhysRevA.93.021601. URL https://journals.aps.org/pra/abstract/10.1103/PhysRevA.93.021601
165. S. Klaiman, A.I. Streltsov, O.E. Alon, Chem. Phys. **482**, 362 (2017). DOI 10.1016/j.chemphys.2016.07.011
166. S. Klaiman, A.I. Streltsov, O.E. Alon, J. Phys. Conf. Ser. **999**, 12013 (2018). DOI 10.1088/1742-6596/999/1/012013. URL https://iopscience.iop.org/article/10.1088/1742-6596/999/1/012013
167. A.U.J. Lode, P. Molignini, R. Lin, M. Büttner, P. Rembold, C. Lévêque, M.C. Tsatsos, L. Papariello. UNIQORN:Universal Neural-network Interface for Quantum Observable Readout from {N}-body wavefunctions, https://gitlab.com/UNIQORN/uniqorn (2021). URL https://gitlab.com/UNIQORN/uniqorn
168. A.U.J. Lode, R. Lin, M. Büttner, P. Rembold, C. Lévêque, M.C. Tsatsos, P. Molignini, L. Papariello. Uniqorn data set (2021). URL https://drive.google.com/file/d/1Du8KRhsITezlMVWEBrLOnDIAfFZOPcEj/view?usp=sharing
169. A.U.J. Lode, R. Lin, M. Büttner, P. Rembold, C. Lévêque, M.C. Tsatsos, P. Molignini, L. Papariello. Uniqorn triple-well data set (2021). URL https://drive.google.com/file/d/1Zqc8wyzeqWrna-7uMJ9XI{_}WFreQBJzDu/view?usp=sharing

Molecules, Interfaces and Solids

In this funding period, the field of molecules, interfaces and solids benefits enormously from the computational resources provided by the High Performance Computing Center Stuttgart and the Steinbuch Centre for Computing Karlsruhe. In what follows, we selected some projects in this area to demonstrate the impact of high performance computing in physics, chemistry, and material science.

The collaborative work by Oelschläger, Klein, Müller and Roth from the Institute of Functional Matter and Quantum Technology and the Graduate School of Excellence advance Manufacturing Engineering at the University Stuttgart is an outstanding example in this respect. The project comprises numerical simulations of selective laser melting with a focus on the additive manufacturing of products by a printer. The main challenge here is to close the gap between manageable system sizes and the industrial scales which still differ by about two orders of magnitude. Therefore the authors perform a proof of principle study to validate the applicability of atomistic molecular dynamic (MD) simulations (based on the classical Newton equation of motions with interactions modelled by embedded-atom potentials) to the problem of powder bed fusion using a laser beam, and demonstrate that all components of the model work quite well. For the time being a mechanism for pore formation could be identified. Furthermore, it could be shown that lower laser velocities favour droplet formation which may cause balling and splashing. Further studies are definitely worthwhile, for example, in order to clarify the effects of different packing densities, of recrystallisation, or of a variable floor.

Another MD study is devoted to the influence of solutes on the tensile behaviour of polyamide6. The simulations performed by Verestek and Schmauder from the IMWF University of Stuttgart mainly address the mechanical properties of the dry as well as the saturated (with water, methanol or ethanol) bulk material. Here, a clear trend with increased weakening is observed from dry to the solutes, where the impact increases when passing over from water to methanol and ethanol. Thereby the weakening is not caused by the growing distance between the amide groups. Tensile test simulations in the triaxial state show that an increased crystallinity comes from the reorganisation of fibril-like structures. Moreover differences in the Young's modulus are observed, where methanol has the highest reduction.

The Theoretical Physics and Materials Research groups at the Justus-Liebig University Gießen led by S. Sanna investigated the lattice dynamics of a Si(553)-Au surface from first principles by means of a DFT-based atomistic simulation using the double-Au strand model and the rehybridised model. In particular, a theoretical characterisation of the surface-localised phonon-mode is given both above and below the phase transition temperature. Thereby the comparison with the experiments allows to assign the spectral features to the calculated displacement pattern. The strong coupling between the electronic and phononic systems observable in Raman scattering could be related to the phonon activated charge transfer between the Au chain and the Si step edge. Modifying the Si(553)-Au nanowire system by atomic deposition or doping would be an interesting direction of future studies.

Chemical functionalisation of semiconductor surfaces, that is, combining established silicon-based technology with the structural richness of organic chemistry, is a promising route to widen the application range of semiconducting devices. This is why the group of Ralf Tonner at the university of Leipzig investigates the reactivity of organic molecules on semiconductor surfaces, using density functional theory as implemented in the VASP (Vienna Ab Initio Simulation Package) code, which makes very efficient use of HLRS' computational resources. The contribution investigates how pyrazine, methyl benzylazide and a cyclooctyne derivative adsorb on and react with a silicon or germanium surface. Furthermore, computational protocols for even more efficient treatments have been explored, namely how far one can reduce a model system that still leads to meaningful results. Reduced model systems allow for more accurate quantum chemical methods beyond density functional theory, so the reduction of the model is especially important if there is more than one organic layer on top of the silicon surface.

The transformation of our industry from fossil to renewable energy sources will require efficient conversion of solar/wind to chemical energy. Splitting water into hydrogen and oxygen by electric energy will be highly important, since „green" hydrogen is the basis for all sorts of industrial processes and heavy-load transportation. Although the oxygen produced during water splitting is a side product disposed into the atmosphere, oxygen formation is a complicated process and thus the bottleneck for large-scale water electrolysis as well as related processes like carbon dioxide or nitrogen reduction. Phenomenologically this means that one needs significant over-potentials to make oxygen formation sufficiently fast. Among the best catalysts for dioxygen formation are iridium dioxide based materials. While iridium is admittedly a too precious metal to be used massively on an industrial scale, an understanding why it is one of the best catalysts for oxygen formation will possibly produce new ideas for catalysts based on earth-abundant elements. Travis Jones from the Berlin-based Fritz Haber institute of the Max Planck society presents large-scale simulations of dioxygen formation on IrO_2 surfaces, which is a multi-step and multi-electron reaction at an electrochemical interface. As in the previous contribution, density functional methods are being used that can properly describe the bond-breaking and bond-making that occur in chemical reactions. Here, the Quantum ESPRESSO package is used, that also can make good used of the huge amounts of CPUs offered by the HLRS computer. One of the important results is that the oxygen formation rate is determined by the amount of surface oxidation, and that the "kink" (or bend) observed experimentally in the logarithmic Tafel plot (reaction rate as a function of the applied potential) arises from the response of the surface oxidation to the potential rather than from a change in the reaction mechanism. This means the "chemistry" (oxygen-oxygen bond formation) as such does not require a high over-potential, which is instead used to produce the necessary amount of surface oxidation.

We wish to stress that almost all projects supported in this field are of high scientific quality, also those which could not be included in report because of space limitations. This underlines once again the strong need of supercomputing facilities in modern condensed matter based science and technology.

Institut für Physik,
Lehrstuhl Komplexe Quantensysteme,
Ernst-Moritz-Arndt-Universität Greifswald,
Felix-Hausdorff-Str. 6,
17489 Greifswald,
Germany,
e-mail: fehske@physik.uni-greifswald.de

Holger Fehske

Fachbereich Chemie,
Technische Universität Kaiserslautern,
Erwin-Schrödinger-Str. 52,
67663 Kaiserslautern,
Germany,
e-mail: vanwullen@chemie.uni-kl.de

Christoph van Wüllen

Molecular dynamics simulation of selective laser melting

Fabio Oelschläger, Dominic Klein, Sarah Müller and Johannes Roth

Abstract This report deals with the atomistic simulation of selective laser melting (SLM) used to produce additive manufactured objects. After a short introduction into the subject, modifications to basic molecular dynamics simulation code are described which are required to simulate the annealing process. Although the sample sizes studied in this report are already impressively large, scaling of system parameters are required to connect simulation and experiment. First results will be reported and further developments and improvements be described.

1 Introduction

Additive manufacturing (AM) of products with a printer plays an increasing role today. AM means adding layers of material in the manufacturing process while traditionally material is removed from a workpiece by subtractive technologies such as turning or milling. Though there are many different methods for AM, all work according to a similar basic principle: a three-dimensional digital model is dissected by a slicer into layers which are subsequently produced and stacked. In contrast to public

---

Fabio Oelschläger
Institut für Funktionelle Materie und Quantentechnologien, Universität Stuttgart,
e-mail: st151193@stud.uni-stuttgart.de

Dominic Klein
Institut für Funktionelle Materie und Quantentechnologien, Universität Stuttgart,
e-mail: dominic.klein@fmq.uni-stuttgart.de

Sarah Müller
Graduate School of Excellence advanced Manufacturing Engineering, GSaME, Universität Stuttgart,
e-mail: Sarah.Mueller@gsame.uni-stuttgart.de

Johannes Roth
Institut für Funktionelle Materie und Quantentechnologien, Universität Stuttgart,
e-mail: johannes.roth@fmq.uni-stuttgart.de

W. E. Nagel et al. (eds.), *High Performance Computing in Science and Engineering '21*,
https://doi.org/10.1007/978-3-031-17937-2_4

belief where simple plastic parts are printed on commodity printers with no special requirements for precision and stability, we are dealing here with industrial printing of load bearing, metallic power parts. Here the additive manufacturing is far away from working perfectly and being competitive with other methods. Deviation of size and defects are frequent. To understand the defects and failures on an atomistic level, large scale molecular dynamics (MD) simulations on supercomputers are carried out with the simulation code IMD[14] which is well suited for this purpose and has been demonstrated to run effectively with billions of atoms (See Sec. 5). For the simulation of additive manufacturing presented here only a few modifications have to be made, for example the setup of the moving laser beam and the addition of gravitational force.

A main challenge is the gap between manageable simulation sizes and industrial scales which typically differ by one to two orders of magnitude in the current case. In principle the supercomputers are large enough to reach industrial scales but such simulations would require the whole machine for weeks together with sufficient resources for storage, analysis, and visualization. For the time being we have to resort to running smaller simulations and scaling the results suitably.

In the present proof of principle study we demonstrate that all components of the model work quite well. First quantitative results for selective laser melting (SLM) of a single sphere of aluminum are also given.

The report is organized as follows: after an introduction into additive manufacturing by SLM we discuss the problems occurring. Then we present the first results of the simulations. The report ends with performance data collect for our simulations on HLRS Hawk.

2 Additive manufacturing by selective laser melting (SLM)

2.1 Description of the method

The method studied in this work is powder bed fusion using a laser beam, often referred to as Selective Laser melting (SLM, SLM Solutions Group). Fig. 1 shows a printer working according to this principle. The printer consists of two containers with movable bases. A feed container is filled with powder material while the other printing container is more or less empty. The base of the feed container is lifted up if a new layer is to be printed. A drum shifts the dosed material to the printing container while the base of the latter is lowered. Then, a mobile laser scans the desired printing plain and fuses the powder into a solid shape[9]. The printed chamber is typically lowered by 30 μm to 100 μm [18] but thicker layers are also common [17]. The most common materials applied in the SLM method are Ti compounds, especially Ti6Al4V in aerospace industry [1, 17, 18]. Al compounds are used in automobile industry since they are suitable in light-way construction [20, 22].

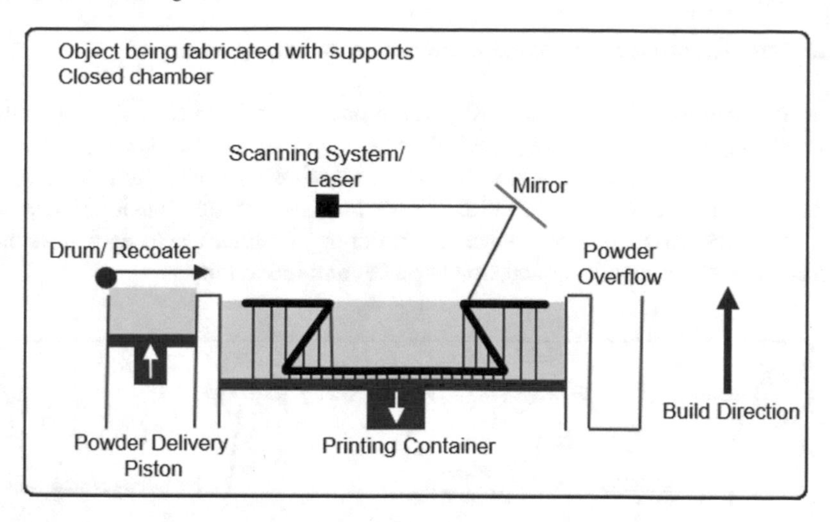

Fig. 1: Schematic presentation of the SLM method. The mobile mirror takes care for the precise positioning of the laser spot. The base of the feed container (left) is lifted while the base of the printing container (center) is lowered.

2.1.1 Defects in manufactured objects

Several defect types can occur during printing: **Pores** are gas inclusions with sizes less that 100 μm[21]. The source may be evaporation of the low-melting metal or protective gas used to avoid oxidation. Without gas a low packing density can be the origin of the pores. The pores are spherical and equidistributed in the sample. The size and shape of the pores has considerable influence on the quality and especially the density of the product.

Lack-of-fusion (LOF) originates predominantly through insufficient energy supply during melting. The width of the molten lane is too small which causes bonding defects and powder inclusions and the different layers are not fused. Consequently LOF occurs predominantly along the lanes and between the layers. The defects increase roughness and disturb the matter flow for the next layer. Thus, they can gradually propagate through several layers. The problem may also be caused by surface cooling which reduces the wettability[21].

After melting, cooling rates reach 10^8 K/s [21]. The temperature gradient and the related strong thermal expansion gradient lead to high stress in the material. **Cracks** are created and propagate through the material. Cracks originate at the surface and propagate into the bulk. Non-spherical pores could also generate cracks[6].

2.1.2 Important parameters and influence of the laser velocity

Many parameters influence the quality of the product in SLM (Fig. 2). Adjustable parameters (Fig. 3) which have great influence on quality are *scanning speed*, *laser power*, *hatch distance or spacing*, *laser wavelength*, *spot size*, and *layer thickness*[15]. Since up to now the studied objects are single powder particles, hatch distance and layer thickness will no longer be taken into account. The laser wavelength plays only an implicit role through its connection to power and reflectivity.

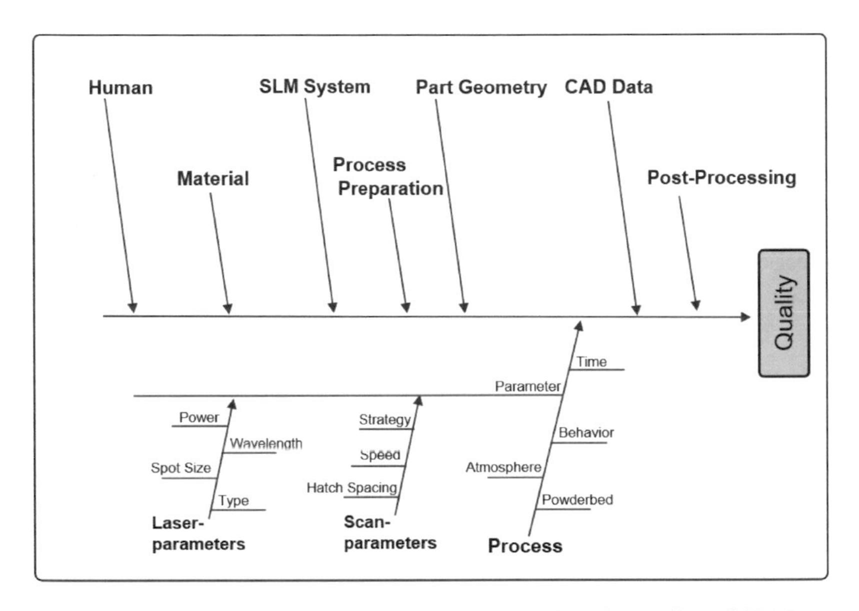

Fig. 2: An overview of the parameters controlling the quality of SLM.

Following the defects discussed in Sec.2.1.1 it is important to melt the powder completely but to avoid supplying too much energy. An optimal matching is required for an optimal outcome. Higher laser velocities for example often require increased laser power in order to guarantee the melting of the material. Sadali et al[15] observe that the pore sizes and the *splashing effect* are reduced with increasing speed. Spherical material agglomeration at the surface are caused by the separation of instable liquid traces through insufficient wetting (*balling effect*). They get fewer and shrink with increasing velocity. Nonetheless, a faster laser velocity does not automatically mean a better result. The number of microcracks for example increase. While these can still be minimized by Hot Isostatic Pressing (HIP), further undesirable effects occur which cannot be removed in post processing[15].

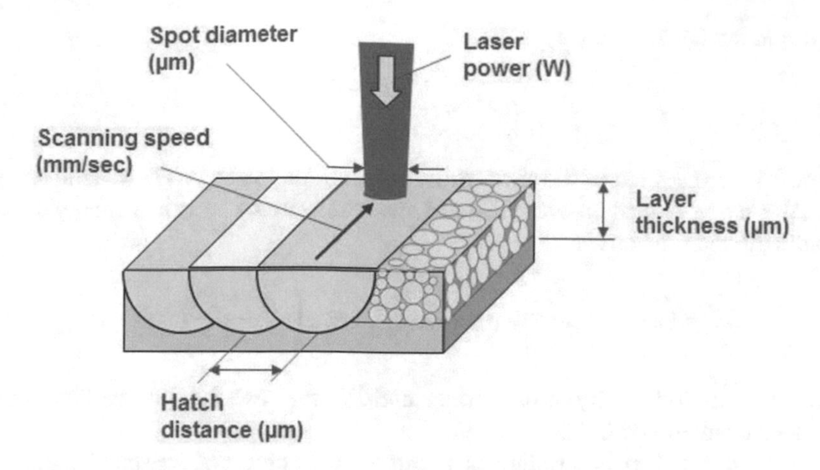

Fig. 3: The most important SLM parameters and the typical orders of magnitude [16].

2.2 Modeling of SLM by molecular dynamics Simulations

In molecular dynamics simulations (MD) the classical Newtonian equations of motions are integrated at discrete time steps, typically fractions of femto-seconds for atomistic systems. Interactions in metals are modeled by embedded-atom potentials (EAM) which incorporate the properties of the free electron gas. To represent a specific material these effective interactions are fitted to the quantum mechanical properties. A linked-cell algorithm can be applied to find interacting atoms since EAM interactions are effectively pairwise and short-ranged. A domain decomposition of the simulation box and message passing can be implemented. Thus, MD ideally scales linearly up to huge sample sizes. The principle setup for the SLM simulations is as follows: the objects which are to be irradiated are placed on a fixed plate which is modeled by immobile atoms. Then a laser beam is moving across the objects with a fixed velocity.

2.2.1 Laser absorption

If matter is irradiated by a laser, beam energy will be absorbed according to the law of Lambert-Beer:

$$I(x) = I(0) \cdot \exp\left(-\mu x\right) \tag{1}$$

where $I(x)$ is the intensity along x and μ is the inverse absorption length. The laser will excite the electrons and the energy will eventually be transformed into thermal energy of the atoms. In the present work the rescaling method has been applied since the time scales of the thermalization processes are shorter than the interaction time scales of the laser. The velocities of the atoms $v(t + \mathrm{d}t) = a(x)v(t)$ are scaled with respect to the Lambert law by

$$a(x)^2 = \frac{E_{\text{kin}}(t + \mathrm{d}t)}{E_{\text{kin}}(t)} \tag{2}$$

where x is the position of the atom with respect to the surface. The kinetic energy $E_{\text{kin}}(t)$ can be summed up over all atoms and results in the absorbed energy $\mathrm{d}E$ per volume $\mathrm{d}V$ and time $\mathrm{d}t$ via

$$\mathrm{d}E = (1 - R) \cdot \mu \cdot \exp\left(-\mu z\right) \cdot \frac{P_{\text{tot}}}{\pi\sigma^2} \cdot \exp\left(-\frac{x^2 + y^2}{\sigma^2}\right) \cdot \mathrm{d}V \cdot \mathrm{d}t \tag{3}$$

where R is the reflectivity of the surface and P_{tot} the absorbed power if a circular Gaussian beam of width σ is assumed.

The derivation here is certainly only correct for an object of constant height. The modification for a sphere and a fixed laser beam could be derived easily, at least numerically. In the present case, however, we are interested in moving laser beams and thus the correct absorption of objects of variable height caused by a moving laser beam together with changing shape during the process can only be modeled by an accompanying ray tracing calculations. Currently, addressing this problem has been postponed since it would complicate the modeling considerably and distract from the proof of principle.

2.2.2 Sample sizes

The size of MD simulation samples have increased tremendously in recent years up to about 10^{13} atoms[2]. But such runs would require full-time simulations on entire top-level supercomputers.

The sizes of powder particle in the SLM process in the range of a few μm [8] is at the limit of current production simulations. For the intended proof of principle simulations the size of the system has been scaled down to a few hundred Å, but will be increased subsequently in the production simulations.

Scale factors have been introduced to connect simulation and experiment. The procedure can be motivated by the work of Glosli et al [7]. This group could demonstrate that the Kelvin-Helmholtz instability in fluids occurs at relatively small particle numbers already. Selective laser melting depends on the behavior of fluids in a similar way. The scaling also includes an increase of the gravitational force.

3 Simulation of SLM

3.1 Preparation of the sample

The initial sample is a crystal with periodic boundary conditions which is equilibrated at 300 K. Then a sphere is cut out of the material and equilibrated under the influence of gravity which causes the sphere to settle down slightly. Thus, the contact plane between the fixed ground and the sphere is enlarged.

3.2 Variation of the laser power at fixed laser velocity

The approximations applied largely exclude the direct transfer of the simulation data to experiment as discussed in Sec. 2.2.1. Therefore the model has been verified predominantly in a qualitative comparison.

The laser velocity is set to a fixed value of 1Å per fs. A number of power values are simulated to obtain a reference simulation for further studies. The first goal is to relate each laser power at this velocity to a melting capability. The criterion is the amount of material molten during simulation.

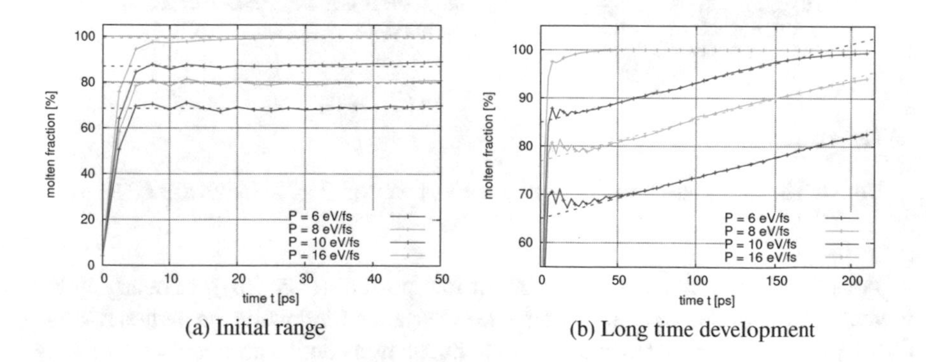

(a) Initial range

(b) Long time development

Fig. 4: Number of particles in the molten phase *vs.* simulation time. The laser beam with velocity $v_{\text{laser}} = 1\text{Å}$ per fs has passed the sphere completely after 10 ps.

Fig. 4 shows the molten fraction of the material *vs.* time for different laser powers P. The short time behavior has been zoomed in on the left (Part (a)). Obviously the laser melts the sample instantaneously and the molten fraction stays constant for a short time. Fitting a constant in this range determines the molten fraction at this laser power.

Table 1: Molten fraction of material.

power P [eV/fs]	molten	increase [%/fs]
6	68.54%	$8.5\cdot10^{-5}$
8	79.49%	$8.6\cdot10^{-5}$
10	86.95%	$8.2\cdot10^{-5}$
16	99.26%	$8.4\cdot10^{-5}$

The data in Tab. 1 show that the sphere is completely molten at a power of $P = 16$ eV/fs. The missing part of $\approx 0.8\%$ results from fluctuations at the beginning. A closer look at Fig. 4 (a) shows that 100% are finally reached. The complete melting is displayed in Fig. 5. The side view in Fig. 5 (b) indicates that the floor is wetted completely as observed in experiment[5]. This is a strong hint that the model is suitable.

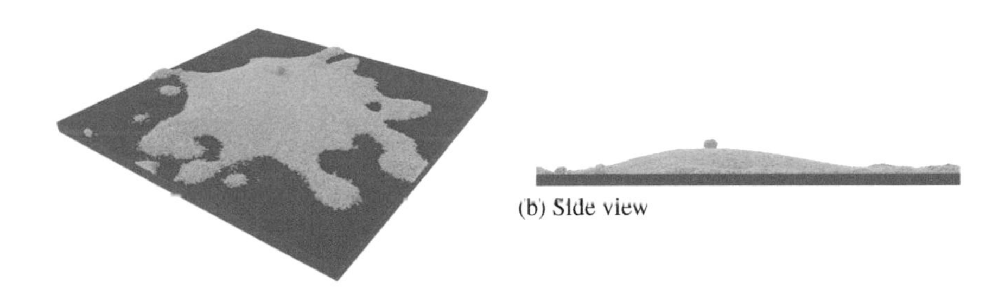

(b) Side view

(a) Perspective view

Fig. 5: The completely molten sphere at a power of $P = 16$ eV/fs after 210 ps.

An interesting phenomenon occurs at low powers (Fig. 4(b)). A nearly linear increase of the molten fraction is visible for a certain period while the molten fraction directly after laser impact is constant up to fluctuations until saturation sets in finally. The rate of increase in this region is nearly the same for all laser powers. The increase of molten material on average is $8.4\cdot10^{-5}$%/fs. The reason is that the fraction molten directly by the laser transfers the heat to the still solid part by heat conduction which is a material property only.

A further phenomenon which is due to the current model limitations is visible in Fig. 6. Shown is a cut through the center of the sphere in the z-x-plane. Red atoms are in the molten part, white atoms are solid. The sphere is not yet completely molten at $P = 10$ eV/fs, but the solid rest shrinks subsequently. Interesting are the voids formed at the bottom in the left figure. The melt from above closes the voids laterally and prevents a spreading of the un-molten nucleus as in Fig. 5. Thus, the partially solid

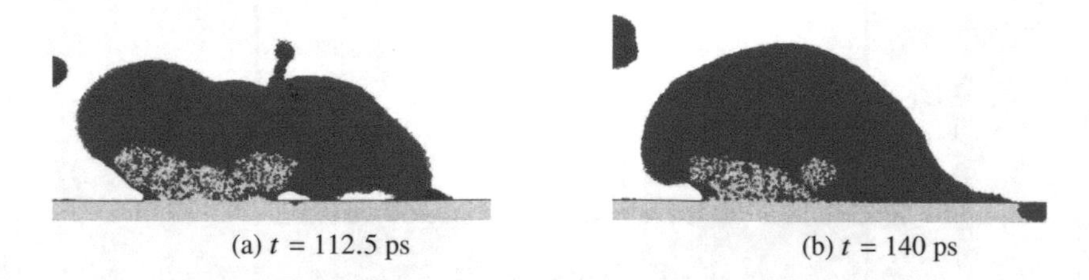

(a) t = 112.5 ps (b) t = 140 ps

Fig. 6: Cuts through the sphere. Voids disappear due to the surrounding vacuum. Inclusions would be visible if the voids were filled with protective gas.

structures lead to void formation. The voids vanish some time later since they are not filled with a gas. The surrounding material has penetrated into the void and closed it. With protective gas the closed void could have formed a pore.

3.3 Behavior at reduced laser velocity

The goal here is to achieve the same melting volume with different parameters. The laser power has to be reduced to obtain the same amount of molten material at lower laser velocity since laser power is applied energy per time. The connection is not a simple proportionality as exemplified by Fig. 7. Obviously, the curves at half power and half velocity do not coincide. The parameter pairs v = 1 Å/fs, P = 8 eV/fs and v = 0.5 Å/fs, P = 5 eV/fs on the other hand show a rather similar behavior at the beginning.

Fig. 8 shows a simulation with higher velocity at the top and one with lower velocity at the bottom. To the right are the same samples at a later time. Details are different although the simulations look rather similar at first. In the shorter time pictures much less small droplets have been formed at lower laser velocities, and this behavior continues until later times. The droplets can be responsible for a number of defects including and splashing or balling. Defects are reduced at higher velocities as mentioned in Sec. 2.1.2 already[15]. The fact is taken as a further sign that the model is suitable for an accurate description of the problem.

4 Summary and outlook

The goal of this work is to validate the applicability of atomistic MD simulations to SLM. The crux is the fact that the scale of the powder particles are a few μm which is a challenge even for big supercomputers. Currently the model is scaled

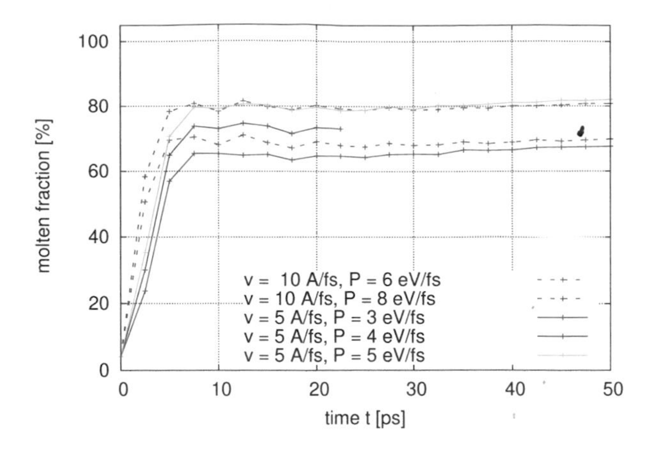

Fig. 7: Fraction of particles belonging to the molten material *vs.* simulation time. Like Fig. 4 (a) for different velocities.

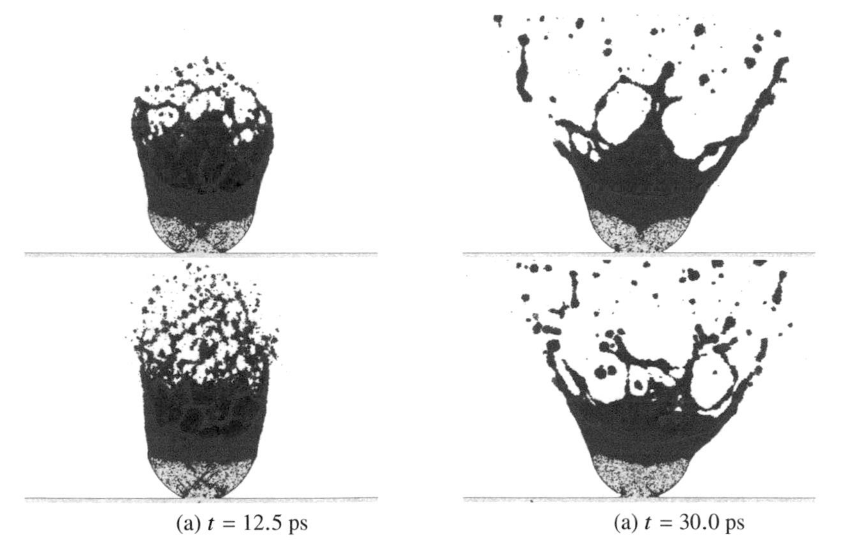

(a) t = 12.5 ps (a) t = 30.0 ps

Fig. 8: Two parameter pairs leading to a similar result. Top: v = 1 Å/fs, P = 8 eV/fs, bottom: v = 0.5 Å/fs, P = 5 eV/fs

down to several hundred Å. The results of these proof of principle simulations have demonstrated that this project should be continued. Hints have been found that the mechanisms at the basis of the observed defects can be studied.

The modeling is not yet complete, but the basic requirements for a successful description of SLM are fulfilled. A mechanism for pore formation could be identified and the disappearance of the pores without protective gas could be observed. It could be shown that lower laser velocities favor droplet formation which has been named as a reason for balling and splashing[15].

Pure aluminum in vacuum has been studied, a material which is rarely used but contained in many compounds as a majority element. Pure metals simplify the identification and discussion of the many open questions.

More realistic simulations are necessary. They include:

- The parameter space of sphere size, laser velocity and laser power should be studied in more detail.
- Several spheres in different configurations should be simulated to observe their interaction. Different packing densities are of highest interest since they may lead to pore formation.
- Recrystallization should also be studied in this context. Then other defects like microcracks also come into play.
- Protective gas like argon should be added to modify the behavior of the pores.
- Cooling process of the fixed floor should be taken into account. Currently the simulations are run in an isolated NVE ensemble which eventually leads to complete melting of the whole setup late after irradiation. This can be avoided by applying an isothermal ensemble which leads to active cooling.
- Up to now the sphere has been placed on solid ground. But typically the un-molten powder lies above an already processed layer and the interaction with a variable floor should also be studied.
- The simulation can be extended to other metals and alloys as desired.

Mesoscopic simulations have been carried out at true sample sizes[13] already. Obviously, they cannot give information about atomistic processes. At the HLRS it should be possible so scale up the simulations to the μm regime where the number of atoms reaches orders of several billions. Ablation simulations of Al[3] and Si[10] have demonstrated already that this is possible for production runs with IMD. And if the whole supercomputer HAWK is available, demonstration runs with even larger sizes should be feasible.

5 Performance of simulation code

In the current reporting period we have successfully carried out molecular dynamics simulations laser ablation simulations on Hawk with

- quasi 1D silicon samples (Si1D),
- quasi 2D silicon samples of small (Si2Dsmall) and big (Si2Dbig) sizes, as well as
- simulations of the selective laser melting of Al samples (SLM).

Further simulation tasks of minor size and already terminated tasks on Hazelhen will not be reported.

5.1 Typical number of applied processors

The typical number (#) of applied processors is displayed in Tab. 2 according to simulation type. Please consider that the SLM simulations have not yet been run in full productive mode.

simulation	wall time	# atoms n	# nodes	# processors	performance $[\mathrm{CPUs}/n\Delta t]$
Si1D	24h	18,121,592	64	8,192	$5.344767 \cdot 10^{-5}$
Si2Dsmall	24h	67,108,864	128	16,384	$1.342413 \cdot 10^{-4}$
Si2Dbig	24h	536,870,912	1024	65,536	$5.174720 \cdot 10^{-4}$
SLM	24h	27,680,214	64	8,192	$1.374238 \cdot 10^{-4}$

Table 2: Typical resources applied and performance achieved in CPU seconds (cpus per atom per simulation-step Δt).

5.2 Parallelization degree

IMD is a typical molecular dynamics simulation code. As such the main computation load is generated in a single main loop where the classical interactions between atoms are calculated. The program achieves parallelization degrees of the order of 90 and more percent, depending on the simulated experimental setup. The main loop can be fully parallelized with MPI due to its short range nature and distributed to any number of processors. It has been demonstrated (for example on JuGene) that IMD shows weak scaling up to the full machine size as long as the system is sufficiently homogeneous.

5.3 Scaling

In "real" simulations the samples are not homogeneously filled with atoms but contain empty space around spheres for example. The simulation box is spatially decomposed and distributed to various nodes and thus some processors may idle. Together with the unavoidable communication load between increasing numbers of processor some performance loss and scaling decay occurs (Fig. 9) especially for large simulations. Applying dynamical load balancing implemented in IMD, performance can be improved to some extent.

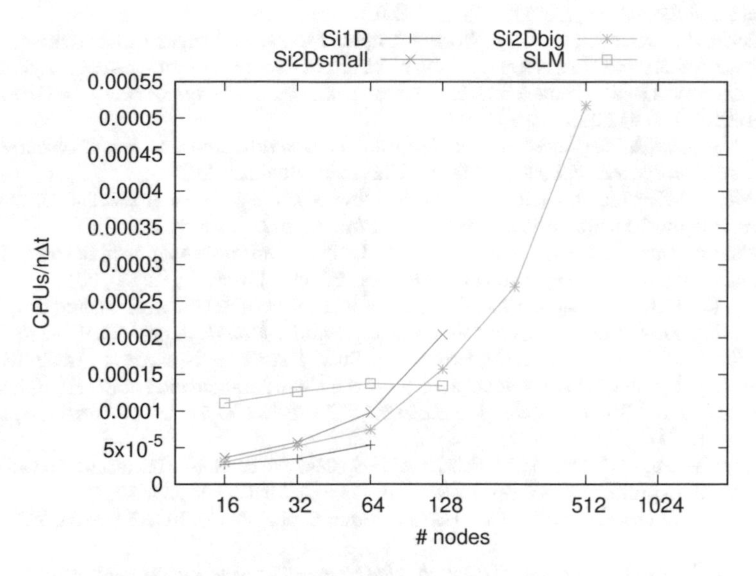

Fig. 9: Scaling of the benchmarked simulation types.

5.4 Consumed computing resources

Since July 2020 a total of 452,758 node-hours have been spent on HPE Apollo (Hawk). The largest share has been used for the ongoing PhD thesis of D. Klein [10, 11] studying laser ablation of silicon (about 407,948 node-hours, including a share for the SLM study). F. Oelschläger [12] has spent 14,738 node-hours for the study of SLM presented here. K. Vietz [19] used 19,685 node-hours for the study of Al-Ni-alloys under laser ablation. The remaining part was spent by E. Eisfeld [3,4] at the completion of his thesis on the molecular dynamics simulation of Al with plasma effects.

Acknowledgements Dominic Klein is supported by the Hans-Böckler-Stiftung. Sarah Müller is supported by the Deutsche Forschungsgemeinschaft DFG.

References

1. E. Brandl, A. Schoberth, and C. Leyens. Morphology, microstructure, and hardness of titanium (Ti-6Al-4V) blocks deposited by wire-feed additive layer manufacturing (ALM). *Materials Science and Engineering: A*, 532:295–307, 2012.
2. W. Eckhardt. Scientists Achieve World's Largest Molecular Dynamics Simulation, 2013. `https://gauss-centre.eu/results/materials-science-and-chemistry/article/scientists-achieve-worlds-largest-molecular-dynamics-simulation/`, abgerufen am 28.04.2021.
3. E. Eisfeld. *Molekulardynamische Simulationen der Laserablation an Aluminium unter Einbeziehung von Plasmaeffekten*. PhD thesis, Universität Stuttgart, 2020.
4. E. Eisfeld, D. Förster, D. Klein, and J. Roth. Laser ablation of al under short and ultrashort pulses: Insights from atomistic simulations. *J. Phys. D*, 2022. submitted.
5. H. Eskandari Sabzi and P. E. J. Rivera-Díaz-del Castillo. Defect Prevention in Selective Laser Melting Components: Compositional and Process Effects. *Materials*, 12(22), 2019.
6. C. Galy, E. Le Guen, E. Lacoste, and C. Arvieu. Main defects observed in aluminum alloy parts produced by SLM: From causes to consequences. *Additive Manufacturing*, 22:165–175, 2018.
7. J. N. Glosli, D. F. Richards, K. J. Caspersen, R. E. Rudd, J. A. Gunnels, and F. H. Streitz. Extending stability beyond CPU millennium: a micron-scale atomistic simulation of Kelvin-Helmholtz instability. In *SC '07: Proceedings of the 2007 ACM/IEEE Conference on Supercomputing*, pages 1–11, 2007.
8. J. Hajnys, M. Pagac, J. Mesicek, J. Petru, and F. Spalek. Research of 316l metallic powder for use in SLM 3d printing. *Advances in Materials Science*, 20(1):5–15, 2020.
9. F. Horsch. *3D-Druck für alle*. Carl Hanser Verlag GmbH & Co. KG, 2 edition, 2014. pp. 114,119.
10. D. Klein. *Computer simulations of laser ablation from simple metals to complex metallic alloys*. PhD thesis, Universität Stuttgart, 2021.
11. D. Klein, E. Eisfeld, and J. Roth. Molecular dynamics simulations of the laser ablation of silicon with the thermal spike model. *J. Phys. D*, 54, 2021. submitted.
12. F. Oelschläger. Molekulardynamische Simulation von selektivem Laserschmelzen. Bachelorarbeit, Universität Stuttgart, 2021.
13. C. Panwisawas, C. Qiu, M. J. Anderson, Y. Sovani, R. P. Turner, M. M. Attallah, J. W. Brooks, and H. C. Basoalto. Mesoscale modelling of selective laser melting: Thermal fluid dynamics and microstructural evolution. *Computational Materials Science*, 126:479–490, 2017.
14. J. Roth, E. Eisfeld, D. Klein, S. Hocker, H. Lipp, and H.-R. Trebin. IMD – the ITAP molecular dynamics simulation package. *The European Physical Journal Special Topics*, 227(14):1831–1836, 2019.
15. M. F. Sadali, M. Z. Hassan, F. Ahmad, H. Yahaya, and Z. A. Rasid. Influence of selective laser melting scanning speed parameter on the surface morphology, surface roughness, and micropores for manufactured Ti6Al4V parts. *Journal of Materials Research*, 35(15):2025–2035, 2020.
16. M. Saunders. X marks the spot - find ideal process parameters for your metal AM parts, 2017. `https://www.linkedin.com/pulse/x-marks-spot-find-ideal-process-parameters-your-metal-marc-saunders`, abgerufen am 24.04.2021.
17. X. Shi, S. Ma, C. Liu, C. Chen, Q. Wu, X. Chen, and J. Lu. Performance of High Layer Thickness in Selective Laser Melting of Ti6Al4V. *Materials*, 9(12), 2016.
18. B. Song, S. Dong, B. Zhang, H. Liao, and C. Coddet. Effects of processing parameters on microstructure and mechanical property of selective laser melted Ti6Al4V. *Materials & Design*, 35:120–125, 2012.
19. K. Vietz. Simulation der Laserbearbeitung von Aluminium-Nickel-Materialien. Bachelorarbeit, Universität Stuttgart, 2021.
20. Q. Yan, B. Song, and Y. Shi. Comparative study of performance comparison of AlSi10Mg alloy prepared by selective laser melting and casting. *Journal of Materials Science & Technology*, 41:199–208, 2020.

21. B. Zhang, Y. Li, and Q. Bai. Defect Formation Mechanisms in Selective Laser Melting: A Review. *Chinese Journal of Mechanical Engineering*, 30(3):515–527, 2017.
22. J. Zou, Y. Zhu, M. Pan, T. Xie, X. Chen, and H. Yang. A study on cavitation erosion behavior of AlSi10Mg fabricated by selective laser melting (SLM). *Wear*, 376-377:496–506, 2017.

Molecular dynamics investigations on the influence of solutes on the tensile behavior of Polyamide6

Wolfgang Verestek, Johannes Kaiser, Christian Bonten and Siegfried Schmauder

Abstract Although PA6 shows remarkable properties, when in contact with specific media, the strength is reduced. Here we report about molecular dynamics (MD) simulations of PA6 with solutes and their influence on the mechanical properties of the bulk material. For this, four selected scenarios have been simulated. More precisely, four cases are defined: the "dry" case without any solute, "saturated with water" at 5.8 mass %, "saturated with ethanol" at 12.4 mass % and "saturated with ethanol" at 12.3 mass %. The simulated nano tensile tests show differences in the Young's modulus and the ultimate tensile stress, where methanol shows the highest reduction.

Wolfgang Verestek
IMWF University of Stuttgart, Pfaffenwaldring 32, 70569 Stuttgart, e-mail: `wolfgang.verestek@imwf.uni-stuttgart.de`

Johannes Kaiser
IKT University of Stuttgart, Pfaffenwaldring 32, 70569 Stuttgart, e-mail: `johannes.kaiser@ikt.uni-stuttgart.de`

Christian Bonten
IKT University of Stuttgart, Pfaffenwaldring 32, 70569 Stuttgart, e-mail: `christian.bonten@ikt.uni-stuttgart.de`

Siegfried Schmauder
IMWF University of Stuttgart, Pfaffenwaldring 32, 70569 Stuttgart e-mail: `siegfried.schmauder@imwf.uni-stuttgart.de`

W. E. Nagel et al. (eds.), *High Performance Computing in Science and Engineering '21*,
https://doi.org/10.1007/978-3-031-17937-2_5

1 Introduction

Polyamide 6 (PA6), among other Polymers, has sparked scientific and industrial interest in commercial research and academic laboratories due to its remarkable properties. Strengthening can happen at relatively low filler concentrations, for example with 4.7 mass %-layered silicates the Young's modulus and the strength of the material can be doubled [1].

Although PA6 shows remarkable properties, when in contact with specific media, the strength is reduced. Here we report about molecular dynamics (MD) simulations of PA6 with solutes and their influence on the mechanical properties of the bulk material. For this, four selected scenarios have been simulated. More precisely, the following four, experimentally based [2] cases are defined:

1. Dry: no solute
2. Saturated with water: 5.8 mass % water
3. Saturated with methanol: 12.4 mass % methanol
4. Saturated with ethanol: 12.3 mass % ethanol

2 Model

In the following the model creation, equilibration and the performed simulations will be described.

2.1 Model creation

Polyamide 6 can be produced by a condensation reaction of monomers or by a ring opening polymerization. For PA6 this results in amide groups (-N(H)C(=O)-) with five CH_2s between each of them. The resulting Polymer with polymerization grade n is depicted in Fig. 1. The solutes investigated in this contribution are shown in Fig. 2.

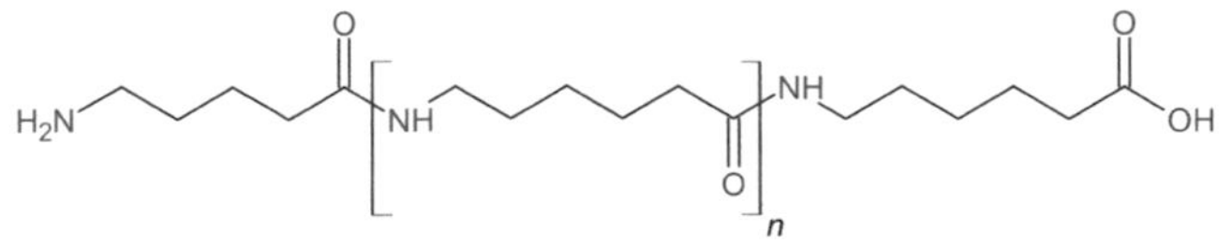

Fig. 1: Polyamide 6 with polymerization grade n. Nitrogen is depicted in Blue, Oxygen in Red. Hydrogens attached to nitrogen and Oxygen are shown, Hydrogens attached to Carbon are not shown for clarity.

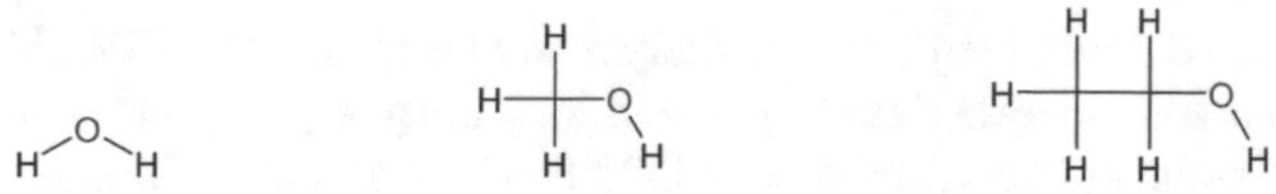

Fig. 2: Solutes. From left to right: Water, Methanol, Ethanol

The three-dimensional periodic models for the tensile test simulations were created with EMC [3] and contain approx. 0.1 mio. atoms. For all models 20 molecules of PA6 were created with a polymerization grade n (see Fig. 1) ranging from 210 to 300 in steps of 10. This results in two molecules for each polymerization grade. The resulting models are summarized in table 1. All models have an initial density of 1.05 g/dm^3 and are thermalized and relaxed during equilibration, see Sec. 2.2. To produce the pure PA6 models without solute the corresponding molecules were simply removed, resulting in a lower initial density, and the equilibration procedure was applied.

Table 1: Model setup

Solute	Mass fraction	Solute molecules
None	0.0 mass %	0
Water	5.8 mass %	1985
Methanol	12.4 mass %	2561
Ethanol	12.3 mass %	1769

2.2 Equilibration

For simulation the open source MD code LAMMPS [4] was used and the 2nd generation Polymer Consistent Force Field (PCFF) [5] with an cutoff of 9.5 Å was applied. The equilibration consists of an initial energy minimization with the conjugate gradient method as well damped dynamics to relax overlapping and very close atoms. This is followed by an NVE ensemble for 10.000 steps with a limited displacement of 0.1 Å/step and a time step width of 0.25 fs to allow for further relaxation.

To allow higher chain mobility a soft-modified version of the PCFF potential was utilized [6] with n = 2, $\alpha_{LJ} = 0.5$ and $\alpha_C = 10$. Two simulations with each 50.000 steps in the NVT ensemble with a starting temperature of 700K and a final temperature relaxed the chain morphology further. The first simulation used $\lambda = 0.8$ and the second one used $\lambda = 0.9$ [6]. To allow larger time steps the rRESPA-Algorithm [7] was employed with 3 levels (bonds, pair and long range/kspace) and scaling factors 2 and 2 resulting in an inner time step of 0.25 fs for bond, 0.5 fs for pair and an outer time step of 1.0 fs for long range interactions.

Finally a sequence of NPT and NVT simulations, each 100.000 steps, was applied. The NVT simulations had a starting temperature of 700 K and an end temperature of 300 K. The NPT simulations had starting and end temperature set to 300 K and a target pressure of 0 bar and isotropic volume scaling. Again, for larger time steps the rRESPA-Algorithm was employed with the same settings as before, resulting in a simulated time of 0.1 ns per simulation run. The final, equilibrated structures are shown for the dry case in Fig. 3 as well as in Fig. 4 for models with solutes.

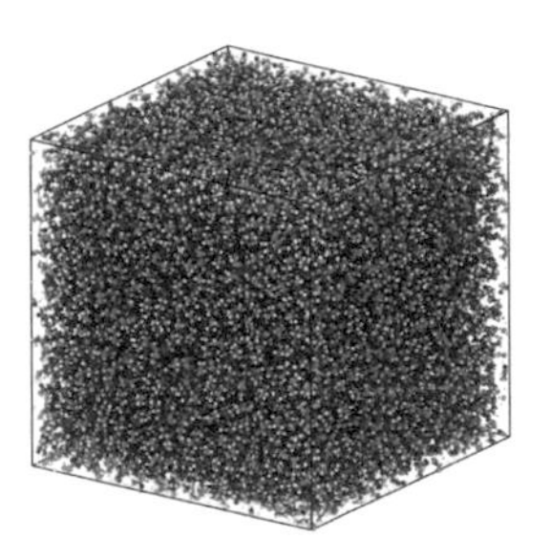 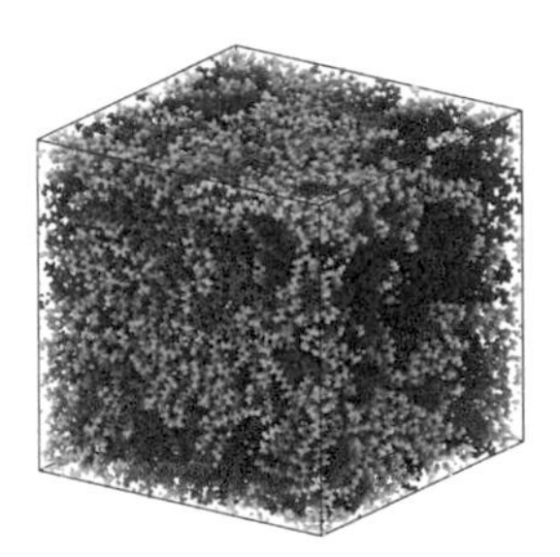

Fig. 3: Equilibrated structure for PA6 without solutes. On the left side, the typical color coding is applied to the Atoms (H: White, C: Grey, N: Blue, O: Red). On the right side the color coding is due to the Molecule ID and allows an better understanding of the molecular morphology.

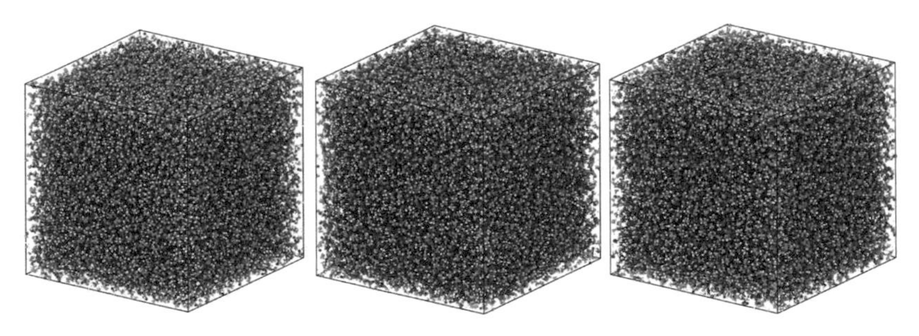

Fig. 4: Equilibrated structures of PA6 with solutes. The typical atomic color coding is applied (H: White, C: Grey, N: Blue, O: Red) plus additionally the solute molecules are colorized in Orange. From left to right the solute is water, methanol and ethanol with mass fractions of approx. 6 and 12 mass % respectively, see Tab. 1.

2.3 Tensile test simulation

As before, the tensile test simulations for a temperature of 300 K were performed with the polymer consistent force field, but the cutoff was increased to 12 Å. Tensile test simulations were done in two different setups. The first setup was simulated in the NPT ensemble with a target pressure of 0 bar in lateral direction and a relaxation time of 0.1 ps for the temperature and 1 ps for the pressure. A second set of simulations was performed in the NVT ensemble with the same settings. Due to the NVT ensemble lateral contraction is not allowed and results effectively in a triaxial stress state. Again, the rRESPA-scheme was employed to allow larger time steps with an inner time step (bonds) of 0.375 fs, a middle time step (pair) of 0.75 fs and an outer time step (long range) of 1.5 fs. For each equilibrated model, tensile tests were simulated. Deformation was applied each time step with true strain rates of 1.0e-5 and 1.0e-6 1/fs. To reduce the noise of the stress-strain curves, the pressure was averaged over 100 time steps (150 fs) before becoming a data point for the plot. To get better, less morphology dependent stress-strain curves, each model was deformed in x, y and z direction, see Fig. 5, and an average resulting stress is computed.

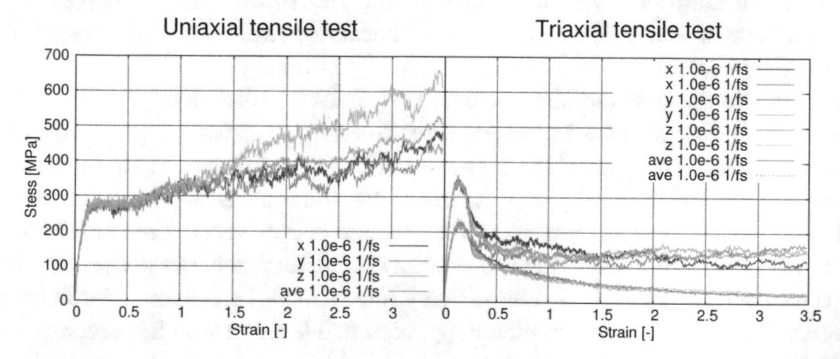

Fig. 5: Stress-strain curves for PA6 without solute for the uniaxial stress state (NPT) and the Triaxial stress state for a true strain rate of 1.0e-6 1/fs for deformation in x, y and z direction and their average. The lower diagram for triaxial stress state additionally shows the averaged stresses orthogonal to the loading direction with dashed lines.

For the uniaxial case the stress goes up nearly linear in the beginning until plastic deformation starts and reaching the ultimate tensile stress. After reaching the a plateau strain hardening takes place and leads to higher stresses than the ultimate tensile stress. One should consider here, that the plots show true stress-strain curves. Transitioning to engineering stress-strain curves would show lower stresses for higher deformation grades as the contraction for the uniaxial case is neglected. It is obvious that the triaxial stress state results in a somewhat higher ultimate tensile stress. But after reaching the ultimate tensile stress the orthogonal stresses decay nearly to zero whereas the stress in loading direction reaches a plateau at which the disentanglement of the molecular chains happens.

3 Results

The uniaxial model, Fig. 6, stays more compact compared to the triaxial case where a big void is already seen in the first picture of Fig. 7. Due to additional degree of freedom in the lateral direction the molecules in the uniaxial model get aligned in loading direction. This is not only true for the voids at higher strains, but also for the bulk phase at lower strains. The molecules align in both cases at higher strains when spanning a void and look like "nano" fibrils.

In Fig. 8 to 11 the simulated stress-strain curves are shown. For the case of the dry PA6 without any solute distinct ultimate tensile stress points can be seen for the uniaxial case for all simulated strain rates. This is followed by a little bit lower yield plateau before strain hardening takes place and reaching a stress of approximately 490 MPa at a strain of 350 %. It is worth mentioning, that the stress-strain curves for both strain rates nearly lie on top of each other in the strain hardening regime after leaving the plateau.

For the triaxial case the ultimate tensile stresses are also clearly visible. This is true for the higher stresses in loading direction as well as for the lower stresses in lateral direction. The ultimate tensile stresses in loading direction are separated by approx. 60 and 100 MPa, respectively, where the strain rate is lower by a factor of 10 and 100. In contrast to this, the maximum stresses in transversal direction show only a very slight reduction due to the lower strain rate. After reaching the ultimate tensile stress the lateral stresses decay towards very low values whereas the stress in loading direction decays only to a certain, nearly horizontal level, at which disentanglement of the molecules takes place.

The tensile test simulation for PA6 with Water are shown in Fig. 9. Compared to the dry, uniaxial case three differences are noticeable. Firstly, for the higher strain rate of 1.5e-5 1/fs the ultimate tensile stress is clearly visible followed by a plateau, whereas for the lower strain rates the plateau vanishes and the ultimate tensile stress is at about 227 MPa and 221 MPa. Secondly, the ultimate tensile stress for all strain rates are lower than those for the dry case. And lastly, the slope of the strain hardening is higher compared to the uniaxial case in Fig. 8 and shows a higher stress, more than 700 MPa at 350 % strain compared to approx. 500 MPa for the dry case. For the triaxial load case the behavior is comparable at a load level approx. 30 MPa lower.

The tensile test simulation for PA6 with methanol are shown in Fig. 10. For the uniaxial case the slope in the hardening regime is higher than in the dry case, comparable to the water loaded case. But the stress level is lower than that for water with approx. 600 and 650 MPA for the two strain rates at 350 % strain. Also the

Fig. 6: Simulated tensile test for dry PA6 with lateral contraction for a true strain rate of 1.0e-6 1/fs after 25, 50 and 100 % simulation time and 45.5 %, 111.7 %, 208.0 % and 348.2 % strain respectively. The color coding is the same as before (H: White, C: Grey, N: Blue, O: Red). Visualization with Ovito [8].

plateau for the higher strain rate of 1.5e-5 1/fs is less pronounced. In general the ultimate tensile stresses are lower than those for the dry case and the water case for all stress states and strain rates.

The uniaxial stress-strain curves for ethanol, Fig. 11 look very similar to those of methanol. But the plateau for the strain rate of 1.0e-5 1/fs is a little bit more distinct than that for methanol. In contrast the slope of the strain hardening regime is less steep and results in lower stresses of approx. 580 MPa and 550 MP. For the triaxial case, the curves are very similar to the methanol case. Generally, the ultimate tensile stresses are slightly higher for ethanol than for methanol, but still lower than for the water case.

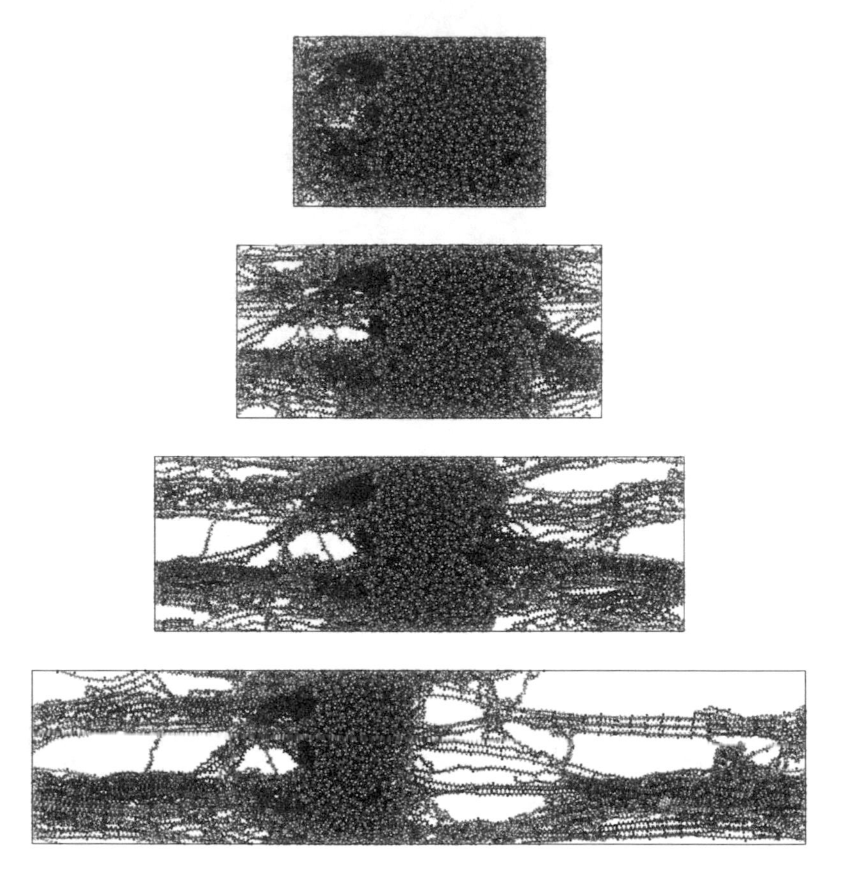

Fig. 7: Simulated tensile test for dry PA6 without lateral contraction for a true strain rate of 1.0e-6 1/fs after 25, 50 and 100 % simulation time and 45.5 %, 111.7 %, 208.0 % and 348.2 % strain respectively. The color coding is the same as before (H: White, C: Grey, N: Blue, O: Red). Visualization with Ovito [8].

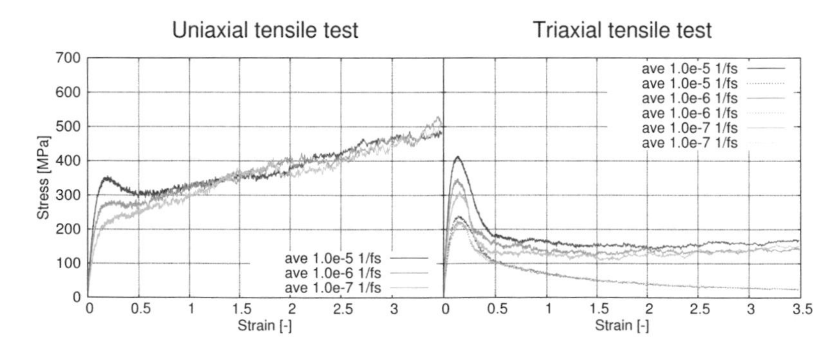

Fig. 8: Simulated tensile test for dry PA6 with and without lateral contraction for true strain rates 1.0e-5, 1.0e-6 and 1.0e-7 1/fs.

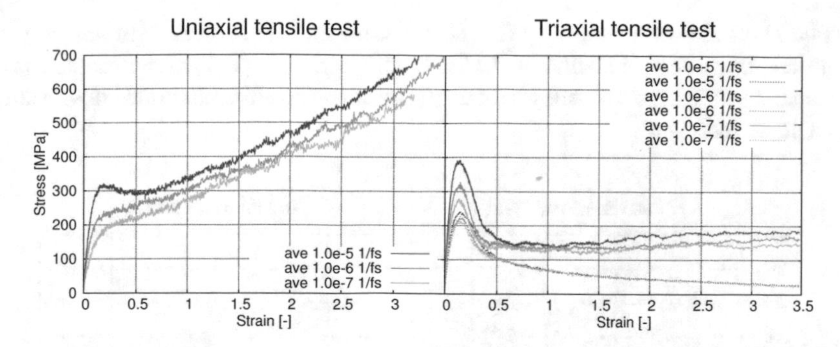

Fig. 9: Simulated tensile test for PA6 with 5.8 mass % water with and without lateral contraction for true strain rates 1.0e-5, 1.0 e-6 and 1.0 e-7 1/fs.

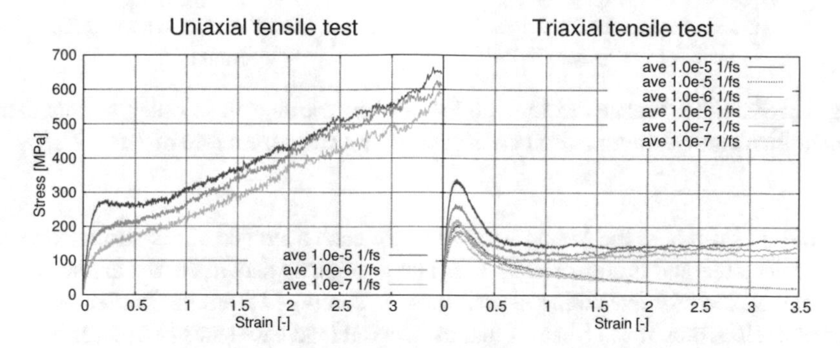

Fig. 10: Simulated tensile test for PA6 with 12.4 mass % methanol with and without lateral contraction for true strain rates 1.0e-5, 1.0 e-6 and 1.0 e-7 1/fs.

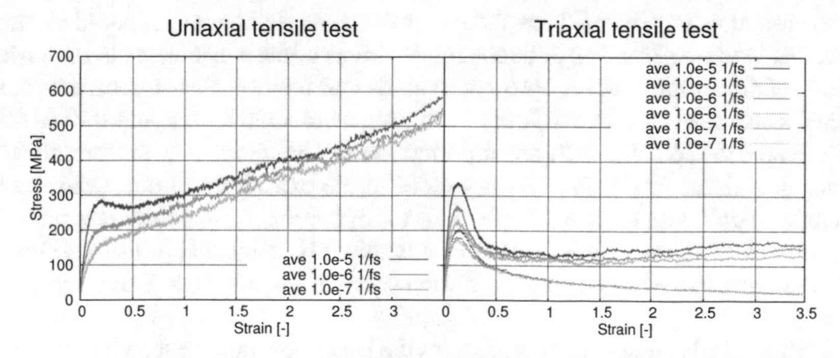

Fig. 11: Simulated tensile test for PA6 with 12.3 mass % ethanol with and without lateral contraction for true strain rates 1.0e-5, 1.0 e-6 and 1.0 e-7 1/fs.

Figure 12 shows a summary for the dry case and all solutes with and without contraction for a strain rate of 1.0e-7 1/fs. To enable a more detailed representation the stress-strain curves are only shown up to 50% strain. The aforementioned observations can also be seen here.

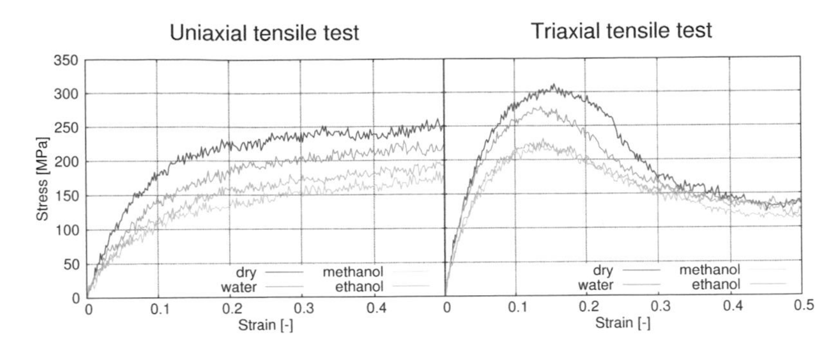

Fig. 12: Simulated tensile test for PA6 for the dry case as well as water, methanol and ethanol with and without lateral contraction for a true strain rate of 1.0 e-7 1/fs.

Starting from here the Young's modulus for each averaged stress-strain curve has been calculated and is shown in Tab. 2. The previously shown curves are somewhat fluctuating, therefor the Young's modulus was computed from the Bézier smoothed curves at 1% strain. It can be seen, that the triaxial load case leads to a stiffer behavior than the uniaxial one. Furthermore a clear trend among the solutes can be observed. Regardless of the load case, dry PA6 has the highest Young's modulus followed by water and ethanol. Methanol shows the lowest Young's modulus. Additionally, experimental values from [2] are shown for the quasi static case. It should be stated here, that in the experiments different compositions of the constituents led to various grades of crystallinity in the PA6 compounds and thereby also to a certain range. Therefor only max and mean values of one composition (90% Ultramid B40 (BASF), 10% SelarPA3426 (DuPont)) are reported here. It is interesting to note, that the Young's modulus of dry PA6 is very close to the experimental one, whereas the Young's moduli with solutes differ by a factor of approx. 2.5-4.5 with respect to the experimental maximum values. Part of this might be subjected to the well known strain rate and size effect. Also the diffusion speed of the solutes in PA6 might play a role.

Having calculated the Young's modulus the next step would be the yield stress. As there is no clear and pronounced yield stress visible in the plots above $R_{p0.2}$ as well as R_{p2} have been computed and are listed in Tab. 3

The ultimate tensile stresses for the different simulations are summarized in Tab. 4 and Fig. 13. From Fig. 13 a clear trend is visible. Each solute lowers the yield stress. water has the least effect. Methanol and ethanol have a similar influence whereas methanol reduces the yield stress a little bit more than ethanol.

Table 2: Young's modulus

strain rate 1/fs	Dry MPa	Water MPa	Methanol MPa	Ethanol MPa
uniaxial				
1e-5	4730	4450	3730	3800
1e-6	3930	3470	2840	3110
1e-7	3000	2550	1890	2130
triaxial				
1e-5	7040	7170	5730	6100
1e-6	6150	6010	4950	5070
1e-7	5530	5320	4290	4410
Experiment	2700/2530	985/480	434/398	664/558

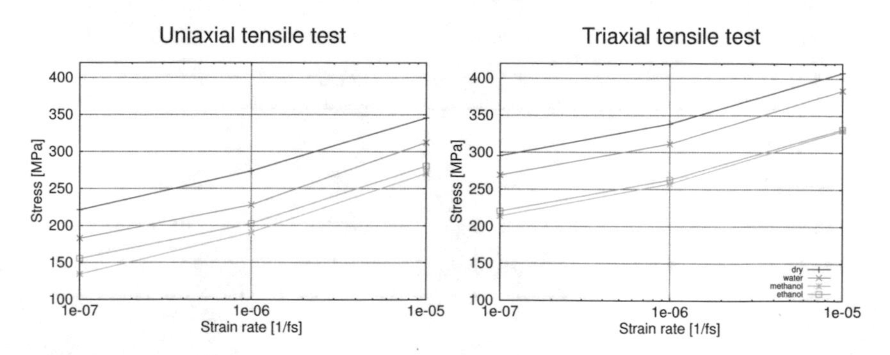

Fig. 13: Ultimate tensile stresses for different solutes, stress states and strain rates.

The reduction of the yield stress is caused by the solute molecules diffusing through the polymer and being attracted by the amide groups. Most of the solute molecules were found to be close to an amide functional group, either the negatively charged oxygen of the solute aiming at the hydrogen of the nitrogen or the doubly bonded oxygen of the amide group attracting a hydrogen of the hydroxyl group of the solute. Due to these hydrogen bridges between the amide group and the solute the chain mobility is increased. There are two reasons for the increased chain mobility. Firstly, the solute effectively shields some amide groups inside the PA6 and prevent these amide groups from building hydrogen bridges with other amide groups. This shielding of the amide groups allow PA6 molecules slide along each other more easily without being pinpointed by hydrogen bridges between amide groups. One

Table 3: Yield stress

strain rate 1/fs	Dry MPa	Water MPa	Methanol MPa	Ethanol MPa
$R_{p0.2}$ uniaxial				
1e-5	131.8	101.4	88.0	87.2
1e-6	87.1	74.9	62.4	61.3
1e-7	64.8	49.6	47.0	40.5
$R_{p0.2}$ triaxial				
1e-5	171.4	149.2	134.1	130.4
1e-6	153.2	122.4	106.6	111.8
1e-7	111.4	104.0	82.4	88.3
R_{p2} uniaxial				
1e-5	253.4	222.1	194.3	200.6
1e-6	195.2	152.8	130.2	130.4
1e-7	143.2	97.8	81.6	86.5
R_{p2} triaxial				
1e-5	334.5	303.9	269.9	265.6
1e-6	277.2	242.5	208.7	208.5
1e-7	224.0	203.1	159.9	171.0

could assume that this shielding happens by the solute molecules getting close to the amide groups and by widening the average distance of the hydrogen bonds between the amide groups. However, the partial radial distribution function (pRDF) for the O-H hydrogen bond of the amide group, see Fig. 4 does not show a strong influence. Two points are worth mentioning. Firstly, for the first peak around 1.8 Å one can see that water leads to a slightly lower probability in the pRDF. Secondly, the second peak at approx. 3.2 Å has two nearly overlapping groups, namely dry/water and methanol/ethanol, whereas the second group with methanol and ethanol shows a slightly higher probability. This is counter intuitive regarding the results of the Young's modulus or the ultimate tensile stress and leads to the conclusion that the shielding is not a simple widening of the hydrogen bonds between the amide groups.

Finally one other interesting observation can be seen when comparing the pRDF for the amide groups over the course of the tensile test simulation for the two different stress states, see Fig. 15. While for the uniaxial stress state the pRDF increases just

Table 4: Ultimate tensile stresses

strain rate 1/fs	Dry MPa	Water MPa	Methanol MPa	Ethanol MPa
uniaxial				
@ 20% strain				
1e-5	345.5	312.3	270.6	280.4
1e-6	273.6	227.9	190.7	202.6
1e-7	221.4	182.9	134.4	155.5
triaxial				
@ 13% strain				
1e-5	407.2	383.4	329.3	331.1
1e-6	338.4	311.4	257.5	262.7
1e-7	295.3	269.3	213.9	220.4

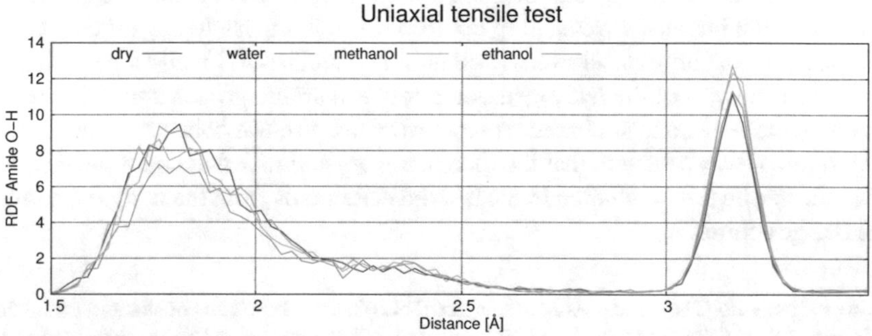

Fig. 14: Partial radial distribution function for O-H in the amide group.

slightly, the pRDF increases nearly linearly for the triaxial stress state and indicates an increasing crystallinity. When examining Fig. 7 one can easily see, that during the later stages of the tensile test, the fibril like structures reorganize in a crystalline way. This explains the increase in the pRDF.

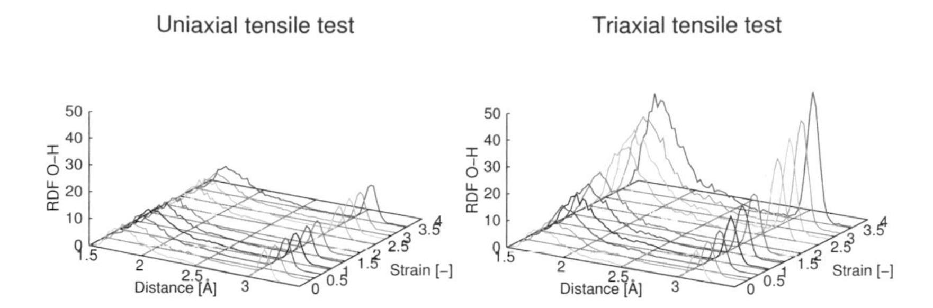

Fig. 15: Partial radial distribution function for O-H in the amide group during the tensile test simulation for the uniaxial and triaxial stress state.

4 Summary

In this work amorphous, dry PA6 as well as PA6 saturated with water, methanol and ethanol have been investigated by means of molecular dynamics. For this, nano tensile tests have been simulated for uniaxial and triaxial stress states and the corresponding Young's moduli, $R_{p0.2}$, R_{p2} and ultimate tensile stresses have been investigated. A clear trend with increased weakening could be seen from dry to the solutes, where water had the least impact followed by ethanol and methanol. Finally a partial radial distribution analysis of the hydrogen bond between amide groups showed, that the weakening mechanism is not based on increasing the distance between amide groups. Furthermore it was revealed, that the increasing crystallinity that could be observed during the tensile test simulation in the triaxial state stems from the reorganization of fibril like structures.

Acknowledgements The authors want to thank the DFG (German Research Foundation) for funding within the project Schm 746/186-1 and BO 1600/30-1. The simulations were performed at the computer clusters Hawk and Hazelhen at the HLRS, University of Stuttgart.

References

1. Usuki, A., Kojima, Y., Kawasumi, M., Okada, A., Fukushima, Y., Kurauchi T. and Kamigaito, O.: Synthesis of nylon 6-clay hybrid. In: Journal of Materials Research **8** 5, 1179–1184 (1993) doi: 10.1557/JMR.1993.1179
2. Schubert, M.: "Einfluss der Blendmorphologie auf das Bruchverhalten von konditionierten PA6-Blends" In: Master Thesis, 2019, IKT Univeristät Stuttgart
3. in't Veld, P. J.: EMC: Enhanced Monte Carlo - A multi-purpose modular and easily extendable solution to molecular and mesoscale simulations. `http://montecarlo.sourceforge.net/emc/Welcome.htmlCited26June2020`
4. Plimpton, S.: Fast Parallel Algorithms for Short-Range Molecular Dynamics. In: Journal of Computational Physics **117** 1, 1–19 (1995) doi: 10.1006/jcph.1995.1039

5. Sun, H., Mumby, S. J., Maple, J. R. and Hagler, A. T.: An ab Initio CFF93 All-Atom Force Field for Polycarbonates. In: Journal of the American Chemical Society **116** 7, 2978–2987 (1994) doi: 10.1021/ja00086a030
6. Beutler, T. C., Mark, A. E., van Schaik, R. C., Gerber, P. R., van Gunsteren, W. F.: Avoiding singularities and numerical instabilities in free energy calculations based on molecular simulations. In: Chemical Physics Letters **222** 6, 529–539 (1994) doi: 10.1016/0009-2614(94)00397-1
7. Tuckerman, M. and Berne, B.J.: Reversible multiple time scale molecular dynamics. In: Journal of Chemical Physics **97** 3, 1990–2001 (1992). doi: 10.1063/1.463137
8. Stukowski, A.: Visualization and analysis of atomistic simulation data with OVITO - the Open Visualization Tool. In: Modeling and Simulation in Materials Science and Engineering **18** 1, 015012 (2010). doi: 10.1088/0965-0393/18/1/015012

Dynamical properties of the Si(553)-Au nanowire system

Mike N. Pionteck, Felix Bernhardt, Johannes Bilk, Christof Dues, Kevin Eberheim, Christa Fink, Kris Holtgrewe, Niklas Jöckel, Brendan Muscutt, Florian A. Pfeiffer, Ferdinand Ziese and Simone Sanna

Abstract The lattice dynamics of the Si(553)-Au surface is modeled from first principles according to different structural models. A multitude of surface-localized phonon modes is predicted. As a general rule, low-energy modes are associated to vibrations within the Au chain, while high-energy modes are mostly localized at the Si step edge. The presence of model specific displacement patterns allows to identify the structural models compatible with the measured spectra at low and at room temperature. Our atomistic models within density functional theory allow to assign spectroscopic signatures available from the literature to displacement patterns, and to explain the activity of nominally Raman silent modes.

1 Introduction

One-dimensional (1D) physics is characterized by a multitude of exotic phenomena such as spin-charge separation, triplet superconductivity, and Luttinger liquid behavior [1–5]. In order to study such phenomena, 1D systems have to be realized. In this respect, self assembled metallic atomic chains on stepped semiconductor surfaces are regarded as the most promising quasi-1D model systems [6–12]. Different realizations of atomic nanowires have been reported in the last decade, that demonstrate their high reproducibility, stability and tunability [9, 13–22].

C. Dues, K. Holtgrewe, S. Sanna and F. Ziese,
Institut für Theoretische Physik and Center for Materials Research (LaMa), Justus-Liebig-Universität Gießen, Heinrich-Buff-Ring 16, 35392 Gießen, Germany,
e-mail: `simone.sanna@theo.physik.uni-giessen.de`

F. Bernhardt, J. Bilk, K. Eberheim, C. Fink, N. Jöckel, B. Muscutt, F.A. Pfeiffer, M.N. Pionteck
Institut für Theoretische Physik, Justus-Liebig-Universität Gießen, Heinrich-Buff-Ring 16, 35392 Gießen, Germany

W. E. Nagel et al. (eds.), *High Performance Computing in Science and Engineering '21*,
https://doi.org/10.1007/978-3-031-17937-2_6

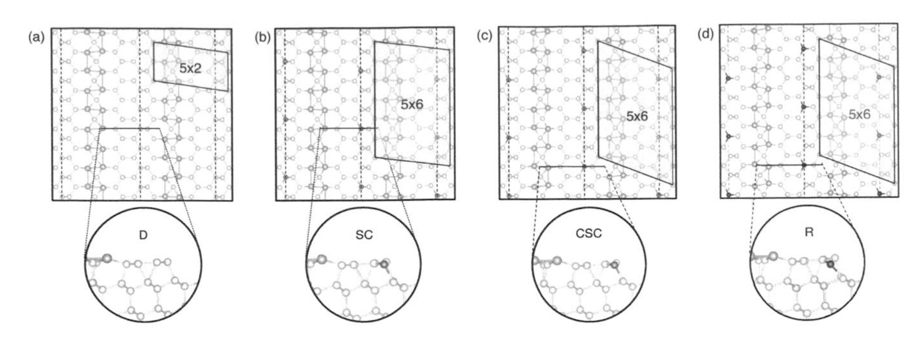

Fig. 1: Structural models of the Si(553)-Au surface. (a) Double Au strand model (D) [16], (b) Spin-chain model (SC) [9], (c) centered spin-chain (CSC) [25]. Spin-polarized atoms are represented in blue (spin up) and red (spin down). (d) Rehybridized model (R) [26]. Si atoms vertically displaced from the step edge are shown in black. Surface unit cells are highlighted.

Among the atomic chains, the Si(553)-Au surface is one of the most studied systems [9, 15, 16, 23–27], as it was proposed to feature an antiferromagnetic spin ordering [9] and to undergo an order-disorder type phase transition below 100 K [28], which is argument of debate [27, 29, 30]. The Si(553) surface is tilted by 12.5° in $[11\bar{2}]$ direction with respect to the (111) plane. It features 14.8 Å wide (111) nanoterraces, separated by steps of double atomic height along the $[1\bar{1}0]$ direction. Evaporation of 0.48 ML Au onto this surface generates one Au chain per terrace as shown in Fig. 1. However, the exact structure of this system has been controversely discussed. Double Au strands and a Si honeycomb chain as shown in Fig. 1 (a), are common to all models. However, while the Krawiec model features a planar Si step edge, in the model proposed by Erwin *et al.* [9] and refined by Hafke *et al.* [25] [Fig. 1 (b) and (c)], the step edge is spin polarized, and every third step edge atom is shifted downwards by 0.3 Å. Recently, Braun *et al.* [26] proposed a diamagnetic, $sp^2 + p$ rehybridized model [Fig. 1 (d)], in which every third edge atom is 0.8 Å lower than the others.

In this manuscript, we provide a thorough theoretical characterization of the surface-localized phonon modes of the Si(553)-Au system, above and below the phase transition temperature. Thereby, the vibrational properties of the system are calculated from DFT-based atomistic simulations for all the proposed structural models. This requires a computational power beyond that of local compute cluster. The comparison with available experimental data [30] allows to assign the measured spectral features to the calculated displacement patterns. The spectra calculated for the double Au strand [16] and the $sp^2 + p$ rehybridized [26] models are compatible with the measured spectra, at 30 K and 300 K, respectively.

2 Methodology

The VASP package [31, 32] implementing the DFT offers highly customizable parallelization schemes to exploit the full capability of massively parallel supercomputers such as the HPE Apollo 9000 system (Hawk) at the HLRS. In order to separate the discussion about the computational setup and the discussion about the algorithm performances, we divide the present section in two parts.

2.1 Computational details

DFT calculations have been performed with the Vienna ab initio Simulation Package (VASP) [31, 32]. Projector augmented waves (PAW) potentials [33] with projectors up to $l = 1$ for H, $l = 2$ for Si and $l = 3$ for Au are used. A number of 1 ($1s^1$), 4 ($3s^2 3p^2$), and 11 ($5d^{10} 6s^1$) valence electrons is employed for H, Si, and Au atoms, respectively. Plane waves up to an energy cutoff of 410 eV build up the basis for the expansion of the electronic wave functions. The silicon surfaces are modeled with asymmetric slabs consisting of 6 Si bilayers stacked along the [111] crystallographic direction, the surface termination including the Au chains, and a vacuum region of about 20 Å. H atoms saturate the dangling bonds at the opposite face of the slabs. These atoms as well as the three lowest Si bilayers are frozen at their bulk position in order to model the substrate, while the remaining atoms are free to relax. The atomic positions are relaxed until the residual Hellmann–Feynman forces are lower than 0.005 eV/Å. $4\times9\times1$ ($4\times27\times1$) Monkhorst–Pack k-point meshes [34] are employed to perform the energy integration in the Brillouin zone of the supercell of 5×6 (5×2) periodicity.

The calculated phonon frequencies depend strongly on the computational approach. Indeed, they depend both directly and indirectly (through the resulting structural differences) on the employed xc-functional. In order to estimate the dependence of the phonon eigenvalues on the computational approach, we have calculated the vibrational frequencies of the double strand model as proposed by Krawiec [16] within the PBEsol [35], LDA [36] and GGA-PBE [37] approach.

According to the actual knowledge of the Si(553)-Au system, the D model is considered as a candidate for the description of the RT structure, while the SC, CSC, and R models are considered for the description of the LT phase. However, all the structures are modeled within DFT at 0 K and thermal lattice expansion is neglected.

While a limited number of surface localized phonons are experimentally detected, the frozen-phonon slab calculations lead to 246 vibrational modes for the double Au chain model by Krawiec [16] and 738 modes for the spin-polarized [9, 25] and rehybridized models [26]. In order to achieve a comparison with the experimental results, we discard the phonons, whose atomic displacement vectors are localized by less than an arbitrarily chosen threshold of 40% in the two topmost atomic layers. Yet, the number of calculated phonon modes is still too high to allow for a frequency-based assignment of the calculated eigenmodes to the measured spectral features. To achieve this task, we compute the Raman scattering efficiency. Raman spectra are generated as described in detail in Refs. [38, 39]. As the electronic structure is self-consistently calculated according to the phonon displacement, phonon induced charge transfer and modifications of the electronic structure are both accounted for. Theoretical spectra are constructed adding Lorentz functions centered at the calculated phonon frequencies, with height corresponding to the calculated Raman efficiency and experimental width. Spectra are calculated for an experimental laser frequency of 647 nm, after consideration of the DFT underestimation of 0.5 eV of the fundamental band gap of the Si bulk.

2.2 Computational performance

The algorithm implemented in VASP can be parallelized with MPI, OpenMP and a hybrid of MPI and OpenMP. However, our tests show that the code performs best with pure MPI parallelization, i.e. one MPI task for each physical core. In addition, VASP offers several parallelization levels that can be specified by VASP-internal input parameters. The main parallelization levels are determined by KPAR, i.e. the number of k-points treated in parallel, and then NPAR, i.e. the number of bands treated in parallel in the matrix diagonalization. According to our tests, these two parameters influence massively the walltime of a calculation. In particular, the parallelization with respect to the k-points improves the performance because it divides the problem into independent chunks, whose demand for inter-communication is restricted. Therefore, performance can be maximized by setting KPAR as high as possible, i.e. setting KPAR to the number of k-points or the number of nodes, whichever is smaller. In general, performance is improved when KPAR is a divisor of both the number of k-points and the number of nodes. On the other hand, k-point parallelization is memory intensive, since the whole problem must be copied to each k-point group. This results in a proportional behaviour between RAM usage and the number of k-points treated in parallel. Based on shared memory, NPAR requires less RAM, however it influences the walltime less than KPAR. NPAR should also be a divisor of bands and number of cores to avoid idle time. The best configuration of KPAR and NPAR depends on the host system. Hence, we need to test the best configuration for HPE Apollo 9000 (Hawk) and to formulate a rule for setting KPAR and NPAR on Hawk. Then we need to check the performance of the code with the best setting.

Since the scaling increases with the problem size, we choose a model system representing the $LiNbO_3(0001)$ surface (z-cut), modeled by a large supercell containing 512 atoms, 3072 bands and 32 k-points for the performance test. Self-consistent minimization of the electronic energy is performed, which is the main framework of the DFT. The calculations are terminated after six minimization steps. Since the first electronic step contains the non-scalable setup of the initial charge, it is excluded from the timing considerations. The loop times for all other steps (step 2 to 6) are averaged to the target value T_{LOOP}. This procedure makes the performance test extrapolatable to real calculations, which perform 30 to 50 electronic steps and, hence, in which the first step with non-scalable portion carries only a small weight. For the tests, we use one MPI task for each physical core, i.e. 128 MPI tasks per node. The left panel of Fig. 2 shows the average walltimes T_{LOOP} of the performance tests with different configurations of KPAR and NPAR as a function of the number of MPI tasks. The green dots denote the average walltime of the best configuration for the respective number of MPI tasks. The tests are in agreement with the experience that setups with higher KPAR require less wall time. However, because of the high memory demand of the problem, each k-point group requires two nodes at the minimum, which limits KPAR to the half of the total number of nodes for this problem. Furthermore, the performance tests yield the rule that NPAR is approximately best set to the square root of the number of MPI tasks employed for each k-point group.

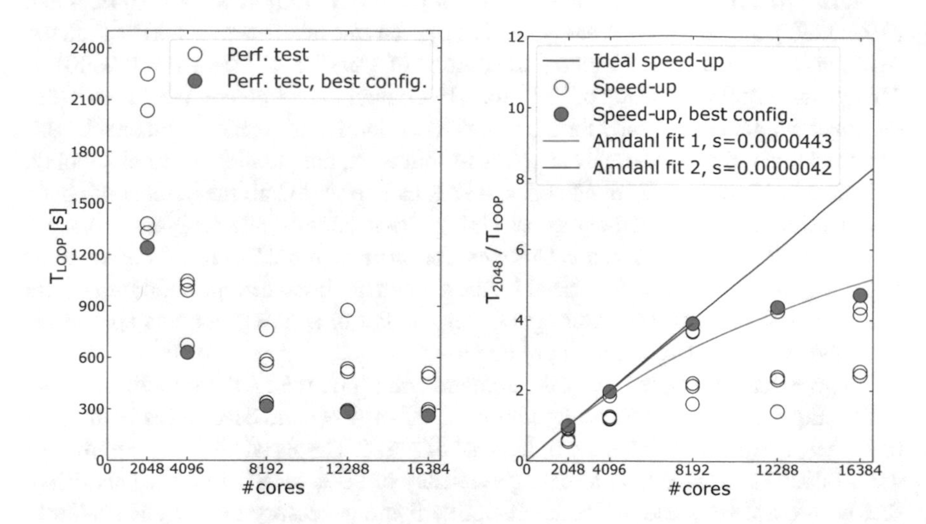

Fig. 2: Results of the performance test; left panel: test results for different parallelization setups, right panel: speed-up of the best guess and Amdahl fits (“Amdahl fit 1”: fit for 2048, 4094, 8192, 12288 and 16384 cores; “Amdahl fit 2”: fit for 2048, 4094 and 8192 cores).

Subsequently, we evaluate the scalability of the VASP code on Hawk. As we use 16 nodes at 128 cores at the minimum, we calculate the speed-up as the ration between the loop walltime for 2048 and N employed cores

$$S_{2048}(N) = \frac{T_{2048}}{T_{\text{LOOP}}(N)}. \tag{1}$$

The right panel of Fig. 2 shows the speed-up as a function of the number of employed cores. The green dots again indicate the best configuration of KPAR and NPAR for the respective number of cores and are fitted to the speed-up by an analytic function. This function is derived from Amdahl’s law, which is defined as the speed-up with respect to one MPI task

$$S(N) = \frac{T_1}{T_{\text{LOOP}}(N)} = \frac{1}{s + \frac{p}{N}}, \tag{2}$$

where $s, p \in (0, 1)$ with $p = s - 1$ denote the serial and parallel portions of the code. Thus the speed-up in (1) can be fit with the serial part s as parameter in accordance with Amdahl's law

$$S_{2048}(N) = \frac{T_{2048}}{T_{\text{LOOP}}(N)} = \frac{T_{2048}}{T_1} \frac{T_1}{T_{\text{LOOP}}(N)} = \frac{S(N)}{S(2048)} = \frac{s + \frac{1-s}{2048}}{s + \frac{1-s}{N}}. \tag{3}$$

At first, the fit is done for the speed-up of the best configurations at 2048, 4094, 8192, 12288 and 16384 cores and plotted as a green line labelled "Amdahl fit 1" in the right panel of Fig. 2. The obtained serial portion of "Amdahl fit 1" is at $s = 0.0000443$. This yields a scale efficiency of 73 % for 8192 cores, i.e. 64 nodes. We observe that the speed-up flattens abruptly for more than 8192 cores. This can be explained by our test conditions. Since we use 32 k-points for our tests, the parallelization of k-points is limited to 32 groups. Up to 64 nodes, KPAR can be set to half the number of nodes (the maximum due to memory constraints). Consequently, the problem can scale with 2048, 4094 and 8192 cores. However, for more than 8192 cores, the number of k-point groups cannot be increased further, resulting in an abrupt flattening of the speed-up. For calculations featuring a larger number of k-points, a linear speedup is expected even for a larger number of cores.

To investigate the speed-up in the domain with optimal KPAR-parallelization, we fit also Eq. (3) with the best configurations at 2048, 4094 and 8192 cores as the blue line labeled "Amdahl fit 2" in the right panel of Fig. 2. The serial portion drops then to $s = 0.0000042$ translating to a scaling efficiency of 96 % for 8192 cores. This shows that we are able to scale our code efficiently if the necessary memory is available and enough k-points can be used. Thus, optical calculations with a high number of k-points are particularly well suited to exploit the full capability of Hawk.

According to our experience, the queue time is significantly longer for jobs longer than 4 hours. Therefore, we have optimized our jobs for walltimes of maximum 4 hours.

3 Results

3.1 Phonon modes at room temperature

The vibrational properties of the Si(553)-Au system at 300 K are computed with the structural model proposed by Krawiec [16] [see Fig. 1 (a)]. Although, according to recent investigations, the Si(553)-Au surface at 300 K oscillates between the D, R, and

SC/CSC phases, the system is for the vast majority of the time in the D configuration [27], which is therefore employed to describe the high temperature phase of the Si(553)-Au surface.

Among the calculated phonon modes a considerable fraction is Raman silent. The calculated phonon spectra shown in Fig. 3 closely reproduce the measured spectra [30]. Solely the peaks of moderate intensity predicted at about 150 cm^{-1} in the crossed polarization are not experimentally observed. However, the assignment of phonon modes in this spectral region is difficult because of the overlap with broad and intense bulk phonons. Generally, both the frequency and the relative intensity of the spectral features are in satisfactorily agreement with the experiment, although the most intense vibrational signatures in the (yx) crossed configuration are somewhat red shifted within PBEsol in comparison with the experiment. The good agreement

Table 1: Raman frequencies (in cm^{-1}) measured at 300 K [30] and calculated (0 K frozen phonon calculations performed with the D model). PBEsol calculated frequencies are listed (Theo.), along with the highest and lowest frequency calculated with other XC-functionals. Char. and Loc. indicate whether the phonon has Au or Si character, and the surface localization of the atomic displacement vectors, respectively. Modes with calculated Raman efficiency below 1% of the main peak are not listed.

z(yy)-z					z(yx)-z				
Exp.	Theo.	Theo. Min-Max	Char.	Loc.	Exp.	Theo.	Theo. Min-Max	Char.	Loc.
37.5±0.3	35.4	35.1−36.4	Au	75%	43.3±0.9	44.2	44.2−46.1	Au	58%
48.3±0.1 {	48.9	47.7−52.9	Au+Si	93%	49.7±0.4	52.7	52.7−59.8	Au	91%
	52.7	52.7−59.8	Au	91%					
60.1±0.1	64.9	64.9−66.5	Au+Si	68%	43.3±0.9	44.2	44.2−46.1	Au	58%
68.4±0.3	69.8	66.1−72.9	Au	92%	49.7±0.4	52.7	52.7−59.8	Au	91%
85.5±0.1	84.2	84.1−86.2	Au+Si	88%	61.6±0.5	62.4	61.7−64.6	Au	91%
100.8±0.2	109.4	109.4−113.1	Si	74%	73.9±1.1	65.4	65.4−67.6	Si	47%
121.3±0.4	122.0	122.0−129.4	Si	55%	87.4±0.9	79.7	79.7−86.2	Si	57%
134.1±1.4	131.9	131.8−141.1	Si	50%	94.0±0.2 {	82.1	82.1−86.1	Au	77%
						84.4	84.4−94.9	Au	77%
147.4±0.7	140.2	140.2−144.0	Si	47%	101.7±0.2	96.8	96.2−98.8	Au	72%
158.4±4.7	165.1	165.1−167.5	Si	55%	113.5±0.5	111.4	111.4−115.3	Si	52%
172.6±1.9	169.8	169.7−171.3	Si	51%					
392.2±0.1	386.2	379.8−388.3	Si	40%					
413.2±0.1	411.5	410.1−415.7	Si	50%					

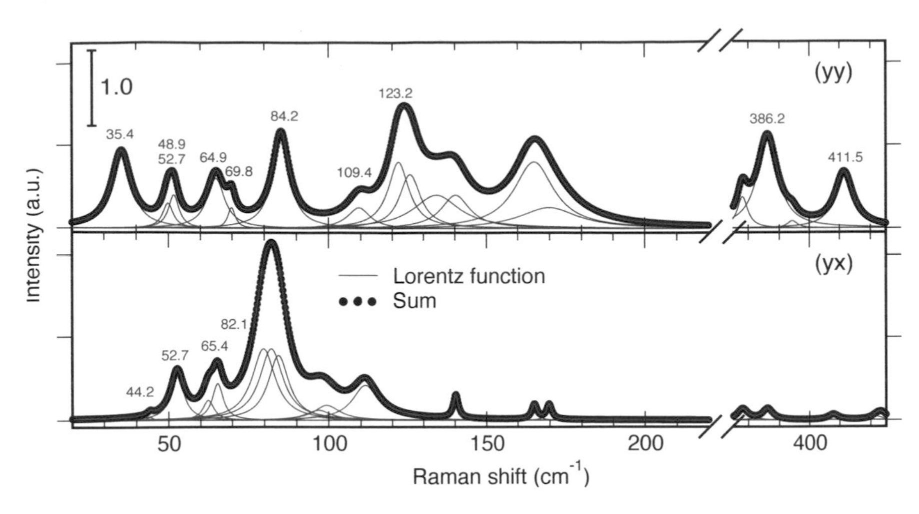

Fig. 3: Raman spectra of the Si(553)-Au surface calculated within DFT-PBEsol for the (yy) and (yx) polarization with the structural model –D– by Krawiec *et al.* [16].

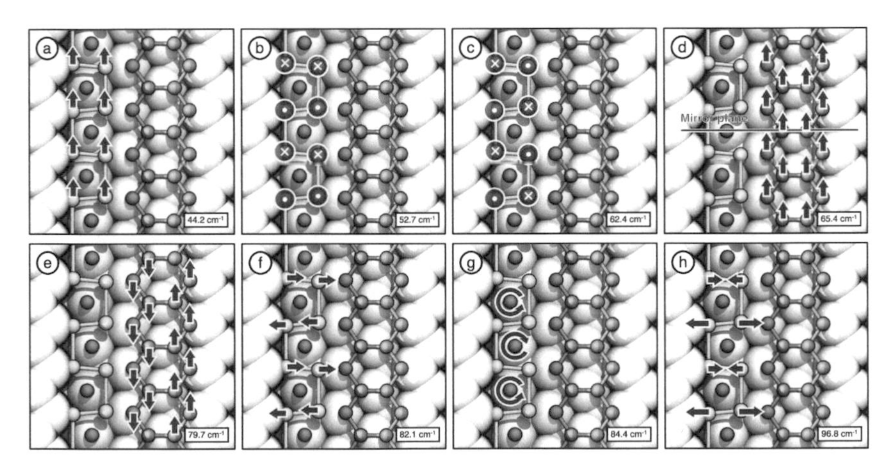

Fig. 4: Schematic representation of the displacement pattern of selected phonon modes that are Raman active in the crossed polarization according to the double Au strand model [16] of the RT phase.

between measured and calculated spectra allows to assign the calculated phonon displacement patterns on the basis of energy, symmetry, and Raman intensity. The result of this procedure is shown in Tab. 1.

Displacement patterns of Raman active modes are represented in Figs. 4–5, in which upwards and downwards vertical displacements are represented by dots and crosses, respectively. The calculated displacement patterns reflect the polarization dependence of the Raman spectra. The modes detectable in parallel polarization are symmetric with respect to the mirror symmetry plane perpendicular to the Au wires shown in Fig. 4 (d) and Fig. 5 (d). The modes that are Raman active in crossed polarization do not possess this symmetry. This relationship between selection rules and phonon symmetry is valid, assuming deformation potentials as the origin of the Raman scattering.

More in detail, in the crossed configuration six modes are clearly visible, along with low intensity signatures. All the corresponding displacement patterns (shown in Fig. 4) break the local symmetry of the terrace, as expected for modes that are Raman active in crossed polarization. Further modes of moderate intensities at higher frequencies are predicted but not experimentally observed (due to the overlap with bulk resonances) and will not be discussed in detail.

In parallel polarization, more modes are Raman active. While at lower frequencies the Au related modes dominate, above 100 cm^{-1} only Si modes are found. This is expected, due to the much higher mass of the Au atoms with respect to Si. Among this modes, the overlap of two close phonons at 48.9 and 52.7 cm^{-1}, with similar displacement pattern, represented in Fig. 5 (b), is experimentally observed as a single peak measured at 48.3 cm^{-1}. Both modes perform the same seesaw movement within the Au-chain, however, the first one features a more pronounced in-phase movement of the step edge, which increases the phonon effective mass and lowers its frequency.

These two modes break the local terrace symmetry and should not be Raman active in parallel polarization, assuming deformation potentials as the origin of the Raman scattering. However, a further Raman mechanism –scattering at charge density fluctuations– must be considered for these modes. Charge fluctuations between the step edge and the Au chain occur when atomic displacements modify the Au-Au bond length and the relative height of the step edge atoms, as revealed by previous studies [20, 21, 27]. Scattering at charge density fluctuations, which is well known,

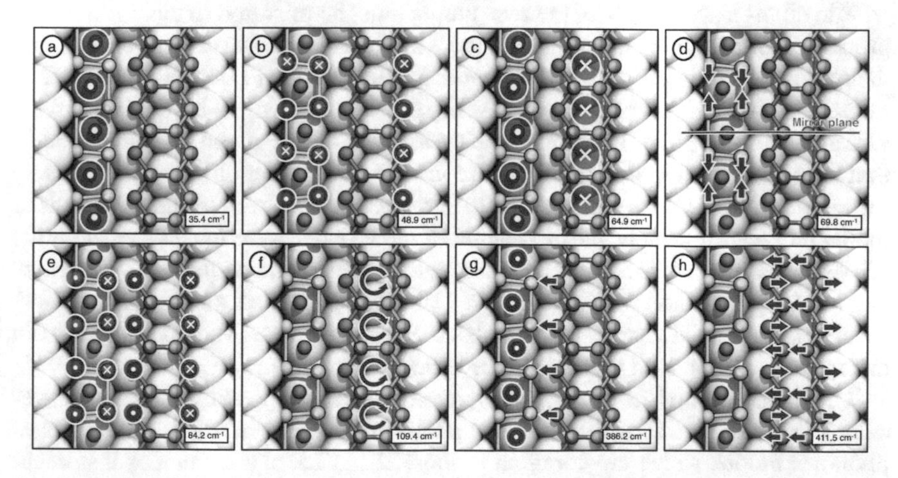

Fig. 5: Schematic representation of the displacement pattern of phonon modes that are Raman active in the parallel polarization according to the double Au strand model [16] of the RT phase.

e.g., for highly doped semiconductors [40], is traced to the modifications of the adsorption edge due to the phonon-induced charge redistribution and had not been previously observed for quasi 1D systems. This mechanism contributes to the Raman scattering only in parallel configuration, and occurs for all modes associated with strong charge fluctuations. Yet, it is generally difficult to distinguish and quantify the relative contributions of scattering at charge density fluctuations and deformation potentials to the total Raman intensity, as both occur in parallel polarization. However, for the mode in Fig. 5 (b), no deformation potential scattering can take place in parallel polarization, due to the phonon symmetry. Therefore only scattering at charge density fluctuations can be responsible for the Raman intensity of this mode.

The close agreement between the theoretical and experimental results strongly suggests that the double Au strand model [16] correctly describes the Si(553)-Au system.

3.2 Phonon modes below and above the phase transition

The Si(553)-Au system is supposed to undergo an order-disorder type structural transition starting below 100 K [27]. According to the actual knowledge of the system, the double Au strand model of (5×2) periodicity [16] describes the high symmetric RT phase, while the (centered) spin-chain [9, 25] and rehybridized [26] models of (5×6) periodicity have been proposed for the description of the lower symmetric LT phase. Between RT and LT the morphology of the step edge of the Si terraces fluctuates among the configurations of the two phases, establishing an interplay (via charge transfer) with the Au chain that continuously enhances the chain dimerization from RT to LT [27].

The different symmetry of the two phases must be mirrored in their vibrational properties. In particular, modes associated with the Au chain are expected to modify their frequency from RT to LT, while modes localized at the Si step edge cannot exist in the same form in the two phases. To verify this assumption, we have calculated the vibrational properties of the Si(553)-Au system at LT with all the structural models that have been proposed in the literature for the description of this phase, namely the spin-chain [9] and centered spin-chain model [25], as well as the rehybridized model [26]. Unfortunately, the calculation of the Raman scattering efficiency for the LT structural models is an exceptionally demanding task, due to both system size and number of phonons (396 atoms and 738 vibrational modes for each structure). However, the knowledge of the calculated Raman frequencies and displacement patterns can still be used to interpret the experimental data.

The comparison of the displacement patterns and corresponding eigenfrequencies calculated with the RT and LT models allows to discriminate three categories of phonons: modes which are common to the RT and LT phases, modes that occur both in the RT and in the LT phase, however with a different frequency, and modes that exist either only in the RT or only in the LT structure. Most of the frequencies

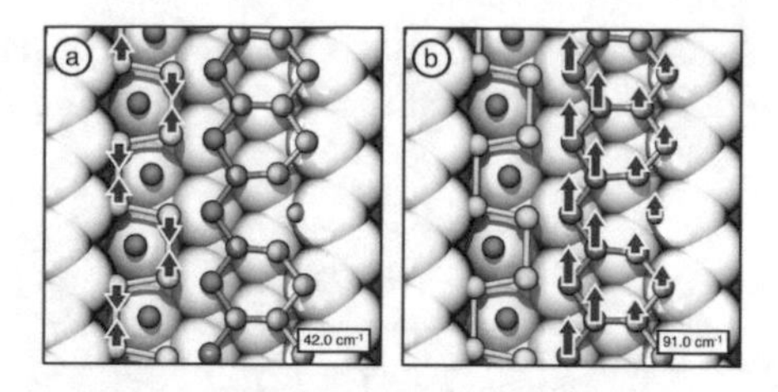

Fig. 6: Schematic representation of the displacement pattern of the dimerization mode and of the honeycomb translation mode. The displacement patterns are calculated according to the rehybridized model[26] of the LT phase.

calculated with the RT model can be identified in all three candidate models within a few cm^{-1}. However, important exceptions are found, corresponding exactly to the measured LT-RT differences [30].

The Au dimerization mode shown in Fig. 6 (a), exists in both the RT and LT structure, yet at different frequencies. This mode is predicted at $18.8\pm5\,cm^{-1}$ with the RT model, and is thus not experimentally accessible. However, it becomes much harder ($42.0\,cm^{-1}$) within the rehybridized model, which features a more pronounced dimerization. As this mode shortens the Au-Au bond length, it requires more energy for a strongly dimerized Au chain. This suggests that the peak observed at LT at $40.5\,cm^{-1}$ is the overlap of a weakly temperature dependent mode at $37.5\,cm^{-1}$ and the dimerization mode at about $42\,cm^{-1}$. This would both explain the observed modifications in intensity and frequency, and also be in agreement with a previous interpretation [27].

The second exception is represented by a further mode with strong frequency shift, the mode calculated at $69.8\,cm^{-1}$ shown in Fig. 5 (d). Similarly to the dimerization mode, this mode shortens the Au-Au bond length and becomes much harder by about $8\,cm^{-1}$ for the rehybridized model.

Another difference between the HT and LT phase is the mode associated to the displacement pattern calculated with the rehybridized model and displayed in Fig. 6 (b). This mode is not existent in the RT structure, but closely related to the chain translation modes predicted at $65.4\,cm^{-1}$ and at $79.7\,cm^{-1}$ [Fig. 4 (d) and (e)] at RT. This mode is a translation of the HC chain, which is strongly hindered by the LT modification of the step edge pinning the outer atoms and therefore making this mode much harder.

Finally, the mode measured at $413\,cm^{-1}$ and associated to the theoretically predicted step edge mode at $411.5\,cm^{-1}$ [see Fig. 5 (h)] is only a phonon mode of the RT phase. Due to the different symmetry of the Si step edge in the structural models associated to the LT and RT phases, this lattice vibration has no low temperature counterpart. In the rehybridized model, this mode is decomposed into local vibrations of the step edge, as previously pointed out by Braun *et al.* [27].

Table 2: Raman frequencies of selected modes discussed in the text calculated within PBEsol according to different structural models representing the low temperature structure.

Exp.	R	SC	CSC
40.5±0.2	42.0	8.7	16.7
82.7±1.4	77.3	69.4	70.7
99.1±1.5	91.0	65.0	64.8

To summarize, the rehybridized model can well explain the observed temperature shifts. On the contrary, in the (centered) spin-chain model [9, 25] the dimerization is not as pronounced as in the rehybridized model [26] and therefore the frequency shifts with respect to the high temperature model cannot be reproduced. For example, the dimerization mode [see Fig. 6 (a), and Tab. 2] is predicted by DFT-PBEsol at 8.7 and 16.7 cm^{-1}, for the spin-chain and centered spin-chain models, respectively. This value is far from the value of 42.0 cm^{-1} calculated with the rehybridized model and assigned to the peak measured at 40.5 cm^{-1}. Similarly, the mode that couples the Au dimers [see Fig. 5 (d)] calculated at 68.9 cm^{-1} for the RT structure, does not significantly shift at LT in the SC and CSC model (69.4 and 70.7 cm^{-1}, respectively) and cannot explain the spectral feature measured at 82.7 cm^{-1}. Thus, the comparison of the calculated vibrational properties with the measured spectra yields a strong argument for the rehybridized model for the description of the low-temperature phase.

4 Conclusions

The dynamical properties of the Si(553)-Au surface have been studied by DFT. The atomistic calculations satisfactorily reproduced the measured spectra, if the double Au strand model and the rehybridized model are used for the simulation of the RT and LT phase, respectively. We propose an assignment of the mode specific phonon displacement patterns to the measured spectral signatures. Furthermore, a pronounced temperature dependence of different modes in the LT phase and in the RT phase is interpreted as a signature of a structural phase transition. As a general feature, modes that modify the Au-Au bond length and modes localized at the Si step edge are significantly influenced by the phase transition.

Modes which involve the Si step edge atoms either disappear or are shifted to much higher energies at LT. In particular, the mode predicted at 411.5 cm^{-1}, associated to transversal vibrations of the Si step edge, disappears at LT, where thermal fluctuations are frozen and the surface structure shows a higher order.

The phonon activated charge transfer between the Au chain and the Si step edge, which is responsible for the observed order-disorder phase transition [27], leads to Raman scattering by charge density fluctuations and allows for a direct observation of the strong coupling between electronic and phononic systems on the Si(553)-Au surface.

The knowledge of the vibrational properties of the Si(553)-Au surface is moreover a prerequisite for future investigations of the Si(553)-Au system modified by atomic deposition or doping, which is often performed to control the coupling with higher dimensions.

Acknowledgements We gratefully acknowledge financial support from the Deutsche Forschungsgemeinschaft in the research unit FOR1700 (project SA 1948/1-2) FOR2824, project SA 1948/2-1 and FOR5044, (project SA 1948/3-1). The Höchstleistungrechenzentrum Stuttgart (HLRS) is gratefully acknowledged for grants of high-performance computer time.

References

1. S. Kagoshima, H. Nagasawa, T. Sambongi, *Electron Correlation and Magnetism in Narrow-Band Systems*, *Springer Series in Solid-State Sciences*, vol. 72 (Springer, Berlin, Heidelberg, 1988). DOI 10.1007/978-3-642-83179-9. URL http://dx.doi.org/10.1007/978-3-642-81639-0{_}15{%}0Ahttp://link.springer.com/10.1007/978-3-642-83179-9http://link.springer.com/10.1007/978-3-642-83179-9
2. G. Grüner, Rev. Mod. Phys. **60**(4), 1129 (1988). DOI 10.1103/RevModPhys.60.1129. URL https://link.aps.org/doi/10.1103/RevModPhys.60.1129
3. J.M. Luttinger, J. Math. Phys. **4**(9), 1154 (1963). DOI 10.1063/1.1704046. URL http://link.aip.org/link/JMAPAQ/v4/i9/p1154/s1{&}Agg=doihttp://scitation.aip.org/content/aip/journal/jmp/4/9/10.1063/1.1704046
4. K. Schönhammer, *Luttinger Liquids: basic concepts* (Springer Netherlands, Dordrecht, 2004), *Physics and Chemistry of Materials with Low-Dimensional Structures*, vol. 25, chap. 4, pp. 93–136. DOI 10.1007/978-1-4020-3463-3_4. URL http://www.springerlink.com/content/p6ru1315m33g7685http://link.springer.com/10.1007/978-1-4020-3463-3{_}4
5. T. Giamarchi, *Quantum Physics in One Dimension* (Clarendon Press, Oxford, 2007)
6. C. Zeng, P. Kent, T.H. Kim, A.P. Li, H.H. Weitering, Nature Materials **7**(7), 539 (2008). URL https://www.nature.com/articles/nmat2209
7. P.C. Snijders, H.H. Weitering, Rev. Mod. Phys. **82**, 307 (2010). DOI 10.1103/RevModPhys.82.307
8. H. Weitering, Nat. Phys. **7**(10), 744 (2011). DOI 10.1038/nphys2074. URL http://dx.doi.org/10.1038/nphys2074
9. S.C. Erwin, F.J. Himpsel, Nat. Commun. **1**(5), 58 (2010). DOI 10.1038/ncomms1056. URL http://www.ncbi.nlm.nih.gov/pubmed/20975712
10. J. Aulbach, J. Schäfer, S.C. Erwin, S. Meyer, C. Loho, J. Settelein, R. Claessen, Phys. Rev. Lett. **111**(13), 137203 (2013). DOI 10.1103/PhysRevLett.111.137203. URL http://link.aps.org/doi/10.1103/PhysRevLett.111.137203
11. C. Brand, H. Pfnür, G. Landolt, S. Muff, J.H. Dil, T. Das, C. Tegenkamp, Nat. Commun. **6**, 8118 (2015). DOI 10.1038/ncomms9118
12. K. Holtgrewe, S. Appelfeller, M. Franz, M. Dähne, S. Sanna, Phys. Rev. B **99**, 214104 (2019). DOI 10.1103/PhysRevB.99.214104. URL https://link.aps.org/doi/10.1103/PhysRevB.99.214104

13. J.N. Crain, J. McChesney, F. Zheng, M.C. Gallagher, P. Snijders, M. Bissen, C. Gundelach, S.C. Erwin, F.J. Himpsel, Phys. Rev. B **69**(12), 125401 (2004). DOI 10.1103/PhysRevB.69.125401
14. I. Barke, F. Zheng, T.K. Rügheimer, F.J. Himpsel, Phys. Rev. Lett. **97**(22), 226405 (2006). DOI 10.1103/PhysRevLett.97.226405. URL http://link.aps.org/doi/10.1103/PhysRevLett.97.226405
15. J. Aulbach, S.C. Erwin, R. Claessen, J. Schäfer, Nano Lett. **16**(4), 2698 (2016). DOI 10.1021/acs.nanolett.6b00354. URL http://pubs.acs.org/doi/abs/10.1021/acs.nanolett.6b00354
16. M. Krawiec, Phys. Rev. B **81**(11), 115436 (2010). DOI 10.1103/PhysRevB.81.115436. URL http://link.aps.org/doi/10.1103/PhysRevB.81.115436
17. I. Miccoli, F. Edler, H. Pfnür, S. Appelfeller, M. Dähne, K. Holtgrewe, S. Sanna, W.G. Schmidt, C. Tegenkamp, Phys. Rev. B **93**, 125412 (2016). DOI 10.1103/PhysRevB.93.125412. URL https://link.aps.org/doi/10.1103/PhysRevB.93.125412
18. C. Braun, C. Hogan, S. Chandola, N. Esser, S. Sanna, W.G. Schmidt, Phys. Rev. Materials **1**, 055002 (2017). DOI 10.1103/PhysRevMaterials.1.055002
19. F. Edler, I. Miccoli, J.P. Stöckmann, H. Pfnür, C. Braun, S. Neufeld, S. Sanna, W.G. Schmidt, C. Tegenkamp, Phys. Rev. B **95**(12), 125409 (2017). DOI 10.1103/PhysRevB.95.125409. URL http://link.aps.org/doi/10.1103/PhysRevB.95.125409
20. Z. Mamiyev, S. Sanna, C. Lichtenstein, T. and. Tegenkamp, H. Pfnür, Phys. Rev. B **98**(24), 245414 (2018). DOI 10.1103/PhysRevB.98.245414
21. C. Hogan, E. Speiser, S. Chandola, S. Suchkova, J. Aulbach, J. Schäfer, S. Meyer, R. Claessen, N. Esser, Phys. Rev. Lett. **120**, 166801 (2018). DOI 10.1103/PhysRevLett.120.166801. URL https://link.aps.org/doi/10.1103/PhysRevLett.120.166801
22. M. Tzschoppe, C. Huck, F. Hötzel, B. Günther, Z. Mamiyev, A. Butkevich, C. Ulrich, L. Gade, A. Pucci, J. Phys. Condens. Matter **31**(19), 195001 (2018). DOI 10.1088/1361-648X/ab0710
23. S. Polei, P.C. Snijders, S.C. Erwin, F.J. Himpsel, K.H. Meiwes-Broer, I. Barke, Phys. Rev. Lett. **111**(15), 156801 (2013). DOI 10.1103/PhysRevLett.111.156801. URL http://link.aps.org/doi/10.1103/PhysRevLett.111.156801
24. S. Polei, P.C. Snijders, K.H. Meiwes-Broer, I. Barke, Phys. Rev. B **89**(20), 205420 (2014). DOI 10.1103/PhysRevB.89.205420. URL http://link.aps.org/doi/10.1103/PhysRevB.89.205420
25. B. Hafke, T. Frigge, T. Witte, B. Krenzer, J. Aulbach, J. Schäfer, R. Claessen, S.C. Erwin, M. Horn-von Hoegen, Phys. Rev. B **94**(16), 161403 (2016). DOI 10.1103/PhysRevB.94.161403. URL http://link.aps.org/doi/10.1103/PhysRevB.94.161403
26. C. Braun, U. Gerstmann, W.G. Schmidt, Phys. Rev. B **98**, 121402 (2018). DOI 10.1103/PhysRevB.98.121402. URL https://journals.aps.org/prb/pdf/10.1103/PhysRevB.98.121402
27. C. Braun, S. Neufeld, U. Gerstmann, S. Sanna, J. Plaickner, E. Speiser, N. Esser, W.G. Schmidt, Phys. Rev. Lett. **124**, 146802 (2020). DOI 10.1103/PhysRevLett.124.146802. URL https://link.aps.org/doi/10.1103/PhysRevLett.124.146802
28. F. Edler, I. Miccoli, H. Pfnür, C. Tegenkamp, Phys. Rev. B **100**, 045419 (2019). DOI 10.1103/PhysRevB.100.045419. URL https://link.aps.org/doi/10.1103/PhysRevB.100.045419
29. B. Hafke, C. Brand, T. Witte, B. Sothmann, M. Horn-von Hoegen, S.C. Erwin, Phys. Rev. Lett. **124**, 016102 (2020). DOI 10.1103/PhysRevLett.124.016102. URL https://link.aps.org/doi/10.1103/PhysRevLett.124.016102
30. J. Plaickner, E. Speiser, C. Braun, W.G. Schmidt, N. Esser, S. Sanna, Phys. Rev. B **103**, 115441 (2021). DOI 10.1103/PhysRevB.103.115441. URL https://link.aps.org/doi/10.1103/PhysRevB.103.115441
31. G. Kresse, J. Furthmüller, Computational Materials Science **6**(1), 15 (1996). DOI https://doi.org/10.1016/0927-0256(96)00008-0
32. G. Kresse, J. Furthmüller, Phys. Rev. B **54**, 11169 (1996). DOI 10.1103/PhysRevB.54.11169
33. P.E. Blöchl, Phys. Rev. B **50**, 17953 (1994). DOI 10.1103/PhysRevB.50.17953
34. H.J. Monkhorst, J.D. Pack, Phys. Rev. B **13**, 5188 (1976). DOI 10.1103/PhysRevB.13.5188

35. J.P. Perdew, A. Ruzsinszky, G.I. Csonka, O.A. Vydrov, G.E. Scuseria, L.A. Constantin, X. Zhou, K. Burke, Phys. Rev. Lett. **100**, 136406 (2008). DOI 10.1103/PhysRevLett.100.136406. URL https://link.aps.org/doi/10.1103/PhysRevLett.100.136406
36. D.M. Ceperley, B.J. Alder, Phys. Rev. Lett. **45**, 566 (1980). DOI 10.1103/PhysRevLett.45.566. URL https://link.aps.org/doi/10.1103/PhysRevLett.45.566
37. J.P. Perdew, K. Burke, M. Ernzerhof, Phys. Rev. Lett. **77**, 3865 (1996). DOI 10.1103/PhysRevLett.77.3865. URL https://link.aps.org/doi/10.1103/PhysRevLett.77.3865
38. E. Speiser, N. Esser, B. Halbig, J. Geurts, W.G. Schmidt, S. Sanna, Surface Science Reports **75**(1), 100480 (2020). DOI https://doi.org/10.1016/j.surfrep.2020.100480. URL http://www.sciencedirect.com/science/article/pii/S0167572920300017
39. S. Sanna, S. Neufeld, M. Rüsing, G. Berth, A. Zrenner, W.G. Schmidt, Phys. Rev. B **91**, 224302 (2015). DOI 10.1103/PhysRevB.91.224302. URL https://link.aps.org/doi/10.1103/PhysRevB.91.224302
40. G. Abstreiter, M. Cardona, A. Pinzcuk, *Light Scattering by free carrier excitations in semiconductors, in: Light scattering in Solids IV: Electronic Scattering, Spin Effects, SERS and Morphic Effects*, *Topics in Applied Physics*, vol. 54 (Springer Verlag, 1984)

Reactivity of organic molecules on semiconductor surfaces revealed by density functional theory

Fabian Pieck, Jan-Niclas Luy, Florian Kreuter, Badal Mondal and Ralf Tonner-Zech

Abstract The present report discusses the reactivity and properties of organic molecules on mainly semiconductor surfaces by means of density functional theory. For pyrazine on Ge(001), a benzylazide on Si(001) and a cyclooctyne derivate on Si(001) adsorption modes and possible reaction paths are presented. The charge transfer effect between the inorganic and organic interface is presented with the example of corroles on Ag(111). Approaches towards more realistic and efficient models are discussed in the third chapter, while the scaling of large simulations on the resources of the High-Performance Computing Center Stuttgart is addressed in a final chapter.

1 Introduction

The increasing demand for consumer electronics is driving the development of cheaper and increasingly powerful microelectronic devices. To satisfy these demands device miniaturization is a core strategy, new device architectures are developed and even the use of germanium as alternative elemental semiconductor material, besides silicon is promoted [1].

To enlarge the property and application range of semiconducting devices the formation of organic-semiconductor and metal-organic interfaces needs to be understood. The basic idea for the first approach is to combine the advantages of the mature silicon-based technology with the chemical richness and diverse electronic properties of organic compounds [43, 71, 84]. While Si(001) is still the technologically most important semiconductor surface for the organic functionalization [33, 46, 53, 58], also the Ge(001) surface is considered [7, 37]. Both surfaces have in common the formation of surface dimers, that act as the fundamental reactive unit of the surface

Ralf Tonner-Zech
Wilhelm-Ostwald-Institut für Physikalische und Theoretische Chemie, Linnéstr. 2, 04103 Leipzig, Germany, e-mail: Ralf.Tonner@uni-leipzig.de

W. E. Nagel et al. (eds.), *High Performance Computing in Science and Engineering '21*,
https://doi.org/10.1007/978-3-031-17937-2_7

[7, 37, 75]. Since these pristine semiconductor surfaces are generally highly reactive [78] the critical step is the attachment of the first organic layer to obtain uniform and defect-free interfaces. A promising approach is the use of tailored bifunctional molecular building blocks [17]. These building blocks can be attached to the substrate by 1,3-dipolar [36] and [2 + 2] cycloaddition reactions [44, 52], while a second functional group is reserved for the attachment of further organic molecules.

To obtain well-defined structures, bifunctional building blocks showing orthogonal reactivity for attachment and further functionalization are required. So far, only a small number of bifunctional molecules, often restricted to a fixed combination of the two functional groups, are promising for the attachment on Si(001) [14, 28, 66, 68, 83].

Here, cyclooctyne has been found to be an excellent platform molecule [44, 52]. While the attachment to the Si(001) surface is accomplished by the strained triple bond, additional organic layers can be grown on the first layer [38] when bifunctional anchor molecules, such as ethynyl-cyclopropyl-cyclooctyne [33] or methyl-enol-ether-functionalized cyclooctyne [19] are used. The approach of combining organic molecules with semiconducting surfaces opens a huge range of possible applications in the area of biosensors [12, 74, 76], microelectronics [2, 22, 25, 56, 74, 77, 79], batteries [8], organic thin film growth [10, 62, 84], organic solar cells and organic light emitting diodes [45, 85].

Enabling the organic functionalization in these applications, requires an understanding of the fundamental reactivity of organic molecules on semiconducting surfaces and efficient screening approaches for potential new molecules. Here, computational approaches by means of density functional theory (DFT) can significantly contribute. Still, modelling of organic layers on semiconducting surfaces remains challenging due to large system sizes often with large conformational freedom, various types of bonding interactions involved and complex chemical environment including solvents and defects. Therefore, developing small and efficient models is a desirable step.

The next sections address various aspects of our theoretical work regarding the functionalization of semiconductor surfaces. In the first section we answer fundamental questions concerning the reactivity and selectivity of organic molecules on semiconducting surfaces at the examples of pyrazine on Ge(001) [63] and benzyl-azide [24] as well as a cyclooctyne derivate [19] on Si(001). For a full understanding of devices, we also need an in-depth understanding of the metal-organic interfaces that appear for example when electrodes are attached to organic semiconductors. We are thus also extending our investigations in this direction. With the electronic properties of a corrole on Ag(111) [86], a metal-organic interface is studied in the second section to understand charge transfer effects, which also occur on semiconducting surfaces. The third section discusses developments towards more realistic and efficient surface models [38, 39], while in a final section our scaling benchmark for HAWK is presented.

2 Reactivity and selectivity of organic molecules on Si(001) and Ge(001)

2.1 Pyrazine on Ge(001)

The behavior of heteroaromatic molecules on semiconductors is of particular interest to understand the role of aromaticity on surface reactions, which can serve as the guiding principle for surface modification [29, 72, 73]. While pyrazine was already studied on a silicon surface [30, 31, 34, 48, 49, 67, 82], only the related molecule pyridine was studied on a germanium surface [5] and no knowledge of pyrazine on Ge(001) was available.

Figure 1 shows the most stable adsorption structures for pyrazine on Ge(001). Interestingly, structures where the molecule bonds to the surface with one or two nitrogen atoms (1N, 2N, Figure 1a and 1b) as well as with two carbon atoms (2C, Figure 1c) are observed. In terms of bonding energies 1N and 2N are with -120 and $-115\,\text{kJ}\cdot\text{mol}^{-1}$ favored over 2C with $-82\,\text{kJ}\cdot\text{mol}^{-1}$. Several organic compounds, e.g. benzene, are known to be able to form covalent bonds with carbon atoms to semiconductor surfaces [20, 26]. Additional structures were found, but with a bonding energy of up to $-13\,\text{kJ}\cdot\text{mol}^{-1}$ they are significantly less stable. In addition, 1N, 2N and 2C adsorption modes could be verified by comparing calculated XPS peaks with experimental results [63].

Besides the most stable adsorption structure, also their occurrence dependent on temperature or coverage is of interest. To derive this, adsorption paths and reaction paths for the conversion of 2N to 1N and 1N to 2C were calculated (Figure 2). In agreement with experimental data, the pyrazine molecule is initially able to adsorb

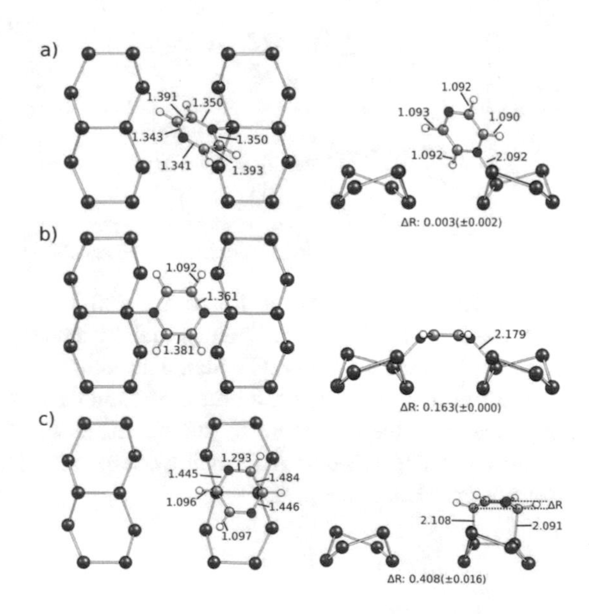

Fig. 1 Top and side view of the optimized structures for the three most stable adsorption structures of pyrazine on Ge(001). 1N (a), 2N (b) and 2C (c). Bond length and out of plane displacement (ΔR) in Å. Reprint from reference [63].

to the 1N and 2N structure. While the reaction barrier between 1N and 2N is with $7\,\text{kJ}\cdot\text{mol}^{-1}$ negligible, the barrier between 1N and 2C is with $89\,\text{kJ}\cdot\text{mol}^{-1}$ more pronounced. Consequently, pyrazine can constantly interconvert between 1N and 2N even at lower temperatures (180 K), while a larger fraction of molecules in the 2C state is only observed at elevated temperatures (423 K) [63].

For the study of multiadsorption effects, the surface model had to be doubled in size. Several arrangements of pyrazine in the 1N and 2N mode were calculated. Based on the change in bonding energy, it was visible that high surface coverage can only be achieved with adsorbates in the 1N structure, while the intermolecular repulsion for molecules in the 2N structure is too large to form densely packed layers.

To deepen our understanding of the reactivity of pyrazine, the loss of aromaticity was further studied by means of nuclear-independent chemical shifts (NICS) and the harmonic oscillator model of aromaticity (HOMA), while the bonding situation was determined by the calculation of partial charges and the application of the energy decomposition method for extended systems (pEDA). Here, in agreement with the out-of-plane displacement (ΔR) a significant loss of aromaticity could be observed for the 2N and 2C structure [63]. Analysis of the partial charges in combination with the pEDA could identify the N-Ge bonds in the 1N and 2N structure as dative bonds

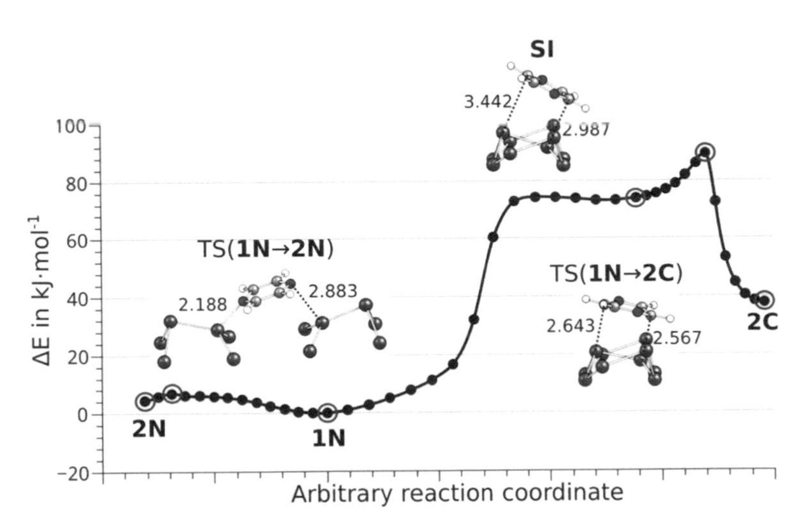

Fig. 2: Reaction path from 1N to 2N and 1N to 2C with the transition states TS(1N→2N) and TS(1N→2C) and a shallow intermediate (SI) on the energy plateau between 1N and TS(1N→2C). Stated structures are indicated along the path with colored circles. In addition, a blue circle indicates that a barrierless adsorption for pyrazine from the gas phase to that structure was found. The bond length of the formed bonds is stated in Å. Bonding energy are given relative to the 1N structure. Reprint from reference [63].

resulting in a minor charge transfer to the surface. The 2C structure showed a stronger surface to molecule donation and could be identified as the product of an inverse electron demand Diels–Alder reaction.

Overall, these results showed that pyrazine adsorbs on Ge(001) forming a mix of reaction products. We could explain the experimentally observed temperature and coverage dependence, while thoroughly analyzing the bonding situation.

2.2 Methyl-substituted benzylazide on Si(001)

For the organic functionalization of silicon, molecules easily reacting with the first organic layer are needed. Here, azides are often used mainly due to reactions such as alkyne-azide couplings used in click chemistry [32]. While the intention is to use azides as a building block for the second organic layer, also their behavior towards a clean Si(001) surface has to be understood since it poses an important side reaction. Studies of the unsubstituted benzylazide on Si(001) revealed an intermediate state involving the intact molecule and the elimination of N_2 to reach the final state with the molecule attached to silicon via the remaining nitrogen atom [6, 36].

For the methyl-substituted benzylazide the computed adsorption and reaction profile on Si(001) is shown in Figure 3. The molecule initially adsorbs without a barrier into a stable [3 + 2] cycloaddition product with a bonding energy of $-263\,\mathrm{kJ{\cdot}mol^{-1}}$.

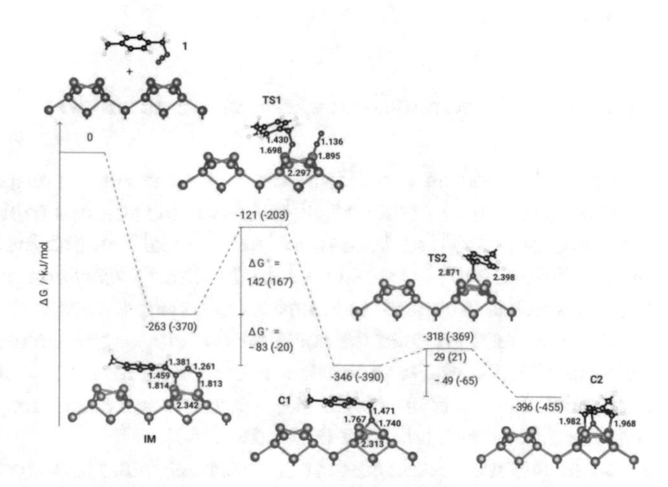

Fig. 3: Adsorption and reaction profile of the methyl-substituted benzylazide on Si(001). Gibbs energies (electronic energies) relative to the free molecule and surface. HSE06 energies with thermodynamic corrections based on PBE-D3. Reprint from reference [24].

The structure consists of a five-membered ring with nearly symmetric Si-N bonds. This intermediate state (IM) can further decompose via N-N bond cleavage and a high barrier of $\Delta G^{\ddagger} = 142\,\mathrm{kJ{\cdot}mol^{-1}}$ toward the first adsorption configuration (C1).

This reaction is thermodynamically favored by $-83\,\mathrm{kJ{\cdot}mol^{-1}}$ and driven by the loss of N_2. A second structure (C2) is accessible via a low-lying transition state structure with a barrier of $29\,\mathrm{kJ{\cdot}mol^{-1}}$. C2 is more stable by $50\,\mathrm{kJ{\cdot}mol^{-1}}$ in comparison to C1. The gain in stability stems from an additional interaction of the benzyl moiety with the silicon dimer row, similar to the butterfly structure observed for benzene on Si(001) [65,69].

Based on calculated STM images and experimental STM and XPS data, the IM state could be identified at low temperatures (50 K), while the C1 and C2 states are observed at 300 K. In agreement with the calculated barrier the adsorbate can switch between these states at 300 K [24].

Besides the structures discussed, additional structures and reaction paths were investigated including other intermediate states showing a single N atom bound to the electrophilic surface silicon atom or structures being bonded only via the tolyl ring. Also, a final state comprising of a Si_{Dimer}-N-$Si_{Subsurface}$ ring and the additional dissociative adsorption of the methyl group [11] were investigated. However, these possibilities could be rejected due to too high barriers or unfavorable thermodynamics. From the DFT investigations, we thus conclude that the [3 + 2] cycloaddition product IM is the most likely species observed at low temperatures, which converts to adsorption configurations C1 and C2 at higher temperatures, exhibiting three membered Si-N-Si rings.

2.3 Methyl enol ether functionalized cyclooctyne on Si(001)

While azides are well suited as a building block for the second organic layer, cyclooctyne derivates selectively react on Si(001) over the strained triple bond of the cyclooctyne ring [33,53,58] and are therefore an ideal building block for the first layer. This chemoselectivity is traced back to the direct adsorption pathway on Si(001) [44,51,58], which is in contrast to almost all other organic adsorbates [15,35]. The unused second functional group of the substituted cyclooctynes can thus be used for further reactions [33,53,58]. A promising cyclooctyne derivate is the methyl enol ether functionalized cyclooctyne (MEECO), since the enol ether group could be employed in click reactions with tetrazine derivatives [42].

Figure 4 shows the four most stable adsorption structures found for the cyclooctyne derivate. Shown are only structures of the Z-isomer of MEECO, since this isomer is found to bind more strongly to the surface than the E-isomer, which is the more stable isomer in the gas phase. For both isomers the same adsorption structures were found and we therefore assume that both react in the same manner with a Si(001) surface [19].

Figure 4a shows the only structure, where the adsorbate is not cleaved into two fragments (*intact*). This structure resembles the on-top structure of cyclooctyne [52] showing the same bonding energy using the PBE functional. This supports previous findings that exchanging the functional group averted from the triple bond does not alter the bonding to the substrate [53].

The structure in Figure 4b (*ether-cleavage*) is obtained by breaking the C-O bond at the ether group. The mechanism of this reaction strongly resembles a previously observed S_N2-type attack [50]. This breaking of the C-O bond proceeds via an intermediate state with the intact ether group being datively bonded to the surface dimer [19]. The bonding energy is with $-599\,\text{kJ}\cdot\text{mol}^{-1}$ considerable larger than for the *intact* structure and similar to the energy of an ether-functionalized cyclooctyne [53]. Alternatively, the ether-cleavage reaction can be performed across the dimer row leading to structure in Figure 4c (*methoxy dissociation*). This structure shows with $-637\,\text{kJ}\cdot\text{mol}^{-1}$ the largest bonding energy.

The formation of a methylene group (Figure 4d, *aldehyde*) leads to the fourth relevant adsorption mode. With a bonding energy of $-440\,\text{kJ}\cdot\text{mol}^{-1}$ this structure is also thermodynamically favored over the *intact* structure. These four adsorption modes agree with experimental XPS data, which furthermore show, that most of the methyl enol ether groups stay intact and are therefore available for a reaction with a second organic building block [19]. Additional adsorption modes like an adsorption via the ether oxygen atom, the enol double bond or both the double and triple bond were also investigated but clearly less stable.

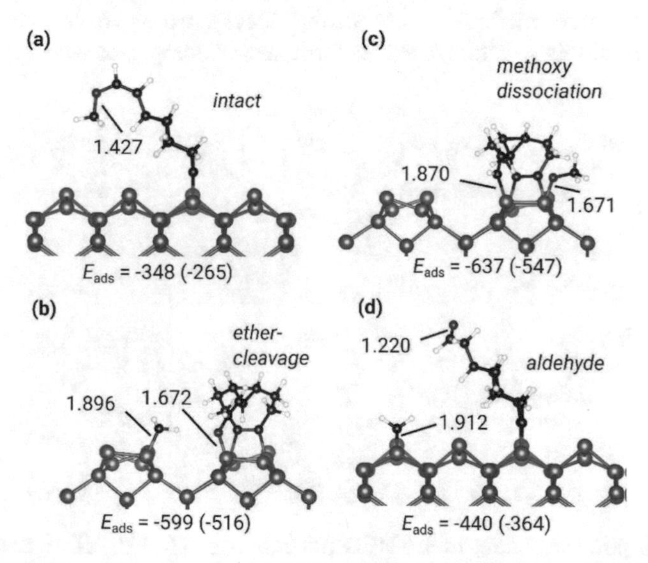

Fig. 4: Most stable adsorption modes of the cyclooctyne derivate. Bonding energies (Gibbs energies) at HSE06-D3 level of theory in $\text{kJ}\cdot\text{mol}^{-1}$. Bond length in Å. Reprint from reference [19].

Finally, also the reaction barriers for the conversion *intact* to *ether-cleavage* and the formation of *methoxy dissociation* were calculated. The obtained barriers of 79 and 102 $kJ \cdot mol^{-1}$, respectively, agree well with experimental observations that for low temperatures a reduced reactivity for the enol ether group of the MEECO molecules is observed. In addition, blocking neighboring surface dimers at high coverage can further reduce the reactivity of the second functional group.

In conclusion, MEECO molecules adsorb chemoselective via the strained triple bond of cyclooctyne at 150 K on Si(001). Increasing the substrate temperature promotes the formation of additional products via reactions of the enol ether group. Therefore, low temperature and high coverages enable MEECO as an excellent molecular building block for the first organic layer.

3 Interface formation of corroles on Ag(111)

While the previous sections focused on the reactivity and chemoselectivity of organic molecules on semiconducting surfaces, this section presents intriguing electronic effects observed in the interface formation of a corrole on Ag(111).

Corroles belong to the class of cyclic tetrapyrroles and are structurally closely related to porphyrins and phthalocyanines [47]. The interface chemistry of tetrapyrroles is of fundamental interest [3,21,80], especially their possibility to coordinate metal atoms [9,13,21,40] and the role of aromaticity [54,70,81]. In contrast to porphyrins, corroles contain three methine bridges and one direct pyrrole-pyrrole bond (see Figure 5). As a result of this structure, corroles form neutral complexes with trivalent metal

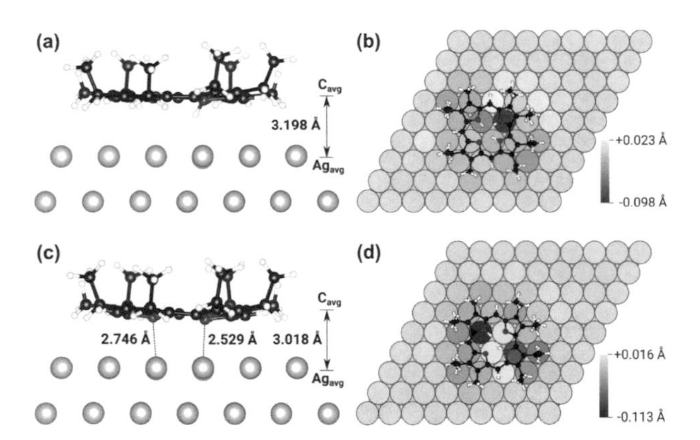

Fig. 5: Adsorption structures of 3H-HEDMC (a,b) and 2H-HEDMC (c,d) on Ag(111). C_{avg} and Ag_{avg} indicate the average high of carbon atoms in the corrole ring and silver atoms of the first layer. Color scheme in (b) and (d) indicates the vertical displacement of surface atoms due to the presence of an adsorbate. Reprint from reference [86].

ions. Furthermore, the contracted corrole macrocycle provides a tighter coordination environment for the metal ion compared to porphyrins, resulting in an additional stabilization of higher oxidation states [4, 18].

In this study 2,3,8,12,17,18-hexaethyl-7,13-dimethylcorrole (3H-HEDMC, Figure 5) and the parent molecule 3H-C (see Figure 6), in which all ethyl groups are replaced by hydrogen atoms, were investigated. In all calculations only the most stable tautomer of the corroles has been used. In the gas phase 3H-HEDMC and 3H-C show a nonplanar structure, in which also the hydrogen atoms of the N-H groups are pointing out of the plane formed by the nitrogen atoms. In Figure 5 the on-surface structure of 3H-HEDMC is shown. Here, all alkyl groups are pointing away from the surface and the out-of-plane angle for one hydrogen atom is even more pronounced. For both, 3H-HEDMC and 3H-C, large bonding energies are obtained with −403 and $-267\,\text{kJ}\cdot\text{mol}^{-1}$, respectively. The interaction with the surface solely stems from dispersion interactions, while the electronic contributions are even repulsive. This nicely explains why different adsorption sites (bridge, hollow, on-top) show similar adsorption energies [86].

The hydrogen atom of the N-H group, which is already pointing towards the surface, is easily transferred to the surface indicated by the small reaction barrier of $76\,\text{kJ}\cdot\text{mol}^{-1}$. This barrier nicely agrees with the experimentally observed onset temperature of 180 K [86]. The N-H dissociation relieves the steric stress in the corroles leading to the planar structures 2H-HEDMC (see Figure 5) and 2H-C. Furthermore, the adsorption heights decrease by around 0.2 Å while the bonding energies increase to −337 (2H-C) and $-457\,\text{kJ}\cdot\text{mol}^{-1}$ (2H-HEDMC). Noteworthy, this increase in the bonding energies is also based on now attractive electronic interactions. Furthermore, this reaction is driven by the increase in entropy due to the combination and release of H_2.

Experimental data indicate a substantial charge transfer contributing to the stabilization of 2H-HEDMC [86]. Charge density difference plots for 3H-C and 2H-C are shown in Figure 6, whereby charge accumulation is shown in blue and charge depletion in red. For 3H-C the most prominent feature is a charge accumulation

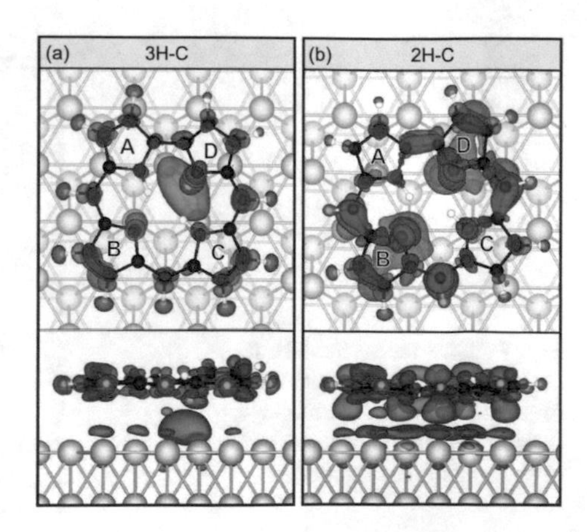

Fig. 6 Charge density difference plots of 3H-C (a) and 2H-C (b) on Ag(111). Charge accumulation shown in blue and charge depletion in red. An isosurface value of $0.0001\,e\cdot\text{Å}^{-3}$ was used. Adapted reprint from reference [86].

directly below the hydrogen atom, which is already directed towards the surface. Larger changes in the electron density are observed for 2H-C. Here, the molecule receives a significant amount of charge from the surface. The distribution of this charge at the molecule resembles the shape of the SOMO of the gas phase molecule. The charge flow to the adsorbate can be quantified by Hirshfeld partitioning to be $-0.08\,e$ for 3H-C and $-0.43\,e$ for 2H-C. Furthermore, an analysis of the spin density indicates that the charge transfer from the surface into the singly occupied molecular orbital molecule almost completely quenches the spin [86]. Finally, the effect on the aromaticity was studied by means of NICS and HOMA. Here, a significant higher aromatic character was observed for the 2H-C anion in comparison to the neutral molecule.

This suggests that the charge transfer from the surface to the adsorbed 2H-HEDMC can be described as aromaticity-driven; by accepting electron density from the metal surface, the molecule gains aromatic character and thus aromatic stabilization. Such interfacial charge transfer effects play a prominent role in the context of charge injection in organic electronic devices [16, 27, 64].

4 Developing accurate and efficient computational models

4.1 The introduction of a hierarchical model system

Studying several organic adsorbates, or assuming a complete first layer and modelling the reactions resulting in the formation of the second layer is a computationally highly expensive task. Therefore, we aim at deriving an accurate and efficient model to study the formation of hybrid interfaces and their growth with DFT.

For our benchmark, three model reactions of interest for the formation of the second layer were selected: With a variant of the azide-alkyne 1,3-dipolar cycloaddition (AAC) [60] a prototypical click reaction with excellent performance [46] was selected.

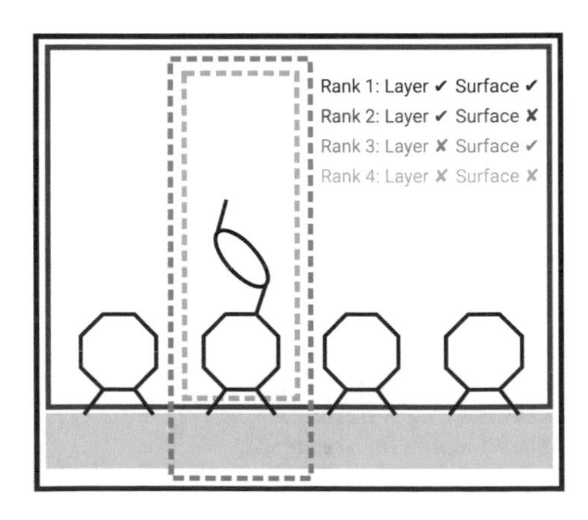

Fig. 7 Hierarchical models to study an interface between an organic layer and a semiconductor surface. For details see text. Reprint from reference [38].

A possible alternative to the AAC reaction is the acyl-chloride mediated esterification (ACE) [61] due to its smaller reaction barriers. As a third reaction an inverse electron demand Diels–Alder (IEDDA) reaction utilizing a tetrazine derivative as the diene [42] was selected. Here, we considered only the first reaction step.

For our model systems we assume an ideal first layer of organic molecules (details see [38]) on the Si(001) surface. Our approach assumes that for the reaction towards the second layer, bond cleavage and formation are predominantly local phenomena. Therefore, the idea is to remove parts of the system, which are not directly involved in the reactions of interest. In Figure 7, the four proposed hierarchical models are sketched. The complete system composed of the surface slab and the complete first organic layer is used as the reference model (rank 1 model). Removing the substrate leads to the rank 2 model. If necessary, dangling bonds are saturated with hydrogen atoms. This model is lacking electronic interactions with the substrate, while template effects and intra-layer-effects are still included. In comparison to the rank 1 model speedups by a factor of 2 to 6 are observed for the AAC reaction.

Instead of removing the substrate, one could remove all molecules of the first organic layer, which do not participate in the reaction with the new adsorbate. This leads to the rank 3 model. Here, electronic interactions with the substrate are included, while the intra-layer-effects are neglected. Still, a minor speedup of a factor 1.5 to 2 is achieved. For the simplest possible system, the substrate and the first organic layer is removed (rank 4 model). While this model lacks all environmental influences, it shows with a factor of 50 the greatest speedup, can be carried out with molecular quantum chemistry methods and might still be accurate enough to sample the reaction between two organic molecules.

All model systems were tested with the three reactions of interest. Furthermore, different density functionals and even CCSD(T) and MP2 calculations were performed for the rank 4 model [38]. Figure 8 shows exemplary results for the AAC reaction. It shows that reaction energies and barriers are only slightly affected. Furthermore, two dominant structural effects can be identified: As shown by the inset in Figure

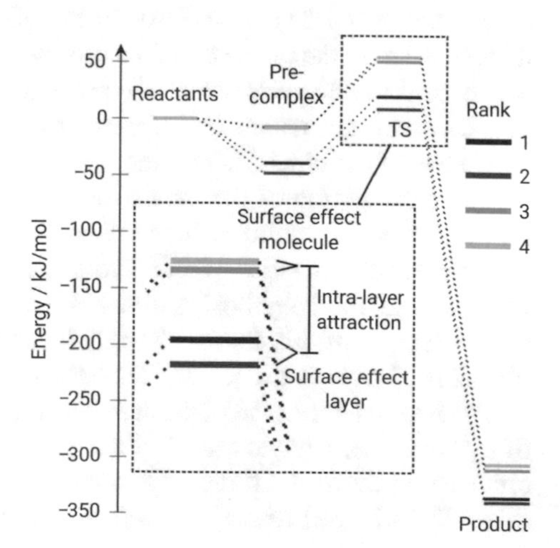

Fig. 8 Reaction energies of the azide-alkyne cycloaddition (AAC). Reprint from reference [38].

8, a clustering of the values for rank 1, 2 and rank 3, 4 are observed. Removing the organic layer and therefore the intra-layer effects leads to the large change from rank 1, 2 to rank 3, 4. Ranks 3 and 4 are higher in energy since they lack the attractive dispersion interactions with the organic layer. Removing the surface results in only a minor change in energy due to the presence of steric repulsion between the silicon surface and the first layer of molecules.

The benchmark of different density functionals for all reaction energies, reaction barriers and models revealed that rank 2 always performs very well (mean error smaller than 5 kJ·mol^{-1}). In contrast, the model ranks 3 and 4 showed larger deviations and for optB88 even larger than 10 kJ·mol^{-1}. Also including solvent effects, does not alter the overall picture.

In conclusion, in terms of efficiency a rank 1 model should only be used in case highly accurate results are necessary. A rank 2 model is commonly suited to study the chemistry within the organic layers. The rank 3 model is a good choice to extract adsorption and reaction energies for the adsorbate-surface interactions. Finally, the gas phase model (rank 4) can be used for screening molecular building blocks and benchmark of DFT approaches.

4.2 Reactivity of a Si(001) dimer vacancy

The previous sections worked with ideal Si(001) or Ge(001) surfaces. In the literature, some studies concerning the influence of step edges exists [41, 57]. Although, the formation of defects can be suppressed [23], we believe that they could be used to increase selectivity of surface reactions. By incorporating surface defects in our models, we are able to quantify and understand changes in reactivity with respect to the pristine surface. Furthermore, the investigation of nonideal surfaces will help to create more realistic computational models of organic-semiconductor interfaces.

In Figure 9 the structure and the formation energy of common defects of a Si(001) surface are shown. The presence of defects allows for additional structural motive and thereby increase the richness of the adsorption chemistry. In our study, we focused on the most favorable point defect, the bonded dimer vacancy (DV) [59]. To study this defect the surface model was increased to a 8 x 4 supercell with eight silicon layers where one of the dimers is missing.

The introduction of the dimer vacancy has some consequences on the atomic and electronic structure of the surface [39]. The defect itself relaxes by forming bonds between the silicon defect atoms as indicated by the bond length of 2.88 Å. Furthermore, the neighboring dimer is flipped. By calculating and visualizing the band structure of this slab, a strongly localized defect state is observed inside the former band gap. Consequently, the band gap is narrowed by 0.15 eV.

To reveal the influence of the defect on the reactivity of the surface, the adsorption of cyclooctyne, acetylene and ethylene was studied. Interestingly, acetylene and ethylene prefer to bond directly to the defect atoms leading to bonding energies of −302 and −321 kJ·mol^{-1}, respectively. However, the most reactive molecule cyclooctyne prefers to adsorb on-top of a silicon dimer leading to a bonding energy of $-307\,\text{kJ}\cdot\text{mol}^{-1}$ (healing the defect: $-250\,\text{kJ}\cdot\text{mol}^{-1}$). As supported by pEDA, an adsorption of cyclooctyne in the tight space of the DV results in a larger Pauli repulsion and smaller orbital interactions between the adsorbate and surface, which significantly lowers the bonding energy [39].

For acetylene and ethylene we also searched for direct adsorption paths resulting in a healing of the defect. Since direct adsorption paths to these final states were not found, also adsorption paths to other structures and paths connecting the different adsorption modes were calculated. The complete reaction network for acetylene is shown in Figure 10. Following the reactions with the lowest barriers (highlighted in red), the most likely steps to reach the thermodynamically most stable structure (pedestal) would start with a direct adsorption in the pre-bridge-E structure. This structure can easily overcome the barrier of $5\,\text{kJ}\cdot\text{mol}^{-1}$ to reach the bridge-E structure. Although a direct connection to the pedestal structure exists, another intermediate step to the bridge structure is more likely due to the smaller barrier of $154\,\text{kJ}\cdot\text{mol}^{-1}$. The last step shows with $170\,\text{kJ}\cdot\text{mol}^{-1}$ the largest barrier and therefore the rate determining

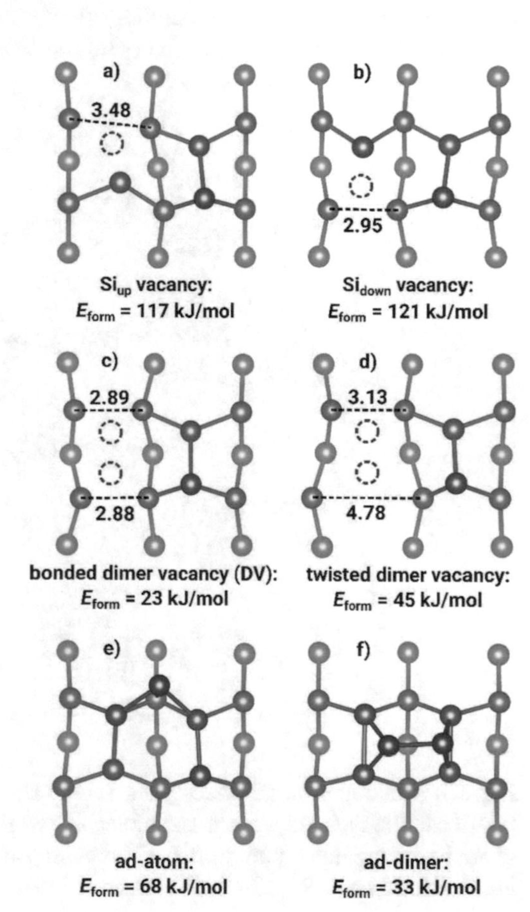

Fig. 9 Common defects of a Si(001) surface. Bond length in Å. Reprint from reference [39].

step. Overall, due to the missing direct adsorption and pronounced barriers for a reaction with the defect, the DV can be concluded to be less reactive than silicon dimers on a pristine surface.

5 Scaling of computational methods

Structural optimizations and reaction path calculations were performed with the software package "Vienna Ab initio Simulation Package" (VASP). The Software is licensed by the VASP Software GmbH located in Vienna, Austria (`https://www.vasp.at/`, `office@vasp.at`). The best code performance on HAWK is obtained by using the latest version of the Intel Fortran compiler together with the HPE MPI. Basic numeric libraries like (SCA)LAPACK, BLAS or FFTW are also used to push the performance.

VASP comprises three levels of parallelization. The least intense level in terms of MPI communication is the parallel calculation of the wave function at different k-points. The second level is composed of the calculation of bands, while in the third

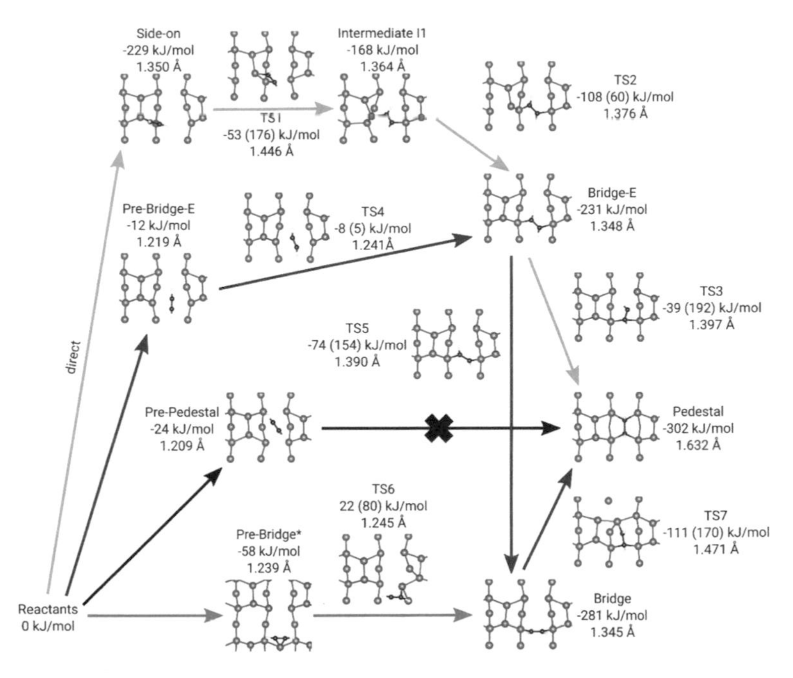

Fig. 10: Reaction network of acetylene on the DV-defect. Bonding energies are stated relative to the isolated surface and molecule, while reaction barriers (in brackets) are given to the previous minimum. C-C bond length given in Å. The most likely path is highlighted in red. Reprint from reference [39].

level the calculation is parallelized over the basis set (plane waves). Due to the high MPI communication in the treatment of the plane wave, this parallelization is usually restricted to the physical socket, i.e. 64 cores for HAWK. For molecular dynamics (MD) and nudged-elastic band (NEB) calculations a very good code scaling was already shown on Cray XC40 (Hazel Hen) at HLRS for up to 1536 and 3072 cores, respectively [55].

In Figure 11 the scaling benchmark of our automatization routine, which is further developed in this project, is shown. The surface reaction network of GaH_3 on GaP(001) is used as a benchmark system. Every calculation was restricted to a wall clock time of 10 minutes, where the first 40 seconds were consumed for initialization. Every instance of VASP had to deal with a system containing 120 atoms in the unit cell ($H_{32}Ga_{50}P_{34}GaH_3$), 366 explicitly treated electrons, 5 k-points, 260 bands per k-point and ca. 107,000 plane waves per band. Consequently, the VASP parallelization over k-points was enabled where possible. Requirements for file IO and memory per core are negligible.

As expected, our automatization routine shows a nearly linear scaling behavior up to 29 nodes due to the trivial parallelization over multiple VASP instances. The usage of more and more nodes for a single VASP instance, more than 29 total used nodes, leads only to a minor decrease in the scaling compared to the ideal scaling. This trend will persist even for larger numbers of cores due to the very good scaling behavior of VASP [55]. Furthermore, a superlinear scalability is observed for 88 used nodes. Here, the reduced memory requirements due to the separation of the bands to multiple nodes overcompensates the increased MPI communication. All in all, with the combination of a high-performing DFT code and our automatization routine we are now able to efficiently use in a single calculation significant amounts of the computational resources provided by HAWK in a massively parallel fashion.

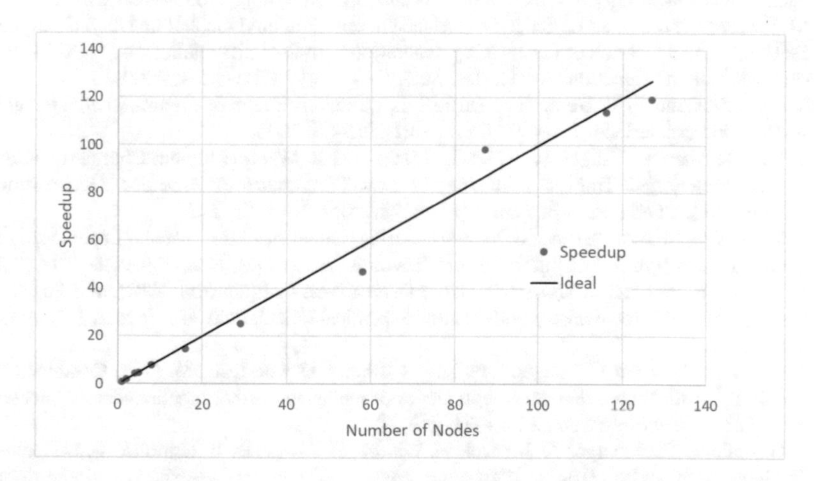

Fig. 11: Scaling behavior of our automatization routine using VASP 5.4.4 on HAWK at HLRS. For details on the calculation see text.

References

1. International Roadmap for Devices and Systems (IRDS) **2020**. https://irds.ieee.org, Last accessed June 2021.
2. S. V. Aradhya and L. Venkataraman. Single-molecule junctions beyond electronic transport. *Nat. Nanotechnol.*, 8(6):399–410, 2013.
3. W. Auwärter, D. Écija, F. Klappenberger, and J. V. Barth. Porphyrins at interfaces. *Nat. Chem.*, 7(2):105–120, 2015.
4. I. Aviv-Harel and Z. Gross. Aura of Corroles. *Chem. - A Eur. J.*, 15(34):8382–8394, 2009.
5. S.-S. Bae, S. Kim, and J. Won Kim. Adsorption Configuration Change of Pyridine on Ge(100): Dependence on Exposure Amount. *Langmuir*, 25(1):275–279, 2009.
6. S. Bocharov, O. Dmitrenko, L. P. Méndez De Leo, and A. V. Teplyakov. Azide Reactions for Controlling Clean Silicon Surface Chemistry: Benzylazide on Si(100)-2 x 1. *J. Am. Chem. Soc.*, 128(29):9300–9301, 2006.
7. J. M. Buriak. Organometallic Chemistry on Silicon and Germanium Surfaces. *Chem. Rev.*, 102(5):1271–1308, 2002.
8. Q. Cai, B. Xu, L. Ye, Z. Di, S. Huang, X. Du, J. Zhang, Q. Jin, and J. Zhao. 1-Dodecanethiol based highly stable self-assembled monolayers for germanium passivation. *Appl. Surf. Sci.*, 353:890–901, 2015.
9. M. Chen, H. Zhou, B. P. Klein, M. Zugermeier, C. K. Krug, H.-J. Drescher, M. Gorgoi, M. Schmid, and J. M. Gottfried. Formation of an interphase layer during deposition of cobalt onto tetraphenylporphyrin: a hard X-ray photoelectron spectroscopy (HAXPES) study. *Phys. Chem. Chem. Phys.*, 18(44):30643–30651, 2016.
10. R. G. Closser, D. S. Bergsman, L. Ruelas, F. S. M. Hashemi, and S. F. Bent. Correcting defects in area selective molecular layer deposition. *J. Vac. Sci. Technol. A Vacuum, Surfaces, Film.*, 35(3):031509, 2017.
11. F. Costanzo, C. Sbraccia, P. Luigi Silvestrelli, and F. Ancilotto. Proton-transfer reaction of toluene on Si(100) surface. *Surf. Sci.*, 566-568:971–976, 2004.
12. Y. Cui, Q. Wei, H. Park, and C. M. Lieber. Nanowire Nanosensors for Highly Sensitive and Selective Detection of Biological and Chemical Species. *Science*, 293(5533):1289–1292, 2001.
13. K. Diller, A. C. Papageorgiou, F. Klappenberger, F. Allegretti, J. V. Barth, and W. Auwärter. In vacuo interfacial tetrapyrrole metallation. *Chem. Soc. Rev.*, 45(6):1629–1656, 2016.
14. M. Ebrahimi and K. Leung. Selective surface chemistry of allyl alcohol and allyl aldehyde on Si(100)2x1: Competition of [2+2] CC cycloaddition with O-H dissociation and with [2+ 2] CO cycloaddition in bifunctional molecules. *Surf. Sci.*, 603(9):1203–1211, 2009.
15. M. A. Filler and S. F. Bent. The surface as molecular reagent: organic chemistry at the semiconductor interface. *Prog. Surf. Sci.*, 73(1-3):1–56, 2003.
16. Y. Gao, Y. Shao, L. Yan, H. Li, Y. Su, H. Meng, and X. Wang. Efficient Charge Injection in Organic Field-Effect Transistors Enabled by Low-Temperature Atomic Layer Deposition of Ultrathin VO_x Interlayer. *Adv. Funct. Mater.*, 26(25):4456–4463, 2016.
17. S. M. George, B. Yoon, and A. A. Dameron. Surface Chemistry for Molecular Layer Deposition of Organic and Hybrid Organic-Inorganic Polymers. *Acc. Chem. Res.*, 42(4):498–508, 2009.
18. A. Ghosh. Electronic Structure of Corrole Derivatives: Insights from Molecular Structures, Spectroscopy, Electrochemistry, and Quantum Chemical Calculations. *Chem. Rev.*, 117(4):3798–3881, 2017.
19. T. Glaser, J. Meinecke, C. Länger, J.-N. Luy, R. Tonner, U. Koert, and M. Dürr. Combined XPS and DFT investigation of the adsorption modes of methyl enol ether functionalized cyclooctyne on Si(001). *ChemPhysChem*, 22(4):404–409, 2021.
20. S. Gokhale, P. Trischberger, D. Menzel, W. Widdra, H. Dröge, H.-P. Steinrück, U. Birkenheuer, U. Gutdeutsch, and N. Rösch. Electronic structure of benzene adsorbed on single-domain Si(001)-(2x1): A combined experimental and theoretical study. *J. Chem. Phys.*, 108(13):5554–5564, 1998.
21. J. M. Gottfried. Surface chemistry of porphyrins and phthalocyanines. *Surf. Sci. Rep.*, 70(3):259–379, 2015.

22. R. Har-Lavan, O. Yaffe, P. Joshi, R. Kazaz, H. Cohen, and D. Cahen. Ambient organic molecular passivation of Si yields near-ideal, Schottky-Mott limited, junctions. *AIP Adv.*, 2(1):012164, 2012.
23. K. Hata, T. Kimura, S. Ozawa, and H. Shigekawa. How to fabricate a defect free Si(001) surface. *J. Vac. Sci. Technol. A Vacuum, Surfaces, Film.*, 18(4):1933–1936, 2000.
24. J. Heep, J.-N. Luy, C. Länger, J. Meinecke, U. Koert, R. Tonner, and M. Dürr. Adsorption of Methyl-Substituted Benzylazide on Si(001): Reaction Channels and Final Configurations. *J. Phys. Chem. C*, 124(18):9940–9946, 2020.
25. G. Hills, C. Lau, A. Wright, S. Fuller, M. D. Bishop, T. Srimani, P. Kanhaiya, R. Ho, A. Amer, Y. Stein, D. Murphy, Arvind, A. Chandrakasan, and M. M. Shulaker. Modern microprocessor built from complementary carbon nanotube transistors. *Nature*, 572(7771):595–602, 2019.
26. W. A. Hofer, A. J. Fisher, G. P. Lopinski, and R. A. Wolkow. Adsorption of benzene on Si(100)-(2x1): Adsorption energies and STM image analysis by ab initio methods. *Phys. Rev. B*, 63(8):085314, 2001.
27. M. Hollerer, D. Lüftner, P. Hurdax, T. Ules, S. Soubatch, F. S. Tautz, G. Koller, P. Puschnig, M. Sterrer, and M. G. Ramsey. Charge Transfer and Orbital Level Alignment at Inorganic/Organic Interfaces: The Role of Dielectric Interlayers. *ACS Nano*, 11(6):6252–6260, 2017.
28. M. Hossain, Y. Yamashita, K. Mukai, and J. Yoshinobu. Selective functionalization of the Si(100) surface by switching the adsorption linkage of a bifunctional organic molecule. *Chem. Phys. Lett.*, 388(1-3):27–30, 2004.
29. H. G. Huang, J. Y. Huang, Z. H. Wang, Y. S. Ning, F. Tao, Y. P. Zhang, Y. H. Cai, H. H. Tang, and G. Q. Xu. Adsorption of nitrogen-containing aromatic molecules on Si(111)-7x7. *Surf. Sci.*, 601(5):1184–1192, 2007.
30. H. G. Huang, Z. H. Wang, and G. Q. Xu. The Selective Formation of Di-σ N-Si Linkages in Pyrazine Binding on Si(111)-7x7. *J. Phys. Chem. B*, 108(33):12560–12567, 2004.
31. S. C. Jung and M. H. Kang. Adsorption structure of pyrazine on Si(100): Density-functional calculations. *Phys. Rev. B*, 80(23):235312, 2009.
32. H. C. Kolb, M. G. Finn, and K. B. Sharpless. Click Chemistry: Diverse Chemical Function from a Few Good Reactions. *Angew. Chemie Int. Ed.*, 40(11):2004–2021, 2001.
33. C. Länger, J. Heep, P. Nikodemiak, T. Bohamud, P. Kirsten, U. Höfer, U. Koert, and M. Dürr. Formation of Si/organic interfaces using alkyne-functionalized cyclooctynes-precursor-mediated adsorption of linear alkynes versus direct adsorption of cyclooctyne on Si(0 0 1). *J. Phys. Condens. Matter*, 31(3):034001, 2019.
34. H.-K. Lee, J. Park, I. Kim, H.-D. Kim, B.-G. Park, H.-J. Shin, I.-J. Lee, A. P. Singh, A. Thakur, and J.-Y. Kim. Selective Reactions and Adsorption Structure of Pyrazine on Si(100): HRPES and NEXAFS Study. *J. Phys. Chem. C*, 116(1):722–725, 2012.
35. T. R. Leftwich and A. V. Teplyakov. Chemical manipulation of multifunctional hydrocarbons on silicon surfaces. *Surf. Sci. Rep.*, 63(1):1–71, 2008.
36. T. R. Leftwich and A. V. Teplyakov. Cycloaddition Reactions of Phenylazide and Benzylazide on a Si(100)-2 x 1 Surface. *J. Phys. Chem. C*, 112(11):4297–4303, 2008.
37. P. W. Loscutoff and S. F. Bent. REACTIVITY OF THE GERMANIUM SURFACE: Chemical Passivation and Functionalization. *Annu. Rev. Phys. Chem.*, 57(1):467–495, 2006.
38. J.-N. Luy, M. Molla, L. Pecher, and R. Tonner. Efficient hierarchical models for reactivity of organic layers on semiconductor surfaces. *J. Comput. Chem.*, 42(12):827–839, 2021.
39. J.-N. Luy and R. Tonner. Organic Functionalization at the Si(001) Dimer Vacancy Defect-Structure, Bonding, and Reactivity. *J. Phys. Chem. C*, 125(10):5635–5646, 2021.
40. H. Marbach. Surface-Mediated in Situ Metalation of Porphyrins at the Solid-Vacuum Interface. *Acc. Chem. Res.*, 48(9):2649–2658, 2015.
41. A. Mazzone. Acetylene adsorption onto Si(100): a study of adsorption dynamics and of surface steps. *Comput. Mater. Sci.*, 35(1):6–12, 2006.
42. J. Meinecke and U. Koert. Copper-Free Click Reaction Sequence: A Chemoselective Layer-by-Layer Approach. *Org. Lett.*, 21(18):7609–7612, 2019.
43. X. Meng. An overview of molecular layer deposition for organic and organic-inorganic hybrid materials: mechanisms, growth characteristics, and promising applications. *J. Mater. Chem. A*, 5(35):18326–18378, 2017.

44. G. Mette, M. Dürr, R. Bartholomäus, U. Koert, and U. Höfer. Real-space adsorption studies of cyclooctyne on Si(001). *Chem. Phys. Lett.*, 556:70–76, 2013.
45. L. Miozzo, A. Yassar, and G. Horowitz. Surface engineering for high performance organic electronic devices: the chemical approach. *J. Mater. Chem.*, 20(13):2513, 2010.
46. N. Münster, P. Nikodemiak, and U. Koert. Chemoselective Layer-by-Layer Approach Utilizing Click Reactions with Ethynylcyclooctynes and Diazides. *Org. Lett.*, 18(17):4296–4299, 2016.
47. S. Nardis, F. Mandoj, M. Stefanelli, and R. Paolesse. Metal complexes of corrole. *Coord. Chem. Rev.*, 388:360–405, 2019.
48. W. K. H. Ng, S. T. Sun, J. W. Liu, and Z. F. Liu. The Mechanism for the Thermally Driven Self-Assembly of Pyrazine into Ordered Lines on Si(100). *J. Phys. Chem. C*, 117(30):15749–15753, 2013.
49. T. Omiya, H. Yokohara, and M. Shimomura. Well-Oriented Pyrazine Lines and Arrays on Si(001) Formed by Thermal Activation of Substrate. *J. Phys. Chem. C*, 116(18):9980–9984, 2012.
50. L. Pecher, S. Laref, M. Raupach, and R. Tonner. Ether auf Si(001): Ein Paradebeispiel für die Gemeinsamkeiten zwischen Oberflächenwissenschaften und organischer Molekülchemie. *Angew. Chemie*, 129(47):15347–15351, 2017.
51. L. Pecher, S. Schmidt, and R. Tonner. Modeling the Complex Adsorption Dynamics of Large Organic Molecules: Cyclooctyne on Si(001). *J. Phys. Chem. C*, 121(48):26840–26850, 2017.
52. L. Pecher, C. Schober, and R. Tonner. Chemisorption of a Strained but Flexible Molecule: Cyclooctyne on Si(001). *Chem. - A Eur. J.*, 23(23):5459–5466, 2017.
53. L. Pecher and R. Tonner. Computational analysis of the competitive bonding and reactivity pattern of a bifunctional cyclooctyne on Si(001). *Theor. Chem. Acc.*, 137(4):48, 2018.
54. M. D. Peeks, T. D. W. Claridge, and H. L. Anderson. Aromatic and antiaromatic ring currents in a molecular nanoring. *Nature*, 541(7636):200–203, 2017.
55. F. Pieck, J.-N. Luy, N. L. Zaitsev, L. Pecher, and R. Tonner. Chemistry at Surfaces Modeled by Ab Initio Methods. HLRS project report, **2019**.
56. S. R. Puniredd, S. Jayaraman, S. H. Yeong, C. Troadec, and M. P. Srinivasan. Stable Organic Monolayers on Oxide-Free Silicon/Germanium in a Supercritical Medium: A New Route to Molecular Electronics. *J. Phys. Chem. Lett.*, 4(9):1397–1403, 2013.
57. M. Raschke and U. Höfer. Influence of steps and defects on the dissociative adsorption of molecular hydrogen on silicon surfaces. *Appl. Phys. B Lasers Opt.*, 68(3):649–655, 1999.
58. M. Reutzel, N. Münster, M. A. Lipponer, C. Länger, U. Höfer, U. Koert, and M. Dürr. Chemoselective Reactivity of Bifunctional Cyclooctynes on Si(001). *J. Phys. Chem. C*, 120(46):26284–26289, 2016.
59. N. Roberts and R. J. Needs. Total energy calculations of missing dimer reconstructions on the silicon (001) surface. *J. Phys. Condens. Matter*, 1(19):3139–3143, 1989.
60. V. V. Rostovtsev, L. G. Green, V. V. Fokin, and K. B. Sharpless. A Stepwise Huisgen Cycloaddition Process: Copper(I)-Catalyzed Regioselective Ligation of Azides and Terminal Alkynes. *Angew. Chemie Int. Ed.*, 41(14):2596–2599, 2002.
61. F. Ruff and Ö. Farkas. Concerted S_N2 mechanism for the hydrolysis of acid chlorides: comparisons of reactivities calculated by the density functional theory with experimental data. *J. Phys. Org. Chem.*, 24(6):480–491, 2011.
62. T. Sandoval and S. Bent. Adsorption of Multifunctional Organic Molecules at a Surface: First Step in Molecular Layer Deposition. In *Encycl. Interfacial Chem.*, pages 523–537. Elsevier, 2018.
63. T. E. Sandoval, F. Pieck, R. Tonner, and S. F. Bent. Effect of Heteroaromaticity on Adsorption of Pyrazine on the Ge(100)-2x1 Surface. *J. Phys. Chem. C*, 124(40):22055–22068, 2020.
64. J. C. Scott. Metal-organic interface and charge injection in organic electronic devices. *J. Vac. Sci. Technol. A Vacuum, Surfaces, Film.*, 21(3):521–531, 2003.
65. K. W. Self, R. I. Pelzel, J. H. G. Owen, C. Yan, W. Widdra, and W. H. Weinberg. Scanning tunneling microscopy study of benzene adsorption on Si(100)-(2x1). *J. Vac. Sci. Technol. A Vacuum, Surfaces, Film.*, 16(3):1031–1036, 1998.
66. Y. X. Shao, Y. H. Cai, D. Dong, S. Wang, S. G. Ang, and G. Q. Xu. Spectroscopic study of propargyl chloride attachment on Si(100)-2x1. *Chem. Phys. Lett.*, 482(1-3):77–80, 2009.

67. M. Shimomura, D. Ichikawa, Y. Fukuda, T. Abukawa, T. Aoyama, and S. Kono. Formation of one-dimensional molecular chains on a solid surface: Pyrazine/Si(001). *Phys. Rev. B*, 72(3):033303, 2005.
68. B. Shong, T. E. Sandoval, A. M. Crow, and S. F. Bent. Unidirectional Adsorption of Bifunctional 1,4-Phenylene Diisocyanide on the Ge(100)-2x1 Surface. *J. Phys. Chem. Lett.*, 6(6):1037–1041, 2015.
69. P. L. Silvestrelli, F. Ancilotto, and F. Toigo. Adsorption of benzene on Si(100) from first principles. *Phys. Rev. B*, 62(3):1596–1599, 2000.
70. M. Stępień, N. Sprutta, and L. Latos-Grażyński. Figure Eights, Möbius Bands, and More: Conformation and Aromaticity of Porphyrinoids. *Angew. Chemie Int. Ed.*, 50(19):4288–4340, 2011.
71. P. Sundberg and M. Karppinen. Organic and inorganic-organic thin film structures by molecular layer deposition: A review. *Beilstein J. Nanotechnol.*, 5:1104–1136, 2014.
72. F. Tao, S. L. Bernasek, and G.-Q. Xu. Electronic and Structural Factors in Modification and Functionalization of Clean and Passivated Semiconductor Surfaces with Aromatic Systems. *Chem. Rev.*, 109(9):3991–4024, 2009.
73. F. F. Tao and S. L. Bernasek. Chemical Binding of Five-Membered and Six-Membered Aromatic Molecules. In *Funct. Semicond. Surfaces*, pages 89 – 104. John Wiley & Sons, Inc.: Hoboken, NJ, USA, 2012.
74. A. V. Teplyakov and S. F. Bent. Semiconductor surface functionalization for advances in electronics, energy conversion, and dynamic systems. *J. Vac. Sci. Technol. A Vacuum, Surfaces, Film.*, 31(5):050810, 2013.
75. P. Thissen, O. Seitz, and Y. J. Chabal. Wet chemical surface functionalization of oxide-free silicon. *Prog. Surf. Sci.*, 87(9-12):272–290, 2012.
76. B. Wang, J. C. Cancilla, J. S. Torrecilla, and H. Haick. Artificial Sensing Intelligence with Silicon Nanowires for Ultraselective Detection in the Gas Phase. *Nano Lett.*, 14(2):933–938, 2014.
77. R. A. Wolkow. CONTROLLED MOLECULAR ADSORPTION ON SILICON: Laying a Foundation for Molecular Devices. *Annu. Rev. Phys. Chem.*, 50(1):413–441, 1999.
78. J. T. Yates. Surface chemistry of silicon-the behaviour of dangling bonds. *J. Phys. Condens. Matter*, 3(S):S143–S156, 1991.
79. J. T. Yates Jr. A New Opportunity in Silicon-Based Microelectronics. *Science*, 279(5349):335–336, 1998.
80. T. Yokoyama, S. Yokoyama, T. Kamikado, Y. Okuno, and S. Mashiko. Selective assembly on a surface of supramolecular aggregates with controlled size and shape. *Nature*, 413(6856):619–621, 2001.
81. Z. S. Yoon, A. Osuka, and D. Kim. Möbius aromaticity and antiaromaticity in expanded porphyrins. *Nat. Chem.*, 1(2):113–122, 2009.
82. A. Yu, Y. Qu, K. Han, and G. He. DFT Investigations About Pyrazine Molecules on Si(100)-2x1 Surface. *Chem. Res. Chinese Univ.*, 23(4):444–451, 2007.
83. Y. P. Zhang, J. H. He, G. Q. Xu, and E. S. Tok. Selective Attachment of 4-Bromostyrene on the Si(111)-(7x7) Surface. *J. Phys. Chem. C*, 115(31):15496–15501, 2011.
84. H. Zhou and S. F. Bent. Molecular Layer Deposition of Functional Thin Films for Advanced Lithographic Patterning. *ACS Appl. Mater. Interfaces*, 3(2):505–511, 2011.
85. H. Zhou and S. F. Bent. Fabrication of organic interfacial layers by molecular layer deposition: Present status and future opportunities. *J. Vac. Sci. Technol. A Vacuum, Surfaces, Film.*, 31(4):040801, 2013.
86. M. Zugermeier, J. Herritsch, J.-N. Luy, M. Chen, B. P. Klein, F. Niefind, P. Schweyen, M. Bröring, M. Schmid, R. Tonner, and J. M. Gottfried. On-Surface Formation of a Transient Corrole Radical and Aromaticity-Driven Interfacial Electron Transfer. *J. Phys. Chem. C*, 124(25):13825–13836, 2020.

Electro-catalysis for H_2O oxidation

Travis Jones

Abstract Electrocatalysts facilitate two fundamental processes – electron transfer and chemical bond formation/rupture – to convert renewable electrical energy into chemical fuels. Doing so requires the protons and electrons supplied by the oxygen evolution reaction. The reaction steps constituting the oxygen evolution reaction are assumed to result in an electrochemical mechanism qualitatively distinct from the purely chemical ones familiar from thermal catalysis. Such an electrocatalytic mechanism is often thought to be well-described by an exponential dependence of rate on applied overpotential, Tafel's law, requiring the electrochemical bias to act directly on the reaction coordinate. The aim of the ECHO project has been to test this assumption for the oxygen evolution reaction by combining experimental efforts with density functional theory based modeling of the electrified solid/liquid interface for an important class of electrocatalysts, iridium (di)oxide. Leveraging the computing power of Hawk enabled the use of computational models with explicit solvent to compute realistic surface phase diagrams of IrO_2 and test its role on the kinetics of the oxygen evolution reaction on those surfaces under fixed bias and fixed charge conditions. The results suggest the oxygen evolution reaction is mediated by oxidative charge rather than the action of the electrochemical bias on the reaction coordinate.

Travis Jones
Department of Inorganic Chemistry, Fritz-Haber-Institut der Max-Planck-Gesellschaft, Faradayweg 4-6, 14195, Berlin, Germany, e-mail: `trjones@fhi-berlin.mpg.de`

W. E. Nagel et al. (eds.), *High Performance Computing in Science and Engineering '21*,
https://doi.org/10.1007/978-3-031-17937-2_8

1 Introduction

Increasing renewable energy deployment is creating new techno-economic challenges for chemical and energy sectors. Among these challenges is the fact that wind and solar are non-dispatchable sources, that is, they cannot continuously meet energy demand owing to their intermittent nature. Energy storage is then required to ensure a continuous supply of electricity from such sources. Chemical energy conversion is poised to play a prominent role in overcoming this challenge, while simultaneously opening new avenues for sustainable chemical production, by transforming the green electricity from intermittent sources into chemical fuels [4, 8, 11, 31]. Such technologies rely on electrocatalysts to facilitate the transformation of electrical energy into chemical bonds. And while a range of electrocatalytic processes are being explored to produce chemical fuels and feedstocks – e.g. CO_2 reduction, N_2 reduction, and water splitting – these transformations require the protons and electrons supplied by the oxygen evolution reaction (OER) [4, 8, 13, 27, 31, 33, 34, 36, 37]. Catalyzing the OER is, however, a significant bottleneck. Even with high-performance catalysts the slow electrochemical kinetics of the OER limit electrochemical conversion efficiencies, and under the acidic working conditions of the electrolyzers capable of coping with the intermittency of renewable sources, only iridium-based materials combine OER activity and operational stability [4, 5, 20, 33]. This has made understanding the OER mechanism, especially over iridium-based materials, a central focus in electrocatalysis research [20, 27, 33, 36, 37].

In the simplest and most general description, the OER involves the transfer of four electrons across the electrode/electrolyte interface, the breaking of O-H bonds, and the formation of an O-O bond, giving the following stoichiometry:

$$2H_2O \rightarrow 4H^+ + 4e^- + O_2. \qquad (1)$$

Thermodynamically the reaction proceeds at 1.23 V on the standard hydrogen electrode (SHE) scale, though appreciable OER rates require applied potentials in excess of this limiting value. Understanding the origin of such overpotentials (η), and finding ways to minimize them, has been a major driver of research into the OER. The complexity of the electrified solid/liquid interface together with the large variety of mechanistic possibilities associated with the transfer of four electrons [16] and the chemical bond making/breaking steps [15] involved in the OER, however, has severely impeded progress towards this goal. This complexity is illustrated schematically in the exemplary reaction pathway shown in Figure 1. As a multistep heterogeneously-catalyzed reaction, several intermediates and transition states should be expected along the reaction coordinate. The energies of these states are likely to be closely linked to the nature of the surface, much like traditional thermal catalysis. Unlike thermal catalysis, however, the reactant and products, including the protons released during the OER, are solvated in the electrochemical double layer. Mass transport and concentration effects can also ad to the complexity of the OER. While these later aspects can be minimized experimentally, the intrinsic complexity of the electrocatalytic kinetics and the corresponding (unknown) double layer structure cannot. To avoid a complete description of this complexity, most efforts to understand OER kinetics rely on the validity of the phenomenological Butler–Volmer kinetic theory [18, 27, 33, 36, 37], which requires the applied electrochemical potential to act directly on the reaction coordinate in accord with the well-developed theory of outer-sphere electron transfer [1, 32]

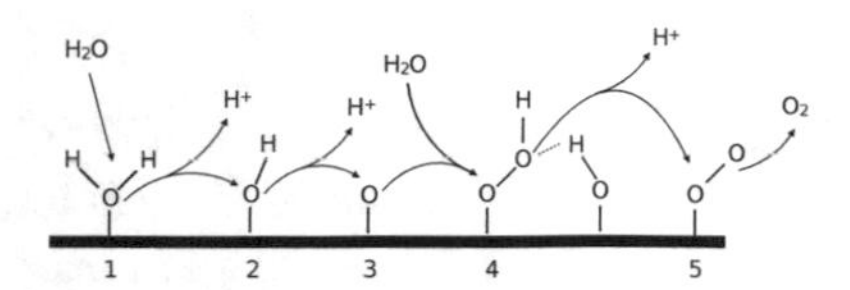

Fig. 1: Schematic reaction pathway of the OER showing water adsorption (1), and its oxidation through OH_{ads} formation (2), O_{ads} (3), OOH_{ads} formation via water nucleophilic attack of O_{ads} (4), and the formation of adsorbed O_2 (5).

When Butler–Volmer kinetic theory holds, the OER rate depends exponentially on the applied overpotential. This dependence can be expressed using the slope of the plot of η versus the log of the OER current, a Tafel plot. An example of a measurement on IrO_2 calcined at 450 °C is shown in Figure 2. Here, the plot of $\eta = a \log i + b$ produces straight lines, in agreement with the exponential dependence of rate on η predicted by Butler–Volmer theory. Within this picture, the Tafel slope, a, takes on characteristic values linked to the nature of the rate determining step under the measurement conditions and can thus be used to derive mechanistic insights [1, 18, 32, 35]. The change in slope from 46 mV/dec to 75 mV/dev in Figure 2 would then indicate a change in mechanism when crossing from the low to high current regime. This type of Tafel slope analysis forms much the experimental foundations of our mechanistic understanding of the OER [18, 27, 28, 33, 36, 37], though there is little experimental evidence demonstrating the validity of Butler–Volmer theory during the OER.

The experimental challenges associated with probing the electronic and atomic structure of the electrified solid/liquid interface have led to a crucial reliance on *ab initio*simulations to address mechanistic questions related to the OER [33, 36, 37]. And while early computational studies have helped shape the current field, these studies focused on the thermodynamics of possible reaction intermediates in vacuum [29, 33, 36, 37], which cannot address the relevant electrocatalytic performance at non-zero overpotentials owing to the lack of a double layer and electrochemical bias. More recent studies have aimed to improve the thermodynamic description by including solvent effects using ice-like models [9], though it is unclear if such models accurately capture the solvation energy of the solid/liquid interface, and a purely

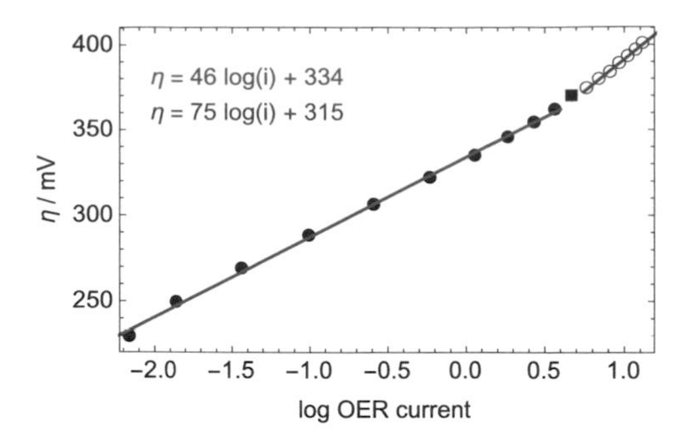

Fig. 2: Experimental Tafel plot of IrO_2 calcined at 450 °C, data from Ref. [21]. The Tafel slope can be seen to change from 46 mV/dec at low overpotentials to 75 mv/dec at high overpotentials, consistent with a change in reaction mechanism within a Butler–Volmer picture.

thermodynamic description does not directly address kinetics. Further efforts have aimed to identify kinetic effects [19, 25], though these too avoid the use of explicit solvent and therefor do not have an atomic scale description of the double layer, which can be circumvented by assuming Butler–Volmer theory holds.

In the ECHO project we aim to move beyond a Butler–Volmer description of the OER by combining density functional theory (DFT) molecular dynamics using explicit solvent with synchrotron based X-ray spectroscopy to identify how the actives phase(s) of OER electrocatalysts evolve as a function of electrochemical bias and by modeling the OER kinetics on these surfaces under constant potential and constant charge conditions. Over this reporting period we have succeeded in developing models of the surface phases present on IrO_2 from the OER onset up to $\eta \approx 400$ mV. We have also been able to compute the activation energies associated with O-H bond breaking and O-O bond formation on these surfaces under constant charge and constant electrochemical bias conditions. Through these efforts were able to show the principle role of the applied overpotential is to modify the surface chemistry of the electrocatalyst rather than to accelerate the OER by direct action on the reaction coordinate. This new picture of the OER is described in Ref. [21] together with the experimental verification. Herein the focus will be on the main computational results.

2 Methods

DFT calculations were performed with the Quantum ESPRESSO package [10] using the PBE exchange and correlation potential [24] and pseudopotentials from the standard solid-state pseudopotentials database [26]. In the case of iridium, several pseudopotentials show similar performance, and the projected augmented wave dataset from the PSLibrary [6] was used owing to improved SCF convergence of the projected augmented wave dataset. All simulations were performed with spin polarization using a wave function cutoff of 60 Ry and a charge density cutoff of 540 Ry. To compute surface phase diagrams symmetric slabs of (110) terminated rutile-type IrO_2 were employed using a (1×2) supercell. The two slides of the slab were separated by 18 water molecules. The phase diagram was constructed by considering the possible permutations of O or OH on the 4 total oxygen sites on the (110) (1×2) surface, 2 μ_2 sites and 2 μ_1 sites. Of these, only five terminations were found to be stable. Denoting each oxygen site on one side of the symmetric slabs the stable surfaces include: i) μ_2-OH/μ_2-OH and μ_1-OH/μ_1-OH; ii) μ_2-O/μ_2-OH and μ_1-OH/μ_1-OH; iii) μ_2-O/μ_2-O and μ_1-OH/μ_1-OH; iv) μ_2-O/μ_2-O and μ_1-O/μ_1-OH; v) μ_2-O/μ_2-O and μ_1-O/μ_1-O (see Figure 4 in Results). Each surface was equilibrated in water for 20 ps using a 1 fs timestep at 350 K (the elevated temperature was used to account for the PBE induced over structuring of water). A Berendsen thermostat controlled the ionic temperature with $dt/d\tau = 1/50$ during the equilibration. The surface (electrochemical) potential vs. pH phase diagram was generated using the computational hydrogen electrode [22]. Configurational entropy for equivalent adsorption sites and the zero-point energies associated with μ_1-OH and μ_2-OH, taken from calculations on surfaces relaxed in vacuum, were included. The surface energies were computed using molecular dynamics snapshots. In an effort to reduce the influence of the error in the energy of bulk water in these snapshots [14] on the computational phase diagram, we defined an interfacial energy following Ref. [38]. That is:

$$\langle E_{int} \rangle = \frac{1}{2N} \sum_{i=1}^{N} \left(E_{tot,i} - E_{H_2O,i} - E_{surf,i} \right) . \tag{2}$$

Here the factor of 1/2 accounts for the two sides of slab and N is the number of snapshots considered in the sum. $E_{tot,i}$ represents the total energy of the solvated slab in the ith snapshot. $E_{H_2O,i}$ is the total energy of the water in the ith snapshot in the absence of the slab. $E_{surf,i}$ is the energy of the slab in the ith snapshot without water. These energies were all computed using a (4×4) **k**-point mesh with Marzari–Vanderbilt smearing ($\sigma = 0.02$ Ry) [17]. The sum in the equation above was continued until the energy convergence of the interfacial energy contribution was better than 0.1 eV. This required less than 32 snapshots per hydrogen coverage. The hydrogen coverage dependence on applied potential was computed by interpolating between the computed hydrogen coverages using a quadratic function to capture the solvent induced Frumkin behavior we found to be associated with hydrogen adsorption.

Minimum energy paths were computed by starting with an molecular dynamics snapshot of the μ_2-O/μ_2-O and μ_1-O/μ_1-OH surface and retaining 2 water bilayers on one side of the slab while introducing $\approx$ 15 Å of vacuum to separate periodic images. Note, while this is not required for constant charge simulations, the constant bias simulations require vacuum; to facilitate comparison between the two approaches we employed the same cell dimensions for both constant charge and bias simulations. After introducing the vacuum layer, μ_1-OH was removed from the non-solvated side of the slab and the bottom two layers of atoms on this side were fixed to their bulk coordinates. The fixed charge minimum energy paths were computed using zero net charge in the simulation cell along the entire length of the path. The fixed bias minimum energy path simulations were performed using the effective screening medium method [2, 23], with the potential of the neutral interface in the initial state and biases 0.1 V to 0.5 V anodic of this zero charge potential. The climbing image nudged elastic band method [12] was used with 8 images per path to locate transition states with a single climbing image/transition state per path. The paths were considered to be converged when the force on each image was below 0.05 eV/Å.

The possible role of finite temperature effects was explored using metadynamics performed at 350 K [3]. As with the minimum energy path simulations, these metadynamics simulations were performed under both constant charge and constant bias conditions on asymmetric slabs with a solvation layer on one side. Here, however, the full solvation layer was preserved during the molecular dynamics simulations, which used the parameters noted above, and $\approx$ 15 Å of vacuum to separate periodic images. To keep the water confined near the surface during the simulations, a potential wall was placed at about $\approx$ 12 Å above the surface. Free energy barriers were computed by biasing the O—O distance of oxygen involved in O-O coupling using Gaussian kernels stored on a grid. Each free energy barrier was computed through three separate simulations, each of which employed 8 parallel walkers to explore the free energy landscape.

The metadynamics and minimum energy path calculations proved to be the most computationally demanding aspects of the work. A similar parallelization strategy was used for each. In the metadynamics simulations DFT based molecular dynamics was performed while biasing a collective variable describing the reaction coordinate, where multiple replicas of the systems, in this work 8, with different starting configurations were used to explore the free energy surface. To allow sufficient time for the simulation to sample the free energy landscape the system must was biased slowly. In this work the total simulation time exceeded 10 ps in each case, which required tens of thousands of DFT force evaluations. Similarly, for minimum energy path simulations using the nudged elastic band algorithm, a minimum energy path was first discretized into a series of images along the path by interpolating the atomic rearrangement between reactant and product. The forces on the atoms in each of the (8) images were then computed. The atoms were allowed to relax using the computed forces that are tangent to the initial path, whereas fictitious spring forces were applied along the path to keep the images from collapsing into a common minimum [12]. As the images relaxed to the minimum energy path, the highest energy image was allowed to feel the full force; however, the force component normal to the path was inverted to drive the highest energy replica towards the transition state. To make these problems computationally tractable efficient parallelization was required

In this work the systems of interest are solvated IrO_2 surfaces containing approximately 110 atoms, 300 Kohn–Sham states, and 16 total **k**-points and 8 replicas. Figure 3 shows scaling data for this system. Owing to the low communication between replicas in both metadynamics and minimum energy path calculations, parallelization over replicas results in trivial linear scaling, though limited by the small number of

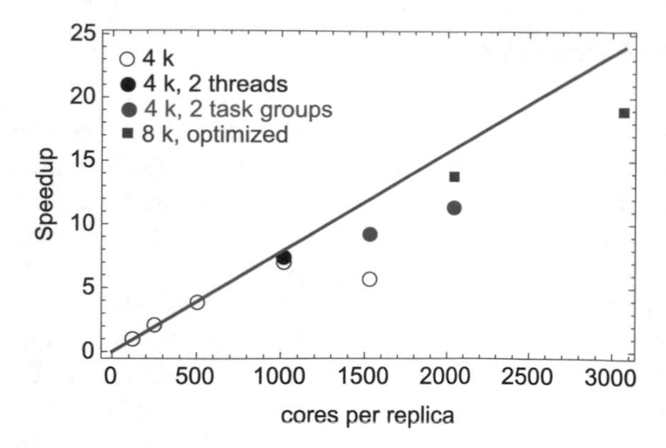

Fig. 3: Scaling data for a single replica of the simulation cell used in this work. As indicated in the figure legend, data is shown for both a coarse and fine **k**-point mesh together with the effect of different parallelization strategies.

replicas. It is then essential to develop a parallelization for each replica. To test the scalability of one replica of the system, a 4-layer oxygen terminated IrO_2 slab was used with 18 water molecules as a solvent layer. This system was tested with both a coarse (4×4) **k**-point mesh and a fine (8×8) **k**-point mesh. In both cases the **k**-points were distributed over n_{pool} pools of CPUs. The communication requirement between these pools is very low and leads to trivial linear scaling that is limited by the small number of **k**-points. Within each pool, n_{PW} CPUs work on a single **k**-point, and the wave-function coefficients are distributed over the n_{PW} CPUs. From previous work it was found this level of parallelization is limited. To further extend the scaling, linear algebra operations needed for subspace diagonalization were also distributed. These operations are performed on $\approx N_b \times N_b$ square matrices, where N_b is the number of Kohn–Sham states, and were thus distributed on a square grid of n_d^2 cores, with $n_d^2 \leq n_{PW}$. Figure 3 shows this parallelization strategy is acceptable up to 1024 cores per replica, where the efficiency is 88%. This efficiency can be improved slightly by employing 2 OpenMP threads to give an efficiency of 92%. For the same system, scaling can be furthered by dividing the n_{PW} cores into n_{task} task groups, where each task group handles the FFT over N_b/n_{task} states. As seen in Figure 3, using two task groups allows the scaling to be extended to at least 2048 cores per replica, though this comes with a drop in efficiency to 71%. As all problems require 8 replicas, 1024 cores per replica is preferred as each problem can then be fit on 64 nodes. Increasing the **k**-point mesh to improve the energy convergence results in a larger number of CPU pools, which increases the scaling when following the aforementioned approach. As with the coarse mesh, 1024 cores yields the most efficient approach and allows the use of 64 nodes for both the minimum energy path and metadynamics simulations.

With this parallelization strategy it became possible to explore the surface phases of IrO_2 and the OER mechanism on Hawk using implicit solvent, as discussed below.

3 Results and discussion

Figure 4a shows an example of the symmetric slab used to compute the surface energies needed to construct the surface (electrochemical) potential vs. pH phase diagram, while Figure 4b-f shows examples the low energy surface terminations found by equilibration using DFT molecular dynamic; solvent has been removed for clarity. Going from Figure 4b to Figure 4f represents an increase in oxidative charge coverage (θ_{h+}) from 0 monolayer (ML) to 1 ML, respectively. While we considered all hydrogen coverages, those shown in Figure 4 represent the lowest energy states. It can be seen that hydrogen is first removed from μ_2-O sites in Figure 4c and d, before hydrogen is lost from μ_1-O sites, Figure 4e and f. Oxidation beyond the 1 ML θ_{h+} surface shown in Figure 4f leads to μ_1-OOH and μ_1-OO, for which we have no experimental evidence [21] and will thus not be considered further in this work.

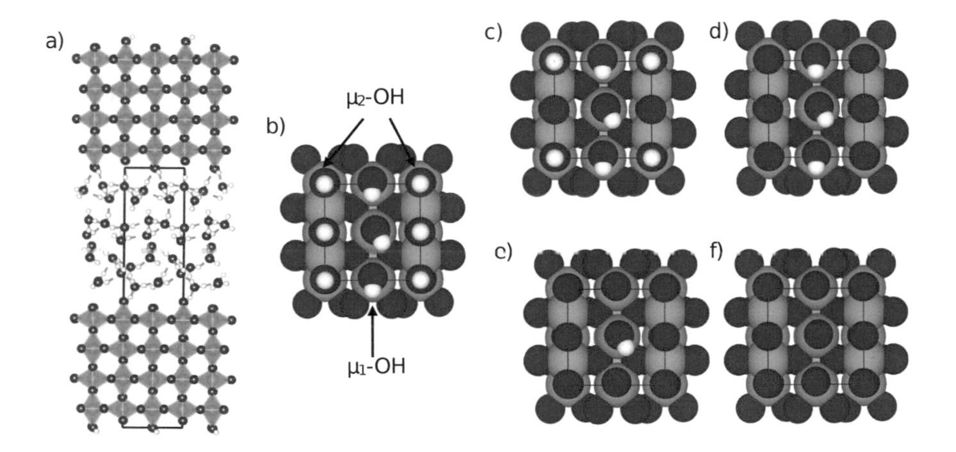

Fig. 4: a) Example of cell used for thermal equilibration. Ir atoms are grey, O as red, and H as white. Low energy terminations of the IrO_2 (110) surface b) μ_2-OH/μ_2-OH and μ_1-OH/μ_1-OH, c) μ_2-O/μ_2-OH and μ_1-OH/μ_1-OH, d) μ_2-O/μ_2-O and μ_1-OH/μ_1-OH, e) μ_2-O/μ_2-O and μ_1-O/μ_1-OH, f) μ_2-O/μ_2-O and μ_1-O/μ_1-O.

The surface energies of the terminations shown in Figure 4 can be used to construct a (electrochemical) potential vs. pH phase diagram following the procedure outlined in the Methods section. Figure 5 shows equilibration in water introduces Frumkin behavior and broadens the surface oxidation window such that it extends over ≈ 1 V. Here it is worth noting only 1/4 ML intervals are shown, though μ_2-OH to μ_2-O oxidation is predicted to begin at ≈ 1 V vs SHE. This onset and the wide surface oxidation window are in good agreement with experimental findings [21]. By comparison, in the absence of explicit solvent surface oxidation is complete by 1.5 V [25], in contrast to experimental findings. From the phase diagram computed with explicit solvent, the active surface under acidic conditions can be expected to

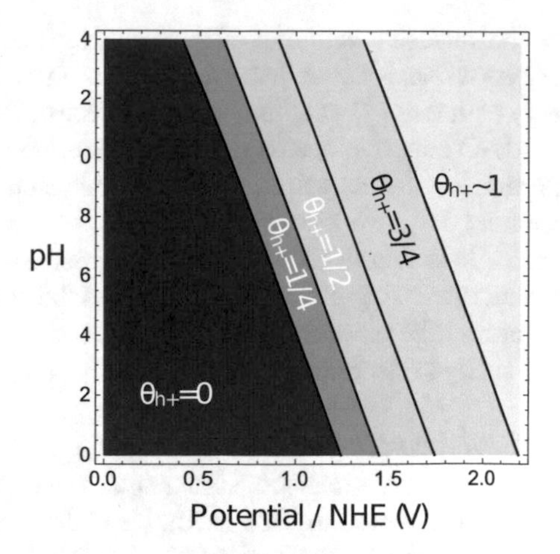

Fig. 5: Computed surface pH vs. potential (normal hydrogen electrode, NHE) phase diagram. Constructed for rutile-type IrO_2 (110) surface. The total oxidative hole coverage with respect to the fully protonated surface indicated.

have an oxidative charge coverage ranging from 1/4 ML to 1 ML depending on the applied bias. Importantly, the oxidative charge coverage is predicted to change over the relevant operating potentials (see Figure 2), and this change in surface oxidation state may be expected to influence the electrocatalytic reaction rate.

Minimum energy path calculations were used to test the effect of oxidative charge on the OER rate by first assuming the elementary steps under acidic conditions can be described as four proton-coupled electron transfers [25, 30]:

$$S + H_2O \rightarrow S - OH + e^- + H^+ \tag{3}$$

$$S - OH \rightarrow S - O + e^- + H^+ \tag{4}$$

$$S - O + H_2O \rightarrow S - OOH + e^- + H^+ \tag{5}$$

$$S - OOH \rightarrow S - OO + e^- + H^+, \tag{6}$$

where S is an empty μ_1 site on the surface.

Of these steps, analysis of experimental Tafel slopes has been used to suggest the second (Equation 4) is rate-limiting [7]. Computational studies using implicit solvent have suggested step 3 (Equation 5) is the rate determining step and the remaining proton-coupled electron transfers are barrierless [25]. In this work, water nucleophilic attack (step 3) was found to be rate-limiting in the presence of the explicit solvent and will be the focus of the remainder of the discussion, see Ref. [21] for more details.

With the rate determining step identified in a realistic model of the electrified solid/liquid interface it becomes possible to explore the validity of Butler–Volmer theory for the OER by focusing on water nucleophilic attack. First consider the activation energy (E_a) for this oxyl-water coupling step on a surface with θ_{h+}=3/4

ML (Figure 4e) at the surface's potential of zero charge (pzc). From Figure 5, this surface oxidation states is consistent with the surface at high overpotential. Computing the minimum energy path for O-O coupling with the electrochemical bias fixed to the surface's pzc shows O-O coupling occurs with the concerted transfer of hydrogen to a μ_2-O site, see Figure 6a. The activation energy for this elementary step is 0.63 eV, and the heat of reaction (ΔH_{rxn}) is near zero. This data is denoted in the plot of E_a vs. ΔH_{rxn} by a green triangle in Figure 6b. The two extra green triangles at $\Delta H_{\mathrm{rxn}} \approx 0$ eV in Figure 6b show that increasing the bias by 0.1 V and 0.5 V to capacitively charge the system while holding the oxidative charge constant at $\theta_{h+} = 3/4$ ML has little impact on E_a. The change introduced by the capacitive charge is indicated by the arrow labeled Q_C.

From Figure 5 it can be seen that constraining θ_{h+} to 3/4 ML to capacitively charge the system by 0.5 V is not realistic. Under such a bias the surface is predicted to oxidize to $\theta_{h+} \approx 1$ ML. Including this oxidative charge yields the total charge, Q_T, which, as shown in Figure 6b, reduces ΔH_{rxn} to $\approx$-0.2 eV. This oxidation charge also lowers E_a from $\approx$0.60 eV to 0.23 eV. Similarly, reducing θ_{h+} from 3/4 to 1/2 ML increases E_a to 0.78 eV, see $Q_{T'}$ in Figure 6b; this latter change cannot be compensated by pure capacitive charging ($Q_{C'}$). Thus, oxidative charge appears to have a stronger influence on the activation energy than capacitive charge.

Another way of viewing the data in Figure 6 is through the Brønsted–Evans–Polanyi (BEP) relationship. The BEP relationship is a linear correlation between E_a and ΔH_{rxn} often observed for chemical bond making and breaking steps [15], and in thermal catalysis in particular. From Figure 6 we can see such a correlation exists for O-O coupling obeys the Brønsted–Evans–Polanyi (BEP) relationship. This relationship holds as changes in ΔH_{rxn} appears to be dominated by oxidative charge rather than purely capacitive charge. Such behavior would be expected when considering classical electrochemical theory, as inner-sphere reaction kinetics are thought to be insensitive to building up the double layer. To test this concept explicitly, minimum energy path calculations were performed without holding the electrochemical potential fixed, that is, a second set of minimum energy path calculations simulations were performed under a fixed charge condition.

The degree to which θ_{h+} alone mediates the BEP relationship can be found by computing the minimum energy paths for O-O coupling using a fixed number of electrons rather than fixing the electrochemical bias as this does not allow capacitive charging. While this approach does not capture the experimental conditions, it does offer a computational means of testing the hypothesis that charge mediates the observed BEP relationship. As expected from classical electrochemical theory, fixing the charge during the minimum energy path calculations does not change the mechanism or break the BEP relationship, see the filled squares in Figure 6b. Thus, the BEP relationship holds and $E_a = E_0 + \alpha\Delta H_{\mathrm{rxn}}$ even in the absence of electron transfer. Moreover, replacing the spectator O/OH species on the surface with adsorbed Cl (an experimentally testable situation) reveals the BEP slope α is primarily controlled by the charge on the ligand rather than chemical nature of the ligand. These results lead to a situation familiar from traditional thermal catalysis: as the surface becomes more reduced or oxidized, E_a increases or decreases, respectively. Thus,

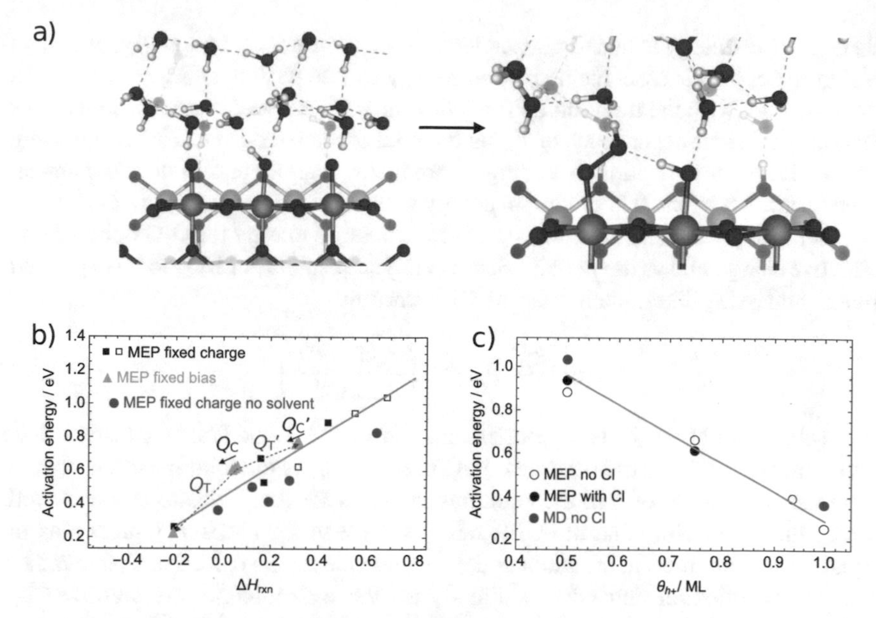

Fig. 6: a) The initial and final states for water-oxyl coupling on the IrO_2 (110) surface. During the reaction a water molecule near the oxyl forms an O-O bond with μ_1-O. During this step hydrogen is transferred from water to a surface mu_2-O through a Zundel-like species. The final state of the elementary step has adsorbed μ_1-OOH on the surface. b) The activation energy computed for O-O coupling along the minimum energy path plotted as a function of the corresponding heat of reaction. The heat of reaction can be seen to become more negative as the oxidative surface charge increases. The green triangles show E_a computed from the minimum energy paths for oxyl-water coupling on surfaces with θ_{h+}=1/2 ML, 3/4 ML, and 1ML (in order of increasing exothermicity) computed under constant potential conditions. In all cases the computed activation energies include reactions on the surfaces with no net charge in the initial state. For θ_{h+} of 1/2 ML and 3/4 ML the green triangles also include results with capacitive charging from 0.1 V to 0.5 V. These results are denoted by the label Q_C for the θ_{h+} = 3/4 ML and the label $Q_{C'}$ for the θ_{h+}= 1/2 ML surface. The small solid arrows under the Q_C and $Q_{C'}$ labels show how the activation energy drops marginally with capacitive charging even as high as 0.5 V. The dashed arrow labeled Q_T shows the effect of allowing the complete charging of the surface/double layer by including oxidative charging, rather than capacitive charging alone, when the θ_{h+} = 3/4 ML surface is biased by an additional 0.5 V. Similarly, $Q_{T'}$ shows the effect of reducing the θ_{h+} = 3/4 ML surface to θ_{h+} = 1/2 ML. By way of comparison, the squares show the computed activation energies as a function of ΔH_{rxn} under constant charge conditions. The unfilled squares show surfaces with adsorbed Cl, which were used to investigate a non-reducible ligand. Circles show the results of minimum energy path calculations without solvent. c) The activation energies computed along the 0 K minimum energy path (labeled MEP) and the activation free energies computed using metadynamics (labeled MD) plotted against the coverage of oxidative charge.

the activation energy is linearly dependent not only on ΔH_{rxn}, but on θ_{h+}, as shown in Figure 6c. In this case, the activation energy can be written as $E_a = \zeta\theta_{h+} + \kappa$ in close analogy with the tradition BEP relationship, where the constants ζ and κ now describe the linear dependence of E_a on the total oxidative charge. This relationship also holds for the activation free energy from metadynamics and so does not appear to be an artifact of the 0 K minimum energy path calculation, see Figure 6c.

Identifying the linear dependence of the activation energy for O-O coupling on oxidative charge allows the per-site electrocatalytic response of IrO_2 to be expressed though an Eyring-like equation for the OER current:

$$i = 4|e|\frac{k_{\text{B}}T}{h}\theta_{\mu_1}\exp\left(-\frac{\zeta\theta_{h+} + \kappa}{k_{\text{B}}T}\right), \tag{7}$$

In the above equation, k_{B} is the Boltzmann constant; h is the Planck constant; T is the temperature; θ_{μ_1} is the coverage μ_1-O species; θ_{h+} is the total oxidative charge coverage. The factor of 4 in the equation accounts for the 4 electrons transferred during the OER. Note that the OER rate response to the electrochemical bias in Equation 7 is captured in the exponential dependence on the oxidative charge rather than the exponential dependence directly on the overpotential, as predicted by Butler–Volmer theory; e.g. $i \propto \exp\left(\eta/k_{\text{B}}T\right)$ in Butler–Volmer theory rather than the $i \propto \exp\left(\theta_{h+}/k_{\text{B}}T\right)$ found in this work.

It is now possible to predict the electrocatalytic response of IrO_2, in particular the Tafel plot, using Equation 7 together with the computed phase diagram. To do so, we can approximate the activation free energy in the Eyring-like equation using the computed activation energies or we can take the computed activation free energies from the metadynamics simulations. Using the BEP relationship computed from the minimum energy reveals good agreement with experiment, see the blue and red fit lines in Figure 7. From these, the computed Tafel plot has a Tafel slope of 39 mV/dec up to 1.58 V, at which point the slope increases to 77 mV/dec. By way of comparison, crystalline IrO_2 has Tafel slopes ranging from 43-47 mV/dec and 71-76 mV/dec over the same potential windows [21], see Figure 2. Using the computed activation free energies also results in good agreement with experiment, with Tafel slopes of 36 mV/dec up to 1.58 V and 69 V/dec above 1.58 V, see the brown and purple fit lines in Figure 7 and the Tafel equations labeled η_{MD}. Now from the calculations, however, the bend in the Tafel slope, irrespective of how the BEP relationship is approximated, can be ascribed to a change in the response of θ_{h+} to the applied electrochemical bias instead of the change in mechanism suggested by Butler–Volmer theory [7]. Here the μ_1-OH species are become depleted at 1.58 V, and, as a result, further increases in the potential beyond this value result in less oxidative charge storage. Beyond describing this behavior, the simulations also reveal the log OER current is linear in θ_{h+}, Figure 7b. This prediction is consistent with experiment, with Figure 7c showing an experimental example of the log OER current response of crystalline IrO_2 as function of the total charge stored in the electrocatalyst. As noted above, the total charge is dominated by the oxidative contribution, θ_{h+}; note, however, experimentally the active surface area of this sample is unknown so the measured charge cannot be converted directly into θ_{h+}. Regardless, the good agreement between the computed and measured trends suggests bond rupture/formation is slow compared to electron transfer. As a result, the OER rate displays BEP behavior and falls outside the scope of Butler–Volmer theory.

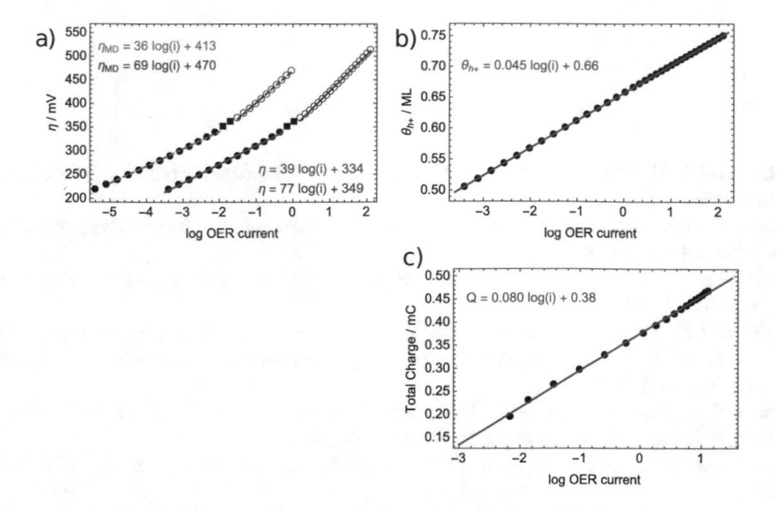

Fig. 7: a) The computed Tafel plot of IrO_2. The Tafel plots computed using the activation energies from the minimum energy path calculation are fit with blue and red fit lines and the corresponding Tafel equations are shown in the lower right. The Tafel plots computed using the activation free energies are fit with brown and purple fit lines and the corresponding Tafel equations are shown in the upper left and denoted η_{MD}. b) The computed θ_{h+} vs. log OER current plot of IrO_2. c) The measured total charge vs. log OER current plot of IrO_2 calcined at 450 °C, data from Ref. [21].

4 Summary

By leveraging the excellent scalability of Quantum ESPRESSO on Hawk it was possible to calculate the surface (electrochemical) potential vs. pH phase diagram of the (110) surface of rutile-type IrO_2 in the presence of explicit solvent. Doing so revealed solvent induced Frumkin behavior that broadens the surface oxidation windows across the range of potentials relevant for the OER under acidic conditions. Minimum energy path calculations and metadynamics performed in the presence of explicit solvent under constant bias and constant charge conditions revealed this oxidative charging controls the OER rate, rather than the direct action of the bias on the reaction coordinate commonly assumed. These results predict bi-linear Tafel plots and a linear response of log OER current to total charge, both of which are confirmed by experiment. Recognizing this role of oxidative charge in OER electrocatalysis demonstrates inner-sphere chemistry controls electrocatalytic OER and provides a fundamental link between thermal- and electrocatalysis. This will enable the tools and concepts developed to understand and engineer traditional catalysts to be applied to their electrochemical counterparts.

References

1. A. J. Bard and L. R. Faulkner. *Electrochemical Methods: Fundamentals and Applications*. Wiley, 2nd edition, 2001.
2. N. Bonnet, T. Morishita, O. Sugino, and M. Otani. First-principles molecular dynamics at a constant electrode potential. *Phys. Rev. Lett.*, 109:266101, 2012.
3. G. Bussi and A. Laio. Using metadynamics to explore complex free-energy landscapes. *Nat. Rev. Phys*, 2:200–212, 2020.
4. A. Buttler and H. Spliethoff. Current status of water electrolysis for energy storage, grid balancing and sector coupling via power-to-gas and power-to-liquids: A review. *Renew. Sus. Energy Rev.*, 82:2440–2454, 2018.
5. M. Carmo, D. L. Fritz, J. Mergel, and D. Stolten. A comprehensive review on PEM water electrolysis. *Inter. J. Hydrogen Energy*, 38:4901–4934, 2013.
6. A. D. Corso. Pseudopotentials periodic table: From H to Pu . *Comput. Mater. Sci.*, 95:337 – 350, 2014.
7. L. De Faria, J. Boodts, and S. Trasatti. Electrocatalytic properties of ternary oxide mixtures of composition $Ru_0.3Ti_{(0.7?x)}Ce_xO_2$: oxygen evolution from acidic solution. *J. Appl. Electrochem.*, 26:1195–1199, 1996.
8. P. De Luna, C. Hahn, D. Higgins, S. A. Jaffer, T. F. Jaramillo, and E. H. Sargent. What would it take for renewably powered electrosynthesis to displace petrochemical processes? *Science*, 364, 2019.
9. J. A. Gauthier, C. F. Dickens, L. D. Chen, A. D. Doyle, and J. K. Nørskov. Solvation Effects for Oxygen Evolution Reaction Catalysis on $IrO_2(110)$. *J. Phys. Chem. C*, 121:11455–11463, 2017.
10. P. Giannozzi, S. Baroni, N. Bonini, M. Calandra, R. Car, C. Cavazzoni, D. Ceresoli, G. L. Chiarotti, M. Cococcioni, I. Dabo, A. Dal Corso, S. de Gironcoli, S. Fabris, G. Fratesi, R. Gebauer, U. Gerstmann, C. Gougoussis, A. Kokalj, M. Lazzeri, L. Martin-Samos, N. Marzari, F. Mauri, R. Mazzarello, S. Paolini, A. Pasquarello, L. Paulatto, C. Sbraccia, S. Scandolo, G. Sclauzero, A. P. Seitsonen, A. Smogunov, P. Umari, and R. M. Wentzcovitch. QUANTUM ESPRESSO: a modular and open-source software project for quantum simulations of materials. *J. Phys.: Condens. Matter*, 21(39):395502, 2009.
11. T. M. Gür. Review of electrical energy storage technologies, materials and systems: challenges and prospects for large-scale grid storage. *Energy Environ. Sci.*, 11:2696–2767, 2018.
12. G. Henkelman, B. P. Uberuaga, and H. Jónsson. A climbing image nudged elastic band method for finding saddle points and minimum energy paths. *J. Chem. Phys.*, 113(22):9901–9904, 2000.
13. C. Hu, L. Zhang, and J. Gong. Recent progress made in the mechanism comprehension and design of electrocatalysts for alkaline water splitting. *Energy Environ. Sci.*, 12:2620–2645, 2019.
14. D. T. Limmer, A. P. Willard, P. Madden, and D. Chandler. Hydration of metal surfaces can be dynamically heterogeneous and hydrophobic. *PNAS*, 110(11):4200–4205, 2013.
15. R. A. Marcus. Theoretical relations among rate constants, barriers, and Broensted slopes of chemical reactions. *J. Phys. Chem.*, 72:891–899, 1968.
16. R. A. Marcus. Electron transfer reactions in chemistry. Theory and experiment. *Rev. Mod. Phys.*, 65:599–610, 1993.
17. N. Marzari, D. Vanderbilt, A. De Vita, and M. C. Payne. Thermal contraction and disordering of the Al(110) surface. *Phys. Rev. Lett.*, 82:3296–3299, 1999.

18. Y. Matsumoto and E. Sato. Electrocatalytic properties of transition metal oxides for oxygen evolution reaction. *Mater. Chem. Phys.*, 14:397–426, 1986.
19. J. T. Mefford, Z. Zhao, M. Bajdich, and W. C. Chueh. Interpreting Tafel behavior of consecutive electrochemical reactions through combined thermodynamic and steady state microkinetic approaches. *Energy Environ. Sci.*, 13:622–634, 2020.
20. T. Naito, T. Shinagawa, T. Nishimoto, and K. Takanabe. Recent advances in understanding oxygen evolution reaction mechanisms over iridium oxide. *Inorg. Chem. Front.*, 8:2900–2917, 2021.
21. H. N. Nong, L. J. Falling, A. Bergmann, M. Klingenhof, H. P. Tran, C. Spöri, R. Mom, J. Timoshenko, G. Zichittella, A. Knop-Gericke, S. Piccinin, J. Pérez-Ramírez, B. R. Cuenya, P. S. Robert Schlögl, D. Teschner, and T. E. Jones. Key role of chemistry versus bias in electrocatalytic oxygen evolution. *Nature*, 587:408–413, 2020.
22. J. K. Nørskov, J. Rossmeisl, A. Logadottir, L. Lindqvist, J. R. Kitchin, T. Bligaard, and H. Jónsson. Origin of the overpotential for oxygen reduction at a fuel-cell cathode. *J. Phys. Chem B*, 108(46):17886–17892, 2004.
23. M. Otani and O. Sugino. First-principles calculations of charged surfaces and interfaces: A plane-wave nonrepeated slab approach. *Phys. Rev. B*, 73:115407, 2006.
24. J. P. Perdew, K. Burke, and M. Ernzerhof. Generalized gradient approximation made simple. *Phys. Rev. Lett.*, 77:3865–3868, 1996.
25. Y. Ping, R. J. Nielsen, and W. A. Goddard. The Reaction Mechanism with Free Energy Barriers at Constant Potentials for the Oxygen Evolution Reaction at the IrO_2 (110) Surface. *J. Am. Chem. Soc.*, 139:149–155, 2017.
26. G. Prandini, A. Marrazzo, I. Castelli, N. Mounet, and N. Marzari. Precision and efficiency in solid-state pseudopotential calculation. *npj Comput. Mater.*, 4:72, 2018.
27. T. Reier, H. N. Nong, D. Teschner, R. Schlögl, and P. Strasser. Electrocatalytic oxygen evolution reaction in acidic environments: Reaction mechanisms and catalysts. *Adv. Energy Mater.*, 7:1601275, 2017.
28. T. Reier, M. Oezaslan, and P. Strasser. Electrocatalytic Oxygen Evolution Reaction (OER) on Ru, Ir, and Pt Catalysts: A Comparative Study of Nanoparticles and Bulk Materials. *ACS Catal.*, 2:1765–1772, 2012.
29. J. Rossmeisl, Z.-W. Qu, H. Zhu, G.-J. Kroes, and J. Nørskov. Electrolysis of water on oxide surfaces. *J. Electroanal. Chem.*, 607:83–89, 2007.
30. J. Rossmeisl, Z.-W. Qu, H. Zhu, G.-J. Kroes, and J. Nørskov. Electrolysis of water on oxide surfaces. *J. Electroanal. Chem.*, 607:83–89, 2007.
31. R. S. Schlögl. Sustainable Energy Systems: The Strategic Role of Chemical Energy Conversion. *Topic. Catal.*, 59:772–786, 2016.
32. W. Schmickler and Santos. *Interfacial Electrochemistry.* Springer, 2010.
33. Z. W. Seh, J. Kibsgaard, C. F. Dickens, I. Chorkendorff, J. K. Nørskov, and T. F. Jaramillo. Combining theory and experiment in electrocatalysis: Insights into materials design. *Science*, 355, 2017.
34. L. C. Seitz, C. F. Dickens, K. Nishio, Y. Hikita, J. Montoya, A. Doyle, C. Kirk, A. Vojvodic, H. Y. Hwang, J. K. Nørskov, and T. F. Jaramillo. A highly active and stable $IrO_x/SrIrO_3$ catalyst for the oxygen evolution reaction. *Science*, 353:1011–1014, 2016.
35. T. Shinagawa, A. Garcia-Esparza, and K. Takanabe. Insight on Tafel slopes from a microkinetic analysis of aqueous electrocatalysis for energy conversion. *Sci. Rep.*, 5:13801, 2015.
36. J. Song, C. Wei, Z.-F. Huang, C. Liu, L. Zeng, X. Wang, and Z. J. Xu. A review on fundamentals for designing oxygen evolution electrocatalysts. *Chem. Soc. Rev.*, 49:2196–2214, 2020.
37. N.-T. Suen, S.-F. Hung, Q. Quan, N. Zhang, Y.-J. Xu, and H. M. Chen. Electrocatalysis for the oxygen evolution reaction: recent development and future perspectives. *Chem. Soc. Rev.*, 46:337–365, 2017.
38. D. Wang, T. Sheng, J. Chen, H.-F. Wang, and P. Hu. Identifying the key obstacle in photocatalytic oxygen evolution on rutile TiO_2. *Nat. Catal.*, 1:291–299, 2018.

Materials Science

In materials science or material science challenges, computational methods require extensive computational capacity. Modern simulation methods from this research area are efficiently parallelised and show a good scaling behaviour on high-performance computers. This part presents a selection of materials science projects (Def-2-Dim and Pace3D), from numerous research activities, that provide new insights into the change of materials on an atomic and microstructural scale, by describing the performance of phase formation and damage simulations using HPC systems. The results can be used to derive design instructions for data-driven, accelerated material development.

The first project (Def-2-Dim) elaborates material simulation methods, such as first principal and molecular dynamics calculations, that operate on an atomic scale, so as to investigate the formation of defects in two-dimensional materials, under the influence of electron and ion irradiation. The quantification of defects, obtained through the investment of extensive computational resources, is important to determine their effect on the electronic, optical and catalytic material properties and provides insights into the post-synthetical doping of two-dimensional materials, using transition metal atoms and a dislocation-mediated mechanism.

The second project (Pace3D) introduces a selected and dedicated application of the high-performance materials simulation framework Pace3D to three-dimensional microstructure simulations on a mesoscopic scale. Large-scale simulations on high-performance clusters are performed with a convincing scalability behaviour that enables the resolution of large-scale microstructure sections. Examples of phase evolution in solid-state phase transformations, involving phase and grain interfaces, are presented and discussed.

The third project (DiHu) chosen for this part illustrates the application of materials science simulation methods to describe the multi-scale, biomechanical phenomena of contracting muscles and highlights the demand of high-performance computing techniques and resources. To study the combined process of muscle contraction and electromyography, the research is conducted using the open-source software OpenDiHu, which shows that the new multi-scale approach is highly scalable.

To conclude this part, simulations for the prediction of muscle contractions (DiHu) are presented as a cross-material application. The results demonstrate the computational effort required to capture multi-scale processes and show the transfer of materials science and solid mechanics methods to life science topics.

Britta Nestler
Institute for Applied Materials –
Computational Materials Science (IAM-CMS),
Karlsruhe Institute of Technology (KIT),
Straße am Forum 7, 76131 Karlsruhe, Germany,
e-mail: britta.nestler@kit.edu

Production of defects in two-dimensional materials under ion and electron irradiation: insights from advanced first-principles calculations

Annual report on 'Bundesprojekt' ACID 44172 (Def-2-Dim)

S. Kretschmer, T. Joseph, S. Ghaderzadeh, Y. Wei, M. Ghorbani-Asl and A.V. Krasheninnikov

Abstract The main goal of the research project with the extensive use of GCS computational resources was to study the formation of defects in two-dimensional (2D) materials under electron and ion irradiation using atomistic simulations. The influence of defects on the electronic, optical and catalytic properties of 2D materials has also been investigated. Specifically, the role of electronic excitations in the production of defects under electron irradiation was elucidated, and the types of defect formed in 2D MoS_2 sheets upon cluster impacts were identified. The first-principles calculations provided insights into the post-synthesis doping of 2D materials with transition metal atoms through dislocation-mediated mechanism, and also allowed for the understanding of how the implanted species (e.g., Cl atoms) affect the electronic properties of 2D transition metal dichalcogenides. Also analytical potential molecular dynamics simulations were used to study the behavior of 2D materials under ion and cluster irradiation. The role of adatoms and surface reconstructions in the novel 2D material hematene was addressed as well. In addition to publications in peer-refereed scientific journals (8 papers published, see Refs. [1–8] and 4 manuscripts are currently under review), the obtained results were also disseminated through popular articles at different internet resources, including the GCS webpage.

A.V. Krasheninnikov
Institute of Ion Beam Physics and Materials Research, Helmholtz-Zentrum Dresden-Rossendorf, Bautzner Landstraße 400, 01328 Dresden,
Tel.: +49 351 260 3148
Fax: +49 351 260 0461
e-mail: a.krasheninnikov@hzdr.de

W. E. Nagel et al. (eds.), *High Performance Computing in Science and Engineering '21*,
https://doi.org/10.1007/978-3-031-17937-2_9

1 Introduction

Since the beginning of the 20th century [9, 10], the interaction of energetic particles – electrons and ions – with matter has been the subject of intensive studies. The research in this area has been motivated first by the necessity to assess the irradiation-induced damage in the materials in radiation-harsh environments such as fission/fusion reactors or cosmic space. It was also realized later on that in spite of the damage, irradiation may have overall beneficial effects on the target [11, 12]. A good example is the industrially important ion implantation into semiconductors [11]. An additional motivation to study effects of electron irradiation on materials comes from transmission electron microscopy (TEM). Every day hundreds, if not thousands, of microscopists in the world get insights into material structure and cope with the beam-induced damage: it is often not clear if the observed defects were present in the sample before it was put into the TEM column or whether they are artefacts of the imaging technique.

The past decade has also witnessed an enormous interest in nanosystems and specifically two-dimensional (2D) materials [13] such as graphene or transition metal dichalcogenides (TMDs), which have already shown good potentials [14–16] for nanoelectronics, photonics, catalysis, and energy applications due to a unique combination of electronic, optical, and mechanical properties. It has also been demonstrated that irradiation, especially when combined with heat treatment, can have beneficial effects on 2D systems. For example, the bombardment of graphene with low-energy ions can be used for increasing the Young's modulus of the system [17], implanting dopants [18], or adding new functionalities, like magnetism[19]. Beams of energetic electrons, which nowadays can be focused to sub-Å areas, have been shown to work as cutting and welding tools on the nanoscale [20, 21] or stimulate local chemical reactions. Further developments of irradiation-assisted methods of 2D materials treatment require complete microscopic theory of defect production in these systems. Moreover, 2D materials give a unique opportunity to understand at the fundamental level the energy deposition and neutralization of the highly-charged ions [22, 23] by measuring their initial/final charge states and kinetic energy. On the theory side, the reduced dimensionality makes it possible to carry out computationally very expensive non-adiabatic first-principles calculations [24–26] to assess the interaction of the ions with the target.

However, a growing body of experimental facts indicate that many concepts of energetic particle-solid interaction are not applicable to these systems, as the conventional approaches based on averaging over many scattering events do not work due to their very geometry, as, e.g., in graphene – a membrane just one atom thick – or require substantial modifications. It is intuitively clear that the impact of an energetic particle will not give rise to a collisional cascade, as in bulk solids, but cause sputtering. A different electronic structure and high surface-to-volume ratio also affect the redistribution of the energy deposited by the energetic particle in the system and thus influence defect production. As for TMDs, experiments indicate that defects are produced at electron energies well below the knock-on threshold (the minimum electron energy required to ballistically displace an atom from the material), and the mechanism of damage production (electronic excitations, etching) in these systems is not fully understood at the moment. The conventional theory based on the charge-state-dependent empirical potentials [27] cannot adequately describe the interaction of highly-charged ions with TMDs, as evident, e.g., from Fig. 3 in Ref. [22].

In this project, by combining first-principles calculations with analytical potential molecular dynamics simulations, we studied the formation of defects in 2D materials under electron and ion irradiation using atomistic simulations. The influence of defects on the electronic, optical and catalytic properties of 2D materials has also been investigated. The highlights from our research are presented below.

2 Results

2.1 Production of defects in two-dimensional materials under electron irradiation

We investigated [1] the role of electronic excitations in defect production by using advanced first-principles simulation techniques based on the Ehrenfest dynamics combined with time-dependent density-functional theory and demonstrate that a combination of excitations and knock-on damage in 2D MoS_2 under electron beam can give rise to the formation of vacancies and explain the experimental observations of defects production far below the knock-on threshold. As our first-principles calculations showed, electronic excitations will quickly delocalize in an otherwise unperturbed periodic structure. The situation changes dramatically when the translational symmetry is broken, as is the case for the displacement of one target atom upon momentum transfer from an impinging electron.

The results of Ehrenfest dynamics (ED simulations of a high-energy electron impact into MoS_2 sheet are presented in Fig. 1. The electron was modeled as a classical particle with a precisely defined trajectory, which can give rise to electronic excitations in the target material, as schematically illustrated in Fig. 1(a). The spatial extent of the electronic excitation created in the system immediately after the impact is illustrated in Fig. 1(b). Simulations where exactly one electron is excited with the excitation initially being localized on a sulfur atom are shown in Fig. 1(c) and panel (d) illustrates the spatial extent of the excitation after 1.6 fs as described within the framework of ED.

Based on the results of our simulations, we proposed a possible mechanism for the observed sub-knock-on threshold damage which involves the localization of the excitation at an emergent vacancy site. According to our calculations, this localized electronic excitation then gives rise to a decrease in the displacement threshold as anti-bonding states are occupied at the emergent defect site. Consequently, beam damage may be expected for voltages below the knock-on threshold as observed by the Sub Angstrom Low Voltage Electron (SALVE) microscope operated by our collaborators at Ulm University.

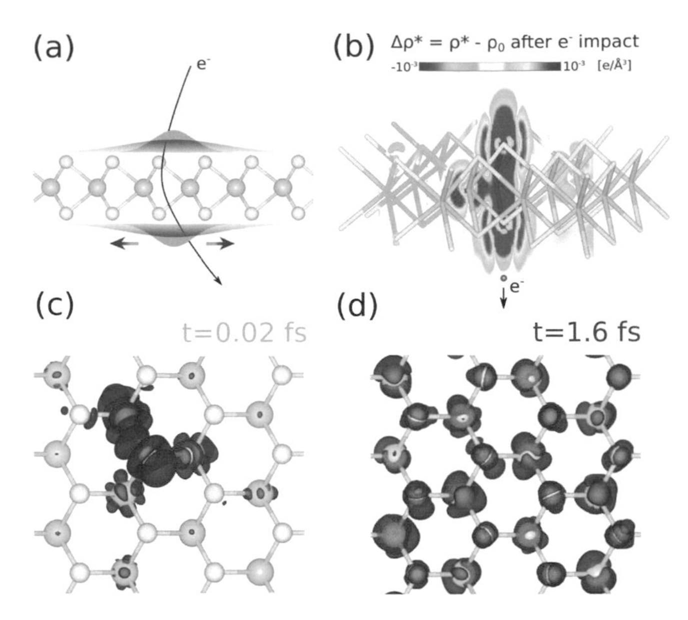

Fig. 1: ED simulations of a high-energy electron impact into MoS_2 sheet. The electron is modeled as a classical particle with a precisely defined trajectory, which can give rise to electronic excitations in the target material, as schematically illustrated in (a). (b) The spatial extent of the electronic excitation created in the system immediately after the impact. (c) Simulations where exactly one electron is excited with the excitation initially being localized on a sulfur atom. (d) The spatial extent of the excitation after 1.6 fs as described within the framework of ED. Reprinted with permission from Ref. [1], Copyright (2020) American Chemical Society.

2.2 Response of supported two-dimensional materials to ion and cluster irradiation

2D materials with nanometer-size holes are promising systems for DNA sequencing, water purification, and molecule selection/separation. However, controllable creation of holes with uniform sizes and shapes is still a challenge, especially when the 2D material consists of several atomic layers as, e.g., MoS_2, the archetypical transition metal dichalcogenide. We used [2] analytical potential molecular dynamics (MD) simulations to study the response of 2D MoS_2 to cluster irradiation. We modelled both freestanding and supported sheets and assess the amount of damage created in MoS_2 by the impacts of noble gas clusters in a wide range of cluster energies and incident angles. We showed that cluster irradiation can be used to produce uniform holes in 2D MoS_2 with the diameter being dependent on cluster size and energy. Energetic clusters can also be used to displace sulfur atoms preferentially from either top or bottom layers of S atoms in MoS_2 and also clean the surface of MoS_2

sheets from adsorbents. Our results for MoS_2, which should be relevant to other 2D transition metal dichalcogenides, suggest new routes toward cluster beam engineering of devices based on 2D inorganic materials.

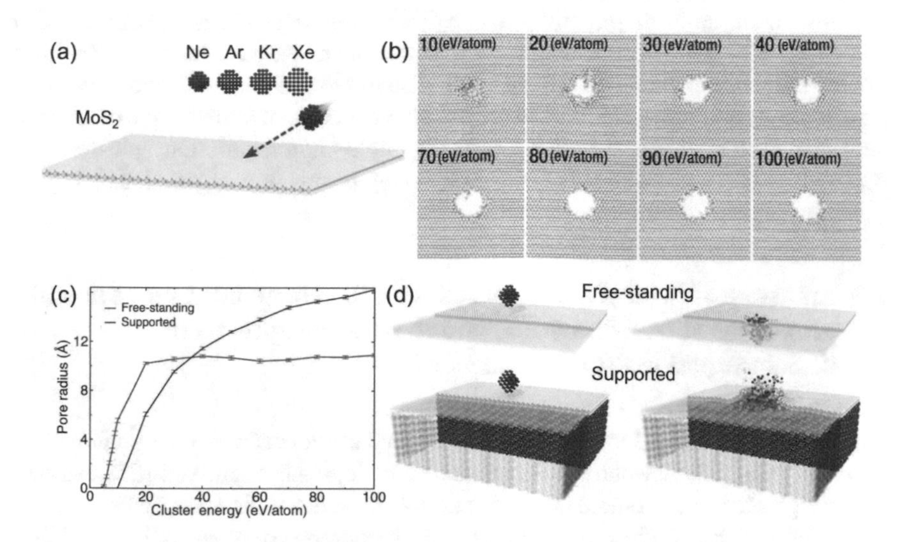

Fig. 2: (a) Simulation setup for irradiation of freestanding MoS_2 monolayers with various noble gas clusters. (b) Defect production in free-standing MoS_2 monolayers after impacts of the Xe_{79} cluster with different initial kinetic energies under normal incidence. (c) Radius of the pore created by the impact as a function of cluster energy. (d) Snapshots from MD simulations of a free-standing and supported MoS_2 monolayer on a SiO_2 substrate.

The effects of ion irradiation on 2D materials have further been studied [8] in the context of Cl ion implantation onto 2D $MoSe_2$, one of the prominent members of the transition metal dichalcogenide materials family. The efficient integration of transition metal dichalcogenides into the current electronic device technology requires mastering the techniques of effective tuning of their optoelectronic properties. Specifically, controllable doping is essential. For conventional bulk semiconductors, ion implantation is the most developed method offering stable and tunable doping. The n-type doping in $MoSe_2$ flakes was experimentally realized by our coworkers through low-energy ion implantation of Cl^+ ions followed by millisecond-range flash lamp annealing. Atomistic simulations at the kinetic Monte Carlo level made it possible to assess the distribution of impurities in the irradiated samples. Density-functional theory calculations were carried out to understand the atomic structure of the irradiated material and assess the effects on the electronic properties. A comparison of the results of the density functional theory calculations and experimental temperature-dependent micro-Raman spectroscopy data indicates that Cl atoms are incorporated into the atomic network of $MoSe_2$ as substitutional donor impurities.

Controlled production of defects in hexagonal boron nitride (h-BN) through ion irradiation has recently been demonstrated to be an effective tool for adding new functionalities to this material such as single photon generation and for developing optical quantum applications. Using atomistic simulations, we studied [7, 28] the response of single-layer and multi-layer h-BN to noble-gas cluster irradiation. Our results quantify the densities of defects produced by noble gas ions in a wide range of ion energies and elucidate the types and distribution of defects in the target. The simulation data can directly be used to guide the experiment aimed at the creation of defects of particular types in h-BN targets for single-photon emission, spin-selective optical transitions and other applications by using beams of energetic ions.

2.3 Influence of defects and impurities on the electronic and chemical properties of two-dimensional materials: insights from density-functional-theory calculations

Highly-doped TMDs. Doping of materials beyond the dopant solubility limit remains a challenge, especially when spatially nonuniform doping is required. In 2D materials with a high surface-to-volume ratio, such as transition metal dichalcogenides, various post-synthesis approaches to doping have been demonstrated, but full control over the spatial distribution of dopants remains a challenge. Post-growth doping of single layers of WSe_2 was performed by our coworkers through adding transition metal (TM) atoms in a two-step process, which includes annealing followed by deposition of dopants together with Se or S. The Ti, V, Cr, and Fe impurities at W sites are identified by using transmission electron microscopy and electron energy loss spectroscopy. The dopants are revealed to be largely confined within nanostripes embedded in the otherwise pristine WSe_2. Density functional theory calculations [6] showed that the dislocations assist the incorporation of the dopant during their climb and give rise to stripes of TM dopant atoms. This work demonstrated a possible spatially controllable doping strategy to achieve the desired local electronic, magnetic, and optical properties in 2D materials.

Exotic, novel 2D Material: hematene. Exfoliation of atomically thin layers from non-van der Waals bulk solids gave rise to the emergence of a new class of 2D materials, such as hematene (Hm), a structure just a few atoms thick obtained from hematite. Due to a large number of unsaturated sites, the Hm surface can be passivated under ambient conditions. Using density functional theory calculations, we investigated [3] the effects of surface passivation with H and OH groups on Hm properties and demonstrate that the passivated surfaces are energetically favorable under oxygen-rich conditions. Although the bare sheet is antiferromagnetic and possesses an indirect band gap of 0.93 eV, the hydrogenated sheets are half-metallic with a ferromagnetic ground state, and the fully hydroxylated sheets are antiferromagnetic with a larger band gap as compared to the bare system. The electronic structure of Hm can be

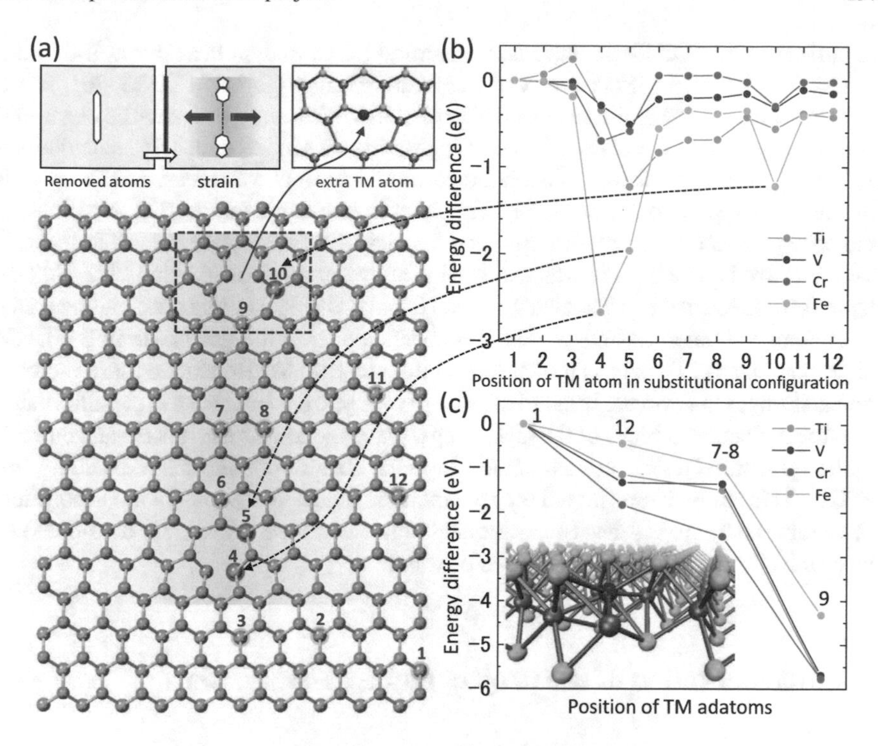

Fig. 3: (a) Atomic model of WSe_2 sheet with a dislocation line created by removing W and Se atoms, as schematically illustrated in the inset, and the resulting strain distribution (orange shading). The green balls represent Se and the gray balls W atoms. b) The energies of transition metal (TM) atom in substitutional (W) positions labeled in (a). It is evident that TM atoms in the substitutional positions prefer to be in the strained areas next to the dislocation cores. c) The relative energies of TM adatoms placed at various positions. The inset shows a configuration when the adatom (red ball) takes the position in the middle of the hollow area, but below Se atoms. Adatoms prefer to be in the strained area and ultimately take the position in the heptagon, with the atomic configuration being shown in (a) in the red frame. Reprinted with permission from Ref. [6], Copyright (2020) Wiley.

further tuned by mechanical deformations. The band gap of fully passivated Hm increases monotonically with biaxial strain, hinting at the potential applications of Hm in electromechanical devices.

Role of mirror twin boundaries in water adsorption. Water adsorption on TMDs and other 2D materials is generally governed by weak van der Waals interactions. This results in a hydrophobic character of the basal planes, and defects may play a significant role in water adsorption and water cluster nucleation. However, there is a lack of detailed experimental investigations on water adsorption on defective 2D materials. By combining low-temperature scanning tunneling microscopy (STM) experiments and DFT calculations, we studied [5] in that context the well-defined mirror twin boundary (MTB) networks separating mirror-grains in 2D $MoSe_2$. These MTBs are dangling bond-free extended crystal modifications with metallic electronic states embedded in the 2D semiconducting matrix of $MoSe_2$. Our DFT calculations indicate that molecular water also interacts similarly weak with these MTBs as with the defect-free basal plane of $MoSe_2$. However, in low temperature STM experiments, nanoscopic water structures are observed that selectively decorate the MTB network. This localized adsorption of water is facilitated by the functionalization of the MTBs by hydroxyls formed by dissociated water. Hydroxyls may form by dissociating water at undercoordinated defects or adsorbing radicals from the gas phase in the UHV chamber. Our DFT analysis indicates that the metallic MTBs adsorb these radicals much stronger than on the basal plane due to charge transfer from the metallic states into the molecular orbitals of the OH groups. Once the MTBs are functionalized with hydroxyls, molecular water can attach to them, forming water channels along the MTBs. This study demonstrated the role metallic defect states play in the adsorption of water even in the absence of unsaturated bonds that have been so far considered to be crucial for adsorption of hydroxyls or water.

3 Implementation of the project and technical data

3.1 Codes employed and their scaling

Three different codes were used to carry out the simulations, GPAW [29], VASP [30,31] and LAMMPS [32]. All these codes are widely used (thousand of users) in atomistic simulations on various platforms with massive parallel architecture, and a good scaling behavior has been demonstrated.

3.1.1 GPAW code

For the Ehrenfest dynamics simulations, we employed the GPAW code (released under GNU Public licence version 3, see `https://wiki.fysik.dtu.dk/gpaw/`) [29]. GPAW is implemented in the combination of Python and C programming languages, where high-level algorithms are implemented in Python, and numerically intensive kernels in C and in numerical libraries. Python has only a small overhead for the calculations, and for example in Cray XT5 GPAW has been measured to execute 4.8 TFLOP/s with 2048 cores, which is 25% of the peak performance. The library requirements for GPAW are NumPy (a fast array interface to Python), BLAS, LAPACK and ScaLAPACK (for DFT calculations only). We modified the GPAW code and the accompanying PAW potentials to enable direct first-principles simulations of the electron impact in the 2D materials. The scaling behavior of the code is presented

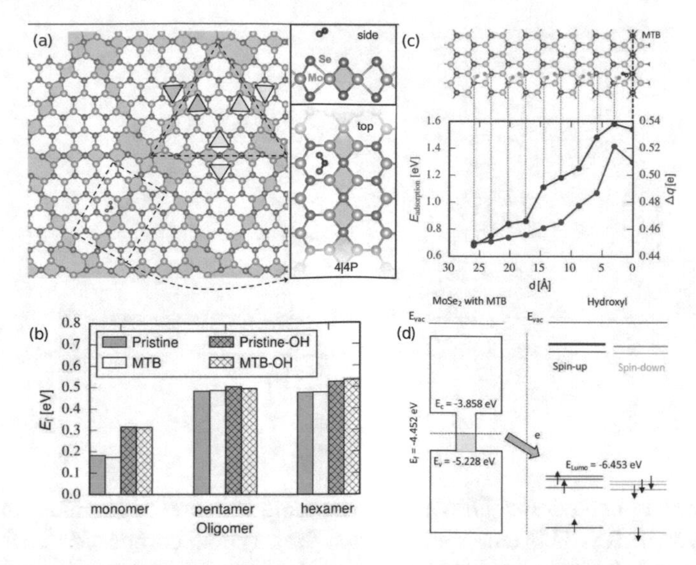

Fig. 4: DFT simulations of water and hydroxyl adsorption on pristine $MoSe_2$ and that with mirror twin boundaries (MTBs). (a–c) Atomic structure of $MoSe_2$ with MTB and adsorbed water molecules, hydroxyl groups and water hexamers attached to the sheet (a) and OH group (b and c). (d) Formation energy of water clusters on top of $MoSe_2$ in the pristine area, next to MTB and OH group. (e) The dependence of adsorption energy of OH group on distance to the MTB defined as the separation between the MTB and the coordinates of the Se atom the OH group is attached to. The plot also shows charge transfer from the MTB into the empty molecular orbitals of the hydroxyl group, as schematically illustrated in panel (f), which presents the electronic structure of MoSe2 with MTB and isolated OH group.

in Fig. 5. Here all the scaling tests were carried out at HLRS facilities (Cray XT5 HazelHen). Typically, these time-dependent DFT calculations were run on 1024 cores.

3.1.2 VASP code

The VASP code [30, 31], which requires a software licence agreement with the University of Vienna, Austria, is currently used by more than 1400 research groups in academia and industry worldwide. At the moment it is the 'standard accuracy reference' in the DFT calculations. We have employed the code for the calculations of the ground state properties of various defective systems and for Born–Oppenheimer molecular dynamics simulations of ion impacts into 2D materials. The scaling behavior of the code is presented in Fig. 6. For these DFT calculations between 128 and 512 cores were used.

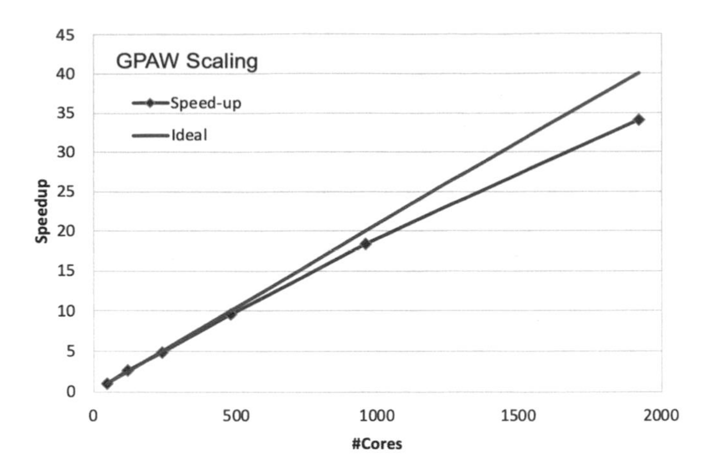

Fig. 5: Scaling behavior of GPAW when running Ehrenfest dynamics simulations on CPUs at HLRS. This data was obtained for a system composed of 36 atoms, 5 K-points and 160 bands were used.

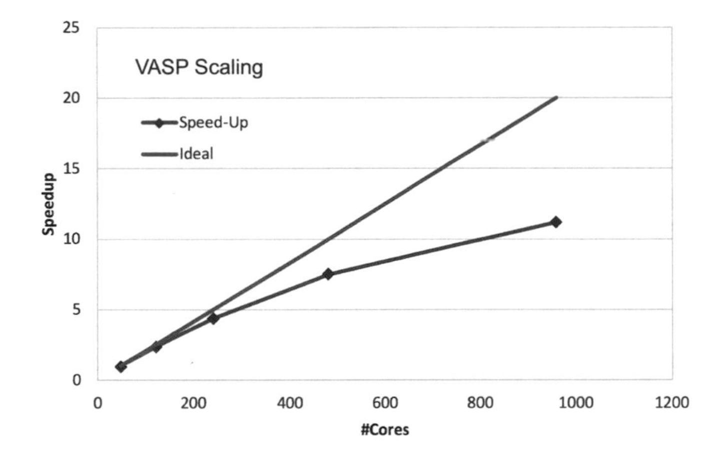

Fig. 6: Scaling behavior of the VASP code when running Ehrenfest dynamics simulations on CPUs at HLRS. This data was obtained for a system composed of 108 atoms, Cut-off energy of 400 eV and 5 K-points.

3.1.3 LAMMPS code

Analytical potential molecular dynamics were carried for large systems comprising a 2D material on a substrate under ion and cluster irradiation using the LAMMPS package [32]. The code has been shown to perform well in several architectures, ranging from x86 clusters to Blue Gene P and Cray supercomputers. LAMMPS has very nice feature called "partition" command which allows it to run replicates under one mpirun. When LAMMPS is run on P processors and this command is not used, LAMMPS runs in one partition, i.e. all P processors run a single simulation. If this command is used, the P processors are split into separate partitions and each partition runs its own simulation. The arguments to the switch specify the number of processors in each partition. Arguments of the form $M \times N$ mean M partitions, each with N processors. We have used a partition command to invoke each impact point to a single processor via the partition command-line switch in the script. The benchmark results are shown in Fig. 7. The figure illustrates the good scalability of ion irradiation simulations for graphene on SiO_2 substrate containing 5101 atoms. The timings given below are for 3000 molecular dynamics steps of each system and 1200 impact points. The LAMPPS simulations were typically run on 512 to 1024 cores.

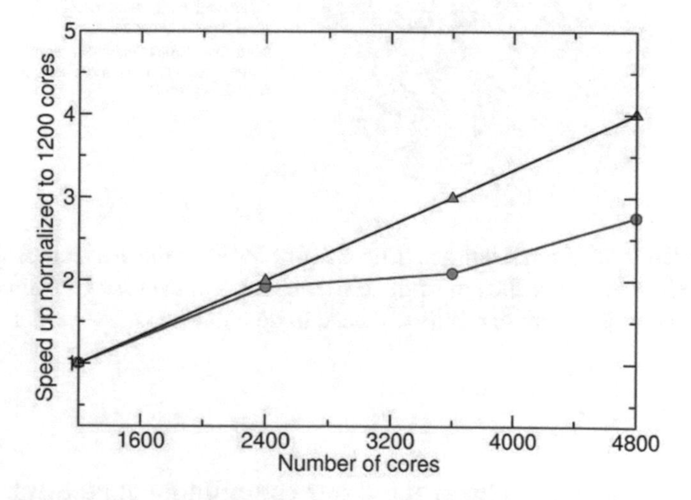

Fig. 7: Scaling behavior of the LAMMPS code when running ion irradiation simulation of graphene deposited on substrate. The speedup is normalized so that at the first data point (1200 processing elements.

3.2 CPU time used

The group has used 28.8 million CPU hours in an extended time frame from March 2020 to April 2021 (total granted CPU time from March 2020 to Feb 2021: 28.2 million CPU hours). The time frame extension was necessary due to delays of some parts of the project which required the collaboration with experimental groups due to the Corona pandemic. Besides, the hiring process of new members of the simulation team took much longer than planned due to delays with getting the visa/work permits. Detailed statistics of the used CPU time down to the level of each sub-project are presented in Fig.8. Note that over one-forth of CPU time is invested in on-going research whose results still need to be published.

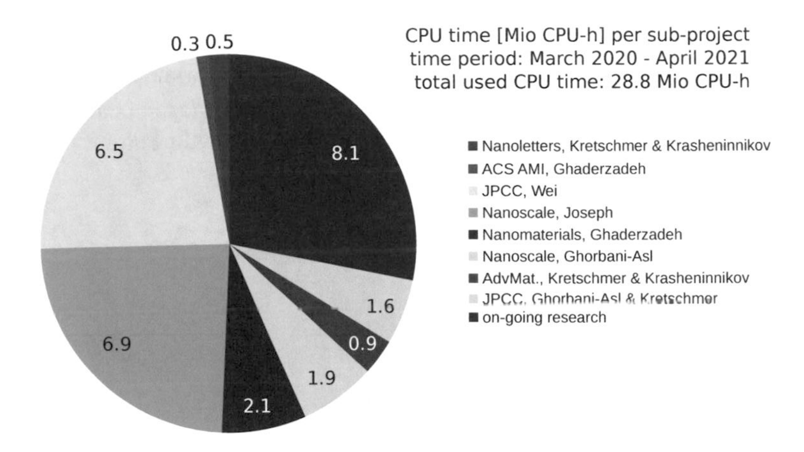

Fig. 8: Detailed CPU time statistics (Mio CPU-h) for the published sub-projects (see Refs. [1–8]) from March 2020 to April 2021. One-forth of the used CPU time is spent on on-going projects whose results still need to be published.

3.3 Popular internet articles about our computational research

Supercomputing Helps Study Two-Dimensional Materials
`https://www.gauss-centre.eu/news/article/supercomputing-helps-study-two-dimensional-materials/`

Natrium statt Lithium: Die Akkus der Zukunft
`https://www.dw.com/de/natrium-statt-lithium-die-akkus-der-zukunft/a-54512116`

Natrium statt Lithium: Die Akkus der Zukunft
https://www.focus.de/perspektiven/nachhaltigkeit/natrium-statt-lithium-natrium-statt-lithium-die-akkus-der-zukunft_id_12306025.html

Simulations Help To Tailor The Properties Of 2D Materials
https://prace-ri.eu/simulations-help-to-tailor-the-properties-of-2d-materials/

Conflict of interest

The authors declare that they have no conflict of interest.

References

1. Silvan Kretschmer, Tibor Lehnert, Ute Kaiser, and Arkady V. Krasheninnikov. Formation of Defects in Two-Dimensional MoS2 in the Transmission Electron Microscope at Electron Energies below the Knock-on Threshold: The Role of Electronic Excitations. *Nano Letters*, 20:2865–2870, 2020.
2. Sadegh Ghaderzadeh, Vladimir Ladygin, Mahdi Ghorbani-Asl, Gregor Hlawacek, Marika Schleberger, and Arkady V Krasheninnikov. Freestanding and Supported MoS 2 Monolayers under Cluster Irradiation: Insights from Molecular Dynamics Simulations. *ACS Applied Materials Interfaces*, 12:37454–37463, 2020.
3. Yidan Wei, Mahdi Ghorbani-Asl, and Arkady V. Krasheninnikov. Tailoring the Electronic and Magnetic Properties of Hematene by Surface Passivation: Insights from First-Principles Calculations. *The Journal of Physical Chemistry C*, 124(41):22784–22792, oct 2020.
4. Janis Köster, Mahdi Ghorbani-Asl, Hannu-pekka Komsa, Tibor Lehnert, Silvan Kretschmer, Arkady V Krasheninnikov, and Ute Kaiser. Defect Agglomeration and Electron-Beam-Induced Local-Phase Transformations in Single-Layer MoTe 2. *The Journal of Physical Chemistry C*, 125:13601–13609, 2021.
5. Jingfeng Li, Thomas Joseph, Mahdi Ghorbani-Asl, Sadhu Kolekar, Arkady V. Krasheninnikov, and Matthias Batzill. Mirror twin boundaries in MoSe2monolayers as one dimensional nanotemplates for selective water adsorption. *Nanoscale*, 13:1038–1047, 2021.
6. Yung-Chang Lin, Jeyakumar Karthikeyan, Yao-Pang Chang, Shisheng Li, Silvan Kretschmer, Hannu-Pekka Komsa, Po-Wen Chiu, Arkady V. Krasheninnikov, and Kazu Suenaga. Formation of Highly Doped Nanostripes in 2D Transition Metal Dichalcogenides via a Dislocation Climb Mechanism. *Advanced Materials*, 33:2007819, 2021.
7. Sadegh Ghaderzadeh, Silvan Kretschmer, Mahdi Ghorbani-Asl, Gregor Hlawacek, and Arkady V Krasheninnikov. Atomistic Simulations of Defect Production in Monolayer and Bulk Hexagonal Boron Nitride under Low- and High-Fluence Ion Irradiation. *Nanomaterials*, 11:1214, 2021.
8. Slawomir Prucnal, Arsalan Hashemi, Mahdi Ghorbani-Asl, René Hübner, Juanmei Duan, Yidan Wei, Divanshu Sharma, Dietrich R T Zahn, René Ziegenrücker, Ulrich Kentsch, Arkady V Krasheninnikov, Manfred Helm, and Shengqiang Zhou. Chlorine doping of MoSe 2 flakes by ion implantation. *Nanoscale*, 13:5834–5846, 2021.
9. E. Rutherford. The scattering of α and β particles by matter and the structure of the atom. *Philos. Mag*, 21(125):669–688, 1911.
10. N Bohr. On the constitution of atoms and molecules. *Philos. Mag*, 26:1–24, 1913.

11. Michael Nastasi, James Mayer, and James K Hirvonen. *Ion-Solid Interactions: Fundamentals and Applications*. Cambridge Solid State Science Series. Cambridge University Press, Cambridge, 1996.
12. Roger Smith, editor. *Atomic and Ion Collisions in Solids and at Surfaces: Theory, Simulation and Applications*. Cambridge University Press, Cambridge, 1997.
13. K S Novoselov, D Jiang, F Schedin, T J Booth, V V Khotkevich, S V Morozov, and A K Geim. Two-dimensional atomic crystals. *Proceedings of the National Academy of Sciences of the United States of America*, 102(30):10451–10453, 2005.
14. J N Coleman, M Lotya, A O'Neill, S D Bergin, P J King, U Khan, K Young, A Gaucher, S De, R J Smith, I V Shvets, s. K Arora, G Stanton, H.-Y. Kim, K Lee, G T Kim, G S Duesberg, T Hallam, J J Boland, J J Wang, J F Donegan, J C Grunlan, G Moriarty, A Shmeliov, R J Nicholls, J M Perkins, E M Grieveson, K Theuwissen, D W McComb, P D Nellist, and V Nicolosi. Two-Dimensional Nanosheets Produced by Liquid Exfoliation of Layered Materials. *Science*, 331:568–571, 2011.
15. B Radisavljevic, A Radenovic, J Brivio, V Giacometti, and A Kis. Single-layer MoS2 transistors. *Nature Nanotechnology*, 6:147–150, 2011.
16. Manish Chhowalla, Hyeon Suk Shin, Goki Eda, Lain-Jong Li, Kian Ping Loh, and Hua Zhang. The chemistry of two-dimensional layered transition metal dichalcogenide nanosheets. *Nature Chemistry*, 5(4):263–275, 2013.
17. G López-Polín, C Gómez-Navarro, V Parente, F Guinea, MI Katsnelson, F Pérez-Murano, and J Gómez-Herrero. Increasing the Elastic Modulus of Graphene by Controlled Defect Creation. *Nat. Phys.*, 11(1):26–31, 2014.
18. U Bangert, W Pierce, D M Kepaptsoglou, Q Ramasse, R Zan, M H Gass, J A den Berg, C B Boothroyd, J Amani, and H Hofsäss. Ion Implantation of Graphene – Toward IC Compatible Technologies. *Nano Lett.*, 13:4902–4907, 2013.
19. R. Nair, M. Sepioni, I-Ling Tsai, O. Lehtinen, J Keinonen, Arkady V. Krasheninnikov, T. Thomson, a. K. Geim, and I. V. Grigorieva. Spin-half paramagnetism in graphene induced by point defects. *Nature Physics*, 8(3):199–202, 2012.
20. A. V. Krasheninnikov and F. Banhart. Engineering of nanostructured carbon materials with electron or ion beams. *Nature Materials*, 6(10):723–733, 2007.
21. A V Krasheninnikov and K Nordlund. Ion and electron irradiation-induced effects in nanostructured materials. *Journal of Applied Physics*, 107(7):071301, 2010.
22. Roland Kozubek, Mukesh Tripathi, Mahdi Ghorbani-Asl, Silvan Kretschmer, Lukas Madauß, Erik Pollmann, Maria O'Brien, Niall McEvoy, Ursula Ludacka, Toma Susi, Georg S. Duesberg, Richard A. Wilhelm, Arkady V. Krasheninnikov, Jani Kotakoski, and Marika Schleberger. Perforating Freestanding Molybdenum Disulfide Monolayers with Highly Charged Ions. *Journal of Physical Chemistry Letters*, 10(5):904–910, 2019.
23. Richard A. Wilhelm, Elisabeth Gruber, Janine Schwestka, Roland Kozubek, Teresa I. Madeira, José P. Marques, Jacek Kobus, Arkady V. Krasheninnikov, Marika Schleberger, and Friedrich Aumayr. Interatomic coulombic decay: The mechanism for rapid deexcitation of hollow atoms. *Physical Review Letters*, 119:103401, 2017.
24. A. V. Krasheninnikov, Y. Miyamoto, and D. Tománek. Role of electronic excitations in ion collisions with carbon nanostructures. *Phys. Rev. Lett.*, 99:016104, 2007.
25. M. Caro, A. A. Correa, E. Artacho, and A. Caro. Stopping power beyond the adiabatic approximation. *Scientific Reports*, 7(1):2618, 2017.
26. M. Ahsan Zeb, J. Kohanoff, D. Sánchez-Portal, A. Arnau, J. I. Juaristi, and Emilio Artacho. Electronic stopping power in gold: The role of d electrons and the H/He anomaly. *Physical Review Letters*, 108(22):225504, 2012.
27. Richard A Wilhelm and Wolfhard Möller. Charge-state-dependent energy loss of slow ions. II. Statistical atom model. *Physical Review A*, 93:052709, 2016.
28. M Fischer, J M Caridad, A Sajid, S Ghaderzadeh, M. Ghorbani-Asl, L Gammelgaard, P Bøggild, K S Thygesen, A V Krasheninnikov, S Xiao, M Wubs, and N Stenger. Controlled generation of luminescent centers in hexagonal boron nitride by irradiation engineering. *Science Advances*, 7:eabe7138, 2021.

29. J Enkovaara, C Rostgaard, J J Mortensen, J Chen, M Dułak, L Ferrighi, J Gavnholt, C Glinsvad, V Haikola, H A Hansen, H H Kristoffersen, M Kuisma, A H Larsen, L Lehtovaara, M Ljungberg, O Lopez-Acevedo, P G Moses, J Ojanen, T Olsen, V Petzold, N A Romero, J Stausholm-Møller, M Strange, G A Tritsaris, M Vanin, M Walter, B Hammer, H Häkkinen, G K H Madsen, R M Nieminen, J K Nørskov, M Puska, T T Rantala, J Schiøtz, K S Thygesen, and K W Jacobsen. Electronic structure calculations with GPAW: a real-space implementation of the projector augmented-wave method. *J. Phys. Condens. Matter*, 22:253202, 2010.
30. G. Kresse and J. Hafner. Ab initio molecular dynamics for liquid metals. *Physical Review B*, 47:558–561, 1993.
31. G. Kresse and J. Furthmüller. Efficiency of ab-initio total energy calculations for metals and semiconductors using a plane-wave basis set. *Computational Materials Science*, 6:15–50, 1996.
32. Steve Plimpton. Fast parallel algorithms for short-range molecular dynamics. *J. Comput. Phys.*, 117:1–19, 1995.

High-performance multiphase-field simulations of solid-state phase transformations using PACE3D

E. Schoof, T. Mittnacht, M. Seiz, P. Hoffrogge, H. Hierl and B. Nestler

Abstract Computational materials science contributes to the accelerated development of new or optimized materials. The phase-field method has established itself as a powerful tool to describe the temporal microstructure evolution during solid-phase transformations. The use of high performance computers allows studying the evolution of large, three-dimensional microstructures incorporating phase- and grain boundary specific behaviors as well as phase transitions. This allows a more realistic representation of the phenomena investigated and lead to more reliable predictions of the microstructural evolution. In this work, current applications of the phase-field method are presented using the PACE3D software package and applied at the ForHLR II supercomputer. Additionally, the scaling behavior is shown when using up to 5041 cores.

1 Introduction

In materials science, understanding the conditions under which certain microstructures are formed during manufacturing is crucial to improving the properties of materials or developing new classes of materials. Complementary to experiments, simulations can help to investigate specific process conditions on the resulting microstructures and thus on the behavior of the material in the application. The phase-field method is an elegant method to describe interface motions numerically, since no explicit interface tracking is necessary [1]. With increasing understanding of this method, it is used for material discovery [2]. Particularly through the possibilities of modern

E. Schoof, T. Mittnacht, M. Seiz, P. Hoffrogge, H. Hierl and B. Nestler
Institute of Applied Materials (IAM), Karlsruhe Institute of Technology (KIT), Straße am Forum 7, 76131 Karlsruhe, Germany

B. Nestler
Institute of Digital Materials, Hochschule Karlsruhe Technik und Wirtschaft, Moltkestr. 30, 76131 Karlsruhe, Germany, e-mail: marco.seiz@kit.edu

W. E. Nagel et al. (eds.), *High Performance Computing in Science and Engineering '21*,
https://doi.org/10.1007/978-3-031-17937-2_10

high-performance computing, realistic processes in the microstructure formation during the production of materials can be investigated [3] and the understanding of critical process parameters can be expanded. The aim of this paper is to present selected results in the simulation of solid-phase transformations made possible by the use of high-performance computers. Three applications are discussed, namely the martensitic transformation, shape-instabilities of finite three-dimensional cementite rods embedded in a ferrite matrix, as well as coarsening and grain growth in solid oxide fuel cells anodes.
This work is structured as follows: In section 2 the basic model of the used multiphase-field framework in presented. The PACE3D framework is described including its performance and scaling behavior in section 3. In section 3 the results of applying the multiphase-field method to simulate solid-state phase transformations and the used cores are discussed.

2 Core phase-field model

The free energy of a system with volume V is generally composed of energy contributions due to interfaces separating physically distinguishable regions and energy contributions present within a region. The latter are referred to as volumetric energy density contributions. In order to parametrize a system, each physically distinguishable region is assigned an order parameter $\phi_\alpha(\mathbf{x}, t)$, which is called phase-field parameter and which is part of the N-tuple $\phi(\mathbf{x}, t)$, where N is the number of all order parameters, t is the time and $\mathbf{x}$ is the location vector. In the multiphase-field context, the free energy can be expressed by

$$\mathcal{F}(\boldsymbol{\phi}, \nabla\boldsymbol{\phi}, \ldots) = \int_V W_{\text{intf}}(\boldsymbol{\phi}, \nabla\boldsymbol{\phi}) + \bar{W}_{\text{bulk}}(\boldsymbol{\phi}, \ldots)\mathrm{d}V. \tag{1}$$

Here, the term $W_{\text{intf}}(\boldsymbol{\phi}, \nabla\boldsymbol{\phi})$ contains the energy contributions of all interfaces and the term $\bar{W}_{\text{bulk}}(\phi, \ldots)$, which can depend on arbitrary quantities like the chemical concentration, contains the volumetric energy contributions. The phase-dependent interpolated volumetric energy density is given by

$$\bar{W}_{\text{bulk}}(\phi, \ldots) = \sum_\alpha h^\alpha(\boldsymbol{\phi}) W^\alpha_{\text{bulk}}(\ldots), \tag{2}$$

whereas $h^\alpha(\boldsymbol{\phi})$ is a normalized interpolation function, which reduces in the simplest case to $h^\alpha(\boldsymbol{\phi}) = \phi_\alpha$ [4]. The effort of a system to minimize the free energy is modeled with a phase-field evolution equation. Two approaches to the formulate the Allen–Cahn evolution equation in the multiphase-field context can be distinguished. Nestler et al. [5] use a Lagrangian parameter λ to ensure the sum constraint $\sum_\alpha \phi_\alpha = 1$:

$$\tau\varepsilon\frac{\partial\phi_\alpha}{\partial t} = -\frac{\delta\mathcal{F}}{\delta\phi_\alpha} - \lambda \tag{3}$$

$$= -\frac{\delta\mathcal{F}}{\delta\phi_\alpha} + \frac{1}{\widetilde{N}}\sum_\beta \frac{\delta\mathcal{F}}{\delta\phi_\beta}, \quad \forall\phi_\alpha, \alpha = 0,\ldots,N. \tag{4}$$

$$\tag{5}$$

Here, $\widetilde{N}$ is the number of locally active phases,

$$\frac{\delta\mathcal{F}(\boldsymbol{\phi},\nabla\boldsymbol{\phi},\ \ldots)}{\delta\phi_\alpha} = \left(\frac{\partial}{\partial\phi_\alpha} - \nabla\cdot\frac{\partial}{\partial\nabla\phi_\alpha}\right)W(\boldsymbol{\phi},\nabla\boldsymbol{\phi},\ \ldots) \tag{6}$$

is the variational derivative of the energy functional $\mathcal{F}$ with respect to the order parameter ϕ_α and the divergence operator is represented by $(\nabla\cdot(\ldots))$. The kinetics of the phase transformations is determined by the relaxation parameter τ and the width of the diffuse interface is set using the parameter ε. The relaxation parameter is weighted locally using the relaxation parameter between two phases $\tau_{\alpha\beta}$ and their volume fractions

$$\tau(\phi) = \frac{\sum_{\alpha<\beta}\tau_{\alpha\beta}\phi_\alpha\phi_\beta}{\sum_{\alpha<\beta}\phi_\alpha\phi_\beta}. \tag{7}$$

Another approach is based on Steinbach [6] and reads

$$\frac{\partial\phi_\alpha}{\partial t} = \frac{1}{\widetilde{N}\varepsilon}\sum_{\beta\neq\alpha}^{\widetilde{N}}\left[M_{\alpha\beta}\left(\Delta W^{\alpha\beta}_{\text{intf}} + \frac{8\sqrt{\phi_\alpha\phi_\beta}}{\pi}\Delta W^{\alpha\beta}_{\text{bulk}}\right)\right], \quad \forall\phi_\alpha, \alpha = 0,\ldots,N, \tag{8}$$

with $\Delta W^{\alpha\beta}_{\text{intf}} = \delta\mathcal{F}_{\text{intf}}/\delta\phi_\beta - \delta\mathcal{F}_{\text{intf}}/\delta\phi_\alpha$ as the contribution the due to curvature minimizing and $\Delta W^{\alpha\beta}_{\text{bulk}} = \delta\mathcal{F}_{\text{bulk}}/\delta\phi_\beta - \delta\mathcal{F}_{\text{bulk}}/\delta\phi_\alpha$ as the driving force due a difference in the volumetric energy densities. Here, the mobility between two phases $M_{\alpha\beta}$ is directly used in the evolution equation. According to Nestler et al. [5], the interfacial energy density

$$W_{\text{intf}}(\boldsymbol{\phi},\nabla\boldsymbol{\phi}) = \varepsilon a(\phi,\nabla\phi) + \frac{1}{\varepsilon}\omega(\phi) \tag{9}$$

consists of the gradient energy density $\varepsilon a(\phi,\nabla\phi)$ and the potential energy density $\omega(\phi)/\varepsilon$ of obstacle type [5, 7]. The interaction of both contributions leads to the formation of a diffuse interface as well as to the representation of the interfacial energy over a volumetric range. This core multiphase-field model can be coupled with additional volumetric energy contribution for variable application scenarios which is detailed in Section 3.

3 The Pace3D framework: Performance and scaling

The models used in this work are implemented in the multiphysics massive parallel Pace3D framework ("Parallel Algorithms for Crystal Evolution in 3D")[8]. To calculate the evolution of the phase-fields and the chemical potentials, the corresponding equations are discretized on a uniform grid with finite differences and an explicit Euler method is employed for the time integration. The mechanical equilibrium condition is solved implicitly to update the stresses and strains using a finite element scheme. Parallelization is integrated employing the Message Passing Interface (MPI) and assigning spatial subdomains to each MPI process (spatial domain decomposition).

Investigations concerning the performance and scaling on the ForHLR II were conducted. Specifically, weak scaling was performed for a baseline solver and an optimized solver employing a thin abstraction layer over single instruction, multiple data (SIMD) intrinsics as well as various other optimizations detailed in [8, 9], e.g. a buffer ensuring that expensive gradients are only calculated once or using the update pattern of the phase-field equation for better vectorization. The domain size as well as the initial and boundary conditions were chosen to be representative of typical simulation domains employed in this work. The results up to 5041 cores are shown in Fig. 3. The plot shows how many lattice updates per second per core were

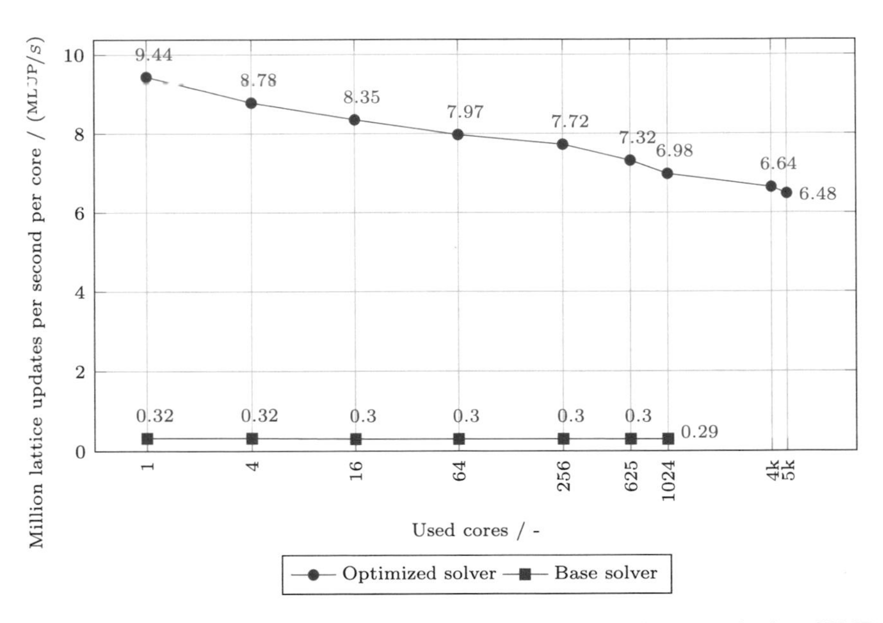

Fig. 1: Weak scaling on the ForHLR II for an optimized solver employing SIMD intrinsics and a baseline solver. While the baseline solver scales almost ideally, its performance makes it unattractive. Compared to this the optimized solver shows reasonable percentages of peak performance with a somewhat worse scalability.

calculated on a whole-application basis, with higher values corresponding to smaller time-to-solution. The optimized version likely became memory-bound within the confines of a single node, as the baseline solver showed almost no drop in MLUP/s whereas the optimized version shows a drop. Scaling beyond one node (starting from 64 cores) for the baseline solver shows almost no drop since the time for a single time step is significantly larger than the time taken for the necessary communication between cores. However, as the optimized solver is about 30 times faster than the baseline solver, it shows a significant efficiency drop as more and more core are added. Strictly speaking there should only be a significant drop between in-node and out-of-node scaling as the network latency is triggered between workers. We suspect the additional drop is due to the network architecture of the ForHLR II since the same scaling studies on the Hazel Hen as well as SuperMUC did not show a significant out-of-node drop in efficiency.

The performance was investigated by counting the floating-point operations executed and correlating these with the runtimes, yielding the floating point operations per second (FLOP/s on a whole-application basis. Comparing the obtained FLOP/s to the values for the Intel Xeon E5-2660 shows that 19 % peak performance was reached for a single core and 14 % for 16 cores.

4 Modeling and simulating solid-state phase transformations using PACE3D

4.1 Martensitic transformation

To precisely control the thermal and mechanical properties of the material, steel is subjected to a defined heat treatment. Depending on the cooling temperature, cooling rate and chemical composition, different microstructures are obtained starting in the (partly) austenitic region. At a rapid cooling rate under the martensite start temperature, martensite is formed from austenite by a diffusionless transformation mechanism which leads to a strengthening of steel.

To model the martensitic transformation in a polycrystalline microstructure, interfacial, chemical and elastic contributions are considered in free energy of the system resulting in

$$\mathcal{F}(\phi, \nabla\phi, \bar{\epsilon}) = \int_V W_{\mathrm{intf}}(\phi, \nabla\phi) + \bar{W}_{\mathrm{elast}}(\phi, \bar{\epsilon}) + \bar{W}_{\mathrm{chem}}(\phi, \mathbf{c}, T)\mathrm{d}V. \tag{10}$$

Here, $\bar{W}_{\mathrm{elast}}$ is the elastic free energy which is given as a function of the phase-field $\boldsymbol{\phi}$ and the elastic strain $\bar{\epsilon}$. The chemical free energy $\bar{W}_{\mathrm{chem}}$ in the context of a constant concentration and temperature depends only on $\boldsymbol{\phi}$. Since the transformation rate of martensite is high, the concentration field is considered to the constant over time. The temporal evolution of the stresses and strains is implicitly calculated based of the assumption of mechanical equilibrium in each discrete time step. The multiphase-field evolution equation. 8 is applied to solve the temporal evolution of the martensitic phases, whereas an additional term is included to ensure a stable interface in the context of high driving forces [10]. For a more detailed description of the underlying models, the reader is referred to [10, 11].

In Fig. 2, results of simulating the martensitic transformation using PACE3D are displayed. Different application scenarios have been investigated, ranging from three-dimensional elastic martensitic transformation in a single-crystal to a two-dimensional elasto-plastic martensitic transformation in a dual-phase microstructure using a phase-dependent plasticity formulation. Starting from a martensitic-free microstructure or a setting with an initial martensitic nucleus within an austenitic grain, martensitic variants nucleate at structural defects like grain boundaries and spread into the austenitic regions. In this process, different variants are formed, which minimize the energy of the system by alternating occurrence of martensitic variants, so that the typical microstructures for lath martensite are formed.

The necessary computational costs depend, among other things, on the number of discretization points, the physical models used, and how far the system is from equilibrium. An overview of typically used cores in different application scenarios can be found in Tab. 4.1. The jump from an elastic to an elasto-plastic simulation reflecting the difference between case IV and case V requires an increase in the necessary resources, since additionally plastic strains have to be calculated. In case IV, the model according to [4] is applied which uses the same plastic fields for all phases. In case V, the models according to [15, 16] are applied which are capable of calculating phase-dependent plastic strains using different nonlinear mechanical constitutive equations for each phase separately. Consequently, the calculation of phase-dependent plastic strains require higher computational costs compared to case IV.

4.2 Shape-instabilities of finite three-dimensional rods

The macroscopic behavior of any metallic material is highly influenced by its microstructural patterns. Rod-shaped patterns, which can be often observed in mesoscopic regions wherein multiple phases are systematically aligned, therefore are of interest in material investigations for many years. To this end, the morphological behavior of three-dimensional finite cementite rods is considered in this work.

To model the diffusion- and curvature-driven shape-instabilities of finite three-dimensional rods, interfacial and chemical contributions are considered in the energy functional of the system resulting in

$$\mathcal{F}(\boldsymbol{\phi}, \nabla\boldsymbol{\phi}, \boldsymbol{c}, T) = \int_V W_{\mathrm{intf}}(\boldsymbol{\phi}, \nabla\boldsymbol{\phi}) + \bar{W}_{\mathrm{chem}}(\phi, \mathbf{c}, T)\mathrm{d}V, \tag{11}$$

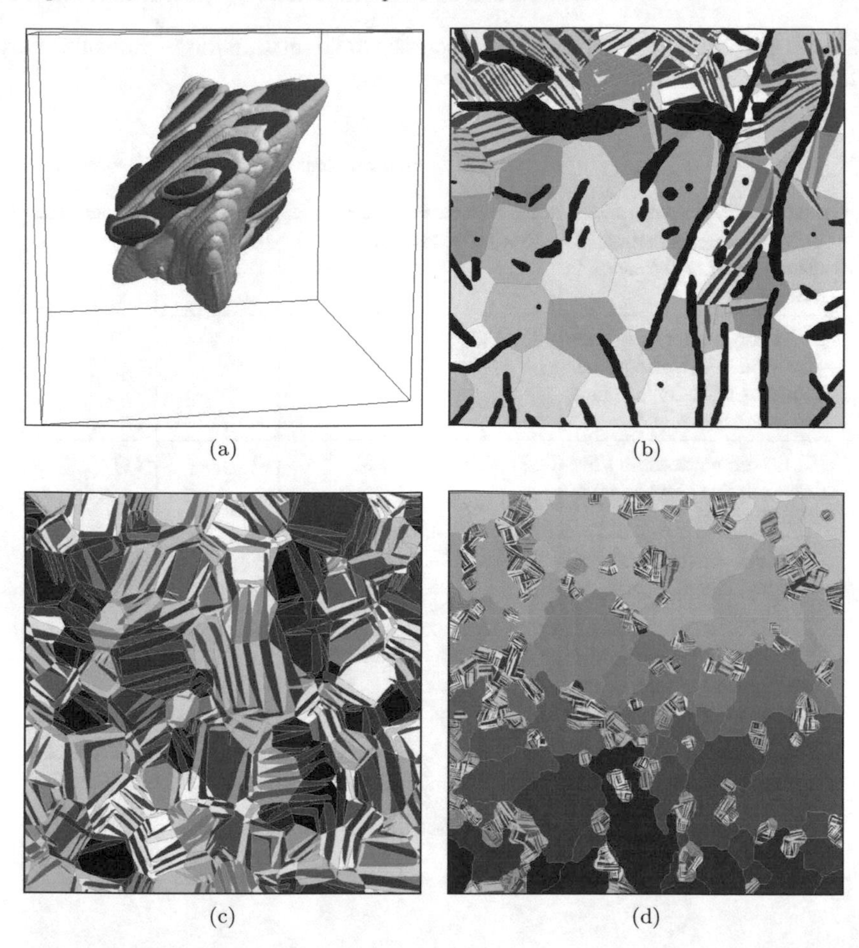

Fig. 2: Resulting microstructures of the martensitic phase transformation using PACE3D. a) Three-dimensional elastic martensitic transformation in a single-crystal [12]. b) Elastic martensitic transformation in a microstructure consisting of austenitic grains and lamellar graphite [13]. Lamellar graphite is shown in black. c) Elastic martensitic transformation in a polycrystalline microstructure [14]. d) Elasto-plastic martensitic transformation in a dual-phase microstructure without phase-dependent plasticity [10]. Ferritic grains are shown in gray scale. In a)-d), Martensitic variants are displayed with colors. In b) and c), austenitic grains are shown in gray scale.

Table 1: Typical used computational resources to simulation the martensitic phase transformation in different application scenarios.

Case	Resolution in voxel cells	Used cores	Duration in hours	Results published in
I: Three-dimensional elastic martensitic transformation in a single-crystal	150 × 150 × 150	16	18	[12]
II: Elastic martensitic transformation in a microstructure consisting of austenitic grains and lamellar graphite	900 × 900	55	48	[13]
III: Elastic martensitic transformation in a polycrystalline microstructure	800 × 800	196	10	[14]
IV: Elastic martensitic transformation in a dual-phase microstructure	4000 × 4000	381	15	[11]
V: Elasto-plastic martensitic transformation in a dual-phase microstructure without phase-dependent plasticity	4000 × 4000	784	38	[10]
VI: Elasto-plastic martensitic transformation in a dual-phase microstructure with phase-dependent plasticity	4000 × 4000	1191	72	[13]

where $\bar{W}_{\text{chem}}$ is dependent of concentration, temperature and the phase-field variable. All the required thermodynamic data leading to diffusion-driven phase evolution are included in that energetic contribution. For an in-depth understanding of the used model, readers are directed to Refs. [5, 17, 18].

Single finite rods have already been analyzed in [19, 20]. To capture the effect of the presence of neighboring rods on the carbon redistribution mechanisms those studies were extended in [21]. Besides an in-depth analysis of the underlying carbon diffusion paths, a critical aspect ratio has been found above which the cementite rods in the lamellar arrangement spheroidize into 2 particles instead of only 1 particle for lower aspect ratios, which is called ovulation. For single isolated rods, the critical aspect ratio is found to be 11, whereas for multiple lamellar arranged rods the critical value alters to 12. Since those previous investigations have been conducted with a fixed interlamellar spacing λ, a variation of that quantity, apart from varying the rods' aspect ratios, is the main object of interest for the present work.

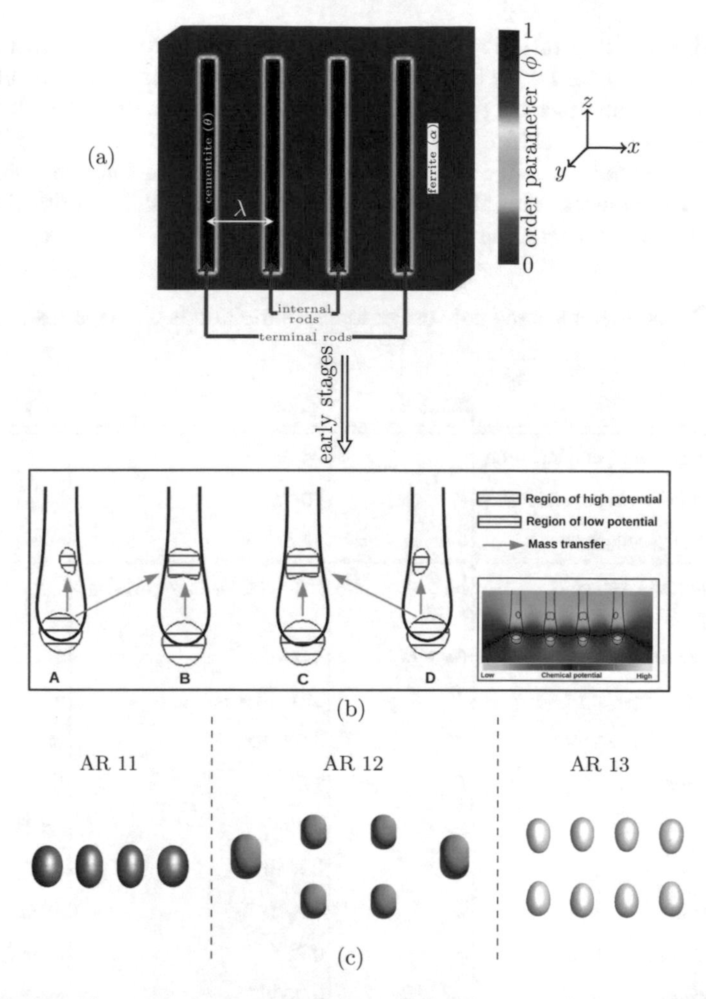

Fig. 3: Temporal morphological evolution. a) Starting simulation setup showing precipitate cementite rods embedded into ferritic matrix in a lamellar arrangement. A cut through the x-z-plane in the middle of the domain is performed to show the longitudinal cross-section of the rods. b) Mass transfer in early stages due to chemical potential fields. c) Final stage morphologies for aspect ratios 11, 12 and 13.

The variation of the interlamellar spacing λ reveals another interesting change in the microstructural evolution, which is shown in Fig. 3c). Hence, apart from the critical aspect ratio obtained in [21], a critical value for the interlamellar spacing is found as well. In this context critical means that there is a particular distance in between the rods in which the shape-instability events of the internal rods highly differ from the ones of the terminal rods. Whilst the internal rods undergo ovulation events, i.e. a break-up of rods, the terminal rods solely experience spheroidization events. This fundamental difference in shape alteration leads to the different amounts of remaining particles for the aspect ratio 12, which represents an intermediate stage between the findings made in [20,21]. Eventually, a small increase of the interlamellar spacing results into a shift of the critical aspect ratio from 12 to 13 for lamellar arranged cementite rods. To conclude, this work figures out an intermediate stage of transformation mechanisms apart from revealing a slightly shifted critical value of the aspect ratio for a variation of the interlamellar spacing.

Table 2: Parameter set and core usage for the simulations of shape-instabilities.

Name	Symbol	Value	Unit
Number of cores per simulation	-	24-36	-
Duration in Hours per simulation	-	70-75	h
Amount of simulations	-	12	-
Resolution in voxel cells	$N_x \times N_y \times N_z$	(130-160) × 45 × (180-200)	-
Discretization spacing	$\Delta x = \Delta y = \Delta z$	1.0×10^{-9}	m
Interlamellar spacing	λ	(30-40) × Δx	m
Interface width parameter	ϵ	3.0 × Δx	m
Temperature	T	973	K
Interfacial Energy	γ_s	0.49	Jm^{-2}[22]
Diffusivity	D	2×10^{-9}	m^2s^{-1}[22]
Molar volume	$\tilde{\Omega}$	7×10^{-6}	m^3/mole
Equilibrium concentration	c_θ^{eq}	0.25	mole fraction
Equilibrium concentration	c_α^{eq}	0.00067	mole fraction

4.3 Coarsening and grain growth in SOFC anodes

Solid oxide fuel cells (SOFCs) are a prominent technology to establish the efficient and environmentally friendly utilization of chemically stored energy. A high durability of fuel cells is key to make this technology also economically feasible by reducing maintenance costs and guaranteeing a long lifetime. One of the most important factors which limits the lifetime of SOFCs is the degradation of the anode. Here, the coarsening of nickel at the relatively high operating temperatures is one of the processes responsible for the degradation [23, Section 2.3]. To gain a deeper understanding of the underlying mechanisms, which govern the coarsening of nickel, we employ multiphase-field simulations to study such process on the micro-scale. An initial polycrystalline microstructure has been artificially generated by a Voronoi diagram, thereby assigning particles statistically uniform to one of the three phases (Nickel, Yttria-stabilized Zirconia (YSZ) and Pores) which results in volume fractions of 34 %, 34 % and 32 % for each of these phases, respectively. Furthermore, the nickel phase is additionally decomposed by a second Voronoi construction using 480 separate order parameters, which results initially in more than 500 nickel grains. The decomposition into grains is done to incorporate the possibility of grain-boundary diffusion and secondly to allow for concomitant thermal grooving, which, albeit subtly, leads to a perturbation of the nickel surface [24,25]. The corresponding initial three-dimensional microstructure is shown in Fig. 4a, which is subject to periodic boundary conditions at either side of the domain.

Please note the clear distinction in the terminology here between *particles* and *grains* of nickel. Whenever the term *particle* is used, we refer to the geometry of the nickel phase as if no information about orientations and GBs was available. Contrarily, a *grain* is a connected region where the orientation does not change, here described as a connected bulk region of a single order parameter. Furthermore, by *coarsening* we mean a relative increase in mean particle size of nickel (again without knowledge about orientations), whereas *grain growth* describes the average increase in grain size due to the annihilation of grains.

We utilize a recently developed extension [25] of the grand-potential model [17,26] which allows to quantitatively render interface diffusion phenomena.

Therefore, the grand-potential functional

$$\mathcal{F}(\boldsymbol{\phi}, \boldsymbol{\nabla\phi}, \boldsymbol{\mu}, T) = \int_V W_{\text{intf}}(\boldsymbol{\phi}, \boldsymbol{\nabla\phi}) + \bar{W}_{\text{chem}}(\phi, \mathbf{c}(\mu), T)\text{d}V, \tag{12}$$

is utilized. It includes an additional dependence on the chemical potentials $\boldsymbol{\mu} = \{\mu_{\text{Ni}}, \mu_{\text{YSZ}}\}$. Note that an isothermal domain is assumed and thus the dependence on temperature T is dropped henceforth. The relaxation parameters are interpolated by Eq. 7. We utilize a set of order parameters $\boldsymbol{\phi} = \{\phi_{\text{Ni}_1}, \ldots, \phi_{\text{Ni}_{480}}, \phi_{\text{YSZ}}, \phi_{\text{Pore}}\}$ to distinguish between a number of nickel grains as well as YSZ and Pores, which are rendered by only a single order parameter each. The chemical contribution to the functional can be written as

$$\begin{aligned} \bar{W}_{\text{chem}}(\phi, \mathbf{c}(\mu)) &= \sum_{\alpha} h_\alpha(\boldsymbol{\phi}) \left[f^\alpha(\mathbf{c}^\alpha(\boldsymbol{\mu})) - \sum_i \mu_i c_i^\alpha(\mu_i) \right] \\ \alpha &\in \{\text{Ni}_1, \ldots, \text{Ni}_{480}, \text{YSZ}, \text{Pore}\}, i \in \{\text{Ni}, \text{YSZ}\}. \end{aligned} \tag{13}$$

Simple parabolic forms of the bulk free energy densities f^α are assumed, which read

$$f^\alpha(\mathbf{c}^\alpha(\boldsymbol{\mu})) = A(c_{\text{Ni}}^\alpha(\mu_{\text{Ni}}) - c_{\text{Ni,eq}}^\alpha)^2 + A(c_{\text{YSZ}}^\alpha(\mu_{\text{YSZ}}) - c_{\text{YSZ,eq}}^\alpha)^2. \tag{14}$$

The flux density which governs the coarsening of nickel is expressed as

$$\mathbf{j}_{\mathrm{Ni}} = -\frac{1}{\epsilon} \sum_{\alpha<\beta} \bar{M}_{\mathrm{Ni}}^{\alpha\beta} \phi_\alpha \phi_\beta \nabla \mu_{\mathrm{Ni}} \tag{15}$$

such that only interface diffusion is considered. The mobilities $\bar{M}_{\mathrm{Ni}}^{\alpha\beta}$ quantify the kinetics of the nickel coarsening. For nickel, only grain-boundary and surface diffusion is assumed, i.e. $\bar{M}_{\mathrm{Ni}}^{\mathrm{YSZ\text{-}Pore}} = \bar{M}_{\mathrm{Ni}}^{\mathrm{Ni}_k\text{-YSZ}} = 0$ for all $k \in \{1, \ldots, 480\}$. Diffusion of YSZ is neglected at operating temperature ($\mathbf{j}_{\mathrm{YSZ}} = 0$), in accordance with experimental findings [27].

The corresponding model parameters are listed in Table 4.3, where the number k in the subscript is omitted from now on since nickel grains share identical properties. We assume equal isotropic interfacial energies between the three phases. The nickel grain-boundaries are assigned a relatively smaller value, the ratio $\gamma_{\mathrm{Ni-Ni}}/\gamma_{\mathrm{Ni\text{-}Pore}}$ is close to the ratio for high-angle GBs determined in a recent and extensive experimental study [24]. The grain-boundary mobility $1/\tau_{\mathrm{Ni\text{-}Ni}}$ is responsible for the rate at which grain growth occurs. Other boundary relaxation parameters are chosen according to [25, p. 12, below Eq. (99)], such that the length scale of attachment kinetics is on the order of the interface width and thus interface diffusion becomes dominant on the length scale of the problem. For the Ni-surface we obtain the characteristic length $l_c = \sqrt{\bar{M}_{\mathrm{Ni}}^{\mathrm{Ni\text{-}Pore}} \pi^2/(32 \cdot 0.8^2)} \approx 0.44 u_l$. The equilibrium compositions $c_{i,\mathrm{eq}}^{\alpha}$ with $i \in \{\mathrm{Ni}, \mathrm{YSZ}\}$ and the thermodynamic prefactor A are so chosen to guarantee sufficient conservation of volume. The parameters are represented in a unit system of the model, where u_l, u_t and u_E are the units of length, time and energy, respectively. The corresponding values in SI-units can be affixed by assuming a surface diffusivity, surface energy, temperature and initial particle size for the system.

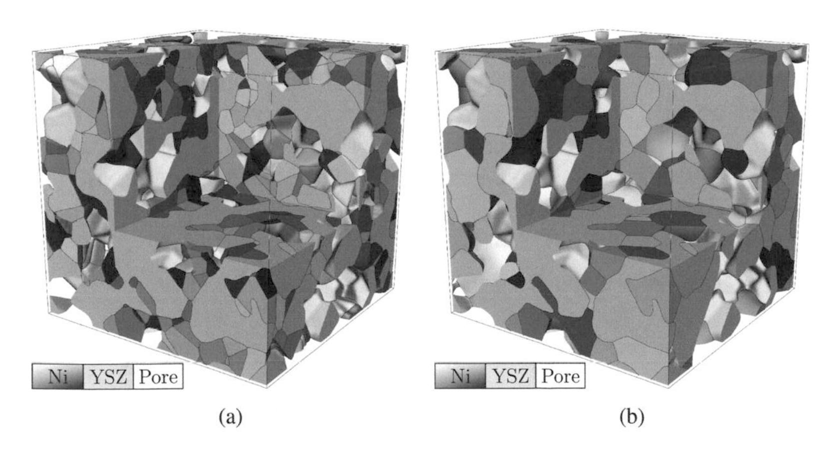

Fig. 4: Initial (a) and final (b) three-dimensional microstructures of the SOFC anode.

Table 3: Parameter set and core usage for the simulation of the SOFC anode.

Name	Symbol	Value	Unit
Number of cores	-	216	-
Duration in Hours	-	144	h
Resolution in voxel cells	$N_x \times N_y \times N_z$	$192 \times 192 \times 192$	-
Discretization spacing	$\Delta x = \Delta y = \Delta z$	1.0	u_l
Interface width parameter	ϵ	4.0	u_l
Relaxation parameter Ni-YSZ	$\tau_{\text{Ni-YSZ}}$	1.0	$u_E u_t / u_l^4$
Relaxation parameter Ni-Pore	$\tau_{\text{Ni-Pore}}$	1.0	$u_E u_t / u_l^4$
Relaxation parameter YSZ-Pore	$\tau_{\text{YSZ-Pore}}$	1.0	$u_E u_t / u_l^4$
Relaxation parameter Ni-GBs	$\tau_{\text{Ni-Ni}}$	100.0	$u_E u_t / u_l^4$
Surface energy Ni	$\gamma_{\text{Ni-Pore}}$	1.0	u_E / u_l^2
Interfacial energy Ni-YSZ	$\gamma_{\text{Ni-YSZ}}$	1.0	u_E / u_l^2
Surface Energy YSZ	$\gamma_{\text{YSZ-Pore}}$	1.0	u_E / u_l^2
Grain boundary energy Ni	$\gamma_{\text{Ni-Ni}}$	0.43	u_E / u_l^2
Equilibrium composition Ni-Ni	$c_{\text{Ni,eq}}^{\text{Ni}}$	0.9	-
Equilibrium composition Pore-Ni	$c_{\text{Ni,eq}}^{\text{Pore}}$	0.1	-
Equilibrium composition YSZ-Ni	$c_{\text{Ni,eq}}^{\text{YSZ}}$	0.1	-
Equilibrium composition Ni-YSZ	$c_{\text{YSZ,eq}}^{\text{Ni}}$	0.1	-
Equilibrium composition Pore-YSZ	$c_{\text{YSZ,eq}}^{\text{Pore}}$	0.1	-
Equilibrium composition YSZ-YSZ	$c_{\text{YSZ,eq}}^{\text{YSZ}}$	0.9	-
Surface diffusion mobility	$\bar{M}_{\text{Ni}}^{\text{Ni-Pore}}$	0.4	$u_l^6 / (u_E u_t)$
GB diffusion mobility	$\bar{M}_{\text{Ni}}^{\text{Ni-Ni}}$	0.4	$u_l^6 / (u_E u_t)$
Thermodynamic prefactor	A	5.0	u_E / u_l^3

During the simulation the microstructure undergoes coarsening of Nickel, which leads to larger and smoother Nickel particles (see Fig. 4b), while the YSZ structure remains invariant. The coarsening of nickel is quantified by means of continuous particle size distributions [28] which were successfully applied in experimental studies on Ni-coarsening in SOFCs [29].

From the individual particle size distributions at different times we calculate the mean particle diameter d_{50} at each time t, which corresponds to the intersection of the distribution with a volume fraction of 50 %. The evolution of the mean particle diameter of nickel is shown in Fig. 7. It is observed, that the particle size increases

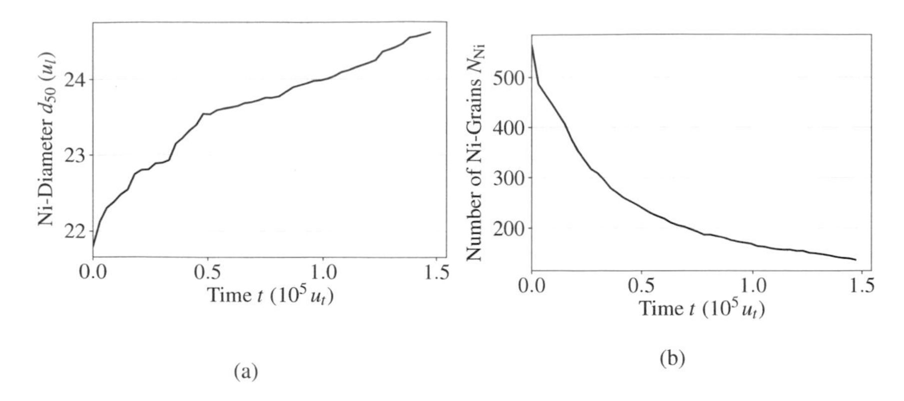

Fig. 5: Evolution of the mean particle diameter of nickel d_{50}, obtained by neglecting the grain structure and calculated by means of continuous particle size distributions [28] (a) and the number of grains N_{Ni} (b) in the SOFC anode.

monotonically with time, while the coarsening rate is largest at early time. The overall increase in particle diameter is about 13 % over the whole simulation time. Inside the nickel particles, the individual grains of the polycrystal are concomitantly growing which causes the number of grains to decay with time (Fig. 5b). From the initial number of 564 grains, only 137 grains persist at the end of the simulation.

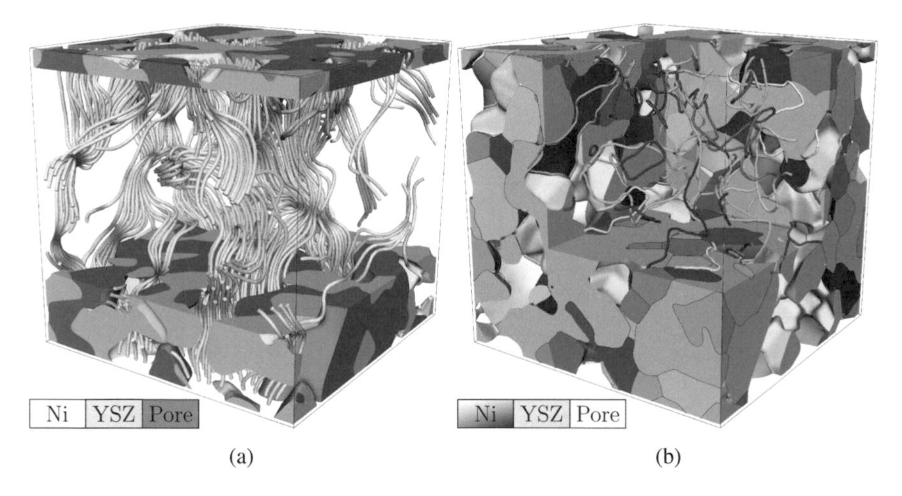

Fig. 6: Tortuous electron pathways inside the nickel structure under an applied voltage difference from top to bottom, where the color of the lines indicates the magnitude of the electric flux density (a) and triple-phase boundaries with each individual segment colored separately (b) in the final state.

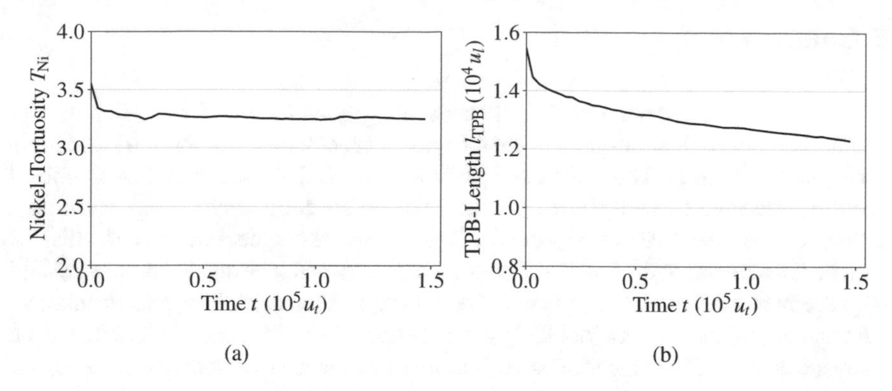

Fig. 7: Evolution of the nickel tortuosity T_{Ni} (a) and triple-phase boundary length l_{TPB} (b) in the SOFC anode.

To gain information about the effective conductivity of the nickel phase, which is relevant for the electron transport, the tortuosity of nickel T_{Ni} has been calculated by applying a voltage difference (Fig. 6a) with respect to two sides of the domain, as in [30]. The tortuosity of the initial microstructure is about $T_{Ni}(t = 0) = 3.5$ and does not change significantly with time (see Fig. 7a), apart from an initial drop, down to a value of about $T_{Ni} \approx 3.25$, which corresponds to a reduction of not more than 8 %. Consequently, the effective conductivity of the nickel network is about 31 % compared to the ideal case of straight transport pathways (corresponding to $T_{Ni} = 1$) which is related to the tortuous nature of the streamlines and in addition to the various bottlenecks where the flux density becomes locally larger in magnitude (see Fig. 6a). On the contrary, the length of the triple-phase boundary (visualized in Fig. 6b) l_{TPB} decreases in a pronounced fashion with time (Fig. 7b). Here, the decrease in l_{TPB} is fastest at early times. Therefore, both the evolution of the particle diameter of Ni as well as the dynamics of TPB length are in agreement with experimental measurements in Ni-YSZ [31] and Ni-CGO anodes [29], at least qualitatively. The overall loss in TPB length, comparing final and initial states of the simulation run, constitutes about 20 %. Since the TPB is the region in the anode at which all of the transport mechanisms (electron, ion and gas transport) are available, a reduced l_{TPB} is unfavorable as it suppresses the oxidization reaction of the fuel gas. Thus, the reduction in l_{TPB} constitutes one of the factors which affect the degradation of the anode material. Please note however, that the assumption of equal interfacial energies (Table 4.3) may not be very accurate. Recent findings [32] show that the interfacial energy between Nickel and YSZ is usually higher than that of the surface energy of YSZ [33]. Together with the surface energies of nickel [34, 35] it is expected that the nickel particles should show dewetting from the YSZ substrate, which is obviously not rendered here. The wetting condition is likely to affect the performance of the anode [36, 37]. Consequently it is probable, that the degradation is underestimated in the current work.

5 Conclusion

In this work, we presented simulation result of applying the multiphase-field model to simulate different solid-phase transformation processes using PACE3D and high-performance facilities. The scaling and performance of a base solver and an optimized version were investigated, showing a factor of 30 speedup after optimization on a single core. However, this performance increase came at a decline in scalability, at least on the network of the ForHLR II, but even at the worst investigated point a 20-fold improvement remained. In the case of modelling martensitic phase transformation, the resulting microstructure including the corresponding stress and strain state can be analyzed whereas the number of used cores increases when more complex models are applied. With regards to the modelling of shape-instabilities, the computational effort in terms of necessary amount of cores also increases slightly with growing aspect ratios and interlamellar spacings. Apart from the computational effort, the simulations of the shape-instabilities for finite cementite rods reveal an intermediate stage for a particular interlamellar spacing in between pure spheroidization or pure ovulation mechanisms in which both modes occur simultaneously. Besides, a shift of the critical value of the aspect ratio is observed when the aforementioned intermediate stage has occurred for the corresponding interlamellar spacing. Furthermore, simulations of an artificially generated three-dimensional and polycrystalline Ni-YSZ SOFC anode are presented. The chosen model is suitable to efficiently delineate herein coarsening by surface and grain-boundary diffusion with concomitant grain growth. The simulations reveal that the Ni coarsening rates in the SOFC anode are fastest at early time and slow down at later times, while, as desired the YSZ-substrate remains invariant. It was shown, that for the assumed parameters, especially equal interfacial energies (wetting angle of 90°), no significant change in the tortuosity of nickel occurs. However, the reduction in triple-phase boundary length, which is also enhanced at early times, might contribute to the degradation of the anode. Further simulations for more realistic wetting angles and experimental microstructures are reserved for a future publication on the coarsening of Nickel in Ni-YSZ SOFC anodes.

Acknowledgements This work was performed on the supercomputer ForHLR II funded by the Ministry of Science, Research and the Arts Baden-Wuerttemberg and by the Federal Ministry of Education and Research. The authors gratefully acknowledge financial support of the parallel code development of the PACE3D package by the Deutsche Forschungsgemeinschaft (DFG) under the grant numbers NE 822/31-1 (Gottfried-Wilhelm Leibniz prize). Furthermore, funding through the coordinated research programme "Virtual Materials Design (VirtMat), project No. 9" and "KNMFi" within the Helmholtz programme "Material System Engineering (MSE), No. 43.31.01" is granted. Research on SOFC was supported by the Federal Ministry for Economic Affairs and Energy (BMWi) within the KerSOLife100 project (Funding No.: 03ET6101A) and the HGF Impuls- und Vernetzungsfonds "Electro-chemical materials for high temperature ion conductors" within the programme "MTET, No. 38.02.01" Part of the work contributes to the research performed at CELEST (Center for Electrochemical Energy Storage Ulm-Karlsruhe) and was funded by the German Research Foundation (DFG) under Project ID 390874152 (POLiS Cluster of Excellence).

References

1. N. Moelans, B. Blanpain, and P. Wollants. An introduction to phase-field modeling of microstructure evolution. *Calphad*, 32(2):268–294, June 2008.
2. M.R. Tonks and L.K. Aagesen. The phase field method: mesoscale simulation aiding material discovery. *Annual Review of Materials Research*, 49:79–102, 2019.
3. J. Hötzer, M. Jainta, P. Steinmetz, B. Nestler, A. Dennstedt, A. Genau, M. Bauer, H. Köstler, and U. Rüde. Large scale phase-field simulations of directional ternary eutectic solidification. 93(0):194 – 204, 2015.
4. D. Schneider, E. Schoof, O. Tschukin, A. Reiter, C. Herrmann, F. Schwab, M. Selzer, and B. Nestler. Small strain multiphase-field model accounting for configurational forces and mechanical jump conditions. *Computational Mechanics*, 61(3):277–295, 2018.
5. B. Nestler, H. Garcke, and B. Stinner. Multicomponent alloy solidification: Phase-field modeling and simulations. *Phys. Rev. E 71, 041609*, 2005.
6. I. Steinbach. Phase-field models in materials science. *Modelling and simulation in materials science and engineering*, 17(7):073001, 2009.
7. J. Hötzer, O. Tschukin, M. Ben Said, M. Berghoff, M. Jainta, G. Barthelemy, N. Smorchkov, D. Schneider, M. Selzer, and B. Nestler. Calibration of a multi-phase field model with quantitative angle measurement. *Journal of materials science*, 51(4):1788–1797, 2016.
8. J. Hötzer, A. Reiter, H. Hierl, P. Steinmetz, M. Selzer, and B. Nestler. The parallel multi-physics phase-field framework Pace3D. *Journal of Computational Science*, 26:1–12, 2018.
9. J. Hötzer. *Massiv-parallele und großskalige Phasenfeldsimulationen zur Untersuchung der Mikrostrukturentwicklung*. PhD thesis, 2017.
10. E. Schoof, C. Herrmann, N. Streichhan, M. Selzer, D. Schneider, and B. Nestler. On the multiphase-field modeling of martensitic phase transformation in dual-phase steel using J_2-viscoplasticity. *Modelling and Simulation in Materials Science and Engineering*, 27(2):025010, 2019.
11. E. Schoof, D. Schneider, N. Streichhan, T. Mittnacht, M. Selzer, and B. Nestler. Multiphase-field modeling of martensitic phase transformation in a dual-phase microstructure. *International Journal of Solids and Structures*, 134:181–194, 2018.
12. E. Schoof. *Chemomechanische Modellierung der Wärmebehandlung von Stählen mit der Phasenfeldmethode*. PhD thesis, Karlsruher Institut für Technologie (KIT), 2020.
13. E. Schoof, C. Herrmann, D. Schneider, J. Hötzer, and B. Nestler. Multiphase-field modeling and simulation of martensitic phase transformation in heterogeneous materials. In *High Performance Computing in Science and Engineering '18*, pages 475–488. Springer, 2019.
14. P. G. Kubendran Amos, E. Schoof, N. Streichan, D. Schneider, and B. Nestler. Phase-field analysis of quenching and partitioning in a polycrystalline fe-c system under constrained-carbon equilibrium condition. *Computational Materials Science*, 159:281–296, 2019.
15. D. Schneider, F. Schwab, E. Schoof, A. Reiter, C. Herrmann, M. Selzer, T. Böhlke, and B. Nestler. On the stress calculation within phase-field approaches: a model for finite deformations. *Computational Mechanics*, 60(2):203–217, 2017.
16. C. Herrmann, E. Schoof, D. Schneider, F. Schwab, A. Reiter, M. Selzer, and B. Nestler. Multiphase-field model of small strain elasto-plasticity according to the mechanical jump conditions. *Computational Mechanics*, 62(6):1399–1412, 2018.
17. A. Choudhury and B. Nestler. Grand-potential formulation for multicomponent phase transformations combined with thin-interface asymptotics of the double-obstacle potential. *Phys. Rev. E*, 85, 2012.
18. T. Mittnacht, P.G. Kubendran Amos, D. Schneider, and B. Nestler. Morphological stability of three-dimensional cementite rods in polycrystalline system: A phase-field analysis. *Journal of Materials Science and Technology*, 77:252–268, 2021.
19. F.A. Nichols. On the spheroidization of rod-shaped particles of finite length. *Journal of Materials Science*, 11(6):1077–1082, 1976.
20. P.G. Kubendran Amos, L.T. Mushongera, Tobias Mittnacht, and Britta Nestler. Phase-field analysis of volume-diffusion controlled shape-instabilities in metallic systems-II: Finite 3-dimensional rods. *Computational Materials Science*, 144:374–385, mar 2018.

21. T. Mittnacht, P.G. Kubendran Amos, D. Schneider, and B. Nestler. Understanding the influence of neighbours on the spheroidization of finite 3-dimensional rods in a lamellar arrangement: Insights from phase-field simulations. In *Numerical Modelling in Engineering*, pages 290–299. Springer, 2018.
22. K. Ankit, A. Choudhury, C. Qin, S. Schulz, M. McDaniel, and B. Nestler. Theoretical and numerical study of lamellar eutectoid growth influenced by volume diffusion. *Acta Materialia*, 2013.
23. San Ping Jiang and Siew Hwa Chan. A review of anode materials development in solid oxide fuel cells. *Journal of Materials Science*, 39(14):4405–4439, 2004.
24. P. Haremski, L. Epple, M. Wieler, P. Lupetin, R. Thelen, and M.J. Hoffmann. A Thermal Grooving Study of Relative Grain Boundary Energies of Nickel in Polycrystalline Ni and in a Ni/YSZ Anode Measured by Atomic Force Microscopy. *SSRN Electronic Journal*, 214, 2020.
25. P.W. Hoffrogge, A. Mukherjee, E.S. Nani, P.G. Kubendran Amos, F. Wang, D. Schneider, and B. Nestler. Multiphase-field model for surface diffusion and attachment kinetics in the grand-potential framework. *Physical Review E*, 103(3):033307, mar 2021.
26. M. Plapp. Unified derivation of phase-field models for alloy solidification from a grand-potential functional. *Phys. Rev. E*, 84:031601, Sep 2011.
27. M. Trini, P.S. Jørgensen, A. Hauch, J.J. Bentzen, P.V. Hendriksen, and M. Chen. 3D Microstructural Characterization of Ni/YSZ Electrodes Exposed to 1 Year of Electrolysis Testing. *Journal of The Electrochemical Society*, 166(2):F158–F167, feb 2019.
28. B. Münch and L. Holzer. Contradicting geometrical concepts in pore size analysis attained with electron microscopy and mercury intrusion. *Journal of the American Ceramic Society*, 91(12):4059–4067, 2008.
29. L. Holzer, B. Iwanschitz, T. Hocker, B. Münch, M. Prestat, D. Wiedenmann, U. Vogt, P. Holtappels, J. Sfeir, A. Mai, and T. Graule. Microstructure degradation of cermet anodes for solid oxide fuel cells: Quantification of nickel grain growth in dry and in humid atmospheres. *Journal of Power Sources*, 196(3):1279–1294, 2011.
30. J. Joos, T. Carraro, A. Weber, and E. Ivers-Tiffée. Reconstruction of porous electrodes by FIB/SEM for detailed microstructure modeling. *Journal of Power Sources*, 196(17):7302–7307, 2011.
31. A. Faes, A. Hessler-Wyser, D. Presvytes, C. G. Vayenas, and J. Van herle. Nickel-Zirconia Anode Degradation and Triple Phase Boundary Quantification from Microstructural Analysis. *Fuel Cells*, 9(6):841–851, dec 2009.
32. H. Nahor, H. Meltzman, and W.D. Kaplan. Ni–YSZ(111) solid–solid interfacial energy. *Journal of Materials Science*, 49(11):3943–3950, jun 2014.
33. A. Tsoga and P. Nikolopoulos. Surface and grain-boundary energies in yttria-stabilized zirconia (YSZ-8 mol%). *Journal of Materials Science*, 31(20):5409–5413, 1996.
34. R. Tran, Z. Xu, B. Radhakrishnan, D. Winston, W. Sun, K.A. Persson, and S.P. Ong. Surface energies of elemental crystals. *Scientific Data*, 3(1):160080, dec 2016.
35. M. Kappeler, A. Marusczyk, and B. Ziebarth. Simulation of nickel surfaces using ab-initio and empirical methods. *Materialia*, 12(March):100675, aug 2020.
36. R. Davis, F. Abdeljawad, J. Lillibridge, and M. Haataja. Phase wettability and microstructural evolution in solid oxide fuel cell anode materials. *Acta Materialia*, 78:271–281, 2014.
37. Z. Jiao and N. Shikazono. Prediction of Nickel Morphological Evolution in Composite Solid Oxide Fuel Cell Anode Using Modified Phase Field Model. *Journal of The Electrochemical Society*, 165(2):F55–F63, 2018.

Bridging scales with volume coupling — Scalable simulations of muscle contraction and electromyography

Benjamin Maier, David Schneider, Miriam Schulte and Benjamin Uekermann

Abstract Measuring the electric signals of a contracting muscle, called electromyography (EMG), is a valuable experimental tool for biomechanics researchers. Detailed simulations of this process can deliver insights on a new level. However, the solution of appropriate biophysically based multi-scale models exhibits high computational loads, which demands for High Performance Computing. At the same time, separate parallelization strategies are required for different spatial scales. Currently, no open-source software is capable of efficiently exploiting supercomputers and handling the complexity of relevant state-of-the-art multi-scale muscle models. We employ a volume coupling scheme to augment the open-source software OpenDiHu and enable coupled simulations of muscle contraction and EMG. We investigate the weak scaling behavior of different components of the multi-scale solver and find that the new approach does not hinder scalability. As a result, we are able to simulate a respective highly resolved scenario with 100 million degrees of freedom on the supercomputer Hawk at the High Performance Computing Center Stuttgart.

1 Introduction

The musculoskeletal system enables humans to perform a variety of vital tasks in everyday life. The capabilities of skeletal muscles satisfy versatile requirements, ranging from forceful to finely controlled actions and from fast and impulsive ones to long, enduring tasks. The functioning of this complex system is governed by the interplay of processes on several scales, including molecular, biochemical processes

Benjamin Maier, David Schneider, Miriam Schulte and Benjamin Uekermann
Institute for Parallel and Distributed Systems, Universität Stuttgart, Universitätsstraße 38, 70569 Stuttgart, Germany,
e-mail: benjamin.maier@ipvs.uni-stuttgart.de,
david.schneider@ipvs.uni-stuttgart.de,
miriam.schulte@ipvs.uni-stuttgart.de,
benjamin.uekermann@ipvs.uni-stuttgart.de

W. E. Nagel et al. (eds.), *High Performance Computing in Science and Engineering '21*,
https://doi.org/10.1007/978-3-031-17937-2_11

at the muscle fiber membranes, macroscopic electrical and mechanical properties on the organ scale, and the neural pathway connecting numerous motor neurons in the spinal cord with hundreds of thousands of muscle fibers in a skeletal muscle.

Detailed, biophysically-based simulations of the neuromuscular system can foster physiological understanding, provide an in-silico laboratory to complement the limited in-vivo studies, and ultimately help to diagnose and treat neuromuscular disorders. For this purpose, we develop simulation software for state-of-the-art multi-scale models of the neuromuscular system. As such detailed models with high temporal and spatial resolutions pose high computational loads, we focus on efficient parallelization and coupling schemes for these multi-physics models in order to exploit the computational power of compute clusters and supercomputers.

1.1 Related work

Relevant multi-scale approaches have been proposed in the literature and combine models of electrophysiology on the one hand, i.e., propagation of electric stimuli along muscle fibers activating the muscle, and continuum mechanics models of muscle contraction on the other hand [11, 13, 22, 23]. They have been implemented using computational software frameworks such as Chaste [19] and OpenCMISS [6].

We base our computations on the multi-scale chemo-electro-mechanical model formulated by the authors of [11, 20, 23] and initially implemented in OpenCMISS. In previous work, we enhanced the software and investigated domain decomposition strategies, which allowed a highly parallel execution of the electrophysiology part of the model [7]. In our new codebase OpenDiHu, we developed tailored solution schemes for the overall model exploiting instruction-level and distributed memory parallelism [15, 17, 18]. OpenDiHu is capable of simulating the multi-scale electrophysiology model with a realistic number of 270,000 muscle fibers for the biceps brachii muscle.

1.2 Research targets

The measurement of electric signals on the skin surface over a contracting muscle, called electromyography (EMG), is an important, non-invasive window to gain insights into the muscle's functioning. However, besides our OpenDiHu code, no open-source software currently exist that are capable of simulating both EMG and muscle contraction using detailed biophysical models. In this work, we present our numeric and algorithmic setup to enable this simulation and demonstrate that our software OpenDiHu is able to conduct the respective detailed simulations by exploiting compute power of the supercomputer Hawk at the High Performance Computing Center Stuttgart.

In such simulations, models of two different scales use discretizations of the three-dimensional (3D) muscle domain: First, the electrophysiological descriptions rely on highly-resolved 3D meshes to accurately represent the finely dosed muscle activation and high-density EMG signals. Second, the solver of the nonlinear solid mechanics formulation usually uses a coarser 3D mesh.

In parallel execution using a domain decomposition approach, the possible number of subdomains in this setup is limited by the small number of elements in the coarser mechanics mesh. Usually, identical partitionings for the fine and coarse 3D meshes are beneficial as the two corresponding models are strongly linked and need to frequently exchange large amounts of data. However, this limits the possible degree of parallelism and prevents High Performance Computing. Therefore, we want to use different partitionings and equip our solid mechanics solver with a coarse mesh and the multi-scale electrophysiology solver with a fine mesh. The high-performance numerical coupling library preCICE [8] is used to glue both partitionings together into a single simulation setup.

1.3 Main contributions of this work

In this work, we evaluate the weak scaling performance of our software OpenDiHu in highly resolved multi-scale simulations of muscular EMG. Further, we present a scheme to bridge the scales between a detailed multi-scale electrophysiology model and an organ-level muscle contraction model, which is also applicable in the area of High Performance Computing. This scheme enables simulations of EMG measurements on a contracting muscle with an unprecedented level of detail.

The remainder of this paper is structured as follows. Section 2 presents the used models and their discretization and solution schemes. Section 2 describes our developed partitioning method to enable the full model for High Performance Computing. Section 3 presents performance results and Sect. 4 concludes this work.

2 Models and solvers

We describe the multi-scale model using the illustration in fig. 1. Figure 1 (a) shows the hierarchical structure of a skeletal muscle. A tendon connects the skeleton and the muscle belly, which consists of dozens of fascicles. Each fascicle contains tens of thousands of muscle fibers and, in each muscle fiber, strings of sarcomeres form the molecular motor, which generates the muscle force.

Figure 1 (b) shows the corresponding representation in the discretized domain, using a 3D finite element mesh (green color) for the macroscopic muscle, numerous embedded 1D meshes (red color) for the muscle fibers, and (0D) points on every fiber (yellow color) to describe the sarcomeres.

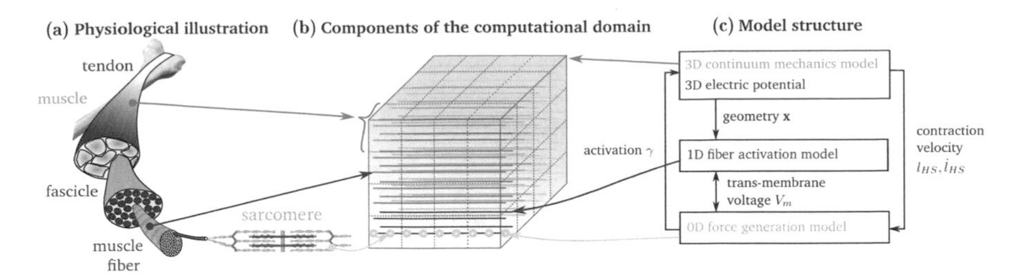

Fig. 1: Components of the implemented multi-scale model and its discretization, comprising the muscle belly, muscle fibers and sarcomeres. The schematic in (c) illustrates the coupled quantities between the four model parts given by different colors.

Figure 1 (c) summarizes the four parts of the multi-scale model and the coupled quantities: (i) the 3D continuum mechanics model (green), (ii) the 3D model of the electric potential (blue color) responsible for the EMG, (iii) numerous instances of the 1D model of electric activation on the muscle fibers (red color), and (iv) numerous instances of the 0D force generation model of the sarcomere (yellow color). The 0D and 1D models are bidirectionally coupled by the trans-membrane voltage V_m, the 3D continuum mechanics model influences all other models by deforming the computational domains, and, furthermore, the 0D model is bidirectionally coupled to the 3D model over the activation γ and the contraction velocity $\dot{l}_{HS}$. In the following sections, we formulate the model equations for the four outlined components of the multi-scale model.

2.1 Continuum mechanics model (3D)

Contraction and deformation of the muscle domain Ω_M are described by a dynamic, incompressible, nonlinear solid mechanics formulation. The quantities of interest are the displacements $\mathbf{u}$, velocities $\mathbf{v} = \dot{\mathbf{u}}$, and stresses in terms of the second Piola–Kirchhoff tensor $\mathbf{S}$. The underlying equations are given by the following balance principles:

$$\text{grad}_{\mathbf{X}}(\mathbf{P}) + \mathbf{B} = \rho_0\,\dot{\mathbf{v}} \quad \text{in } \Omega_M \quad \text{(balance of linear momentum)}, \tag{1}$$

$$\mathbf{S} = \mathbf{S}^\top \quad \text{in } \Omega_M \quad \text{(balance of angular momentum)}, \tag{2}$$

$$\text{div}_{\mathbf{X}}(\mathbf{v}) = 0 \quad \text{in } \Omega_M \quad \text{(incompressibility)}. \tag{3}$$

Here, eq. (1) is the balance of linear momentum with body forces $\mathbf{B}$ in reference configuration and the first Piola–Kirchhoff stress tensor $\mathbf{P} = \mathbf{F}\,\mathbf{S}$ with the deformation gradient $\mathbf{F}$ with respect to referential coordinates $\mathbf{X}$. The symmetry of the stress tensor $\mathbf{S}$ in eq. (2) follows from conservation of angular momentum. Equation (3) follows from conservation of mass assuming a constant density ρ_0 throughout the muscle domain. We additionally specify Dirichlet and Neumann boundary conditions to fix certain displacements and prescribe surface traction forces, respectively.

The constitutive relation between stresses and strains is based on the work of Heidlauf et al. [12]. The model uses a hyperelastic, transversely-isotropic material with an additional active stress term modeling force production in the sarcomeres, which we omit here for the sake of brevity. This term depends on a spatially varying muscle activation level $\gamma \in [0, 1]$, which is computed in the 0D model and then homogenized to the 3D domain. The resulting displacements of the mechanics model deform the 3D domain and the embedded 1D fiber domains such that all other models are influenced by this model of muscle contraction.

2.2 Model of the electric potential (3D)

In the 3D muscle domain Ω_M, extracellular and intracellular spaces are assumed to coexist at every point. Electric conduction in the extracellular space is described by the bidomain equation,

$$\operatorname{div}\big((\boldsymbol{\sigma}_e + \boldsymbol{\sigma}_i)\operatorname{grad}(\phi_e)\big) + \operatorname{div}\big(\boldsymbol{\sigma}_i \operatorname{grad}(V_m)\big) = 0 \quad \text{in } \Omega_M. \tag{4}$$

Here, $\boldsymbol{\sigma}_i$ and $\boldsymbol{\sigma}_e$ are the conductivity tensors in the intra and extracellular spaces, ϕ_e is the electric potential in the extracellular space, and V_m is the trans-membrane voltage, measured between the electric potentials of intra and extracellular spaces. If V_m is known at every point, we can compute ϕ_e by solving the given equation. The resulting value of ϕ_e corresponds to the electric signals measured during intramuscular EMG.

2.3 Model of muscle fiber activation (1D)

The muscle fibers are frequently stimulated by the nervous system at points around their centers. Upon stimulation of a fiber Ω_f^j, an electric spike, called action potential, propagates towards both ends of the fiber. This is described by the monodomain equation, which is a 1D diffusion-reaction equation:

$$\frac{\partial V_m}{\partial t} = \frac{1}{A_m^j\, C_m^j}\left(\sigma_{\text{eff}}\frac{\partial^2 V_m}{\partial s^2} - A_m^j\, I_{\text{ion}}(V_m, \mathbf{y})\right) \quad \text{on } \Omega_f^j. \tag{5}$$

Here, V_m is again the trans-membrane voltage, A_m^j and C_m^j are the surface-to-volume-ratio and the electric capacitance of the membrane of fiber j, respectively, σ_{eff} is the effective conductivity, s is the spatial coordinate along the fiber, and I_{ion} is the current over the membrane. I_{ion} acts as the reaction term in this equation and is computed by the 0D model.

After solving this equation for V_m, we prolong the V_m values from the 1D fiber domains Ω_f^j to the 3D muscle domain Ω_M using tri-linear interpolation and can subsequently compute the model of the 3D electric potential given in eq. (4).

2.4 Subcellular force generation model (0D)

On spatial "0D" points on every muscle fiber Ω_f^j, we solve the dynamics of opening and closing ion channels in the fiber membranes, intracellular calcium dynamics and cross-bridge cycling leading to force production. These dynamic processes are described by a system of differential-algebraic equations:

$$\frac{\partial \mathbf{y}}{\partial t} = G(V_m, \mathbf{y}), \quad I_{\text{ion}} = I_{\text{ion}}(V_m, \mathbf{y}), \quad \gamma = H(\mathbf{y}, \dot{\lambda}_f) \quad \text{on } \Omega_s^i \tag{6}$$

Here, $\mathbf{y}$ is the vector of evolving internal states, from which quantities such as the ionic current I_{ion} and the muscle activation γ can be derived. I_{ion} and γ additionally depend on the trans-membrane voltage V_m given by the 1D model in eq. (5) and the contraction velocity $\dot{\lambda}_f$ in fiber direction computed by the 3D continuum mechanics model, eqs. (1) to (3).

Models with different degrees of detail exist, e.g., the model of membrane potential depolarization by Hodgkin and Huxley [14] with $\mathbf{y} \in \mathbb{R}^3$ or the model proposed by Shorten et al. [24] with $\mathbf{y} \subset \mathbb{R}^{56}$. Subcellular models can be conveniently stored in CellML format [9], an XML-based description language, and can directly be loaded and solved in OpenDiHu. Such CellML models are exchanged among bioengineering researchers via an open-source online model repository[1] established by the Physiome project [21], which hosts a multitude of curated models.

2.5 Discretization and solution

All spatial derivatives in the presented models are discretized using the finite element method and the meshes introduced in fig. 1. The 3D mechanics model uses Taylor-Hood elements, where the displacements $\mathbf{u}$ and velocities $\mathbf{v}$ are discretized using quadratic ansatz functions and a Lagrange multiplier p that enforces incompressibility, identified as the hydrostatic pressure, is discretized using linear ansatz functions. The transmembrane voltage V_m on the 1D fiber domains is also described using linear ansatz functions. The nodes of the 1D meshes are the locations where the 0D model instances are solved. Thus, we need to solve as many 0D model instances as there are fibers multiplied by the number of nodes in each 1D fiber mesh.

[1] https://www.cellml.org/

We solve the overall multi-scale model by a subcycling scheme. After solving one timestep of the two 3D models, we proceed to solve several smaller timesteps of the 0D and 1D models using the Strang operator splitting [25], before we continue to again solve the 3D models. The 3D mechanics model is discretized in time using the implicit Euler method, the 1D model is solved using the Crank–Nicolson method, and, for the 0D model, we use Heun's method. More details can be found in [15, 18].

We perform all computations within our open-source software framework OpenDiHu ("Digital Human Model") [1], where modular time stepping schemes and solvers for the individual parts of the multi-scale model can be combined. OpenDiHu interfaces the numeric solver and preconditioner libraries PETSc [3–5], MUMPS [2], and HYPRE [10]. Tailored optimizations are implemented for the 0D and 1D models, e.g., to explicitly exploit vector instructions using the library 'Vc' [16] or to communicate the data of every 1D model to a single core and then use an efficient linear-complexity Thomas algorithm to solve the respective linear system of equations.

3 Partitioning methods

As mentioned in Sect. 1.2, we first partition the meshes on all domains in an identical way, such that each process owns a distinct part of the overall geometry. Figure 2 (a) and (b) illustrate this concept, showing the same domain decomposition into 16 subdomains for both the 3D mesh (fig. 2 (a)) and the 1D fiber meshes (fig. 2 (b)). While this partitioning approach prevents communication during data mapping between the coupled model parts, it can hinder parallelization. In fact, in fig. 2 (a), the 3D mesh consists of 16 elements with quadratic ansatz functions, as used by the solid mechanics model, and, consequently, a partitioning with more than 16 subdomains is not possible. However, for the computationally intense fiber models we would like to use a higher degree of parallelism, e.g., a partitioning with 1024 processes as shown in fig. 2 (c).

To resolve this issue, we instead split our multi-scale solver implementation in OpenDiHu into two separate programs. The first program solves the 0D and 1D model parts and the 3D electric potential model part, which can be summarized as the multi-scale electrophysiology model. The second program only solves the 3D solid mechanics part. Both programs are coupled using the library preCICE and can, thus, make use of individual domain decompositions.

Figure 3 presents the resulting coupled architecture. The yellow block on the left-hand side represents the first simulation program, which solves the 3D model of the electric potential and the EMG and the 0D/1D fiber models. The program uses finely resolved 3D and 1D fiber meshes that are partitioned identically, here visualized for 16 processes. The data mapping and numerical coupling between the 0D, 1D, and 3D models are performed entirely within OpenDiHu. The right yellow block in fig. 3

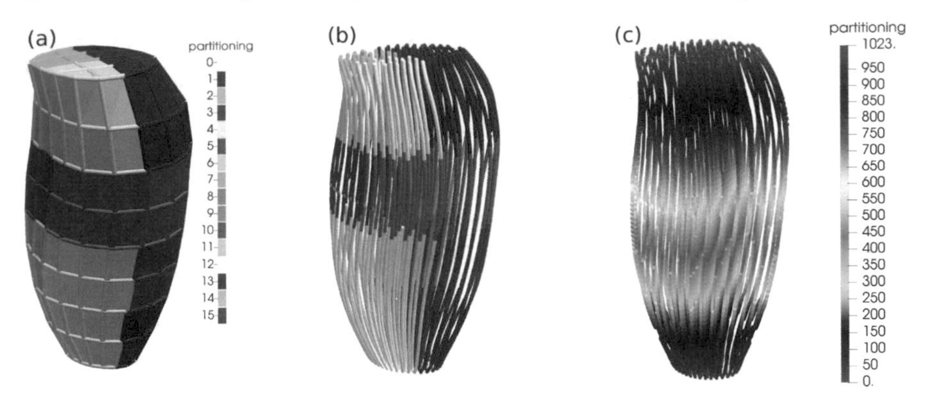

Fig. 2: Partitioning schemes of muscle and fiber domains. (a) Domain decomposition of a 3D mesh for 16 processes, given by the different colors. (b) The identical domain decomposition applied to the 1D meshes of 81 fibers. (c) A different partitioning for the same number of fibers, but for 1024 processes.

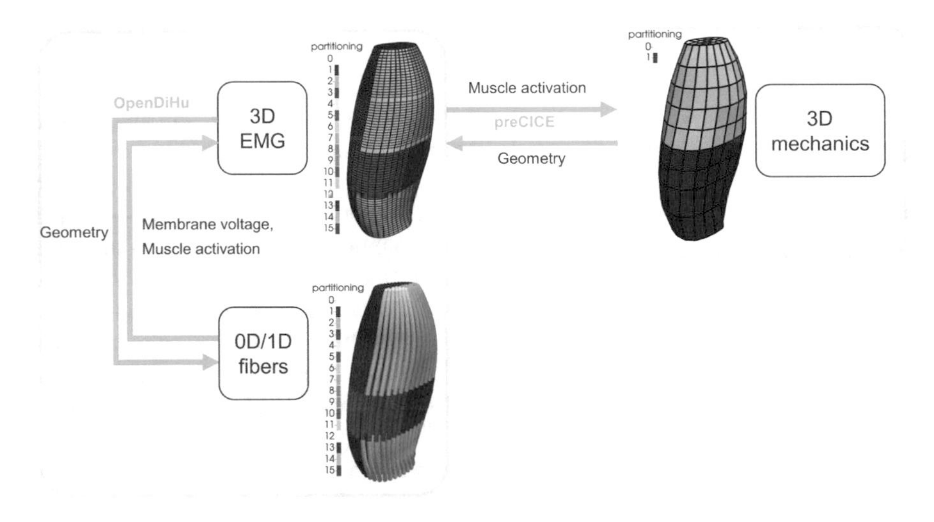

Fig. 3: Overview of the software coupling scheme between two different instances of the solver OpenDiHu using the coupling library preCICE. The left program utilizes more processes and uses finer meshes for the detailed electrophysiology simulations than the program on the right-hand side, which uses a coarser 3D mesh for the mechanics model. The figure shows the value exchanges between the solvers within OpenDiHu and between the programs by blue arrows.

represents the second OpenDiHu program that solely solves the mechanics model. It uses a coarser mesh with different domain decomposition, in the shown example for only two processes.

The bidirectional data mapping and data communication between the two programs is handled by preCICE. The degrees of freedom (dofs) are mapped between the fine 3D mesh on the left-hand side and the coarse 3D mesh on the right-hand side. Note that a direct mapping between the muscle fiber meshes on the left-hand side and the coarse 3D mesh on the right-hand side would not be feasible because of the large number of nodes in the muscle fiber meshes.

We use the serial-explicit coupling functionality of preCICE and employ a consistent mapping with compact polynomial radial basis functions (cf. [8]).

4 Performance results

In this section, we present performance and simulation results for the described multi-scale skeletal muscle model. Section 4.1 begins with a weak scaling investigation, followed by results of our introduced coupling scheme in Sect. 4.2.

4.1 Weak scaling of the electrophysiology solver

In a first step, we study the weak scaling behavior of the multi-scale electrophysiology model, i.e., the 0D, 1D, and 3D electric potential model parts. We increase the problem size by adding more muscle fibers until the realistic number of 270×10^3 fibers for the biceps brachii muscle is reached. At the same time, we increase the number of processes to 26,912. We run the simulation on the supercomputer Hawk at the High Performance Computing Center Stuttgart. Each compute nodes consists of a dual-socket AMD EPYC 7742 processor with 2.25 GHz base frequency and 256 GB RAM per node. We use 64 processes per compute node.

The scenario contains the 0D model of Hodgkin and Huxley [14], which we solve with a timestep width of $dt_{0D} = 10^{-3}$ ms. We simulate a time span of $t_{end} = 2$ ms, corresponding to two invocations of the 3D model solver in the described subcycling scheme. To proportion the time for initialization, we multiply the measured runtimes for all parts except the initialization by the factor 500 such that the results correspond to a simulation time of 1 s.

Figure 4 shows the resulting runtimes of the different solvers and illustrates their main characteristics. It can be seen that the computation of the 0D model contributes most to the total runtime for the scenarios with up to 3600 muscle fibers, whereas the 1D model solver consistently exhibits relatively low runtimes. Further, the 0D and 1D solvers show a perfect weak scaling behavior even for massively parallel simulations

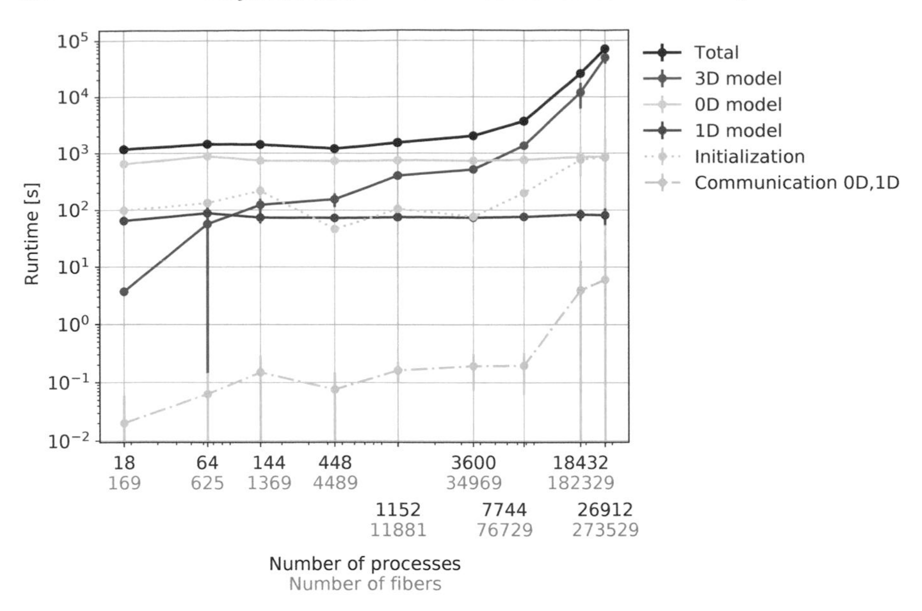

Fig. 4: Weak scaling results for the electrophysiology solvers with runtimes for the different model components (different colors, see legend) and standard deviation over all processes (vertical bars) using between 18 and 26,912 cores and between 169 and 273,529 muscle fibers.

on up to 26,912 cores. The associated communication cost between the 0D and 1D models is negligible. Also, the time for initialization is relatively low compared with the solver runtimes.

We observe an increase in runtime for the 3D model solver, which is due to an increasing number of iterations in the conjugate gradient scheme for larger numbers of degrees of freedom, which we plan to address in future work by implementing a multigrid preconditioner. However, the scaling behavior suffices to be able to run simulations up to the maximum physiologically meaningful resolution of 270,000 muscle fibers of a biceps brachii muscle.

4.2 Performance evaluation of the full multi-scale solver

Next, we perform simulations of the entire multi-scale model of muscular EMG and muscle contraction that was described in Sect. 2. We use the coupling scheme presented in Sect. 2 where the total multi-scale solver is split into two parts, first, the highly parallel multi-scale electrophysiology solver and, second, the solid mechanics solver.

In contrast to the weak scaling study in Sect. 4.1, the simulations here require the more detailed and compute-intense sarcomere model of Shorten et al. [24] to compute the muscle activation for the 3D mechanics model. In addition to the larger number of dofs in this model, a smaller time step width of $dt_{0\mathrm{D}} = 1.25 \times 10^{-5}$ ms has to be used, which further increases the computational work.

Multi-scale Electrophysiology Solver				Mechanics Solver		
processes	fibers	fine 3D mesh nodes	total unknowns	processes	coarse 3D mesh nodes	total unknowns
256	49	6321	4,070,185	1	441	2726
1024	169	21,801	14,037,985	1	129	3959
4096	625	80,625	51,915,625	1	225	1395
9216	1369	176,601	113,715,985	1	225	1395

Table 1: Simulation of muscle contraction and EMG, parameters for different scenarios with varying problem sizes and numbers of processes.

We compute four differently sized scenarios for a time span of $t = 100$ ms and evaluate the runtimes of the individual solvers. Table 1 lists the discretization parameters for the four scenarios. The electrophysiology solver is run with between 256 and 9216 processes on Hawk using 64 processes per compute node, similar to the study in Sect. 4.1. Accordingly, we increase the problem size and consider between 49 and 1369 1D muscle fiber meshes with a constant number of 1481 nodes each. The mesh resolution of the 3D mesh used for the 3D electric potential model is also increased, as listed in table 1. As a consequence, the total number of unknowns increases to over 113 million for the largest scenario.

For all four scenarios, we execute the mechanics solver in serial using a coarser 3D mesh. This is justified, as we are only interested in the macroscopic, uniform deformation of the contracting muscle, in contrast to the spatially higher resolved electric potentials computed by the 3D EMG model. The preCICE library maps and communicates data between the fine and coarse 3D meshes, whose node counts are listed in the third and second last columns in table 1, respectively. The runtime of the coupled mechanics solver without waiting time is below two minutes in every scenario, which is negligible compared to the runtimes of several hours for the electrophysiology solver.

Figure 5 presents the resulting runtimes in the electrophysiology solver. The 0D, 1D and 3D solver parts show characteristics similar to the weak scaling study in Sect. 4.1. In addition, it can be seen that the preCICE mapping contributes negligible additional runtime and exhibits good weak scaling properties. This shows us that our software setup using preCICE can also be used for even larger runs in future work.

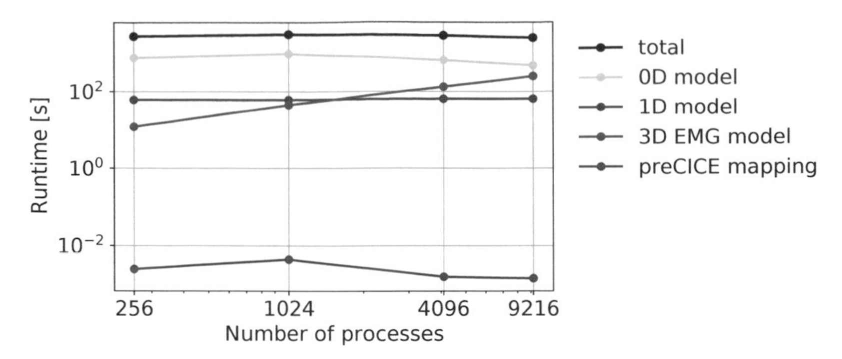

Fig. 5: Weak scaling study of the full multi-scale model of muscular EMG and contraction, including the preCICE volume coupling scheme. The runtimes of the model parts (different colors, see legend) are given for the three scenarios specified in table 1.

4.3 Simulation results of the full multi-scale model

Finally, we present simulation results of a large scenario corresponding to the second last row in table 1. Solving this scenario for a simulation time span of 190 ms takes around 8 h 27 min using 64 compute nodes on Hawk.

The left plot in Figure 6 shows the fine fiber meshes with the trans-membrane potential V_m. Action potentials can be seen spreading from the center of the fibers towards to top and bottom ends of the muscle. The middle plot shows the resulting EMG given by the extracellular potential ϕ_e. The right plot depicts the reference configuration of the solid mechanics mesh by the yellow wireframe and the current deformed mesh colored according to the active Piola–Kirchhoff stress. It can be seen that the three different meshes are identically deformed, as a result of the numerical coupling. Judging from experiences in experimental studies, the simulation results look plausible. For example, the results exhibit a delay between the onset of the muscle activation and the actual shortening of the muscle, which can also be observed in real experiments.

5 Conclusion and outlook

In this work, an existing, detailed multi-scale model of the neuromuscular system was described, which can be used to simulate EMG and muscle contraction. The simulation software OpenDiHu is capable of solving different configurations of this multi-scale model framework using High Performance Computing techniques. However, the combination of the solid mechanics model on the macroscopic organ scale with the electrophysiology models on a smaller spatial scale hindered highly

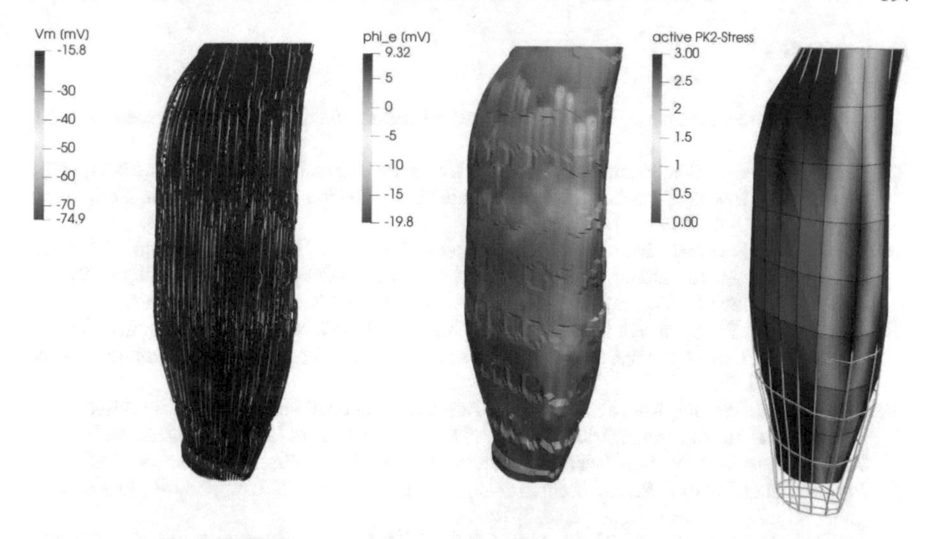

Fig. 6: Simulation result at $t = 110\,\mathrm{ms}$ of the EMG and muscle contraction model with 615 muscle fibers of the biceps brachii muscle. The simulation was computed with 4096 processes on the supercomputer Hawk. The three images depicts the trans-membrane voltage V_m (left), the muscle activation γ (middle) and the active Piola–Kirchhoff stress (right).

parallel executions. In the present paper, we described a scheme to separate the partitionings of these models using the volume coupling functionality of the library preCICE, ultimately resolving the issue.

We investigated the weak scaling behavior of the electrophysiology solver and found good scaling properties for the 0D and 1D model parts, while the 3D part could still be improved. Furthermore, we demonstrated the use of our new coupling scheme with the full multi-scale model and, again, found good weak scaling behavior also for the added coupling part.

Finally, we showed unprecedented simulation results of the combined muscular EMG and muscle contraction model using a fine spatial resolution and a detailed biophysical model with more than 100 million degrees of freedom in total. Such simulations are only possible by exploiting High Performance Computing facilities such as the supercomputer Hawk.

Future work needs to address the mediocre weak scaling behavior of the 3D action potential model, which currently uses an unpreconditioned conjugate gradient solver. Here, the use of a multigrid preconditioner seems promising. Further studies can also extend the simulations to even higher numbers of muscle fibers.

Acknowledgements We thank the Deutsche Forschungsgemeinschaft (DFG, German Research Foundation) for supporting this work by funding – EXC2075 – 390740016 under Germany's Excellence Strategy. Furthermore, we acknowledge the support by the Stuttgart Center for Simulation Science (SimTech).

References

1. OpenDiHu Documentation. https://opendihu.readthedocs.io/en/latest/, 2021. Accessed: 2021-02-12.
2. Patrick Amestoy, Alfredo Buttari, Jean-Yves L'Excellent, and Théo Mary. On the complexity of the block low-rank multifrontal factorization. *SIAM Journal on Scientific Computing*, 39(4):1710–1740, 2017.
3. Satish Balay, Shrirang Abhyankar, Mark F. Adams, Jed Brown, Peter Brune, Kris Buschelman, Lisandro Dalcin, Alp Dener, Victor Eijkhout, William D. Gropp, Dmitry Karpeyev, Dinesh Kaushik, Matthew G. Knepley, Dave A. May, Lois Curfman McInnes, Richard Tran Mills, Todd Munson, Karl Rupp, Patrick Sanan, Barry F. Smith, Stefano Zampini, Hong Zhang, and Hong Zhang. PETSc users manual. Technical Report ANL-95/11 - Revision 3.14, Argonne National Laboratory, 2020.
4. Satish Balay, Shrirang Abhyankar, Mark F. Adams, Jed Brown, Peter Brune, Kris Buschelman, Lisandro Dalcin, Alp Dener, Victor Eijkhout, William D. Gropp, Dinesh Kaushik, Matthew G. Knepley, Dave A. May, Lois Curfman McInnes, Richard Tran Mills, Todd Munson, Karl Rupp, Patrick Sanan, Barry F. Smith, Stefano Zampini, Hong Zhang, and Hong Zhang. PETSc Web page. http://www.mcs.anl.gov/petsc, 2018.
5. Satish Balay, William D. Gropp, Lois Curfman McInnes, and Barry F. Smith. Efficient management of parallelism in object oriented numerical software libraries. In E. Arge, A. M. Bruaset, and H. P. Langtangen, editors, *Modern Software Tools in Scientific Computing*, pages 163–202. Birkhäuser Press, 1997.
6. Chris Bradley, Andy Bowery, Randall Britten, Vincent Budelmann, Oscar Camara, Richard Christie, Andrew Cookson, Alejandro F. Frangi, Thiranja Babarenda Gamage, Thomas Heidlauf, Sebastian Krittian, David Ladd, Caton Little, Kumar Mithraratne, Martyn Nash, David Nickerson, Poul Nielsen, Øyvind Nordbø¸, Stig Omholt, Ali Pashaei, David Paterson, Vijayaraghavan Rajagopal, Adam Reeve, Oliver Röhrle, Soroush Safaei, Rafael Sebastián, Martin Steghöfer, Tim Wu, Ting Yu, Heye Zhang, and Peter Hunter. OpenCMISS: A multi-physics & multi-scale computational infrastructure for the VPH/Physiome project. *Progress in Biophysics and Molecular Biology*, 107(1):32–47, 2011. Experimental and Computational Model Interactions in Bio-Research: State of the Art.
7. Chris P. Bradley, Nehzat Emamy, Thomas Ertl, Dominik Göddeke, Andreas Hessenthaler, Thomas Klotz, Aaron Krämer, Michael Krone, Benjamin Maier, Miriam Mehl, Rau Tobias, and Oliver Röhrle. Enabling detailed, biophysics-based skeletal muscle models on HPC systems. *Frontiers in Physiology*, 9(816), 2018.
8. Hans-Joachim Bungartz, Florian Lindner, Bernhard Gatzhammer, Miriam Mehl, Klaudius Scheufele, Alexander Shukaev, and Benjamin Uekermann. preCICE – a fully parallel library for multi-physics surface coupling. *Computers and Fluids*, 141:250–258, 2016. Advances in Fluid-Structure Interaction.
9. Autumn A. Cuellar, Catherine M. Lloyd, Poul F. Nielsen, David P. Bullivant, David P. Nickerson, and Peter J. Hunter. An Overview of CellML 1.1, a Biological Model Description Language. *SIMULATION*, 79(12):740–747, 2003.
10. Robert D. Falgout and Ulrike Meier Yang. Hypre: A library of high performance preconditioners. In *Proceedings of the International Conference on Computational Science-Part III*, ICCS '02, pages 632–641, Berlin, Heidelberg, 2002. Springer-Verlag.
11. T. Heidlauf and O. Röhrle. Modeling the Chemoelectromechanical Behavior of Skeletal Muscle Using the Parallel Open-Source Software Library OpenCMISS. *Computational and Mathematical Methods in Medicine*, 2013:1–14, 2013.
12. T. Heidlauf and O. Röhrle. A multiscale chemo-electro-mechanical skeletal muscle model to analyze muscle contraction and force generation for different muscle fiber arrangements. *Frontiers in Physiology*, 5(498):1–14, 2014.
13. B. Hernández-Gascón, J. Grasa, B. Calvo, and J. F. Rodríguez. A 3D electro-mechanical continuum model for simulating skeletal muscle contraction. *Journal of Theoretical Biology*, 335:108–118, 2013.

14. A. L. Hodgkin and A. F. Huxley. A quantitative description of membrane current and its application to conduction and excitation in nerve. *The Journal of physiology*, 117(4):500–544, 1952.
15. Aaron Krämer, Benjamin Maier, Tobias Rau, Felix Huber, Thomas Klotz, Thomas Ertl, Dominik Göddeke, Miriam Mehl, Guido Reina, and Oliver Röhrle. Multi-physics multi-scale HPC simulations of skeletal muscles. In Wolfgang E. Nagel, Dietmar H. Kröner, and Michael M. Resch, editors, *High Performance Computing in Science and Engineering ' 20: Transactions of the High Performance Computing Center, Stuttgart (HLRS) 2020*, 2021.
16. Matthias Kretz and Volker Lindenstruth. Vc: A C++ library for explicit vectorization. *Software: Practice and Experience*, 42(11):1409–1430, 2012.
17. Benjamin Maier, Nehzat Emamy, Aaron S. Krämer, and Miriam Mehl. Highly parallel multi-physics simulation of muscular activation and EMG. In *COUPLED PROBLEMS 2019*, pages 610–621, 2019.
18. Benjamin Maier, Dominik Göddeke, Felix Huber, Thomas Klotz, Oliver Röhrle, and Miriam Schulte. OpenDiHu - Efficient and Scalable Software for Biophysical Simulations of the Neuromuscular System (forthcoming). *Journal of Computational Physics*, 2021.
19. Gary R. Mirams, Christopher J. Arthurs, Miguel O. Bernabeu, Rafel Bordas, Jonathan Cooper, Alberto Corrias, Yohan Davit, Sara-Jane Dunn, Alexander G. Fletcher, Daniel G. Harvey, Megan E. Marsh, James M. Osborne, Pras Pathmanathan, Joe Pitt-Francis, James Southern, Nejib Zemzemi, and David J. Gavaghan. Chaste: An open source c++ library for computational physiology and biology. *PLOS Computational Biology*, 9(3):1–8, 03 2013.
20. M. Mordhorst, T. Heidlauf, and O. Röhrle. Predicting electromyographic signals under realistic conditions using a multiscale chemo-electro-mechanical finite element model. *Interface Focus*, 5(2):1–11, February 2015.
21. IUPS Physiome Project. Physiome Model Repository. https://models.physiomeproject.org/, 2020. [Online; accessed 8-December-2020].
22. O. Röhrle, J. B. Davidson, and A. J. Pullan. Bridging scales: a three-dimensional electromechanical finite element model of skeletal muscle. *SIAM Journal on Scientific Computing*, 30(6):2882–2904, 2008.
23. O. Röhrle, J. B. Davidson, and A. J. Pullan. A physiologically based, multi-scale model of skeletal muscle structure and function. *Frontiers in Physiology*, 3, 2012.
24. P. R. Shorten, P. O'Callaghan, J. B. Davidson, and T. K. Soboleva. A mathematical model of fatigue in skeletal muscle force contraction. *Journal of Muscle Research and Cell Motility*, 28(6), 2007.
25. Gilbert Strang. On the construction and comparison of difference schemes. *SIAM Journal on Numerical Analysis*, 5(3):506–517, 1968.

Computational Fluid Dynamics

The following presents a selection of research projects conducted in the field of Computational Fluid Dynamics (CFD) during the reporting period. Numerical simulations were carried out on the new HPE Apollo supercomputer "Hawk" of the High Performance Computing Center Stuttgart (HLRS) as well as on the ForHLR II of the Steinbuch Centre for Computing (SCC) in Karlsruhe. As in previous years, CFD had the strongest demand for supercomputing resources of all disciplines. A share of about 45% of the available computing power of Hawk as well as of ForHLR II were made available to researchers in this field, helping them to expand scientific knowledge in fundamental flow physics or to better understand complex flow behaviour in technical applications. Many important results and new scientific findings have been obtained and remarkable progress has been made. Research has focused on the simulation of fluid dynamic phenomena that can only be captured by extremely fine temporal and spatial discretization, often in combination with high-order methods, as well as on the hardware-related optimization of the employed numerical codes with respect to computational performance.

This year, 36 annual reports were submitted and subjected to a peer review process. Research groups from various institutions across the country took advantage of the supercomputer resources offered by the two centres. From the large number of excellent reports, 11 contributions were selected for publication in this book. The spectrum of projects presented in the following covers fundamental research, investigations into application-oriented problems with industrial relevance as well as further developments of codes and algorithms with respect to the HPC architecture. Challenging fluid dynamic problems are addressed in the fields of bypass transition of compressible boundary layers, gas turbine cooling, hydropower turbines, spray painting processes, and aerodynamic propeller-wing interactions, to name but a few. Different numerical methods, such as Finite Difference, Finite Volume, and Discontinuous Galerkin, were employed using in-house, commercial or open-source codes. The first two articles describe the results of so-called Gauss Large Scale Projects, i.e. these projects were granted more than 100 million core hours per year on Hawk from the Gauss Centre for Supercomputing (GCS).

The contribution of *Borgelt, Hösgen, Meinke and Schröder* from the Institute of Aerodynamics at RWTH Aachen University deals with rim seal flow in an axial turbine stage. The rim seal shall prevent hot gas ingress into the wheel space between the stator and the rotor, which could lead to machine failure. Experiments as well as unsteady RANS simulation had shown that unsteady flow phenomena occur inside the rim seal cavity. However, it had been observed that URANS could not reliably predict the complex unsteady turbulent flow field for each operating condition. Therefore, the authors studied the flow around the rim seal and in the cavity using large eddy simulations (LES). They found that the instantaneous results strongly depend on the cooling gas mass flow rate and are particularly pronounced for lower fluxes. In this case, oscillations of the radial velocity in the rim seal gap occur which lead to an ejection of the cooling gas out of the wheel space and an ingress of hot gas from the main flow, resulting in an reduced cooling effectiveness. Since LES is quite expensive, the authors also applied a zonal RANS/LES method to a generic turbine setup and found good agreement with pure LES, qualifying the hybrid method for future work on the prediction of hot gas ingress in axial turbines.

Ohno, Selent, Kloker and Rist from the Institute of Aerodynamics and Gas Dynamics, University of Stuttgart, present results of their direct numerical simulations (DNS) of bypass transition in a compressible boundary layer induced by isotropic free-stream turbulence. To create the turbulent inlet condition, modes of the continuous spectrum resulting from linear stability analysis were superposed. So far, this approach by Jacobs and Durbin has been used by various researchers for incompressible flow problems. Now, the authors have adapted it for compressible flows. Comparison with DNS data from literature validates the numerical method for quasi-incompressible flows as well as for higher Mach number ($M = 0.7$). In the latter case, the results reveal a clear influence of the compressibility on the transition process. A performance analysis of the applied in-house high-order finite-difference code NS3D shows a near ideal parallel efficiency for both strong and weak scaling up to 1024 nodes on Hawk.

In industrial gas turbines, water is ingested to improve the thermal efficiency by cooling the air before and during compression. However, the interaction between the liquid droplets and the compressor components can lead to a degradation of the structure. To study this interaction in detail, *Schlottke, Ibach, Steigerwald and Weigand* from the Institute of Aerospace Thermodynamics at the University of Stuttgart address the atomization process of a liquid rivulet at the trailing edge of a compressor blade. Up to now, only experimental investigations are known from literature. They carried out DNS with their in-house multiphase flow solver FS3D and compared them with experimental data generated in their own test facility. Additionally, a grid dependence study was performed revealing that the numerical setup is very well suited to reproduce the different atomization processes qualitatively, but even with a very high grid resolution (more than a billion grid cells), some features still cannot be resolved. In a comprehensive performance study on Hawk with different compilers installed on the new system and various optimization options, significant efficiency gains could be achieved with the appropriate settings.

Another working group at the same institute is concerned with the issue of turbine blade cooling, which is mandatory if an increase of system efficiency is to be achieved through higher turbine entry temperatures. Swirl cooling is a promising new technique as it causes high heat transfer rates. On the other hand, high pressure losses occur due to axial flow reversal resulting from vortex breakdown. *Seibold and Weigand* present a numerical study in which they used delayed detached eddy simulations (DDES) with OpenFOAM to analyse the impact of convergent tube geometries on the flow field and the heat transfer in a swirl cooling system. They found that the converging tubes enforce an axial and a circumferential flow acceleration, with the former counteracting the flow reversal. By this, the vortex breakdown can be suppressed. Furthermore, the flow becomes more insensitive to disturbances from the tube outlet. The heat transfer in terms of Nusselt numbers increased significantly compared to a pipe flow without swirl, but shows a strong dependency on the tube geometry. Overall, good agreement between the numerical results and experimental data was achieved. The simulations were performed on the ForHLR II.

Besides swirl cooling, other cooling methods are common to avoid thermal damage to turbine blades, e.g. the use of pin fins or ribs inside the flow channels. In combination with high Reynolds numbers, such complex geometries make the use of LES very time consuming. Thus, for industrial applications, RANS methods are usually chosen despite their lower accuracy. The project of *Wellinger and Weigand* aims at providing very accurate results for various cooling features associated with periodic pin-fin arrays and ribbed channels in order to evaluate the validity of the linear Boussinesq hypothesis. The latter establishes a relationship between the unknown Reynolds stresses and the mean strain rate tensor and is the basis of most existing RANS models. To this end, LES is used to generate accurate data for different test cases, which themselves are validated with experimental or DNS data. The plotted results in terms of the misalignment of the eigenvectors of the Reynolds stress tensor and the strain rate tensor give a very good impression of the areas where the linear Boussinesq hypothesis can be considered valid and where it is violated. The simulations were performed with the commercial STAR-CCM+ code on ForHLR II using about 25,000 cells per core yielding a parallel efficiency of almost 80%.

The next two contributions are from the Numerical Methods group of the Institute of Aerodynamics and Gas Dynamics at the University of Stuttgart and are based on a high-order Discontinuous Galerkin (DG) method developed in the group over the last years. While LES of flows along complex geometries are extremely time consuming and RANS simulations suffer from well-known shortcomings with regard to laminar-turbulent transition and flow separation, hybrid methods have proven to be a valuable compromise for industrial application. This class of methods includes zonal LES, which reduces the computational effort compared to pure LES significantly, but requires the generation of time-accurate turbulent inflow conditions from the RANS solution at the RANS-LES interface. *Kempf, Gao, Beck, Blind, Kopper, Kuhn, Kurz, Schwarz and Munz* present two turbulent inflow methods that they had implemented in their open-source simulation framework FLEXI, a high-order DG Spectral Element method. The inflow methods are a recycling-rescaling anisotropic linear forcing and a synthetic eddy method. They are compared to numerical reference cases for turbulent boundary layers along a flat plate. A particular focus of the report is on the HPC application on the new HAWK architecture, and current performance data are provided.

For the modelling of compressible multi-phase flows, there are two main simulation approaches, the diffuse-interface and the sharp-interface method. A variant of the latter is the high-order level-set ghost-fluid method that has been developed in the Numerical Methods group. The method used exhibits a dynamic variation in the computational workload, and a dynamic load balancing strategy is required to ensure an efficient resource utilization on large-scale supercomputers. *Appel, Jöns, Keim, Müller, Zeifang and Munz* present such a strategy for their numerical framework, which consists of the flow solver FLEXI and an interface-capturing scheme. The strong scaling behaviour was investigated up to 16,384 cores for a generic setup revealing near-ideal parallel efficiency and a significant performance gain compared to the previous, unbalanced simulations. Actually, the authors demonstrate that considerable runtime reductions are achievable even for more complex, realistic scenarios.

The article of *Ye and Dreher* from the Fraunhofer Institute for Manufacturing Engineering and Automation in Stuttgart and *Shen* from the Esslingen University of Applied Science deals with industrial spray painting processes. In order to enhance the appearance of the colour, as, for example, by metallic effect coatings in the automotive industry, effect pigments (small flat flakes) are added to the paint. It is known from comprehensive experimental studies that the orientation of the flakes within the paint layer influences the final metallic effect decisively and strongly depends on the processes during the impact of the paint droplets on solid surfaces. However, numerical investigations have not been reported in literature so far. The authors present for the first time a detailed numerical study of flake orientation during droplet impact on dry and wetted solid walls. For that purpose, a 6-degree-of-freedom model describing the rigid body motion of the flakes was implemented in the commercial CFD-code ANSYS-FLUENT and validated with experimental observations. Subsequently, a parameter study was performed at the HLRS for varying grid resolutions, initial flake positions, Reynolds and Ohnesorge numbers using the Volume-of-Fluid method. The study provides a valuable contribution to a better understanding of the painting processes.

Denev, Zirwes, Zhang and Bockhorn from the Steinbuch Centre for Computing and the Engler-Bunte Institute for Combustion Technology, respectively, at the Karlsruhe Institute of Technology present a new three-dimensional, explicit low-pass Laplace filter for linear forcing. The latter is often used in numerical codes for DNS or LES to force turbulence and maintain it stable throughout the solution time and in the complete computational domain. However, when applied in physical space, it degrades the numerical efficiency, so low-pass filtering of the velocity field is used. The authors provide a new filter that is numerically more efficient than existing ones, has good scaling properties and resolves a larger scale range of turbulence. The new filter and a second improvement, i.e., a particular form of the linear forcing term, have been implemented and successfully tested in OpenFOAM. The report contains a mathematical description of the filter, DNS results, and a performance and efficiency study performed during the installation phase of the new HoreKa cluster, which later replaced ForHLR II at the SCC in Karlsruhe.

Hydropower turbines are expected to play a key role in compensating fluctuations in the power grid. *Wack, Zorn and Riedelbauch* from the Institute of Fluid Mechanics and Hydraulic Machinery at the University of Stuttgart investigated the vortex-induced pressure oscillations in the runner of a model-scale Francis turbine at a far off-design operating point. The focus was on determining the influence of mesh resolution on torque and head at deep part load conditions. The mesh resolution considerably impacts the prediction of the inner-blade vortices and further vertical structures traveling upstream. Refinement in the main flow region as well as in the boundary layer notably improved the results in terms of vortex location and induced pressure fluctuations. For this study, the commercial CFD software ANSYS-CFX was applied using a hybrid RANS/LES approach with an SBES turbulence model. Frequency spectra of the pressure fluctuations are compared with model tests. The code showed good scaling behaviour down to about 36,000 cells per core on Hawk.

The aim of reducing CO2 emission in aviation is the prominent goal of current aeronautical research. One path is the use of full or hybrid electric propulsion systems. The separation of energy supply and propulsive force generation via an electric motor driving a propeller offers new design opportunities, two of which are wing-tip mounted propellers and distributed electric propulsion with many propellers arranged along the wingspan. If properly designed, the mutual interaction between the propeller(s) and the wing can improve the aerodynamic performance of the aircraft. To find an optimal solution, CFD simulations are very helpful, but also very expensive due to the wide, multi-dimensional parameter space. One way to reduce the computational effort is to model the propellers using an Actuator Disc (ACD) or an Actuator Line (ACL) approach instead of fully resolving them. To ensure a physically correct prediction of the actual interference effects when using these models, *Schollenberger, Firnhaber Beckers and Lutz* from the Institute of Aerodynamics and Gas Dynamics at the University of Stuttgart provide in their report a comparison and validation with fully resolved simulations and experimental data known from literature. With the input data for the ACD and the ACL methods derived from steady-state 3D simulations from a single blade, the authors can demonstrate that both approaches can predict the interactions between propeller and wing in both directions with sufficient accuracy making them well suited for design studies. The TAU-code of the German Aerospace Center was applied for this study showing an ideal scaling behaviour between 50,000 and 10,000 grid points per core on Hawk.

The selection of projects presented reflects the continuing progress of high performance computing in the field of CFD. Numerical simulations on supercomputers like those at the HLRS and the SCC are crucial to gain insight into often complex, time-dependent flow physics. Some of these flow phenomena occur only on very small temporal and spatial scales and can only be brought to light by an extremely fine discretization of the temporal and/or physical domains. The sustained increase in computational power at both HPC centres, recently realised by the replacement of the Hazel Hen with the Hawk supercomputer at the HLRS, enables the tackling of increasingly complex fluid dynamic problems, including unsteady and transient processes, within reasonable timeframes. This also facilitates industrial design processes, where the use of high-fidelity numerical methods on modern supercomputers helps to enhance the reliability and the efficiency of the simulations, mitigate development time, risks and costs and, thus, increase industrial competitiveness. But, the rise in hardware performance has to be accompanied by the development of new numerical algorithms. Furthermore, to exploit the full potential of the new hardware, code adaptations to the particular HPC architecture are indispensable. A very close cooperation of the researchers with the experts from the HLRS and the SCC is key to successfully accomplishing this demanding task. In this context, the staffs of both computing centres deserve thanks for their individual and custom-tailored support. Without their dedication, it would not have been possible and will not be possible in the future to maintain the high scientific quality we see in the projects.

Institute of Aerodynamics and Gas Dynamics, *Ewald Krämer*
University of Stuttgart,
Pfaffenwaldring 21,
70550 Stuttgart,
Germany,
e-mail: kraemer@iag.uni-stuttgart.de

Analysis of the hot gas ingress into the wheel space of an axial turbine stage

Jannik Borgelt, Thomas Hösgen, Matthias Meinke and Wolfgang Schröder

1 Introduction

The thermal efficiency of gas turbines strongly depends on the turbine inlet temperature. Increasing the inlet gas temperature improves the engine performance, but results in new challenges for the cooling of the turbine material. To avoid hot gas ingress from the main annulus flow into the wheel space between the stator and the rotor disks, which can lead to machine failure, cooling air from the turbine's secondary air system is used to seal the rim seal gap and to cool the wheel space. To increase the thermodynamic efficiency of the turbine, it is necessary to minimize the cooling mass flow rate.

The rim seal flow has been extensively studied over the past years. A recent review of the progress made is given in [8]. Recent experimental studies showed the existence of periodic flow phenomena inside the rim seal cavity [3, 20]. The unsteady flow structures were unrelated to blade passing and expected to have significant impact on the sealing effectiveness. These unsteady phenomena were only present at low cooling gas mass flow rates and vanished abruptly when the cooling gas mass flux was sufficiently increased. The fluctuations were later confirmed by numerical simulations based on the unsteady RANS approach of the same 1.5-stage turbine test rig [13]. Further studies showed that the unsteady flow phenomena, apart from the amount of injected cooling gas, depend on the rotor speed [1] and the rim seal geometry [9].

Although several authors observed unsteady flow phenomena using the unsteady RANS approach, it does not reliably predict the complex unsteady turbulent flow field for each flow condition and rim seal geometry [10, 11]. To further understand

Jannik Borgelt, Thomas Hösgen, Matthias Meinke and Wolfgang Schröder
Institute of Aerodynamics, RWTH Aachen University, Wüllnerstraße 5a, Aachen, 52062, Germany,
e-mail: j.borgelt@aia.rwth-aachen.de, t.hoesgen@aia.rwth-aachen.de

Wolfgang Schröder
JARA Center for Simulation and Data Science, RWTH Aachen University, Seffenter Weg 23, Aachen, 52074, Germany

W. E. Nagel et al. (eds.), *High Performance Computing in Science and Engineering '21*,
https://doi.org/10.1007/978-3-031-17937-2_12

the inherent unsteady nature of the rim seal flow, necessary for the optimization and development of new rim seal designs, more profound methods with fewer modeling assumptions need to be used.

The work presented in this paper is a continuation of previous investigations, where the flow in the axial turbine was predicted by LES for the full 360° circumference [12, 17]. Two rim seal geometries were considered in [17]. Kelvin–Helmholtz type instabilities were identified in a single lip rim seal gap, which interacted with the main annulus flow. Adding a second sealing lip damped these fluctuations and reduced the hot gas ingress. In [12], one important finding was the occurrence of standing acoustic waves inside the wheel space, where the frequencies coincide with acoustic modes of the wheel space. The computational costs of such large scale LES are, however, extremely large.

To reduce the computational costs a computationally less expensive zonal RANS/LES method is developed, with which the flow field in the rim seal gap, dominated by the shear layers of the main annulus flow and the cavity flow, can be resolved by LES, while the flow field radially above the seal gap, i.e. in the hot gas flow, and radially below the seal gap, i.e. in the wheel space, is determined using the RANS approach.

The paper is organized as follows. First, the governing equations and the numerical approach are discussed. Second, the results of [12] are briefly summarized. Third, the zonal RANS/LES approach is presented and validated based on a reduced computational setup. Finally, some conclusions are drawn.

2 Numerical method

2.1 Governing equations

The motion of a viscous, compressible fluid is described by the Navier–Stokes equations. For an arbitrarily moving control volume V with surface A, the conservation equations for mass, momentum, and energy read

$$\frac{\mathrm{d}}{\mathrm{d}t} = \int_{V(t)} Q \,\mathrm{d}V + \oint_{A(t)} \bar{H} \cdot n \,\mathrm{d}A = 0. \tag{1}$$

Here, $Q = [\rho, \rho u, \rho E, \rho Y]^T$ is the vector of the conservative variables with the density ρ, the velocity vector u, the total energy $\rho E = \rho e + \rho u^2/2$, and the concentration of a passive scalar Y. The variable e denotes the specific internal energy. The quantity $\bar{H}$ is the flux tensor and n the outward normal vector on the surface A. The flux tensor in non-dimensional form is given by

$$\bar{H} = \bar{H}^{inv} - \bar{H}^{vis} = \begin{pmatrix} \rho\,(u - u_{\partial V}) \\ \rho u\,(u - u_{\partial V}) + p\bar{I} \\ \rho E\,(u - u_{\partial V}) + \rho u \\ \rho Y\,(u - u_{\partial V}) \end{pmatrix} - \frac{1}{Re_0} \begin{pmatrix} 0 \\ \bar{\tau} \\ \bar{\tau} u + q \\ -\rho D \nabla Y / Sc_0 \end{pmatrix}. \tag{2}$$

where $u_{\partial V}$ is the velocity of the control volume's surface, p is the pressure, and the unit tensor is given by $\bar{I}$. Additionally, D is the mass diffusion coefficient of the passive scalar Y, and $Sc_0 = 1$ is the Schmidt number. The Reynolds number and the speed of sound are expressed by

$$Re_0 = \left(\rho_0 a_0 l_{ref}\right) / \eta_0\,, \quad \text{and} \quad a_0 = \sqrt{\gamma p_0 / \rho_0} \tag{3}$$

with the ratio of specific heats $\gamma = 1.4$ and the characteristic length l_{ref}. For a Newtonian fluid with zero bulk viscosity, the stress tensor is written

$$\bar{\tau} = (2/3)\,\eta\,(\nabla \cdot u)\,\bar{I} - \eta\left(\nabla u + (\nabla u)^T\right)\,, \text{where} \quad \eta(T) = T^{3/2}\frac{1+S}{T+S}, \tag{4}$$

is determined from Sutherland's law with $S = 111K/T_0$. For constant Prandtl number $Pr_0 = 0.72$, the non-dimensional vector of heat conduction according to Fourier's law is

$$q = \frac{-\eta}{Re_0 Pr_0 (\gamma - 1)} \nabla T. \tag{5}$$

The equations are closed by the equation of state for an ideal gas, which is written in non-dimensional form $\gamma p = \rho T$. On the adiabatic considered walls the no-slip condition is imposed. The pressure at the solid boundaries is determined via a robin-type boundary condition derived from the momentum equation.

The Reynolds averaged Navier–Stokes equations (RANS) are closed using the Boussinesq hypothesis

$$-\overline{u_i' u_j'} = \nu_t \left(\frac{\partial \bar{u}_i}{\partial x_j} + \frac{\partial \bar{u}_j}{\partial x_i}\right) - \frac{2}{3} k \delta_{ij}, \tag{6}$$

where the turbulent viscosity ν_t is determined by the density corrected Spalart–Allmaras turbulence model [7].

2.2 Numerical method

A coupled finite-volume and a level-set solver is used to simulate the flow field. The solvers own subsets of a shared hierarchical Cartesian mesh, which are adapted independently based to the movement of the blade surfaces.

The conservation equations (1) are discretized by a cell-centered finite-volume method. In a LES, only the large scale turbulent structures are resolved. The sub-grid contribution is modeled here by the monotone integrated LES (MILES) approach [6], in which the dissipative part of the truncation error dissipates the energy at the unresolved scales.

A low-dissipation variant of the advection upstream splitting method (AUSM) proposed [15] is used to compute the inviscid flux tensor $\bar{H}^{inv}$ (2). The primitive variables at the cell surfaces are obtained by a second-order accurate MUSCL extrapolation. The viscous flux tensor $\bar{H}^{vis}$ (2) is computed using a central-difference scheme. The cell-centered gradients are computed by a weighted least-square method as described in [21]. The gradients at the cell-surface centroids are determined with the recentering-approach proposed by [2].

The time-integration is performed by an explicit second-order accurate 5-stage Runge–Kutta scheme [21], which is optimized for an efficient integration for cases with moving domain boundaries. The Runge–Kutta coefficients are chosen for maximum stability.

A strictly conservative cut-cell method is used to accurately represent the surfaces of embedded bodies, which are prescribed by a zero level-set contour obtained from the level-set solver. To prevent arbitrarily small time steps, a flux redistribution method is applied in small cut-cells. A detailed description of the cut-cell method and the flux redistribution method is given in [21].

A level-set method is applied to track the movement of embedded bodies. The level-set is a signed-distance function that divides the domain into a solid G^s and fluid G^f part. The surface of the body is given by the zero level-set contour Γ. To preserve the resolution of the initial level-set surface while moving the bodies, the kinematic motion level-set approach described by [18] is used. In this method a reference level-set field is generated from a high-resolution stl at the start of the simulation and used to interpolated the level-set at later time steps. Each time step the grid cell from which the level-set was interpolated is stored and serves to efficiently find the correct grid cell for interpolation in the following time step. In a previously implemented method, the reference level-set remains unchanged throughout the entire simulation. Hence, for parallel operations, the reference grid cell can be located on an arbitrary MPI rank, and after a certain rotor rotation a global MPI communication is required until a full rotor rotation is completed and the level-set arrives again at the reference location. Since this global communication strongly hampers the performance for a large number of MPI ranks, the previous method was modified. In the new version, the reference level-set field is reconstructed from the stl each time the reference grid cell moves onto a neighboring MPI rank. Before the reconstruction the coordinated of the stl are rotated according to the current rotor location. For a large number of MPI ranks, the cost for the level-set reconstruction are significantly lower compared to the idle time of individual MPI ranks generated by the global communication.

2.3 Zonal RANS/LES method

The zonal RANS/LES or embedded LES method is a hybrid turbulence modeling method, where the governing equations for LES or RANS are solved in segregated predefined computational domains. The domains are linked via an overlapping region, in which the transition from RANS to LES and vice versa occurs. The aim of the zonal RANS/LES method is to reduce the domain size of the computationally more expensive LES approach to the flow region around the rim seal of the axial turbine.

2.3.1 Synthetic turbulence generation (STG)

To ensure a smooth transition between the RANS domain located upstream of the rim seal gap and the downstream LES domain, the time averaged RANS values are used to determine the solution at the LES inflow. The time averaged pressure of the LES domain is prescribed at the outflow of the RANS domain to ensure the correct propagation of information from the LES domain the upstream RANS domain. The velocity fluctuations at the LES inflow are generated by the reformulated synthetic turbulence generation (STGR) method [19]. This method is a synthetic eddy method (SEM), which composes a turbulent velocity field by a superposition of eddies in a virtual volume.

2.3.2 Reconstruction of the eddy viscosity (RTV)

The coupling of the LES domain around the rim seal with the RANS domain located further downstream, requires a reconstruction of the eddy viscosity for the Spalart–Allmaras model from the LES solution. Here, the methodology of [14] is used. In addition, the time averaged pressure of the downstream located RANS domain is used as an outflow condition for the embedded LES domain. The temporal average of the flow quantities in the LES domain is calculated by a moving average. The turbulent kinetic energy k and the specific rate of dissipation ω are computed based on Bradshaw's hypothesis using the time averaged and instantaneous flow quantities of the LES domain.

3 Computing resources

The LES for the full circumference, i.e., 360° of the axial turbine is performed on 64 compute nodes of the HAWK high-performance computer installed at HLRS. Each node consists of 2 AMD EPYC™ 7742 CPUs. Due to the low flow velocity in the wheel space, it takes about 40 full rotations until the turbulent flow in the wheel space becomes fully developed. Depending on the operating conditions, one degree of rotor rotation can take between 8 and 15 minutes computing time on 8192 compute cores.

The LES for the 360° axial turbine is conducted using approx. 450 million grid cells. Therefore, a snapshot of the flow variables requires approximately 58.5 GBytes of disk space.

By using a solution adaptive mesh refinement, the number grid cells on the individual MPI ranks changes considerably due to the rotation of the turbine blades. To prevent load imbalances between the compute cores and to guarantee a high parallel efficiency a dynamic load balancing method is utilized. During the load balancing the parallel subdomains are redistributed among the compute cores [16]. Necessary communication between solvers is minimized by partitioning the shared hierarchical Cartesian grid based on a space-filling curve. The load balancing is based on the individual compute loads of all MPI ranks, measured for 5 time steps after each mesh adaptation. A detailed description of the load balancing method can be found in [16].

4 One-stage axial turbine

4.1 Computational setup

In this section, the setup of the one-stage axial flow turbine also investigated in [12] is described. The geometry is shown in Figure 1. Note that the distance of the stator and rotor rows are increased in axial direction for a better visualization. The turbine stage consists of 30 stator vanes and 62 rotor blades. A double lip rim seal is used to seal the rotor-stator wheel space from the main annulus. More details of this setup can be found in [4, 5].

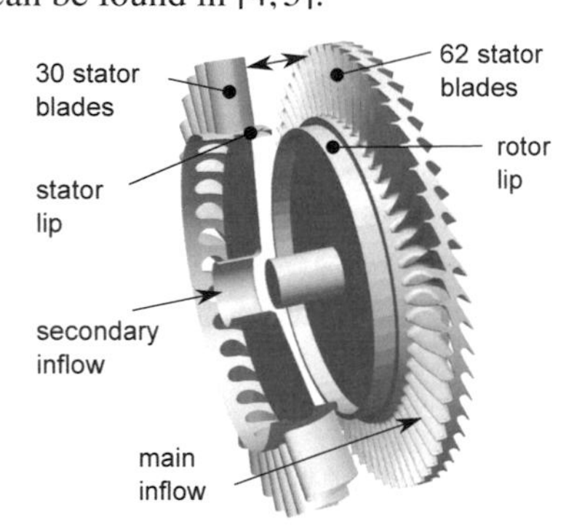

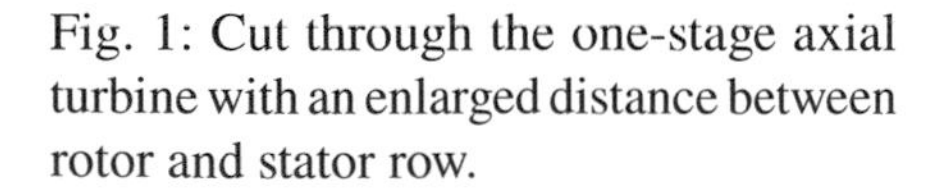

Fig. 1: Cut through the one-stage axial turbine with an enlarged distance between rotor and stator row.

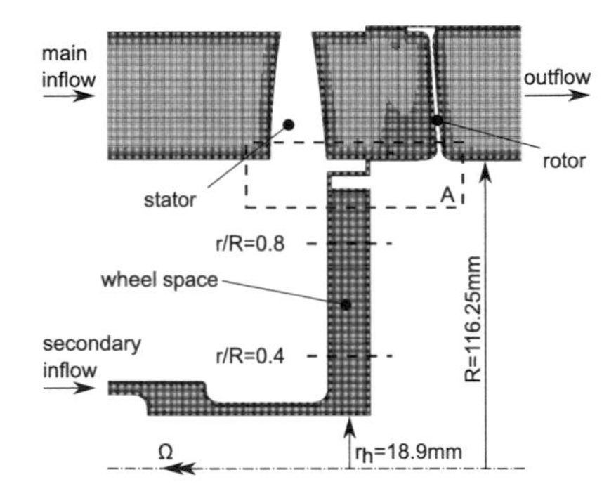

Fig. 2: Dimensions and computational mesh in an axial plane of the axial turbine stage.

The full 360° circumference of the main annulus flow and wheel space is resolved by an adaptive Cartesian mesh with approximately 450 million cells. The subset used by the finite-volume solver comprises approximately 400 million leaf cells of the shared grid, while the level-set solver uses about 450 million leaf cells. Figure 2 shows an axial cut through the mesh. A zoom of the rim seal geometry for the region marked in Figure 2 by the dashed line is displayed in Figure 3.

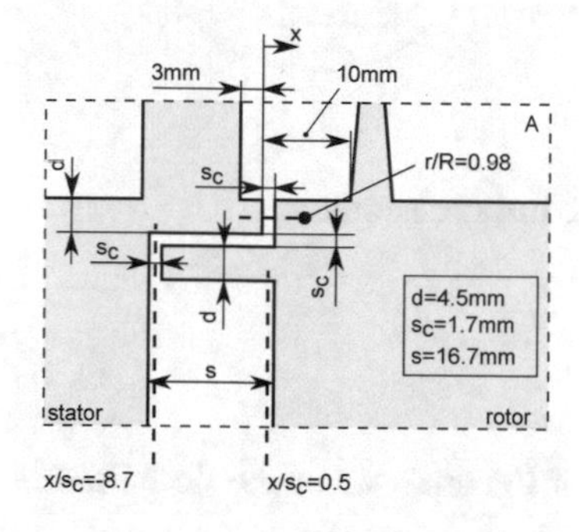

Fig. 3: Schematic view of the rim seal geometry.

Fig. 4: Operating conditions.

	CW1K	CW2K
$Re_{c_1} = \frac{\rho_1 c_1 R}{\mu_1}$	$0.8 \cdot 10^6$	$0.8 \cdot 10^6$
$M_{c_1} = \frac{c_1}{\sqrt{\gamma R T_1}}$	0.37	0.37
$Re_u = \frac{\rho_{cg} \Omega R^2}{\mu_{cg}}$	$0.8 \cdot 10^6$	$0.8 \cdot 10^6$
$cw = \frac{\dot{m}_{cg}}{\mu_{cg} R}$	1000	2000

The operating conditions are defined by the four non-dimensional parameters listed in Table 4. The subscript 1 indicates the flow state 1.5 mm downstream of the stator blades and cg denotes the conditions inside the cooling gas inlet. Two cases with the same flow conditions in the main annulus, i.e., Reynolds and Mach number Re_{c1}, and M_{c1}) and rotor speed Re_u were conducted. The two cases differ in the non-dimensional cooling gas mass flux cw, which is reduced by 50% in case CW1K compared to case CW2K.

At the main and secondary inflow boundaries the density and the three velocity components are prescribed. At the outflow boundary, the static pressure is fixed. To reduce numerical wave reflections sponge layers are used at all in- and outflow boundaries. At the fixed and the moving walls, an adiabatic no-slip condition is prescribed.

4.1.1 Reduced computational setup for zonal RANS/LES method

A simplified two-dimensional generic turbine setup (GT) without rotor blades based on the geometry discussed in section 4.1 is used to test the zonal RANS/LES method. Periodic boundary conditions in the z-direction are used. The investigations are focused on the transition from RANS to LES domains especially for the shear layers. The simplified turbine setup is shown Figure 5.

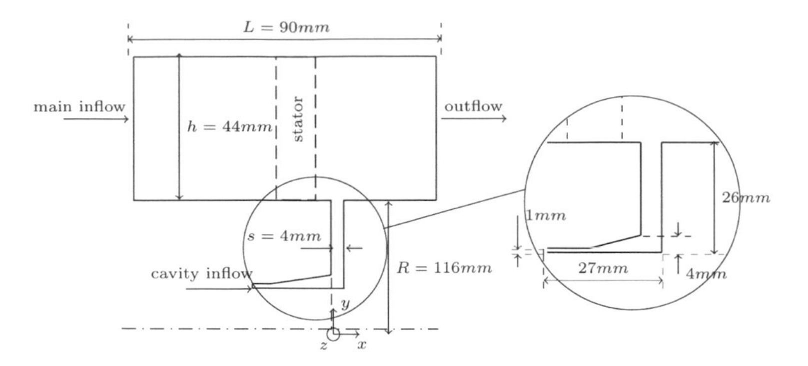

Fig. 5: Schematic of the generic turbine geometry.

4.2 Results

First, the numerical results presented in [12] for the analysis of the flow field in a one-stage axial flow turbine for the two cases with different cooling gas mass flow rates are briefly summarized. The data shown for case CW2K is taken from [17]. For both operating conditions, it takes approximately 40 full rotor rotations until the flow field is fully developed. For the statistical analyses the data of the final 3 rotations are used. The turbulent flow field in the stator hub boundary layer is shown in Figure 6 by the instantaneous contours of the q-criterion. The contours a colored by the absolute Mach number. The position of the rim seal gap is indicated by the black cylinder

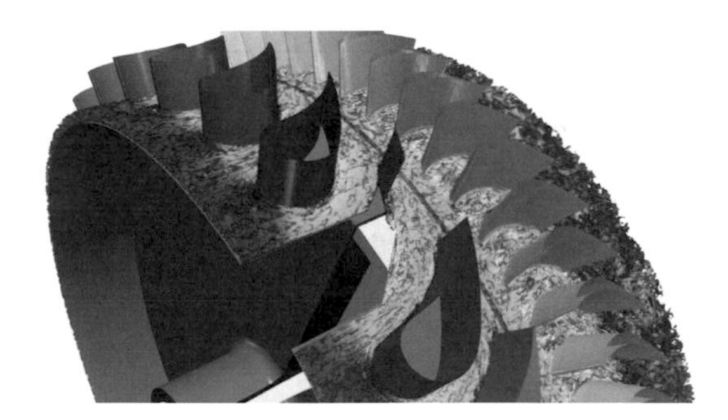

Fig. 6: Instantaneous contours of the q-criterion colored by the absolute Mach number

between the vanes and the blades.

Figure 7 displays the radial distributions of the rms values of the velocity fluctuations in the wheel space at the fixed azimuthal position of $\Theta = 0°$. The distributions in two axial positions are shown, i.e., near the stator wall at $x/s_c = -8.7$ and near the rotor wall at $x/s_c = 0.5$, see Figure 3. While the rms values of the azimuthal and the radial velocity components have somewhat distributions, the pressure fluctuations show large differences for the two cooling mass flow rates. For case CW2K, the pressure rms values are almost constant along the radius. For case CW1K, fluctuations of significantly higher amplitude are visible, which exceed the rms values of case CW2K by more than a factor of two. To identify the reason for the increased pressure fluctuations, the instantaneous flow field inside the wheel space is analyzed. The pressure signals in the wheel space of the 3 full rotor rotations were sampled in 408 radially distributed bins along the stator wall. For these signals cross-correlations are computed relative to the signal at the location $r/R = 0.98$.

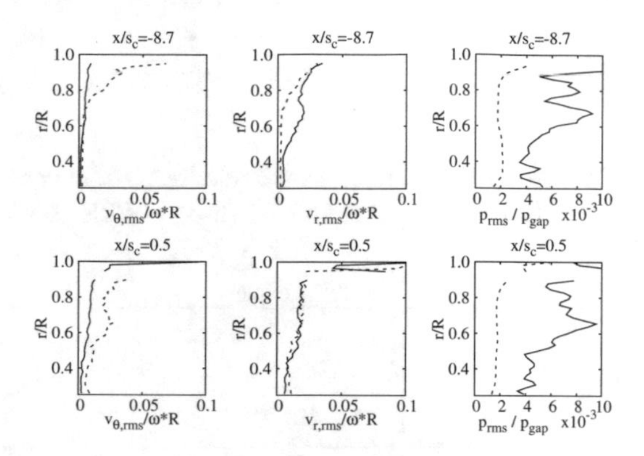

Fig. 7: Radial distributions of the azimuthal and radial velocity components v_θ and v_r and the pressure fluctuations at two axial positions; – CW1K, ... CW2K.

The energy spectra computed from these cross-correlations, shown in Figure 8, reveal a distinct peak at a frequency of approximately 94.33 times the rotor speed n for case CW1K, which is seen along the entire stator wall [12]. The energy spectrum for case CW2K only shows a minor peak at the blade passage frequency (BPF) of $f/n = 62$.

This dominant frequency for case CW1K is also found in the instantaneous flow field in the rim seal gap at $r/R = 0.98$. Since the radial velocity in the rim seal gap determines the hot gas ingress, the signals of the effective radial velocity were recorded along the entire circumference in 1440 equidistantly spaced positions. Using the same procedure as for the energy spectra in Figure 8, the energy spectra of the effective radial velocity fluctuations in the rim seal gap were obtained. The energy spectrum for case CW1K in Figure 9 shows the same dominant frequency of $f/n \approx 94.33$ as the pressure fluctuations for case CW1K in Figure 8 along with additional peaks. Below the dominant frequency $f/n \approx 94.33$ only a broadband of lower amplitudes is observed. This is also apparent for case CW2K, where no distinct frequency can be observed for the effective radial velocity.

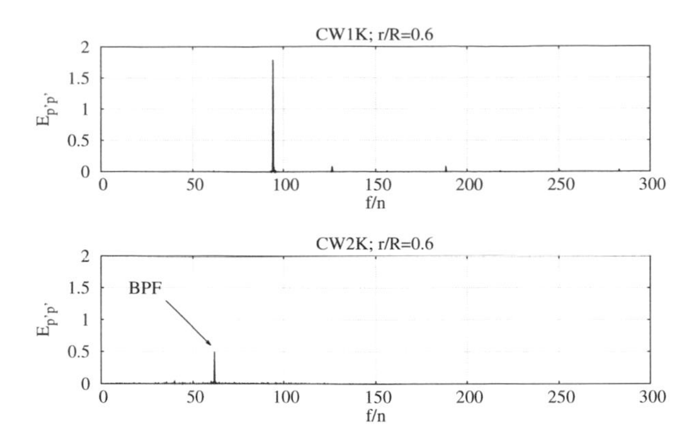

Fig. 8: Energy spectra computed from cross-correlations of the pressure fluctuations along the stator wall in the wheel space; CW1K (top), CW2K (bottom); x/s_c=-8.7.

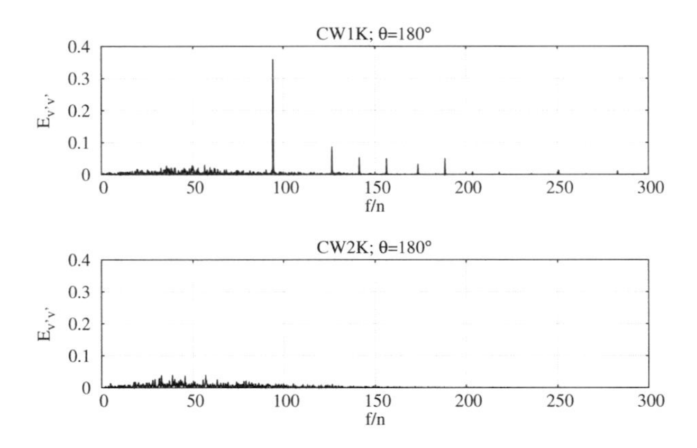

Fig. 9: Energy spectra computed from cross-correlations of the effective radial velocity fluctuations in the rim seal gap; CW1K (top), CW2K (bottom); r/R=0.98.

To explain the origin of the peaks observed in Figure 9, the frequencies are compared to the theoretical acoustic eigenfrequencies of a closed pipe. These frequencies are defined by

$$f = \frac{m}{2 \cdot L} \cdot a, \tag{7}$$

where a is the speed of sound, L the length of the resonator, and m the order of the harmonic. The speed of sound in the wheel space is almost constant and has the value of approximately $a/\omega R = 2.853$. Using the height of the wheel space $L = R - (2 \cdot d + s_c) - r_h = 86.65$ mm the harmonics in Table 1 can be determined.

Table 1: Harmonics of a closed pipe normalized by rotor speed f/n.

m	7	8	11	12	13	16	17	18	21	24
Theory	84.17	96.20	132.27	144.29	156.32	192.39	204.42	216.44	252.51	288.59
LES	79.53	94.33	126.6	141.5	156.3	188.7	203.5	218.3	250.6	283.1
rel. Error	-5.5 %	-1.9 %	4.2 %	-1.9 %	-0.01 %	-1.9 %	-0.5 %	0.9 %	-0.8 %	-1.9 %

Despite the strong simplifications introduced by the 1D model, the comparison of the LES data for CW1K from Figure 9 with the theoretical finding shows a good match. Especially, the 8th harmonic is very close to $f/n \approx 94.33$. The strongest deviations between the theoretical values and the LES results occur for the frequencies close to a multiple of the blade passage frequency. For case CW1K several of the wheel spaces' harmonics are excited, which results in the development of a standing wave inside the wheel space.

Filtering the flow field of the effective radial velocity inside the rim seal gap at the frequency of the dominant 8th harmonic at $f/n = 94.33$, shows how the fluctuations can influence the hot gas ingress, see Figure 10. A square pattern consisting of

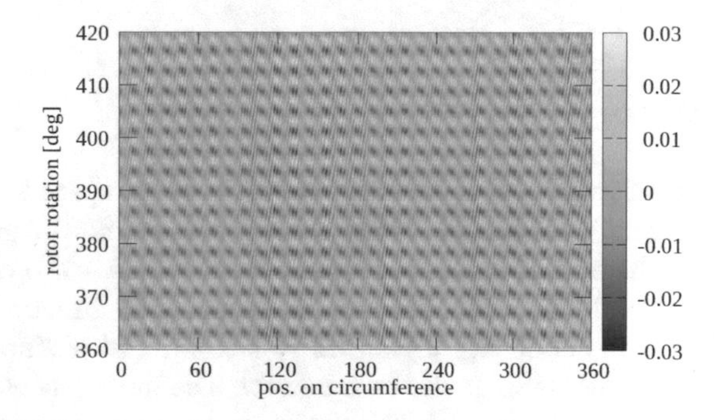

Fig. 10: Filtered effective radial velocity $\tilde{v}_r/\omega R$ inside the rim seal gap at r/R = 0.98; CW1K.

30 fields in the azimuthal direction and in time is evident for the filtered effective radial velocity $\tilde{v}_r$, which correspond to the 30 stator blades. The temporal fluctuation repeats approximately every 3.82° rotor rotation and corresponds to the standing wave inside the wheel space. This wave leads to alternating positive and negative values of the effective radial velocity.

The radial distribution of the time and circumferentially averaged cooling effectiveness for cases CW1K and CW2K is shown in Figure 11. The cooling effectiveness is computed by

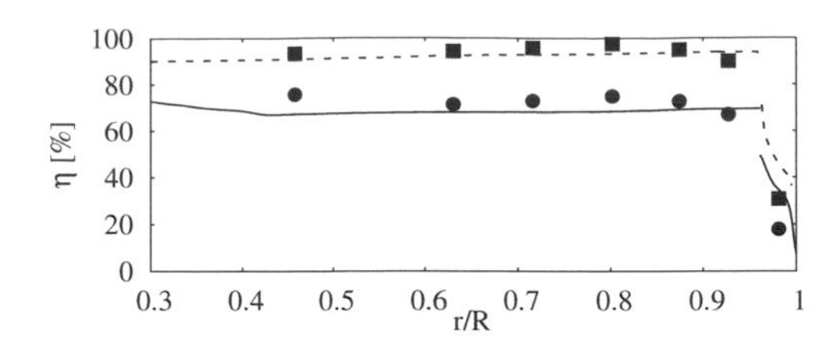

Fig. 11: Radial distribution of the time and azimuthally averaged cooling effectiveness; LES: CW1K (–), CW2K (– – –), exp. data [4]: CW1K (•), CW2K (▪).

$$\eta = \frac{Y(r) - Y_{hg}}{Y_{cg} - Y_{hg}}, \tag{8}$$

where Y is the concentration of a tracer gas mixed into the cooling gas. The concentration is obtained from a passive scalar transport equation. The subscripts hg and cg denote the concentrations at the main flow and the cooling gas inlet. In both cases the LES results match the experimental data from [4] convincingly. The fact that the cooling effectiveness is reduced for case CW1K compared to CW2K shows the importance of accurately resolving the instantaneous flow field.

4.3 Reduced setup

In the following, the comparison of the flow field in the wheel space for the GT setup is discussed for a pure LES and the zonal RANS/LES method. Figure 13 shows the comparison of the instantaneous tracer gas concentration in the wheel space in the axial cut at $x/s = 0.875$, a cross-section in the periodic direction and a radial cross-section inside the boundary layer $r/R = 1.0043$. The position in periodic direction is chosen at the trailing edge of the stator. The instantaneous cooling gas concentration of the zonal RANS/LES setup shows a good agreement with the instantaneous cooling gas concentration of the pure LES.

Figure 13 shows the radial distributions of the cooling effectiveness η and the components of the Reynolds stress tensor $\overline{u'u'}$ and $\overline{v'v'}$ at the axial position $x/s = 0.98$. The zonal RANS/LES setup shows a good agreement in the radial distributions of the cooling effectiveness η compared to the pure LES setup validating the described STG for shear layer flows. A good quantitative agreement of the relevant components of the Reynolds stress tensor $\overline{u'u'}$ and $\overline{v'v'}$ is shown in Figure 13. The STG is therefore capable of generating a physically meaningful turbulent energy spectrum from the RANS domains reducing the extent of the scale resolving LES to a region of interest whilst reducing the computing time for the zonal RANS/LES method by reducing the number of grid cells.

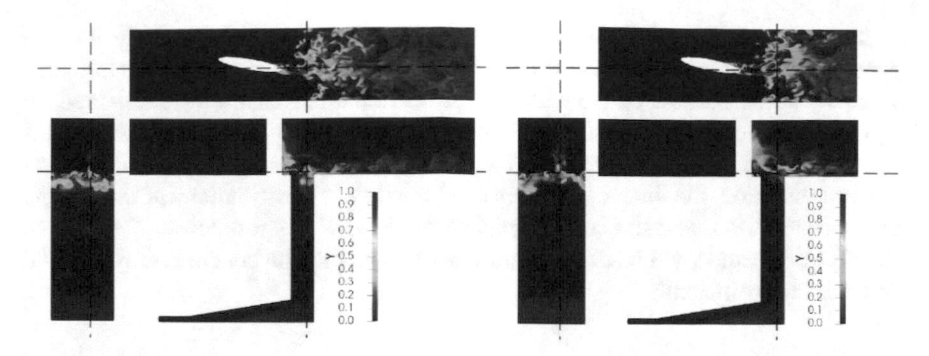

Fig. 12: Distribution of the instantaneous cooling gas concentration inside the wheel space of the (left) pure LES setup and (right) zonal RANS/LES setup for the GT geometry.

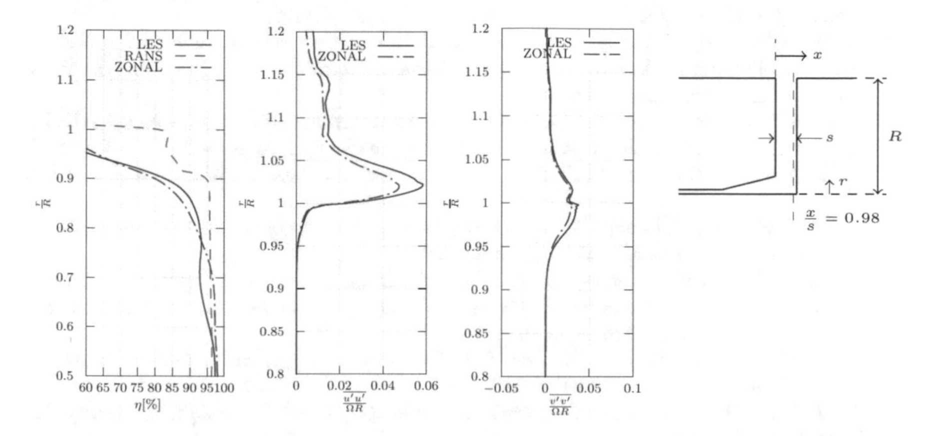

Fig. 13: Radial distributions of the cooling effectiveness η and the components of the Reynolds stress tensor $\overline{u'u'}$ and $\overline{v'v'}$ at the axial position $x/s = 0.98$

5 Conclusion

The flow field in an axial turbine stage was predicted by LES for two cooling gas mass flow rates. After about 40 full rotor rotations a statistically converged flow field in the wheel space was obtained. The analysis showed that the instantaneous results strongly depend on the cooling gas mass flow rate. For the lower cooling gas mass flux (case CW1K), several of the wheel space's harmonics are excited, which generate oscillations of the radial velocity component in the rim seal gap. Especially at the lower frequent oscillations, cooling gas is ejected out of the wheel space, followed by an injection of hotter gas from the main annulus. This leads to a reduced cooling effectiveness compared to the case CW2K.

The presented zonal RANS/LES method employs a synthetic eddy method generating the turbulent fluctuations at the inflow and a reconstruction of the turbulent viscosity at the outflow of the embedded LES domain. Results were obtained for a generic turbine setup and compared to a pure LES setup. The zonal RANS/LES method showed good agreement with the pure LES setup in the instantaneous cooling gas concentration, the time and spanwise averaged radial distributions of the cooling gas concentration, and the components of the Reynolds-stress tensor. This zonal RANS/LES method will be used in future work to predict the hot gas ingress in axial turbines more efficiently.

References

1. P. F. Beard, F. Gao, K. S. Chana, and J. Chew. Unsteady flow phenomena in turbine rim seals. *Journal of Engineering for Gas Turbines and Power*, 139(3), 2016.
2. M. Berger and M. Aftosmis. Progress towards a cartesian cut-cell method for viscous compressible flow. In *50th AIAA Aerospace Sciences Meeting including the New Horizons Forum and Aerospace Exposition*, 2012.
3. D. Bohn, A. Decker, H. Ma, and M. Wolff. Influence of sealing air mass flow on the velocity distribution in and inside the rim seal of the upstream cavity of a 1.5-stage turbine. In *Volume 5: Turbo Expo 2003, Parts A and B*, Turbo Expo: Power for Land, Sea, and Air, pages 1033–1040, 2003.
4. D. Bohn and M. Wolff. Entwicklung von berechnungsansätzen zur optimierung von sperrgassystemen für rotor/stator-kavitäten in gasturbinen, 2001.
5. D. Bohn and M. Wolff. Improved formulation to determine minimum sealing flow – cw, min – for different sealing configurations. In *Volume 5: Turbo Expo 2003, Parts A and B*, Turbo Expo: Power for Land, Sea, and Air, pages 1041–1049, 2003.
6. J. P. Boris, F. F. Grinstein, E. S. Oran, and R. L. Kolbe. New insights into large eddy simulation. *Fluid Dynamics Research*, 10(4):199–228, 1992.
7. S. Catris and B Aupoix. Embedded les-to-rans boundary in zonal simulations. *Aerosp. Sc. Technol*, 4:1–11, 2000.
8. J. W. Chew, F. Gao, and D. M. Palermo. Flow mechanisms in axial turbine rim sealing. *Journal of Mechanical Engineering Science*, 233(23-24):7637–7657, 2019.
9. M. Chilla, H. Hodson, and D. Newman. Unsteady interaction between annulus and turbine rim seal flows. *Journal of Turbomachinery*, 135(5), 2013.
10. J. T. M. Horwood, F. P. Hualca, J. A. Scobie, M. Wilson, C. M. Sangan, and G. D. Lock. Experimental and computational investigation of flow instabilities in turbine rim seals. *Journal of Engineering for Gas Turbines and Power*, 141(1), 10 2018.
11. J. T. M. Horwood, F. P. Hualca, M. Wilson, J. A. Scobie, C. M. Sangan, G. D. Lock, J. Dahlqvist, and J. Fridh. Flow instabilities in gas turbine chute seals. *Journal of Engineering for Gas Turbines and Power*, 142(2), 01 2020.
12. T. Hösgen, M. Meinke, and W. Schröder. Large-eddy simulations of rim seal flow in a one-stage axial turbine. *Journal of the Global Power and Propulsion Society*, 4:309–321, 2020.
13. R. Jakoby, T. Zierer, K. Lindblad, J. Larsson, L. deVito, Dieter E. Bohn, J. Funcke, and A. Decker. Numerical simulation of the unsteady flow field in an axial gas turbine rim seal configuration. In *Volume 4: Turbo Expo 2004*, Turbo Expo: Power for Land, Sea, and Air, pages 431–440, 2004.
14. D. König, M. Meinke, and W. Schröder. Embedded les-to-rans boundary in zonal simulations. *Journal of Turbulence*, 67:1–25, 2010.
15. M. Meinke, W. Schröder, E. Krause, and Th. Rister. A comparison of second- and sixth-order methods for large-eddy simulations. *Computers & Fluids*, 31(4):695–718, 2002.

16. A. Niemöller, M. Schlottke-Lakemper, M. Meinke, and W. Schröder. Dynamic load balancing for direct-coupled multiphysics simulations. *Computers and Fluids*, 199(104437), 2020.
17. A. Pogorelov, M. Meinke, and W. Schröder. Large-eddy simulation of the unsteady full 3d rim seal flow in a one-stage axial-flow turbine. *Flow, Turbulence and Combustion*, 102(1):189–220, 2019.
18. A. Pogorelov, L. Schneiders, M. Meinke, and W. Schröder. An adaptive cartesian mesh based method to simulate turbulent flows of multiple rotating surfaces. *Flow, Turbulence and Combustion*, 100(1):19–38, 2018.
19. B. Roidl, M. Meinke, and W. Schröder. A zonal rans-les method for compressible flows. *Computers and Fluids*, 67:1–15, 2015.
20. Bernd Rudzinski. *Experimentelle Untersuchung des Heißgaseinzuges in die Rotor-Stator-Zwischenräume einer eineinhalbstufigen Turbine für unterschiedliche Dichtkonfigurationen: Zugl.: Aachen, Techn. Hochsch., Diss., 2009*. Verlag Dr. Hut, München, 2009.
21. L. Schneiders, C. Günther, M. Meinke, and W. Schröder. An efficient conservative cut-cell method for rigid bodies interacting with viscous compressible flows. *Journal of Computational Physics*, 311:62–86, 2016.

Direct numerical simulation of bypass transition under free-stream turbulence for compressible flows

Duncan Ohno, Björn Selent, Markus J. Kloker and Ulrich Rist

Abstract Direct numerical simulations (DNS) of bypass transition of a flat-plate boundary-layer subject to free-stream turbulence (FST) were performed. A synthetic turbulent inflow is generated by superposition of modes of the continuous spectrum resulting from linear stability theory (LST). The present article briefly explains the adaption of the method for a fully compressible framework, including the generation of turbulence as well as the treatment of the inflow boundary conditions. The results of a flat-plate flow at three different integral turbulent length scales were compared and validated with results from the literature. In addition, a performance analysis of the codes and methods used was performed on the HLRS *Hawk* supercomputer.

1 Introduction

Fluid flows above a critical ratio of inertia to viscous forces $Re = \tilde{U}_\infty \tilde{l}_{\text{char}}/\tilde{\nu}$ usually undergo a transition from laminar to turbulent state when triggered by suitable perturbations. Even though there is a wide range of low-order models to mimic the effects of this transition it is still often necessary to perform highly resolved numerical simulations of both the onset and the transition process itself. High-fidelity simulations allow on the one hand to gain a deeper understanding of the physical details of the transient and non-linear interactions and on the other hand they enable the development of improved models to predict the impact on industrial-scale applications. This is even more true for high-speed flow where not only increased momentum transfer but also heat flux and a change in thermodynamical properties result from the laminar-turbulent transition.

D. Ohno, B. Selent, M. Kloker and U. Rist
Institut für Aero- und Gasdynamik (IAG), Universität Stuttgart, Pfaffenwaldring 21, D-70550 Stuttgart, Germany, e-mail: {ohno,selent,kloker,rist}@iag.uni-stuttgart.de

W. E. Nagel et al. (eds.), *High Performance Computing in Science and Engineering '21*,
https://doi.org/10.1007/978-3-031-17937-2_13

A common method for generating free-stream turbulence in DNS is the utilization of multiple Fourier-modes scaled to a turbulent spectrum as inflow boundary condition. However, this method is only applicable when the computational domain covers the flow field well ahead of the aerodynamic body. Extending the DNS to the leading edge of a flat-plate or an airfoil results in an increase in computational expenses and may lead to numerical instabilities.

Jacobs & Durbin [1] described for the first time a method for generating artificial free-stream turbulence, while taking the wall boundary condition at the inflow of the simulation domain into account. In this approach, modes of the continuous spectrum resulting from linear stability analysis are used instead of Fourier-modes. These modes are a solution of the underlying linear stability problem of the laminar baseflow including the boundary-layer. Introducing these modes with small amplitudes at the inflow boundary satisfies the conservation equations, since the modes are a solution of the linearized Navier-Stokes equations. In recent years, intensive studies were carried out using this method, including investigations on 2D flows with pressure gradient by Zaki & Durbin [2], 3D flows by Schrader *et al.* [3], corner flows by Schmidt & Rist [4] or in an airfoil flow by Ohno *et al.* [5].

However, most of the previous studies were conducted for incompressible flow problems. The present paper presents an adaption of this method for compressible flows, using results from Brandt *et al.* [6] for validation.

2 Numerical methods

Direct numerical simulation

All simulations are conducted with a revised version of the high-order in-house DNS code *NS3D* [7] which solves the three-dimensional compressible Navier-Stokes equations in cartesian coordinates. Spatial derivatives are discretized with 8th-order explicit finite differences, while an explicit 4th-order 4-step Runge–Kutta scheme is used for time integration.

The solution vector consists of the numerical fluxes, i.e. $\mathbf{Q} = [\rho, \rho u, \rho v, \rho w, E]^T$. Here, ρ is the density, u, v, w are the velocity components in x-y-z coordinates—representing streamwise, wall-normal and spanwise directions, respectively—and E is the total energy per volume. Furthermore, the thermodynamic variables temperature T and pressure p are derived from the solution vector. Dimensional variables are denoted with $\tilde{\bullet}$. The quantities in both, DNS and the linear stability solver (including frequencies and wavenumbers), are nondimensionalized using free-stream values $\tilde{U}_\infty$, $\tilde{\rho}_\infty$, $\tilde{T}_\infty$ and a characteristic length $\tilde{l}_{\text{char}}$.

Generation of free-stream turbulence

The continuous modes were calculated using an in-house solver for the compressible Navier-Stokes equations linearized about a steady baseflow [8] at the inflow location of the DNS domain. The resulting linear system is solved directly by LAPACK [9] routines for eigenvalue problems.

The complex solution vector $\hat{\mathbf{q}} = (\hat{\rho}, \hat{u}, \hat{v}, \hat{w}, \hat{T})^T$ contains the 1D eigenfunctions of the flow field variables in y-direction of each mode. While $\hat{u} = \hat{v} = \hat{w} = \hat{T} = 0$ is applied at the wall, a homogeneous Neumann boundary condition $\partial\hat{\mathbf{q}}/\partial y = 0$ is used at the free-stream. Examples for the shape of these eigenfunctions are depicted in Fig. 2(a). Evidently, the development of these modes is periodic in the free-stream, but rapidly dampened in the boundary-layer due to shear sheltering [10].

In order to create a broad spectrum of disturbances, multiple modes need to be selected from the wavenumber domain α-γ-β, representing the wavenumbers in the spatial directions x, y and z, respectively. The free-stream modes move with the free-stream velocity $c = U_\infty = \omega/\alpha$. Considering Taylor's hypothesis of frozen turbulence, the streamwise wavenumber is therefore connected to the frequency $\alpha_r \approx \omega_r$ for both spatial and temporal analysis in the non-dimensional formulation with $U_\infty = 1$. The spanwise wavenumber β is always a real number and can be considered as a parameter. The corresponding (complex) wall-normal wavenumber γ can be determined using the dispersion relation

$$\gamma^2 = iRe_{\delta^*} \cdot (\alpha U_\infty + \beta W_\infty - \omega) - \alpha^2 - \beta^2 \tag{1}$$

which is given in Mack [8] or Schrader *et al.* [3]. Using temporal theory—where α is a real number—and assuming $\omega_r = \alpha$ for free stream modes, the theoretical complex eigenvalue

$$\omega = \alpha - \frac{i}{Re_{\delta^*}}(\alpha^2 + \beta^2 + \gamma_r^2 - \gamma_i^2) \tag{2}$$

with

$$\gamma_i = \frac{Re_{\delta^*}}{2\gamma_r}(\alpha(1 - U_\infty) - \beta W_\infty) \tag{3}$$

can be calculated. For 2D-baseflows with $U_\infty = 1$ and $W_\infty = 0$, $\gamma_i = 0$ applies, simplifying Eq. (2). With this relation, the complex eigenvalue ω of the required mode m can be searched for in the spectrum of the stability solution at α_m and β_m with the desired wall-normal wavenumber γ_m.

The disturbances generated by superposition of individual modes can be described with the modal ansatz

$$\mathbf{q}'(x, y, z, t) = \sum_{m=1}^{M} A_m \hat{\mathbf{q}}_m(y) \cdot \mathrm{e}^{i(\alpha_m x + \beta_m z - \omega_m t)} \tag{4}$$

where A_m represents the amplitude coefficient. While the wall-normal wavenumber γ_m is determined by the eigenfunction $\hat{\mathbf{q}}_m(y)$, a Fourier ansatz is chosen for the disturbances in spanwise direction with the wavenumber β_m thus satisfying periodicity. Aforementioned Taylor's hypothesis of frozen turbulence allows streamwise wavenumbers to be set via frequency $\alpha_m \approx \omega_m$. In cases where the modes are prescribed in more than one point of the flow direction x, it is necessary to transform the temporal amplification rate ω_i to the spatial one α_i by applying Gaster's transformation [11]

$$\frac{\omega_i^{\text{temporal}}}{\alpha_i^{\text{spatial}}} = -\frac{\partial \omega_r}{\partial \alpha_r}. \tag{5}$$

This transformation guarantees the correct growth of the modes within the forcing region.

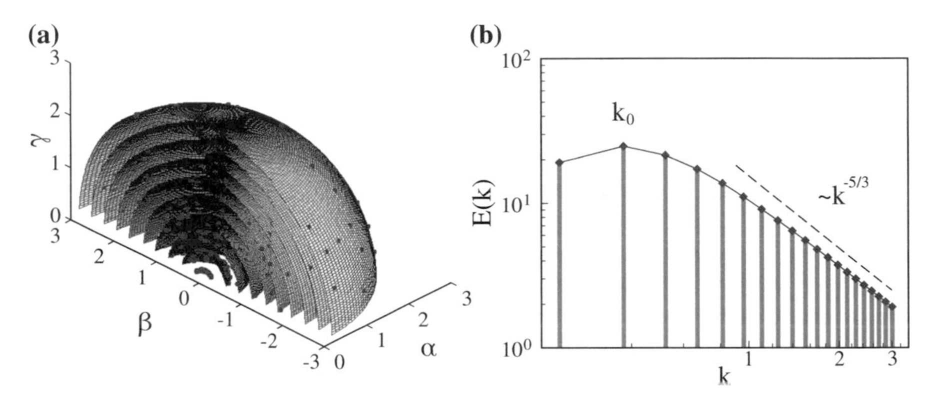

Fig. 1: (a) Selection of free-stream modes in the α-γ-β wavenumber domain on 20 spheres (only 10 depicted) with identical wavenumber magnitudes. (b) Exemplary von Kármán energy spectrum; the columns represent the energy of each sphere.

To ensure the isotropy of the free-stream turbulence, the wavenumber vector $\mathbf{k}_m = [\alpha_m, \gamma_m, \beta_m]^T$ must fulfill certain constraints. Similar to the approach of Brandt *et al.* [6], free-stream modes are distributed in the wavenumber domain α-γ-β over concentric spheres with identical wavenumber magnitudes $k = \sqrt{\alpha^2 + \gamma^2 + \beta^2}$. This is achieved by random rotation of a dodecahedron in the wavenumber domain, where the 20 vertices represent the wavenumber vectors $\mathbf{k}_m$. An example with 20 spheres, including the randomly distributed wavenumbers, is depicted in Fig. 1(a). For the wall-normal and streamwise wavenumbers, the condition $\alpha > 0$ and $\gamma > 0$ applies, thus a limitation is applied.

Finally, the amplitudes of all modes A_m in Eq. (4) need to be defined with the spectral energy distribution $E(k)$. In most investigations known from literature, the von Kármán spectrum is applied, which is defined as

$$E(k) = \frac{2}{3} \frac{a(kL)^4}{(b + (kL)^2)^{17/6}} Lq. \tag{6}$$

Here, the constants $a = 1.606$ and $b = 1.350$ are used and the turbulent kinetic energy is defined as $q = \frac{3}{2}Tu^2$ with the turbulence intensity Tu. The parameter L denotes the integral length scale. The amplitude coefficient is given by

$$A_m^2(k) = E(k)\frac{\Delta k}{N_{\text{shell}}}, \tag{7}$$

where Δk denotes the difference in wavenumber between two contiguous shells and N_{shell} the number of all spheres. An example of a von Kármán spectrum is depicted in Fig. 1(b). The wavenumber of maximum energy k_0, which can be calculated according to Tennekes & Lumley [12] with the integral length scale $k_0 = \frac{1.8}{L}$, is clearly visible.

Modified boundary conditions

In all simulations the wall boundary condition was chosen to be adiabatic, and periodic boundary conditions were used for the spanwise direction. For compressible subsonic flows, characteristic boundary conditions at the inflow, outflow, and free-stream are advantageous, since acoustic reflections can be prevented, see Giles [14]. These boundary conditions are effective in one spatial direction as well as in time and allow harmonic perturbations to leave the domain while still prescribing the baseflow. In the proximity of the boundary, the disturbance flow field

$$\phi_c' = \phi - \phi_{\text{ref}} \tag{8}$$

is transformed to characteristic variables based on a one-dimensional decomposition of the Euler equations into upstream and downstream travelling waves. For an inflow boundary on the left side of the domain, the characteristic variables can be calculated with

$$\begin{pmatrix} c_1 \\ c_2 \\ c_3 \\ c_4 \\ c_5 \end{pmatrix} = \begin{pmatrix} -a_{\text{ref}}^2 & 0 & 0 & 0 & 1 \\ 0 & 0 & \rho_{\text{ref}} \cdot a_{\text{ref}} & 0 & 0 \\ 0 & 0 & 0 & \rho_{\text{ref}} \cdot a_{\text{ref}} & 0 \\ 0 & \rho_{\text{ref}} \cdot a_{\text{ref}} & 0 & 0 & 1 \\ 0 & -\rho_{\text{ref}} \cdot a_{\text{ref}} & 0 & 0 & 1 \end{pmatrix} \cdot \begin{pmatrix} \rho_c' \\ u_c' \\ v_c' \\ w_c' \\ p_c' \end{pmatrix} \tag{9}$$

where a represents the speed of sound. In case of steady-state boundary conditions, all variables with the subscript ref are set to baseflow values $\phi_{\text{ref}} = \phi_{\text{bf}}$. The characteristic variables represent entropy perturbations (c_1), vorticity perturbations in the spanwise and streamwise directions (c_2, c_3), as well as downstream and upstream travelling sound waves (c_4, c_5). The boundary condition sets the incoming disturbances (c_1-c_4)

to zero and extrapolates the characteristic variable c_5 of the upstream travelling acoustic wave. In a final step, the primitive variables are computed by an inversion of Eq. (9) and added to the current flow field.

For simulations of bypass transition due to free-stream turbulence, the characteristic boundary conditions described above are suitable for the free-stream as well as for the outflow. However, for the inflow condition, they must be manipulated in order to allow the introduction of unsteady disturbances into the domain. In this case, the calculated superposition of all modes from Eq. (4) is added to the baseflow

$$\phi_{\text{ref}}(t) = \phi_{\text{bf}} + \phi'(t) \tag{10}$$

at every (sub)iteration and used for the characteristic boundary condition treatment. This allows disturbances to be introduced into the domain, while at the same time allowing upstream acoustic waves to exit the domain.

An additional approach to reduce reflections and improve numerical stability is to apply sponge zones in front of the boundary conditions. Using equation

$$\frac{\partial \mathbf{Q}}{\partial t} = \left.\frac{\partial \mathbf{Q}}{\partial t}\right|_{\text{NS}} - G(\mathbf{x}) \cdot (\mathbf{Q} - \mathbf{Q}_{\text{ref}}), \tag{11}$$

the time derivative of the conservative numerical fluxes $\mathbf{Q}$ of the unsteady solution is forced to the reference state, with a gain function $G(\mathbf{x})$ specifying the magnitude and spatial distribution. Typically, this function describes the fading from a maximum value of the gain G at the boundaries to the null parameter $G = 0$ in the inner evaluable computational domain. For simulations like presented in this paper it is essential to use such damping zones in the free stream as well as at the outflow to avoid distortions due to the boundary conditions. In those areas, the values of the steady baseflow ϕ_{bf} are used to calculate the conservative reference field $\mathbf{Q}_{\text{ref}}$ in order to reduce disturbances.

However, when using a sponge zone at the inflow, the introduced disturbances must again be taken into account. Analogously to the treatment of the unsteady characteristic boundary conditions, see equation Eq. (8), a primitive variable field can be generated at each time step at the inflow region by adding the perturbation field from Eq. (4) to the baseflow. This field is converted into a conservative field $Q_{\text{ref}}(t)$ at each time step and considered in Eq. (11). With this method, the sponge acts simultaneously as an unsteady forcing region—which also contains the attenuation rates of the modes via the Gaster transformation—as well as a dampening zone for outgoing waves. This approach has also been used successfully for an inflow with a turbulent boundary-layer flow, see Appelbaum *et al.* [15].

3 Numerical simulations

Setup

For further validation and further investigations, a bypass transition scenario in a Blasius boundary-layer flow—simulated by Brandt *et al.* [6]—has been employed. The first three cases presented in this publication, 'Case1', 'Case2' and 'Case3', were reproduced. The cases differ only in the spectrum of the free-stream turbulence (see Tab. 1), but not in the domain dimensions.

The Blasius boundary-layer displacement thickness at the beginning of the computational domain $\tilde{\delta}_0^*$ determines the characteristic length $\tilde{l}_{\text{char}}$ for all presented simulations. In the aforementioned paper, a domain size of $101.3\delta_0^* \leq x \leq 1101.3\delta_0^*$, $0.0 \leq y \leq 100.0\delta_0^*$ and $0.0 \leq z \leq 90.0\delta_0^*$ was employed, with the Reynolds number $Re_{\delta_0^*} = 300$ at the inflow. Therefore, $x = 0$ represents the theoretical location of the flat-plate leading edge.

Table 1: Overview of performed simulations with parameters used to define FST according to Brandt *et al.* [6]; FTT: flow-through time.

	Free-stream turbulence			Simulation properties			
Ma	Label	L/δ_0^*	*Tu*	FTT total	FTT stat.	Iter. total	CPU-h
0.3	Case1	5.0	4.7%	12.4	9.9	7270k	777.0k
0.3	Case2	2.5	4.7%	4.9	2.3	2840k	303.5k
0.3	Case3	7.5	4.7%	12.4	9.9	7270k	777.0k
0.7	Case1	5.0	4.7%	18.3	13.5	3060k	327.1k
0.7	Case2	2.5	4.7%	13.8	8.9	2300k	245.8k
0.7	Case3	7.5	4.7%	18.3	13.5	3060k	327.1k

However, in the present investigations a prolonged computational domain of $101.3\delta_0^* \leq x \leq 1450.6\delta_0^*$ has been used in order to reduce the influence of the outflow boundary and to allow an extended sponge zone. In contrast to the incompressible simulations of the reference, a Mach number must be specified for the compressible simulations with *NS3D*. In the investigations, two Mach numbers, $Ma = 0.3$ and $Ma = 0.7$ have been selected, where the former can be considered as a quasi-incompressible case. The baseflows for both cases were acquired with steady direct numerical simulations using a 2D Blasius boundary-layer solution as boundary and initial conditions.

The free-stream turbulence intensity is set to $Tu = 4.7\%$ for all cases using 800 eigenmodes. At this relatively high turbulence intensity, instantaneous nonlinear interaction occurs, although the low-amplitude single modes are still governed by the linear theory when simulated individually. However, three different integral length scales L are selected to determine the free-stream turbulence: $L = 5.0\delta_0^*$ (Case1), $L = 2.5\delta_0^*$ (Case2) and $L = 7.5\delta_0^*$ (Case3). The von Kármán energy spectrum for $L = 5.0\delta_0^*$ is depicted in Fig. 1(b).

A grid resolution of $1800 \times 180 \times 256$ points in (x-y-z) direction was used. The simulations were decomposed in all spatial directions to $60 \times 6 \times 6 = 1440$ MPI processes. Furthermore, all domains were parallelized in 4 openMP threads in spanwise direction, leading to a total of 5760 processes on 45 nodes. Regarding the computing time, all important information is given in Tab. 1, including the flow-through time (FTT, number of runs through domain) for the entire simulation and for the recording/averaging of flow statistics. It is noted that compressible simulations with lower Mach numbers require a smaller timestep for time integration. For the given flow this means that the case with $Ma = 0.3$ needs about 3.5 times more steps then for the calculation with $Ma = 0.7$ to simulate the same physical time span.

Single free-stream disturbances

To demonstrate the unsteady inlet boundary condition, results of 2D simulations with single free-stream disturbances (FSD) in a shorter domain with $101.3\delta_0^* \leq x \leq 460.8\delta_0^*$ in streamwise direction are presented first. In three separate simulations, free-stream modes—normalized with the maximum of $\hat{u}$—with amplitude $A_m = 0.001$ were introduced at the inflow boundary. The real part of the $\hat{u}$ components of the modes with the same angular frequency of $\omega = 1.0$ (therefore with streamwise wavenumber $\alpha_r = 1.0$), but different wall-normal wavenumbers γ, are shown in Fig. 2(a). The selected wavenumbers lie within the range of the discretized free-stream turbulence spectrum of the performed simulations of bypass transition. For larger γ, the amplitudes of the modes at the boundary-layer edge increase. The modal streamwise amplitude development of the corresponding modes in the simulations is depicted in Fig. 2(b). In accordance with the temporal linear stability analysis in conjunction with Gaster's transformation (dashed lines) the decay rate α_i of the FSD increases with higher wall-normal wavenumbers γ. The boundary conditions described in the previous chapter can be considered validated since the introduced disturbances enter the domain without discontinuity at the inlet and then behave according to the damping rates predicted by LST in downstream direction.

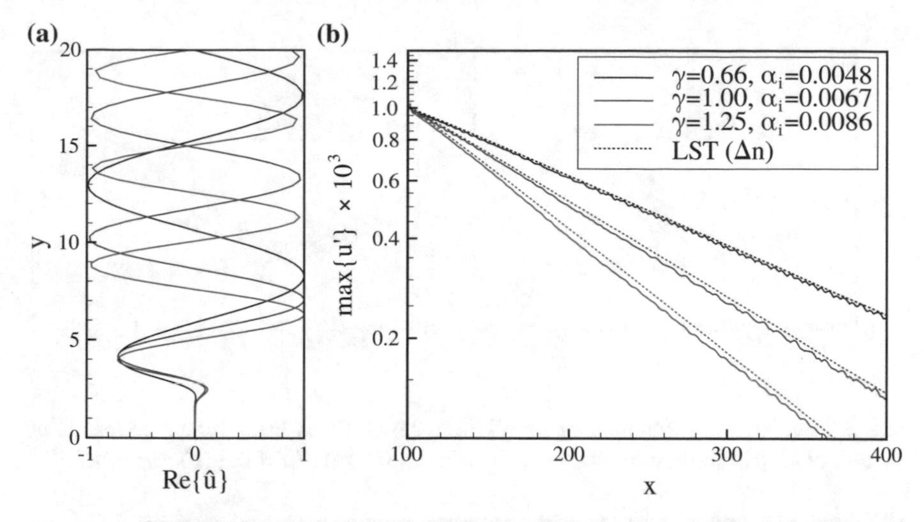

Fig. 2: (a) Shape of eigenmodes from the continuous spectrum. (b) Streamwise development of modal u'-amplitudes for different FSD with $\omega = \alpha_r = 1.0$.

Bypass transition

Fig. 3(a) shows a very good agreement of the transition location indicated by the skin friction coefficient c_f between the reference and the simulations performed with *NS3D* for the low Mach number for Case1 and Case3. Furthermore, the overshoot is resolved very well. Minor differences are caused by slight compressibility effects or differences in the numerical schemes. As in the reference, transition does not occur at all in Case2. However, the c_f curve lies somewhat closer to the analytical laminar solution, which could possibly indicate that the presented adapted method involves less numerical perturbations.

The skin friction coefficient c_f of the simulations with the higher Mach number of $Ma = 0.7$ is depicted in Fig. 3(b). The transition behavior of Case1 and Case3 is similar to the low Mach number case or the incompressible reference, but already exhibits significant differences. The onset of the transition seems to start at similar positions—however, the c_f value does not jump as rapidly to the analytic solution of the turbulent flow. Furthermore, in both cases an overshoot is hardly visible. The skin friction coefficient curve for Case2 is again even closer to the analytical solution. Thus, it can be concluded that there is a clear Mach number effect, which can only be achieved with a numerical setup of this type.

The wall normal distribution of the RMS values of the velocity components, u_{rms}, v_{rms}, and w_{rms} versus the streamwise position (averaged in spanwise direction) can be seen in Fig. 4(a)-(c) for Case3 ($L = 7.5\delta_0^*$) at $Ma = 0.3$. The distributions show very good agreement with the results from the reference [6], where the same contour lines were chosen. In addition, the distribution of the RMS values of the density ρ_{rms} can now be plotted analogously. As expected, ρ_{rms} correlates most closely with the

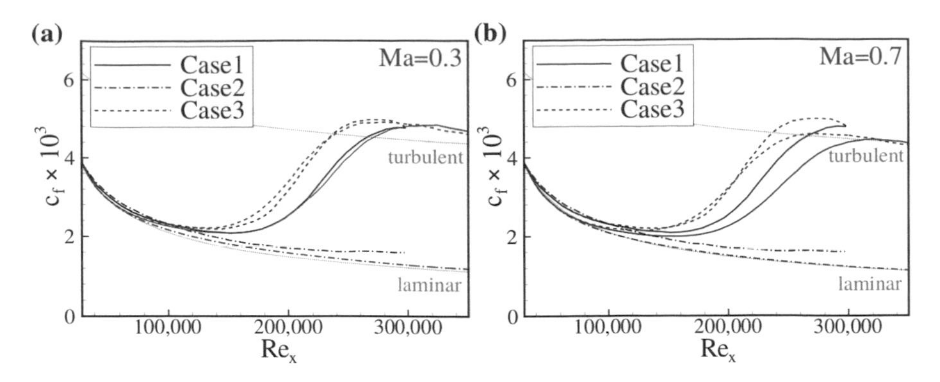

Fig. 3: Skin friction coefficient c_f for all three cases. Black lines: reference results by *Brandt et al.* [6]. Red/blue lines: results with *NS3D*. (a) $Ma = 0.3$ (b) $Ma = 0.7$.

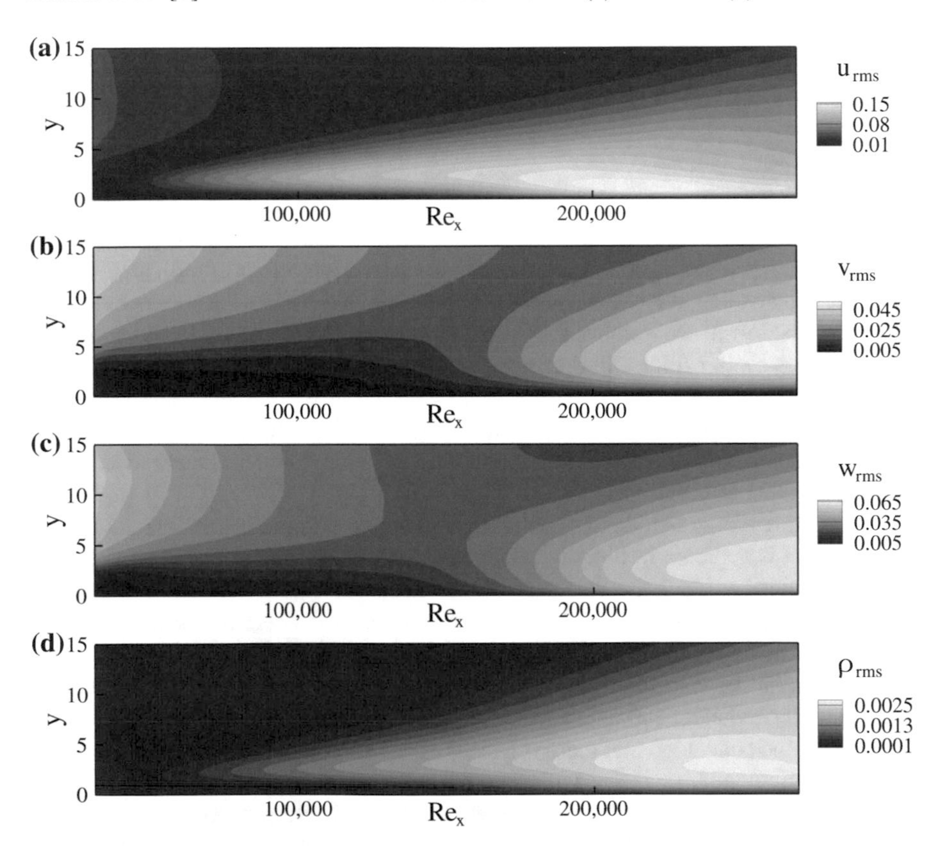

Fig. 4: RMS-values for Case3 ($L = 7.5\delta_0^*$) with $Ma = 0.3$. (a)-(c) velocities u, v, w (d) density ρ.

RMS distribution of u. Fig. 5(a) and Fig. 5(b) are depicting visualizations of the

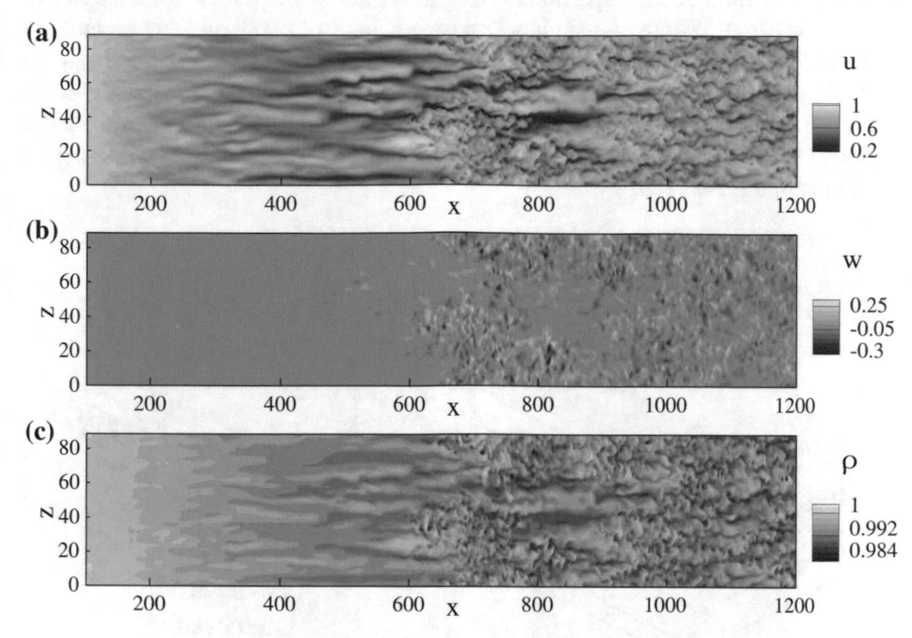

Fig. 5: Instantaneous flowfield at $y/\delta_0^* = 2$ from simulation for Case3 ($L = 7.5\delta_0^*$) with $Ma = 0.3$, (a) Streamwise velocity u (b) Spanwise velocity w (c) Density ρ.

instantaneous streamwise velocity u and spanwise velocity w for Case3 ($L = 7.5\delta_0^*$), respectively. The corresponding plots can also be found in [6], with which they display a very good qualitative agreement. The clearly visible high and low speed streaks—also known as Klebanoff modes—can be regarded as precursors to bypass transition. Secondary instabilities of these modes are leading to patches of turbulence which are finally causing a fully turbulent boundary-layer [13]. Due to the compressible setup, the variation of the density ρ can be shown again in this case, see Fig. 5(c).

4 Performance analysis

The *NS3D* code has been programmed from the start with a strong focus on parallel efficiency and scalability. The finite difference method lends itself quite naturally to techniques for high performance computing and allows for a combination of several techniques such as vectorization of array operations, distributed computing and symmetric multiprocessing.

The application of these methods makes *NS3D* to be equally well suited for both vector and scalar CPU systems. The number of processes used in distributed computing correlates with the number of block structured domains of equal size. The decomposition is realized by Message Passing Interface (MPI) [16] routines and is arbitrary in all three spatial directions as long as the number of points in the respective direction is equal. Within domain blocks a second layer of parallelization is achieved by threading the spanwise coordinate direction using openMP [17] directives. An exemplary decomposition of the computational domain is shown in Fig. 6. The setup of all test cases, i.e. initial and boundary conditions as well as scheme settings, was identical to the one described in §3 except for the domain size which was adapted according to the tests run.

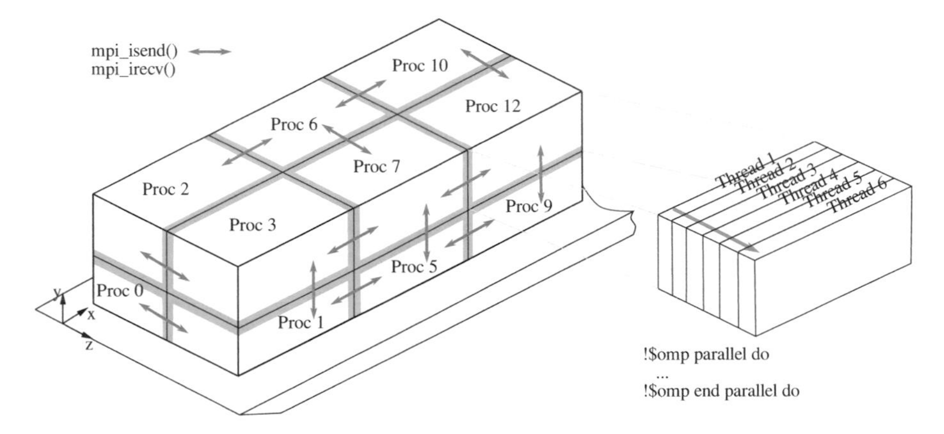

Fig. 6: Schematic of domain decomposition and parallel distribution within *NS3D*

Single node performance

At first a scaling analysis was done on a single node of the *Hawk* system. These tests provided insight in how to optimally distribute processes and threads on the nodes and thus make optimal use of the highly hierarchical structure of the AMD EPYC Rome CPUs.

A mesh of $240 \times 180 \times 256$ points in (x-y-z) direction was used. The simulations were run for 1000 time steps and used 1–128 cores. Three different modes, namely pure distributed computing, pure symmetric multiprocessing and a combination of processes and threads were compared. Fig. 7 shows the speed-up $S = t_1/t_{\mathrm{NP}}$ and parallel efficiency $E = S/\mathrm{NP}$, where $t_\bullet$ is the CPU time and NP is the number of processing elements. For pure MPI parallelization, a linear speed-up and almost ideal efficiency up until 32 processes is achieved. For more than 64 processes no performance gains are obtained. Pure openMP parallelization does not show any significant performance gains for more than 16 cores. The hybrid mode demonstrates a speed-up for up to 128 cores, albeit small beyond 64 cores. It can be concluded that using 32 MPI processes and four openMP threads should generally be the preferred setting for simulations using *NS3D*. This combination ensures fast access on shared L3 cache within a core complex (CCX) (cf. the linear speed-up up to four cores for pure openMP).

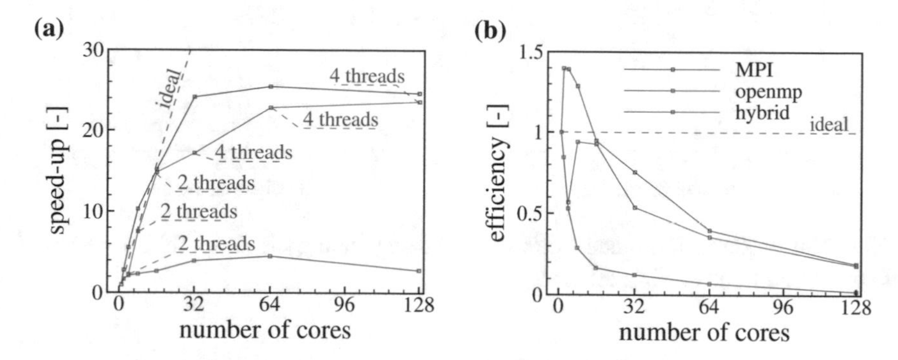

Fig. 7: Single node performance on *Hawk*: (a) Strong scaling speed-up; (b) Strong scaling efficiency

Multi node performance

Multi node scaling tests were subsequently using 32 processes and four threads on each node. Both strong and weak scaling tests were run. Two different grids were used for strong scaling tests in order to avoid simulations with an unfeasible small number of grid points per core. The number of points and the respective distribution for the computations are listed in Tab. 2. Furthermore in order to smooth out distortions due to node errors tests have been run ten times each and averaged. The effectiveness

Table 2: Grid points and parallel distribution

Case / nb. nodes	1	4	16	64	256	1024
a: 480x180x256	4-2-4,4[a]	8-4-4,4	16-4-8,4	x	x	x
b: 3720x240x256	4-2-4,4	x	x	32-4-16,4	64-8-16,4	128-8-32,4

[a] MPIx-MPIy-MPIz,openMP

of the parallel programming techniques used in *NS3D* can be seen in Fig. 8. Strong scaling computations show continuously super-linear speed-up for up to 256 nodes and linear speed-up until 1024 nodes. The efficiency is always above one accordingly.

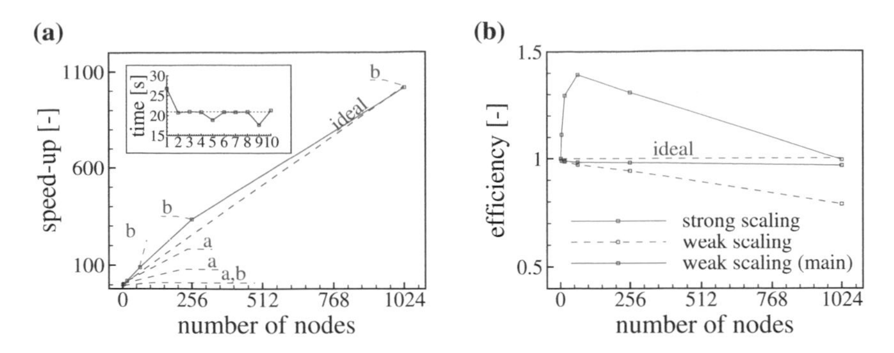

Fig. 8: Multi node performance on *Hawk*: (a) Strong scaling speed-up; (b) Comparison of strong and weak scaling efficiency

The scaling performance varied somewhat especially for the last test case on 1024 nodes because of individual node errors which led to poor performance as can be seen by the inlay in Fig. 8(a). The probability to include a node with either faulty hardware or poor inter-node connection obviously increases when using a number of cores of order 10^5. Weak scaling computations were done with the number of grid points per computational element, i.e. a single node, set to 3686400. For 1–256 nodes the efficiency is almost ideal with a drop of only 6%. Even for 1024 nodes an efficiency of 79% percent can be achieved. This performance drop can be attributed to initialization routines. If the time to complete the main loop is compared only (cf. solid line in Fig. 8(b)) the efficiency is almost ideal for up to 1024 nodes. As the test cases had been run only for 1000 time steps this timer was deemed the more appropriate indicator for performance measurements.

In conclusion it can be said that the historically proven strong parallel performance of *NS3D* can also be uphold on the most recent HLRS system

5 Conclusions

This study presents results of high-fidelity direct numerical simulations of compressible transitional boundary-layer flows. The laminar-turbulent transition is triggered by introducing isotropic free-stream turbulence into the flow through appropriate boundary conditions. The turbulent disturbances were computed from linear stability analysis following an established method which was adapted to include compressibility. It was shown that the disturbances are a realistic representation of isotropic turbulence above walls; disturbance growth followed theoretical predictions and shear sheltering is captured. Very good agreement with reference cases from the literature was achieved for quasi-incompressible flow speeds.

The results of the transition simulations show a clear Mach number dependency of the flow. For identical length scales the onset of transition is delayed for higher Mach number. The velocity fluctuations are in very good agreement with results from literature. Additionally, the fluctuation density is quantified thus allowing for a future estimation of not only wall friction but also wall heating.

In scaling tests the revised version of the *NS3D* solver proved to maintain its strong parallel performance on a massively parallel system such as the *Hawk*. The combination of techniques for distributed computing and symmetric multiprocessing resulted in super-linear and linear speed-up when running on up to 256 and 1024 compute nodes res. Both strong and weak scaling consistently demonstrated near ideal efficiency up to 1024 nodes.

Acknowledgements Duncan Ohno acknowledges funding from the Bundesministerium für Wirtschaft und Energie through the LuFo project "LAINA". Direct numerical simulations were performed on resources provided by the High Performance Computing Center Stuttgart (HLRS) under grant GCS_Lamtur, ID44026.

References

1. Jacobs, R.G., Durbin, P.A.: Simulations of bypass transition. J. Fluid Mech. **428**, 185–212 (2001)
2. Zaki, T.A., Durbin, P.A.: Continuous mode transition and the effects of pressure gradient. J. Fluid Mech. **563**, 357–388 (2006)
3. Schrader, L.U., Brandt, L., Henningson, D.S.: Receptivity mechanisms in three-dimensional boundary-layer flows. J. Fluid Mech. **618**, 209–241 (2009)
4. Schmidt, O., Rist, U.: Numerical investigation of classical and bypass transition in streamwise corner-flow. Procedia IUTAM **14**, 218–226 (2015)
5. Ohno, D., Romblad, J., Rist, U.: Laminar to turbulent transition at unsteady inflow conditions: Direct numerical simulations with small scale free-stream turbulence. New Results in Numerical and Experimental Fluid Mechanics XII (2020)
6. Brandt, L., Schlatter, P., Henningson, D.S.: Transition in boundary layers subject to free-stream turbulence. J. Fluid Mech. **517**, 167–198 (2004)
7. Babucke, A., Linn, J., Kloker, M., Rist, U.: Direct numerical simulation of shear flow phenomena on parallel vector computers. In High Performance Computing on Vector Systems (pp. 229-247). Springer, Berlin, Heidelberg. (2006)
8. Mack, L.M.: Boundary-layer linear stability theory. AGARD Report No. 709 (Special Course on Stability and transition of Laminar Flow), pp. 3/1–81 (1984)
9. LAPACK: Linear Algebra PACKage. http://netlib.org/lapack
10. Hunt, J., Durbin, P.: Perturbed vortical layers and shear sheltering. Fluid Dyn. Res. **24**(6), 375–404 (1999)
11. Gaster, M.: A note on the relation between temporally-increasing and spatially-increasing disturbances in hydrodynamic stability. Journal of Fluid Mechanics. Vol. 14 (1962) 222–224.
12. Tennekes, H., Lumley, J.L.: A First Course in Turbulence. MIT Press, Cambridge (1972)
13. Durbin, P.A.: Perspectives on the phenomenology and modeling of boundary layer transition. Flow Turbul. Combust. **99**(1), 1–23 (2017)
14. Giles, M.B.: Nonreflecting boundary conditions for Euler equation calculations. AIAA journal 28.12 (1990): 2050-2058
15. Appelbaum, J., Ohno, D., Rist, U., Wenzel, C.: DNS of a Turbulent Boundary Layer Using Inflow Conditions Derived from 4D-PTV Data. Experiments in Fluids. (2021) [In review]

16. N.N.: MPI - A message-passing interface standard. Technical Report CS-94-230, University of Tennessee, Knoxville (1994)
17. OpenMP: The OpenMP API Specification for Parallel Programming. http://www.openmp.org

Direct numerical simulation of a disintegrating liquid rivulet at a trailing edge

Adrian Schlottke, Matthias Ibach, Jonas Steigerwald and Bernhard Weigand

Abstract Water ingestion in gas turbines is used in industrial applications to improve the thermal efficiency by cooling the air before and throughout compression. However, this also leads to interactions between the liquid droplets and the compressor parts, which cause a faster degradation of the structure. The current work addresses the numerical investigation of the atomization process at the trailing edge of a compressor blade as there have only been experimental investigations in literature considering the ambient conditions in a gas turbine compressor. Direct numerical simulations (DNS) are carried out using the multiphase flow solver Free Surface 3D (FS3D). Therefore, a numerical setup has been developed to model the trailing edge as a thin plate corresponding to experiments performed at ITLR [22]. Four different cases have been performed to take the experimentally observed atomization processes into account. Additionally, the dependence of the simulation results on the grid resolution and with it the limits to reproduce the experimental findings has been investigated in a grid study. The results show that the developed numerical setup works well and the different atomization processes are reproduced qualitatively, with best results for very high grid resolutions.
Furthermore, an investigation of the available compilers on the new Hawk supercomputer platform reveals that a performance gain up to 36% for the amount of completed calculation cycles per hour is possible compared to the standard compiler setup used as standard practice over the previous years. Optimization options as well as improvement during link time lead to a significant speed-up while simulation results remain unaffected.

Adrian Schlottke
Institute of Aerospace Thermodynamics (ITLR), University of Stuttgart, Pfaffenwaldring 31, 70569 Stuttgart, Germany, e-mail: adrian.schlottke@itlr.uni-stuttgart.de

W. E. Nagel et al. (eds.), *High Performance Computing in Science and Engineering '21*,
https://doi.org/10.1007/978-3-031-17937-2_14

1 Introduction

The ingestion of water droplets into gas turbine compressors, also called Fogging or High-Fogging, is used to improve the thermal efficiency of the turbine by cooling the air before and throughout the compression. However, many of the ingested droplets interact with the compressor parts, e.g. liquid films accumulate on the blade surface and disintegrate at the trailing edge, generating new droplets. The droplets impacting onto the blades lead to erosion and a faster degradation of the material. The focus of this work lies on the numerical investigation of the trailing edge disintegration as an atomization process. The atomization process of different nozzles, e.g. circular, flat, and also prefilming air-blast atomizers has been investigated in literature, both experimentally and numerically. Atomization of liquid jets has been investigated experimentally for a long time and many authors have contributed to the understanding of the process, e.g. the work of Rayleigh [18] and Weber [25] are well known. Dumouchel [2] gives a detailed overview of experimental investigations on jet and sheet atomization including also more recent work as well. Simulations and especially highly resolved Direct Numerical Simulations (DNS) of jet breakup just became feasible with increasing computational power in recent years [4, 19]. However, these can only be performed at supercomputers and therefore Evrard [6] proposed a hybrid Euler-Lagrange method, especially for jet breakup phenomena, to reduce computational efforts.

Although the same morphological atomization behavior occurs for air-assisted jets and for trailing edge disintegration, the limits and transition from one atomization regime to another depend strongly on the application. The atomization process of prefilming air-blast atomizers resembles qualitatively the trailing edge disintegration at compressor blades. Experimental investigations on planar and swirl prefilming air-blast atomizers were performed in [8] and [9] and have shown the dependence of the atomization process and the resulting ejected droplets on the prefilmer geometry and the ambient conditions. Numerically, Koch et al. [15] investigated planar prefilm atomizer using a smoothed particle method and a 2D setup. The findings were validated with experimental results from [9]. Nevertheless, the used 2D setup with the smoothed particle method has some shortcomings, e.g. the small computational domain in longitudinal direction, which disables the representation of the complete liquid breakup process. Additionally, atomization at the trailing edge of a prefilmer and of a compressor blade differ quantitatively due to the influence of the walls within the prefilmer. The small geometrical distance between liquid film surface and top wall influences the liquid film behavior, which is not the case for compressor blades.

Regarding the trailing edge disintegration at compressor blades, there exists a lack of numerical data as only experimental investigations were reported in literature to the best knowledge of the authors. The investigations mainly focus on the statistical quantification of the atomization process of a liquid film at a trailing edge. Kim [14] uses high-speed shadowgraphy imaging to evaluate the ejected droplet diameter, acceleration and disintegration frequency. The experimental setup relates to conditions of large steam turbine blades. Consecutively, Hammit et al. [10] investigate the change in droplet diameter spectra as a function of distance from the trailing edge. A more recent experimental test campaign of Javed et al. [12, 13] investigates the influence of the trailing edge thickness and the blade's angle of attack on the disintegration of a liquid rivulet. It has been shown that at higher ambient velocities, i.e. when breakup is dominated by aerodynamic forces, a larger trailing edge thickness leads to a smaller distribution angle of the ejected droplets. At a constant Weber number, depending on the trailing edge thickness, the ejected droplet diameters correlate with trailing edge thickness. An increasing angle of attack leads to increased droplet sizes due to decreased local air momentum in the region after the trailing edge. Additionally, it is stated that the droplet size distribution is independent of the liquid mass flow rate. So far, the process of trailing edge disintegration under gas turbine conditions has only been investigated experimentally. The experiments focus on an integral observation of the process. This is important for a first understanding of the process, but for a complete prediction, it is important to understand the local process and the underlying physics. For this reason, DNS performed with FS3D will enable the insight into the liquid structure during breakup. As a first step, this work provides the numerical setup and first results for comparison with the findings from detailed experiments performed in [22].

2 Mathematical description and numerical approach

The in-house CFD code Free Surface 3D (FS3D) performs DNS of incompressible multiphase flows and is continuously developed at ITLR for the last 25 years. Several recent studies show the applicability of FS3D to simulate highly dynamic multiphase processes like droplet deformation [20], droplet impacts onto dry and wetted surfaces [1, 7], droplet collisions [17] and atomization of liquid jets [5]. Simulations of such complex problems require high spatial and temporal resolution and thus a high parallel efficiency of the used solver. For this reason, FS3D is parallelized with MPI as well as OpenMP. FS3D solves the governing equations for mass and momentum conservation on finite volumes

$$\partial_t \rho + \nabla \cdot (\rho \mathbf{u}) = 0, \tag{1}$$

$$\partial_t (\rho \mathbf{u}) + \nabla \cdot (\rho \mathbf{u} \otimes \mathbf{u}) = \nabla \cdot (\mathsf{S} - \mathsf{I} p) + \rho \mathbf{g} + \mathbf{f}_\gamma. \tag{2}$$

In equations (1-2), $\mathbf{u}$ denotes the velocity vector, p the static pressure, $\mathbf{g}$ the gravitational acceleration, S the shear stress tensor and I the identity matrix. The term $\mathbf{f}_\gamma$ represents the body force which is used to model surface tension at the interface. The governing equations are solved in a one-field formulation where the different phases are regarded as a single fluid with variable physical properties across the interface. FS3D uses the Volume-of-Fluid (VOF) method by Hirt and Nichols [11] to identify these different phases. An additional variable f is introduced, which is defined as

$$f(\mathbf{x},t) = \begin{cases} 0 & \text{outside the liquid phase,} \\ (0,1) & \text{at the interface,} \\ 1 & \text{inside the liquid phase.} \end{cases} \tag{3}$$

The variable f is then advected using the transport equation

$$\partial_t f + \nabla \cdot \left(f \mathbf{u} \right) = 0. \tag{4}$$

The corresponding f-fluxes are calculated using the PLIC method by Rider and Kothe [21] to maintain a sharp interface. By using the volume fraction f local variables such as the density can be calculated as

$$\rho(\mathbf{x},t) = \rho_l f(\mathbf{x},t) + \rho_g \left(1 - f(\mathbf{x},t)\right). \tag{5}$$

The surface tension forces are modeled by the continuous surface stress (CSS) model by Lafaurie et al. [16]. Further details on numerical implementations and applications of FS3D are given e.g. in Eisenschmidt et al. [3].

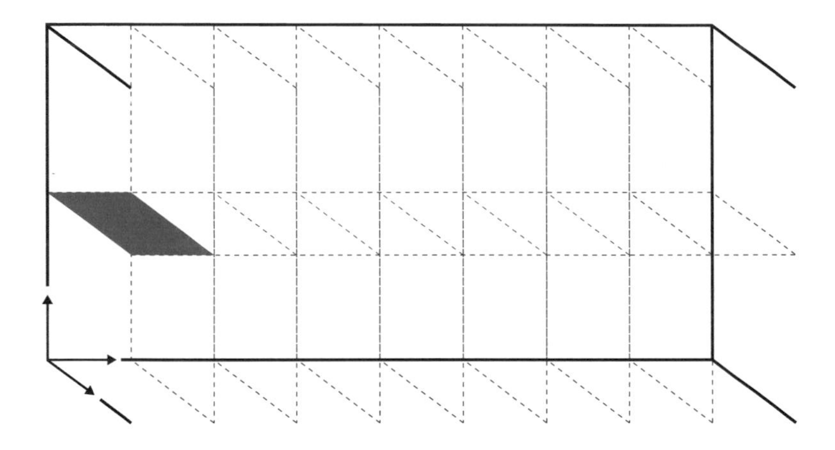

Fig. 1: Computational domain for simulation of a disintegrating liquid rivulet at a trailing edge.

Table 1: Inflow conditions.

Inflow velocity (m/s)		Circle segment (mm)		Re_G (-)	Disintegration regime
Air u_g	Liquid u_l	Height	Width		
11.5	0.103	1.35	2.70	15×10^3	Symmetrical Rayleigh breakup
19.2	0.172	0.81	2.70	25×10^3	Asymmetrical Rayleigh breakup
26.8	0.199	0.70	2.70	35×10^3	Bag breakup
34.5	0.199	0.70	2.70	45×10^3	Bag breakup

3 Computational setup

The simulation of the trailing edge disintegration at a thin plate is performed in a rectangular computational domain as visualized in fig. 1. A rivulet flows on top and over the edge of a thin horizontal plate of length L in the middle of the domain. The dimensions of the domain are $16L \times 4L \times 8L$ in streamwise, spanwise and in plate-normal direction, respectively. In order to increase the spatial resolution, only half of the rivulet is simulated. On the left side of the domain at $x = 0L$ an inflow of both gaseous and liquid phase is applied. The gas flows with velocity u_g through the entire side into the domain whereas liquid flows with a lower velocity u_l on top of the thin plate into the domain with an idealized shape as a circular segment. In the present study, four different settings are investigated. Both velocities as well as the geometrical shape of the inflowing rivulet, defined by the maximum height and width, correspond to experimental findings of Schlottke and Weigand [22]. An overview of the inflow conditions with the corresponding Reynolds numbers of the air flow $Re_G = \rho_g u_g d_h / \mu_g$ is given in table 1. According to the experiments in [22], the characteristic length is defined by the hydraulic diameter $d_h = 32L$ of the experimental flow channel.

As can be seen in the last column of table 1, different disintegration scenarios were investigated: Symmetrical and asymmetrical Rayleigh breakup and bag breakup. In order to accurately reproduce these scenarios and to identify the necessary resolution of the used equidistant Cartesian grid, all cases were simulated with three different grid resolutions: A coarse grid with $512 \times 256 \times 128$ cells ($\Delta x = 78\,\mu m$), a grid with $1024 \times 512 \times 256$ cells ($\Delta x = 39\,\mu m$), and a fine grid with $2048 \times 1024 \times 512$ cells ($\Delta x = 20\,\mu m$).

The novelty of the present setup is the treatment of the thin plate. So far it has not been feasible to represent a non-moving, rigid part with a finite contact angle inside the computational domain with FS3D. To overcome this problem a new methodology for a special treatment of artificial inner boundaries was implemented in FS3D which is explained in the following. The whole computational domain is divided into equally sized blocks. Between the sides of adjacent blocks all available boundary conditions that are implemented in FS3D can be applied. In the present study the domain is divided into $16 \times 2 \times 1$ blocks in x- ,y-, and z-direction which are indicated by the gray dashed lines in fig. 1. For reasons of clearer visualization only half of the blocks in x-direction are shown. For the treatment of the thin plate a no-slip boundary condition with a constant contact angle of 90° is applied between the first two blocks in y-direction at the inlet side. In this way, an idealized thin plate with infinitesimal thickness is reproduced. Besides the inflow and the symmetry conditions at the domain boundary a continuous (homogeneous Neumann) boundary condition is applied at the right side of the domain at $x = 16L$, representing the outlet. All other sides use a free-slip boundary condition. Furthermore, the simulation accounts for gravitational forces with $g = 9.81\,\mathrm{m/s^2}$ in y-direction. The physical properties of gas and liquid are given in table 2.

Table 2: Physical properties of air and water.

	Density ρ (kg/m^3)	Dynamic viscosity μ (Pa s)	Surface tension σ (N/m)
Air	1.190	1.824×10^{-5}	73.0×10^{-3}
Water	998.2	1.000×10^{-3}	

4 Results and discussion

Within this chapter, the results of the DNS will be evaluated and compared to experimental findings from [22]. The discussion of the results is divided into three parts. First, a qualitative, morphological comparison of the breakup process is presented to highlight that the different breakup regimes and their individual features can be reproduced. The second part focuses on the behavior of the ligament at the trailing edge. Especially the geometrical dimensions, i.e. the width, height and maximum length before breakup, are of interest. At last, the ejected droplet diameter distribution will be discussed. The influence of grid resolution will be addressed in each part separately. Regarding the last part it has to be mentioned that the evaluation of droplets that are generated during the course of the simulation is not a trivial task. In all previous studies with FS3D, the counting and characterization of these droplets was always part of the post-processing. Only field data that were written during the simulation could be used. Additionally, loading the data into the memory again is very time-consuming. With the new generation of supercomputers like the HPE Apollo system (Hawk), however, the spatial grid resolution of the computational domain can be further refined. Due to that, the traditional way of droplet evaluation via post-processing has become inefficient and almost unfeasible. For this reason, the whole process of counting and full characterization of existing droplets/ligaments inside the computational domain (volume, position, velocity) has been shifted from the post-processing into the simulation cycle. A new fully parallelized output routine that is based on a Connected-component labeling (CCL) algorithm for identifying separate droplets was implemented in FS3D, which is described in the following. In a first step, all liquid ligaments within a local processor domain are identified by means of a CCL algorithm and they are marked incrementally and in a processor-dependent manner. The properties of each ligament are calculated during the labeling as well. The generated ligament marker field is shared afterwards by using an exchange routine of the solver which is normally used to update the ghost cells of each local processor domain. Due to that, a simple loop over all grid cells at the local domain boundary can be used to check if two differently marked regions belong to the same ligament. In case the markers of two neighboring grid cells differ, this pair of markers is added to a global adjacency list, representing an undirected graph. The connected components of the graph or the liquid ligaments, respectively, can then be identified by using the complete adjacency list and a depth-first search. In a last step, associated ligament characteristics are merged. This leads to a highly efficient way of extracting droplet information from the simulation during runtime, independent from other data that is written and almost independent from the chosen grid resolution.

The comparison of DNS and experimental results at specific points in time is difficult due to the strong statistical behavior of the breakup process. This is mainly affected by the changing initial conditions as each breakup relates on the previous one rather than being independent. This dependence increases for higher ambient gas velocities as the breakup frequency increases and, therefore, there is less time for relaxation between two breakups. Nevertheless, it is important to evaluate if the DNS can reproduce the morphological structures, which were observed experimentally and how they are affected by the grid resolution. Figure 2 shows selected side view snapshots of the trailing edge breakup for two different Re_G and two different breakup regimes, respectively. The experimental images are compared to DNS results for two different grid resolutions. At $Re_G = 15 \times 10^3$, symmetric Rayleigh breakup can be observed. The experimental images show the typical behavior where one large droplet is ejected straight in streamwise direction at breakup, often followed by few small satellite droplets. This can also be seen in the DNS results for both grid resolutions. At $Re_G = 45 \times 10^3$, the bag breakup regime can be observed. Here, the single breakup event cannot be distinguished anymore, but rather the flapping motion at the trailing edge is considered to distinguish between breakups. In the experiment, two states of the bag breakup can be observed: One is the thinning of the lamella right at the trailing edge where the bag is created. The other is the already fragmented liquid bag at the front of the ligament, which creates many small droplets from the lamella and few larger droplets from the rim. The DNS results also show the continuous flapping motion of the ligament. However, the simulation could not reproduce the creation of a bag. This may result from two reasons: One is the grid resolution. Even for the highly resolved case the thickness of the lamella is in the order of the size of a grid cell. Due to that, the interface cannot be accurately reconstructed anymore since both sides of the lamella influence each other. The second reason may be due to the symmetry plane at $z = 0$. In the context of 2D simulations, it has been stated in

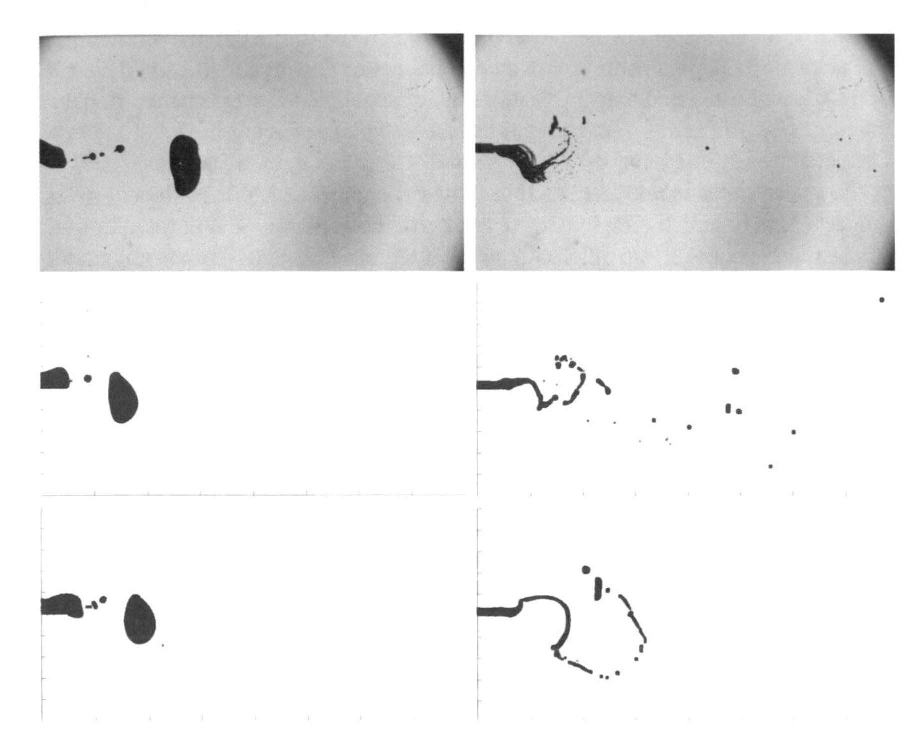

Fig. 2: Instantaneous morphological comparison between DNS results and experimental data for $Re_G = 15 \times 10^3$ and $Re_G = 45 \times 10^3$. Side view.

literature that instabilities in spanwise direction trigger the formation of liquid bags [9, 15]. Although spanwise instabilities are completely ignored in 2D simulations the use of the symmetry plane still suppresses these instabilities drastically.

In the following, a quantitative analysis of the liquid rivulet will be presented. It considers the geometrical dimensions, i.e. width and height of the rivulet close to the trailing edge and the maximum length of the ligament directly before breakup. This quantities are shown in fig. 3 in dependence of Re_G for all simulated cases. The results for the different grid resolutions are depicted side-by-side for better visualization. However, it needs to be mentioned that the simulations are performed at the same inflow conditions, i.e. the same Re_G. The shown results contain the range of evaluated values over time (up and down arrow), as well as the mean value (circle). The mean values for width and height are time averaged whereas the maximum length is averaged by the number of breakups. The dependence of the mean width and height on the grid resolution is rather small and lies in a range of up to 200 µm. While the mean width decreases from $Re_G = 15 \times 10^3$ to $Re_G = 25 \times 10^3$ and than stays constant the mean height is decreasing monotonously for higher Re_G.

This accords well with the experimental findings in [22], both qualitatively and quantitatively. A closer look at the range of width and height reveals a small difference of the results using different fine grids, especially for the cases with high ambient gas flow at $Re_G \geq 35 \times 10^3$. It can be seen that with higher grid resolution the range is increasing. This behavior can be explained by the wavy surface of the liquid rivulet at these Re_G. A sufficient grid resolution is needed to represent these surface waves and their influence on the movement of the liquid rivulet. In contrast, the grid resolution has a strong influence on the maximum length of the ligament before breakup. While for $Re_G = 15 \times 10^3$ the mean values lie quite close together, the difference grows up to 39% at $Re_G = 35 \times 10^3$. It is apparent that the maximum length at breakup diminishes for higher grid resolution. Looking at the range of maximum length, it can be stated that for $Re_G > 25 \times 10^3$ the difference between minimum and maximum value stays quite constant for the middle and high grid resolution with $\Delta \approx 5.0$ mm and $\Delta \approx 5.7$ mm, respectively. The general behavior of the maximum length is the same for all grid resolutions. It elongates from $Re_G = 15 \times 10^3$ to $Re_G = 25 \times 10^3$

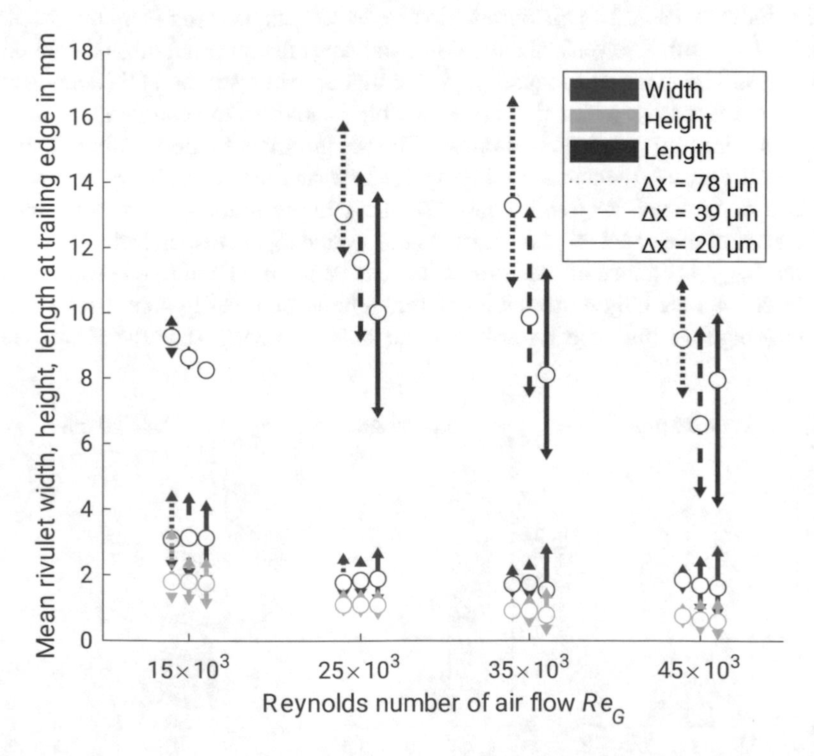

Fig. 3: Mean numerical results of rivulet width and height measured at the trailing edge. Length is defined as the maximum distance between ligament tip and trailing edge. Length is measured at the instance of breakup. Results are shown for the three resolutions used.

and decreases then monotonously. The same qualitative trend is observed in the experiments. However, it needs to be mentioned that the quantitative values are about 50% smaller than the experimental data. A possible reason for this is the inlet condition of the liquid rivulet as the defined velocity strongly affects the liquid's momentum at the trailing edge. Higher velocities at the inlet cause higher momentum at the trailing edge and lead to larger maximal length of the ligament before breakup. As a conclusion, it can be assumed that in case of symmetrical Rayleigh breakup, where only large droplets occur and individual breakups are nearly independent of one another, all used grid resolutions produce comparable results. At other breakup regimes, when the ejected droplets become smaller, more numerous and the influence of small perturbations, e.g. surface waves, is growing, higher grid resolutions are needed to resolve these phenomena adequately.

At last, the ejected droplet spectra will be evaluated and the influence of the grid resolution is shown. Here, only the case at $Re_G = 45 \times 10^3$ is discussed in detail. Figure 4 shows histogram plots for all three used grid resolutions.

The bars represent the probability of the corresponding equivalent spherical droplet diameter where the normalization is performed with the total number of ejected droplets for each case. The histogram bin size is 200 μm, starting from the droplet diameter $d = 0\,\mu\text{m}$. The vertical dashed line indicates the minimal droplet diameter which can still be resolved adequately. With a higher grid resolution this diameter is decreasing, respectively. The first histogram bin is adapted in accordance with the individual minimum resolvable diameter. The results show highest probabilities at the smallest droplet diameters, excluding the first smaller bin at $\Delta x = 78\,\mu\text{m}$ and $\Delta x = 20\,\mu\text{m}$. Although the trend is the same, the grid resolution influences the mean and the maximum ejected droplet diameters d_{mean} and d_{max}, respectively. The mean diameter d_{mean} is defined as the arithmetic mean value of all identified droplets and d_{max} represents the largest droplet found throughout the investigation. Both values are decreasing for a higher grid resolution, see table 3 for details and the comparison

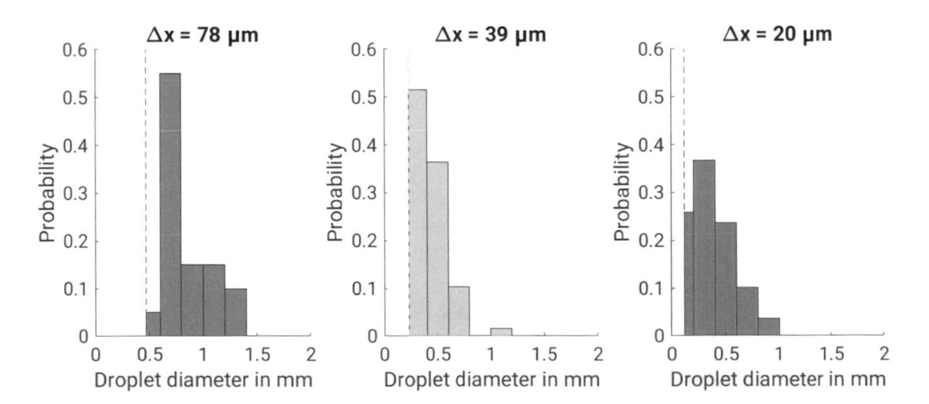

Fig. 4: Histogram of ejected droplets at the trailing edge for all three resolutions at $Re_G = 45 \times 10^3$.

to the experimental values from [22]. The mean diameter in the experiments is even smaller than the computed mean diameter with the highest grid resolution, although it is capable to resolve droplets down to 120 µm. The trend towards smaller ejected droplets, the limitation of the minimum resolvable droplet diameter, and the missing appearance of liquid bags before breakup indicate that even the highest resolution with $\Delta x = 20\,\mu\text{m}$ is not sufficient to predict the ejected droplet diameter distribution.

Table 3: Mean and maximum ejected droplet diameter.

Grid resolution Δx	78 µm	39 µm	20 µm	Experiment [22]
Number of grid cells	16.8 Mio	134.2 Mio	1,073.7 Mio	
d_{mean}	0.83 mm	0.42 mm	0.36 mm	0.21 mm
d_{max}	1.36 mm	1.11 mm	0.97 mm	1.13 mm

At this point, the reader is reminded of the already very high number of grid cells used which was about 1 billion for the highest resolution and is close to the upper limit of feasible computations using FS3D on HAWK. With the presented computational domain further grid refinement is not possible. This leads to the final remarks of this section where the shortcomings of the presented setup shall be addressed shortly. Due to the use of the symmetry plane the spanwise disturbances, which support the evolution of liquid bags, are limited. In comparison to the 2D simulations found in literature, this is still an improvement, as 2D simulations completely suppress interaction in spanwise direction. Furthermore, the trailing ligament is sensitive to the inlet conditions of the liquid. As a larger computational domain is not feasible in terms of computational power, further improvement of the inlet conditions is necessary to match the experimental findings. At last, the results indicate that a grid refinement is needed to enable the simulation of bag breakup and the subsequent droplet diameter distribution. This will only be possible with a smaller computational domain. The authors are aware of the limitations of the presented setup and future studies will focus on the improvement of the simulations. Nevertheless, the results show the successful representation of a thin plate and the atomization at the trailing edge and good qualitative agreement with experimental results.

5 Computational performance

The transition from one generation of high performance computing systems to another, such as from the Cray XC40 system “Hazel Hen” to the new AMD-based HPE Apollo supercomputer “Hawk” at HLRS, always entails new challenges but also possibilities when porting the users’ code and application. This step inevitably requires a good knowledge and thorough understanding of the interactions between the utilized code and the provided hardware, e.g. at core-, socket- and node-level, where the code gets executed and actual computational work is performed. One way of exploiting a possible speedup of the ported application as a first step is to tweak environmental settings and test available compilers and compiler options.

With the successful implementation of a tree structured communication in the implemented multigrid solver in FS3D [23] simulations with more than 8^3 processors are now practicable. This is of paramount importance especially for highly spatially and temporally resolved direct numerical simulations such as for the presented case of atomization of a liquid rivulet at a trailing edge. In the frame of this computational performance study we investigated the available compilers and compiler options to lay the foundation of an optimal setup for subsequent numerical investigations on the HPE Apollo platform. In the following section we present an updated benchmark case and provide a report on the performance analysis in terms of calculation cycles per hour (CPH) measuring the parallel performance of both strong and weak scaling.

5.1 Benchmark case and performance analysis

For the setup of the representative benchmark case we simulate an isolated but stretched droplet similar to fig. 2 resulting from the trailing edge disintegration at $Re_G = 25 \times 10^3$. The reason for the choice of this setup is the symmetrical character which distributes the computational load somewhat evenly making a performance analysis more practicable and comparable. The computational domain consists of an elongated and subsequently oscillating droplet initialized as an ellipsoid (semi-principal axis of $a = b = 1.357$ mm and $c = 0.543$ mm with an equivalent spherical diameter of $d = 2$ mm) in a cubic domain with an edge length of 8 mm. The fluid of the droplet is water at standard ambient conditions $T = 293,15$ K and pressure of one atmosphere similar to the cases presented in [23, 24].

For the analysis of strong and weak scaling we varied the number of processors from 2^3 up to 16^3 which corresponds directly to the initiated MPI-processes as we did not employ hyperthreading or parallelization on loop level with OpenMP. A baseline case for single core performance is added for the weak scaling investigation. For all simulations we tracked the exact time to initialize the simulation and set the walltime to 20 minutes in order to estimate the amount of completed calculation cycles per hour (CPH). To be independent of the interconnected file system and analyze plain computational performance we omitted the output of restart files, integral and field data. The simulations are performed with the latest revision of FS3D employing novel implementations including an improved advection scheme [26], a revision of the balanced CSF algorithm, the implicit viscosity calculation as well as the computation of normal vectors. Note that these numerical settings have been adapted compared to former FS3D benchmark cases and a direct comparison of calculation cycles per hour to previous results is therefore pointless.

Several compilers and optimization options are available within the HPE Apollo Hawk environment. The choices consist of an Intel Compiler (version 19.1.0), a GNU Compiler (version 9.2.0) and an AOCC AMD Optimizing C/C++ Compiler based on LLVM with a Fortran compiler flag (version 2.1.0) with the utilized version denoted in brackets. The optimization options, which can reduce code size and improve performance of the application, can be varied starting from a low level (O2) to a higher optimization (O3) finishing with a more aggressive optimization option (Ofast). Additionally, the technique of Link-Time Optimization (LTO for GNU and AOCC) and Interprocedural Optimization (IPO for Intel)[1] is available allowing the compiler to optimize the code at link time. During this, further rearrangement of the code from separate object files is performed. For the subsequent performance analysis we use a combination of the available compilers and compiler options with the possibility of link-time optimization to assess the optimal compiler setup leading to maximal performance on the new supercomputer system. A version without all optimization (optimization level O0, debug) was not investigated since the computational performance is significantly lower compared to low level optimization and the usage of this binary code is not feasible for the applied benchmark case.

5.1.1 Strong scaling

For the strong scaling performance we used a baseline case with a grid resolution of 512^3 cells while increasing the number of cores from 2^3 up to 16^3. The intermediate steps of $2 \cdot 8^3$ and $4 \cdot 8^3$ MPI-processes are also taken into account although the amount of cells per process is not cubic. As no calculation cycle could be performed for a single core during the allotted walltime this case is disregarded. We pinned processes and threads to CPU cores via *omplace* utilizing 64 of the available 128 cores to increase the available memory bandwidth per core. To have a representative statistical evaluation we performed ten runs for each investigated case spanning a period of several weeks for performance calculations. Simulations which included obvious performance deviations of more than 25% due to instabilities in the system were omitted. The presented results are therefore arithmetic means of the employed walltime, initialization time and the performed calculation cycles.

It has been standard practice in the past, for reasons of compatibility of locally available compilers, to compile with the Intel Compiler using the O2 optimization option. Therefore, we chose this case as the reference case on which all results are referred to. The strong scaling setup is summarized in table 4 along with the results of the peak performance case with 2048 processors.

Figure 5 depicts the estimated cycles per hour for 2^3, 8^3 and $4 \cdot 8^3$ MPI-processes for all regarded compilers and compiler options.

For the cases of 2^3 and 8^3 processes a slight increase in performance from the standard option O2 towards the highest optimization level Ofast can be observed for the GNU and Intel Compiler whereas no effect can be discerned for the AOCC

[1] In further context this option is denoted with a superscript ^{L} for all compilers and compiler options equivalently.

Table 4: Setup for strong scaling and calculation cycles per hour (CPH) at peak performance
(2048 processors) for all compilers and compiler options with Intel O2 as the reference case.

Problem size			512^3			
MPI-processes	2^3	4^3	8^3	$2 \cdot 8^3$	$4 \cdot 8^3$	16^3
Cells per process	256^3	128^3	64^3	$64^2 \cdot 32$	$64 \cdot 32^2$	32^3
Nodes	1	1	8	16	32	64
	Cycles per hour (CPH) at peak performance for $4 \cdot 8^3$ MPI-processes					
Compiler options	O2	O2^L	O3	O3^L	Ofast	Ofast^L
Intel (CPH)	3542	3741	3758	3815	3794	3862
	–	+5.6%	+6.1%	+7.7%	+7.1%	+9.0%
GNU (CPH)	3990	4281	4670	4701	4739	4831
	+12.6%	+20.8%	+31.8%	+32.7%	+33.8%	+36.4%
AOCC (CPH)	3528	3538	3542	3511	3420	3516
	−0.4%	−0.1%	0.0%	−0.9%	−3.4%	−0.7%

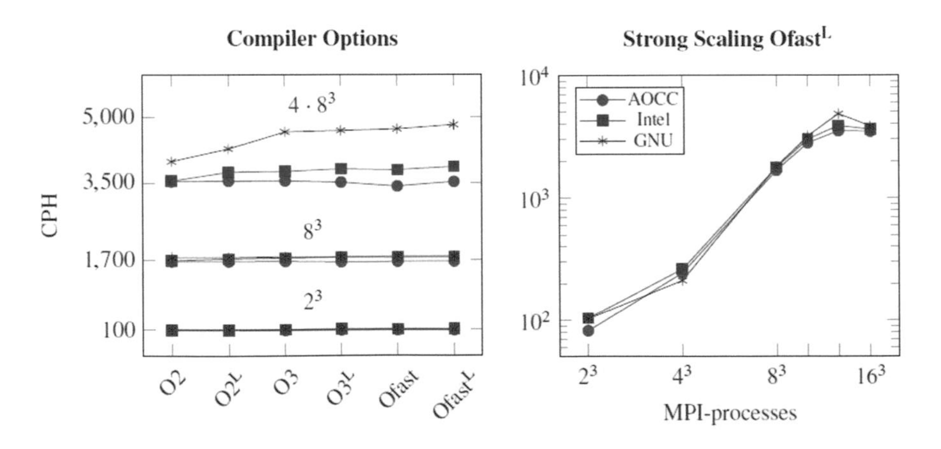

Fig. 5: Comparison of CPH for different compiler options for 2^3, 8^3 and $4 \cdot 8^3$ MPI-processes (constant problem size of 512^3 grid cells) and strong scaling performance for the Ofast^L case.

Compiler. Generally, the best performance is achieved by the GNU Compiler whereas the AOCC Compiler cannot reach the performance of the Intel reference case in any configuration. For the peak performance case of $4 \cdot 8^3$ MPI-processes this trend is more obvious: For both the GNU and the Intel Compiler the performance increases with an increasing level of optimization whereas the AOCC Compiler cannot exploit the potential of code optimization. The GNU Compiler shows a performance boost of about +20% for the Ofast^L option compared to the GNU standard O2 case while the Intel Compiler can still reach an approximate +9% rise. The best performance gain comparing the compilers is achieved with the GNU Compiler which outperforms the Intel reference case with more than +36% comparing the amount of calculated cycles per hour. The link-time optimization can increase the performance for each compiler and compiler optimization option up to 7% but mostly leads to a boost of approximately $1 - 3\%$ depending on compiler and optimization level. A drawback of the linking with LTO/IPO is that it takes a considerable amount of time longer than regular linking. However, this is quickly compensated with a steady use oft the compiled binary in subsequent simulations saving computational resources.

For the analysis of strong scaling we chose the compiler option Ofast^L for the sake of clarity as it revealed the maximum performance gain for all compilers. The estimated cycles per hour show a similar trend as it is expected from previous strong scaling analysis with FS3D. For the HPE Apollo Hawk system the peak performance is achieved with 2048 MPI-processes following a decrease in CPH with 4096 processors. For even more processors (not depicted here) the completed cycles per hour do not decrease drastically but stay constant, however, decreasing the strong scaling efficiency continuously. The significant performance increase for the GNU Compiler is most clearly visible at peak performance. It should be noted that the Ofast optimization level enables all optimization options disregarding strict standards compliance. This could lead to deviations in simulation results originating solely from the compilation of the code. However, for this benchmark case we compared integral and field data and confirmed that the results were binary identical.

5.1.2 Weak scaling

For the weak scaling analysis we kept the cells per processor constant at 64^3 while progressively increasing the number of allotted MPI-processes from one to 16^3. This leads to problem sizes ranging from 64^3 up to 1024^3 grid cells. The setup is summed up in table 5 along with the selected results of 16^3 processors.

Some general trends and findings from the strong scaling analysis are also observed for the weak scaling measurements: Higher optimization levels also lead to a performance gain (see figure 6) where an effective optimization of the code is noticeable for the GNU and the Intel Compiler. Note that the case for the weak scaling with 8^3 processors is already depicted in figure 5 as it is the same setup. The Intel and AOCC Compiler are surpassed by the GNU Compiler for all options and investigated cases for weak scaling by up to 45% for the single and the eight processor case and still 7.5% with 4096 processors.

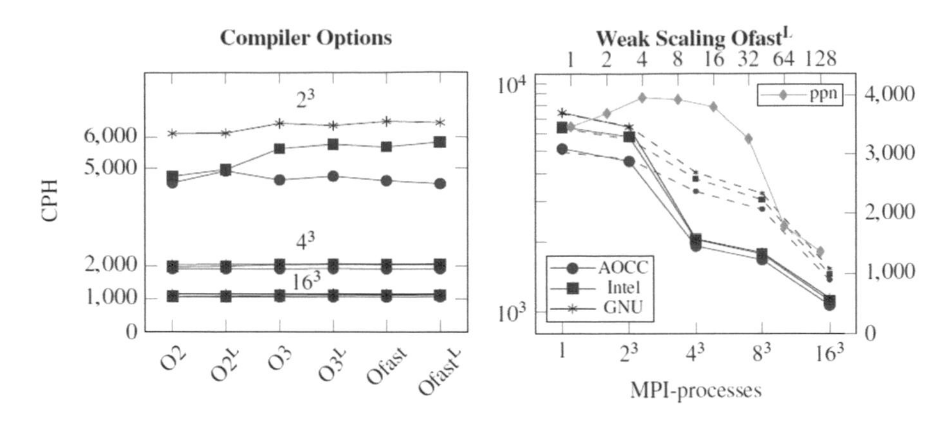

Fig. 6: Comparison of CPH for different compiler options for 2^3, 4^3 and 16^3 MPI-processes (constant cells per process of 64^3) and weak scaling performance for the Ofast^L case. In green: Results for the GNU Ofast^L case with 8^3 processors with a variation of allotted processes per node (ppn).

For the weak scaling, however, the biggest influence of the compiler options can be observed for one and eight MPI-processes and decreases with an increasing amount of MPI-processes. The link-time optimization only leads to a maximum boost of

Table 5: Setup for weak scaling and calculation cycles per hour (CPH) for 4096 processors for all compilers and compiler options with Intel O2 as the reference case.

Cells per process			64^3			
Problem size	64^3	128^3	256^3	512^3	1024^3	
MPI-processes	1	2^3	4^3	8^3	16^3	
Nodes	1	1	1	8	64	
	Cycles per hour (CPH) for 16^3 MPI-processes					
Compiler options	O2	O2^L	O3	O3^L	Ofast	Ofast^L
Intel (CPH)	1061	1074	1094	1109	1110	1109
	–	+1.2%	+3.1%	+4.6%	+4.6%	+4.6%
GNU (CPH)	1140	1142	1140	1142	1141	1140
	+7.5%	+7.6%	+7.4%	+7.7%	+7.5%	+7.5%
AOCC (CPH)	1052	1051	1054	1057	1056	1061
	−0.8%	−1.0%	−0.7%	−0.3%	−0.5%	0.0%

$\approx 1\%$ compared to the non-LTO option. Again, the AOCC compiler cannot compete with the Intel reference case and does not represent a viable option for compiler choices for the FS3D framework.

We chose the Ofast^L compiler option for the weak scaling representation again. A decrease in calculated cycles per hour with a growing number of MPI-processes is expected as this is inevitable for algorithms employing multigrid solvers. A sharp drop of CPH going from eight to 64 processors on a single node can be observed. This steep decrease for the weak scaling efficiency can be attributed to the reduced available memory bandwidth for each processor on a node. An enhancement of the weak scaling performance can be achieved by allotting only 32 processors per node represented by the dashed lines in figure 6. This, however, not only increases the amount of requested nodes but also the required computational resources by a factor of two. The time when node-to-node-level communication becomes the limiting factor is depicted by the green plot in figure 6 depicting the amount of assigned processors per node for the GNU Ofast^L case with 512 processors[2]. The maximum performance is achieved with four processors, however, leaving the remaining processors at idle.

Yet, this issue can be addressed in future endeavors by exploiting OpenMP parallelization on loop level utilizing all the cores within a node. The update of all critical routines in this respect within the FS3D framework is an ongoing process and features considerable potential to further increase the performance on the new HPE Apollo Hawk system. With the implementation of the tree structured communication in the multigrid solver other bottlenecks were identified after the transfer to the new platform. Major parts of the computational load can now be attributed to other parts of the code where subsequent tracing and performance analysis can help to specifically target the implementation of OpenMP in the identified routines.

6 Conclusions

Within this study the atomization process of a liquid rivulet at the trailing edge of a thin plate has been investigated using DNS and the multiphase flow solver FS3D. To do that, the computational domain has been structured into equally sized blocks and a no-slip boundary condition between two blocks has been used to discretize the plate with infinitesimal thickness. To validate the used computational setup simulations were performed in analogy to experiments performed at ITLR. The inflow conditions of the simulations were chosen to represent different occurring atomization regimes. In addition, a grid study revealed the influence and the necessary grid resolution to reproduce the experimental findings adequately. The results show that the setup is generally capable of simulating the disintegration of a liquid rivulet at the trailing edge. Also, the atomization regime changes with the inflow conditions and matches the experimental results qualitatively well. However, the results show that even with the highest grid resolution (with more than one billion cells) the very thin lamella at

[2] For the plot depicting the processors per node (ppn) the appropriate scale for the x-axis is depicted on top, the completed cycles per hour on the right hand side.

bag breakup cannot be resolved. A quantitative comparison of the liquid rivulet's geometry at the trailing edge indicates a strong influence on the inlet condition of the liquid. With the knowledge of the current limitations the authors will focus on the enhancement of the resolution and a more precise specification of the liquid inlet condition to match the experimental findings. The detailed insight into the disintegration process will help to improve the existing empirical models for the prediction of trailing edge disintegration.

Furthermore, several available compilers and compiler options have been tested after the transition and adaption of the FS3D code to the new HPE Apollo Hawk platform at HLRS. The usage of the optimization flags and link-time optimization during compile time led to a significant performance increase compared to the previously utilized O2 optimization. For the FS3D benchmark case the compiling option Ofast with link-time optimization yields the best performance regarding computational cycles per hour. For several years it has been standard practice to use the Intel Compiler. A transfer to the usage of the GNU Compiler can achieve performance gains of up to 36% for strong and 7.5% for weak scaling and should become the standard for future numerical investigations at Hawk. The availability of 128 cores per node engenders new challenges that can be addressed by means of hybrid MPI (on node level) and OpenMP programming (inside each node on loop level). Nearest neighbor communication (namely topology mapping) could also ensure optimal system utilization especially for calculation involving multiple nodes.

Acknowledgements The authors kindly acknowledge the *High Performance Computing Center Stuttgart* (HLRS) for support and supply of computational resources on the HPE Apollo (Hawk) platform under the Grant No. FS3D/11142. In addition, the authors kindly acknowledge the financial support of the Deutsche Forschungsgemeinschaft (DFG, German Research Foundation) under grant number WE2549/36-1, WE2549/35-1 and under Germany's Excellence Strategy - EXC 2075 - 390740016. We also acknowledge the support by the Stuttgart Center for Simulation Science (SimTech).

References

1. Baggio, M., Weigand, B.: Numerical simulation of a drop impact on a superhydrophobic surface with a wire. Physics of Fluids **31**(11), 112107 (2019)
2. Dumouchel, C.: On the experimental investigation on primary atomization of liquid streams. Experiments In Fluids **45**, 371–422 (2008)
3. Eisenschmidt, K., Ertl, M., Gomaa, H., Kieffer-Roth, C., Meister, C., Rauschenberger, P., Reitzle, M., Schlottke, K., Weigand, B.: Direct numerical simulations for multiphase flows: An overview of the multiphase code FS3D. Journal of Applied Mathematics and Computation **272**(2), 508–517 (2016)
4. Ertl, M., Reutzsch, J., Nägel, A., Wittum, G., Weigand, B.: Towards the Implementation of a New Multigrid Solver in the DNS Code FS3D for Simulations of Shear-Thinning Jet Break-Up at Higher Reynolds Numbers, p. 269–287. Springer (2017)
5. Ertl, M., Weigand, B.: Analysis methods for direct numerical simulations of primary breakup of shear-thinning liquid jets. Atomization and Sprays **27**(4), 303–317 (2017)

6. Evrard, F., Denner, F., van Wachem, B.: A hybrid Eulerian-Lagrangian approach for simulating liquid sprays. In: ILASS–Europe 2019, 29th Conference on Liquid Atomization and Spray Systems. Paris, France (2019)
7. Fest-Santini, S., Steigerwald, J., Santini, M., Cossali, G., Weigand, B.: Multiple drops impact onto a liquid film: Direct numerical simulation and experimental validation. Computers & Fluids **214**, 104761 (2021)
8. Gepperth, S., Bärow, E., Koch, R., Bauer, H.J.: Primary atomization of prefilming airblast nozzles: Experimental studies using advanced image proprocess techniques. In: ILASS – Europe 2014, 26th Annual Conference on Liquid Atomization and Spray Systems. Bremen, Germany (2014)
9. Gepperth, S., Koch, R., Bauer, H.J.: Analysis and Comparison of Primary Droplet Characteristics in the Near Field of a Prefilming Airblast Atomizer. In: Proceedings of ASME Turbo Expo 2013: Turbine Technical Conference and Exposition. San Antonio, Texas, USA (2013)
10. Hammit, F., Krzeczkowski, S., Krzyzanowksi, J.: Liquid film and droplet stability consideration as applied to wet steam flow. Forschung im Ingenieurwesen **47**(1) (1981)
11. Hirt, C.W., Nichols, B.D.: Volume of fluid (VOF) Method for the Dynamics of Free Boundaries. Journal of Computational Physics **39**(1), 201–225 (1981)
12. Javed, B., Watanabe, T., Himeno, T., Uzawa, S.: Effect of trailing edge size on the droplets size distribution downstream of the blade. Journal of Thermal Science and Technology **12**(2) (2017)
13. Javed, B., Watanabe, T., Himeno, T., Uzawa, S.: Experimental Investigation of Droplets Characteristics after the Trailing Edge at Different Angle of Attack. International Journal of Gas Turbine, Propulsion and Power Systems **9**(3), 32–42 (2017)
14. Kim, W.: Study of liquid films, fingers, and droplet motion for steam turbine blading erosion problem. Ph.D. thesis, University of Michigan, Michigan, USA (1978)
15. Koch, R., Braun, S., Wieth, L., Chaussonnet, G., Dauch, T., Bauer, H.J.: Prediction of primary atomization using Smoothed Particle Hydrodynamics. European Journal of Mechanics B/Fluids **61**(2), 271–278 (2017)
16. Lafaurie, B., Nardone, C., Scardovelli, R., Zaleski, S., Zanetti, G.: Modelling Merging and Fragmentation in Multiphase Flows with SURFER. Journal of Computational Physics **113**(1), 134–147 (1994)
17. Liu, M., Bothe, D.: Numerical study of head-on droplet collisions at high Weber numbers. Journal of Fluid Mechanics **789**, 785–805 (2016)
18. Rayleigh, L.: On The Instability Of Jets. Proc. London Math. Soc. **10**, 4–13 (1878)
19. Reutzsch, J., Ertl, M., Baggio, M., Seck, A., Weigand, B.: Towards a Direct Numerical Simulation of Primary Jet Breakup with Evaporation, chap. 16, p. 243–257. Springer, Cham (2019)
20. Reutzsch, J., Kochanattu, G.V.R., Ibach, M., Kieffer-Roth, C., Tonini, S., Cossali, G., Weigand, B.: Direct Numerical Simulations of Oscillating Liquid Droplets: a Method to Extract Shape Characteristics. In: Proceedings ILASS–Europe 2019. 29th Conference on Liquid Atomization and Spray Systems (2019)
21. Rider, W.J., Kothe, D.B.: Reconstructing Volume Tracking. Journal of Computational Physics **141**(2), 112–152 (1998)
22. Schlottke, A., Weigand, B.: Two-Phase Flow Phenomena in Gas Turbine Compressors with a Focus on Experimental Investigation of Trailing Edge Disintegration. Aerospace **8**(4), 91 (2021). URL `https://doi.org/10.3390/aerospace8040091`
23. Steigerwald, J., Ibach, M., Reutzsch, J., Weigand, B.: Towards the Numerical Determination of the Splashing Threshold of Two-component Drop Film Interactions. Springer (2022)
24. Steigerwald, J., Reutzsch, J., Ibach, M., Baggio, M., Seck, A., Haus, B., Weigand, B.: Direct Numerical Simulation of a Wind-generated Water Wave. Springer (2021)
25. Weber, C.: Zum Zerfall eines Flüssigkeitsstrahles. Zeitschrift für Angewandte Mathematik und Mechanik **11**, 136–154 (1931)
26. Weymouth, G., Yue, D.K.P.: Conservative Volume-of-Fluid method for free-surface simulations on Cartesian-grids. Journal of Computational Physics **229**(8), 2853–2865 (2010)

Numerical Investigation of the Flow and Heat Transfer in Convergent Swirl Chambers

Florian Seibold and Bernhard Weigand

Abstract Confined swirling flows are a promising technique for cooling applications since they achieve high heat transfer rates. In such systems, however, an axial flow reversal can occur, which corresponds to the axisymmetric vortex breakdown phenomenon.
This report presents a numerical study using Delayed Detached Eddy Simulations (DDES) in order to analyze the impact of convergent tube geometries on the flow field and the heat transfer in cyclone cooling systems. For this purpose, a comparison is drawn for a Reynolds number of $10,000$ and a swirl number of 5.3 between a constant-diameter tube and four convergent tubes. The latter comprise three geometries with linearly decreasing diameters yielding convergence angles of 0.42 deg, 0.61 deg and 0.72 deg, respectively. Additionally, a single tube with a hyperbolic diameter decrease was analyzed.
The results demonstrate that converging tubes enforce an axial and circumferential flow acceleration. The axial flow acceleration counteracts the flow reversal and thus was proved capable of suppressing the vortex breakdown phenomenon. Further, the heat transfer in terms of Nusselt numbers shows a strong dependency on the tube geometry.

Florian Seibold
Institute of Aerospace Thermodynamics (ITLR), University of Stuttgart, Pfaffenwaldring 31, 70569 Stuttgart, Germany, e-mail: florian.seibold@itlr.uni-stuttgart.de

Bernhard Weigand
Institute of Aerospace Thermodynamics (ITLR), University of Stuttgart, Pfaffenwaldring 31, 70569 Stuttgart, Germany, e-mail: bernhard.weigand@itlr.uni-stuttgart.de

W. E. Nagel et al. (eds.), *High Performance Computing in Science and Engineering '21*,
https://doi.org/10.1007/978-3-031-17937-2_15

1 Introduction

In modern gas turbines, the maximum turbine entry temperature limits the overall engine efficiency. As a consequence, this temperature is increased more and more whereby the cooling system for turbine blades becomes one of the most critical parts in the design process. The so called cyclone cooling or swirl cooling is an innovative cooling technique that can be applied for instance in the leading edge of turbine blades [2]. Cyclone cooling systems consist of a vortex chamber, which is an internal flow passage, and one or more tangential inlets that induce the swirling motion. This technique promises high heat transfer rates but is also accompanied by high pressure losses [3]. Chang and Dhir [4] identified two major mechanisms that explain the heat transfer enhancement in such swirl chambers: On the one hand, the high maximum axial velocity close to the wall results in high heat fluxes at the wall. On the other hand, a high turbulence level improves the fluid mixing.

An important phenomenon of swirling flows is the so called vortex breakdown, which is defined as an abrupt change in the structure of the core flow [5]. The vortex breakdown manifests itself in different types: spiral, double-helix and axisymmetric form. The axisymmetric vortex breakdown is characterised by a stagnation point followed by an axial flow reversal [6]. This flow reversal significantly influences the flow field and, thus, also the heat transfer performance of cyclone cooling systems. Although a great deal of research has been conducted on the vortex breakdown phenomenon, it is not yet fully understood. Hence, several explanations exist among which the most popular one is based on the propagation of waves. In this regard, Squire [7] and Benjamin [8] hypothesized the existence of a critical flow state that separates a supercritical state from a subcritical one. In a subcritical state, disturbances can propagate upstream and downstream whereas in a supercritical state only downstream propagation is possible. As a consequence of this categorization, Escudier et al. [9] observed a strong dependency of subcritical flows on the tube outlet conditions. In contrast, supercritical flows showed no influence at all. The authors related this outcome to the axial flow reversal that enables disturbances to propagate upstream and, hence, are able to affect the entire flow field within the swirl tube. Accordingly, the supercritical state represents a more robust flow that is insensitive to disturbances from downstream [10].

In the past, research on swirl chambers was mainly conducted on constant-diameter tubes. Investigations on convergent tubes are rare and mainly focused on the heat transfer enhancement as for instance in [11, 12] or the temperature separation in Ranque–Hilsch vortex tubes [13, 14]. However, there is no detailed study of the flow pattern in convergent vortex tubes and its impact on the heat transfer.

This report aims to summarize selected results from the conducted research on convergent vortex tubes. For this purpose, the flow field and the heat transfer in a constant-diameter tube is analyzed and systematically compared to four convergent geometries with different diameter variations.

2 Geometry

Figure 1 depicts the computational domain of the here presented simulations representing an upscaled generic model of a cyclone cooling system. The fluid enters the domain through two tangential inlets, which feature a height of $h = 5$ mm and a length of 15 times of their hydraulic diameter. After the inlet area with a diameter of $D_0 = 50$ mm, a convergent section follows with an axial length of $20D_0$. The geometry is in accordance with the experimental setup of Biegger et al. [16]. The subscript 0 denotes values at the axial location $z = 0$.

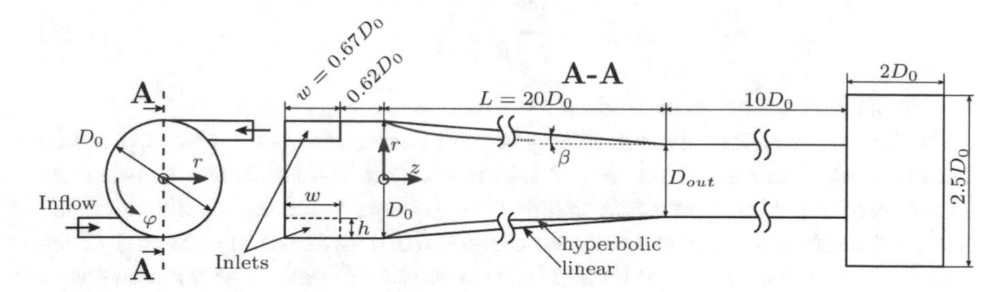

Fig. 1: Sketch of the computational domain [15]

Based on these geometrical constraints, different non-dimensional numbers can be defined. First of all, the local Reynolds number Re is based on the local axial bulk velocity $\overline{u}_z = \frac{2}{R^2} \int_0^R r\, u_z(r,z)\, \mathrm{d}r$ and the local tube diameter D

$$Re = \frac{\overline{u}_z D}{\nu} . \tag{1}$$

Here, ν represents the kinematic viscosity. In addition, an inflow Reynolds number Re_0 can be defined based on the tube inlet diameter D_0 and its corresponding axial bulk velocity $\overline{u}_{z0} = \frac{2}{R^2} \int_0^R r\, u_z(r, z{=}0)\, \mathrm{d}r$ yielding

$$Re = \frac{\overline{u}_{z0} D_0}{\nu} = \frac{4\dot{m}}{\pi D_0 \mu} . \tag{2}$$

Here, $\dot{m}$ and μ denote the mass flow rate and the dynamic viscosity, respectively. The inflow Reynolds number Re_0, defined by Eq. (2), can be evaluated prior to simulation in order to determine the operating point of the device.

Moreover, a dimensionless swirl number S can be defined as the ratio of axial flux of circumferential momentum $\dot{I}_\varphi$ divided by the local tube radius R and the axial flux of axial momentum $\dot{I}_z$ [17]

$$S = \frac{\dot{I}_\varphi}{R\dot{I}_z} = \frac{\int_0^R \rho u_\varphi u_z 2\pi r^2 \mathrm{d}r}{R\int_0^R \rho u_z^2 2\pi r \mathrm{d}r} \,. \tag{3}$$

Here, ρ denotes the fluid density. Equation (3) cannot be evaluated in advance since the velocity profiles are unknown. However, when assuming a uniform velocity at the inlets, a geometrical swirl number S_{geo} can be determined prior to simulation

$$S_{geo} = \frac{\left(R_0 - \frac{h}{2}\right)\pi R_0^2}{R_0\, 2wh} \,. \tag{4}$$

All simulations were conducted for $Re_0 = 10,000$ and $S_{geo} = 5.3$.

In the present report, five different geometries are investigated: One vortex tube with a constant cross-section ($\beta = 0\,\mathrm{deg}$) is used for validation and as reference for comparison. Moreover, three chambers with linearly decreasing diameters are analyzed. these correspond to convergence angles β of 0.42 deg, 0.61 deg and 0.72 deg and area ratios A_{out}/A_0 of $1/2$, $1/3$ and $1/4$, respectively. Finally, one geometry with a hyperbolically decreasing diameter is investigated. The latter enforces a linear increase in the local Reynolds number when assuming a constant bulk density $\overline{\rho}$ yielding

$$D - \frac{D_0}{1 + \frac{z}{L}} \,. \tag{5}$$

The geometrical features are summarized in Tab. 1.

Table 1: Investigated geometries

geometry	D_{out}	A_{out}/A_0	β [deg]
1	50 mm	1	0
2	35.4 mm	$1/2$	0.42
3	28.9 mm	$1/3$	0.61
4	25 mm	$1/4$	0.72
5	25 mm	$1/4$	hyperbolic

3 Numerical setup

The here presented numerical calculations are carried out as compressible Delayed Detached Eddy Simulations (DDES) using the open source code OpenFOAM version 6. The DDES approach was introduced by Spalart et al. [18] as hybrid model that applies a Reynolds Averaged Navier Stokes (RANS) model close to the wall and a Large Eddy Simulation (LES) in the free stream.

Originally, Spalart et al. [19] designed this hybrid approach by modifying the one-equation turbulence model from Spalart and Allmaras [20]

$$\frac{\mathrm{D}\tilde{\nu}}{\mathrm{D}t} = c_{b1}\tilde{S}\tilde{\nu} + \frac{1}{\sigma}\left[\nabla \cdot \left(\left(\nu + \nu_t\right)\nabla\tilde{\nu}\right) + c_{b2}\left(\nabla\tilde{\nu}\right)^2\right] - c_{w1}f_w\left(\frac{\tilde{\nu}}{\tilde{d}}\right)^2 . \tag{6}$$

$\tilde{\nu} = \nu_t/f_{v1}$ represents an effective eddy viscosity and c_{b1}, c_{b2}, c_w and σ denote model coefficients. Further, $\tilde{d}$ is the so-called DES-limiter

$$\tilde{d} = \min\{d; C_{DES}\Delta\} . \tag{7}$$

Here, d represents the wall distance, $\Delta = \max\{\Delta_x; \Delta_y; \Delta_z\}$ denotes the local grid spacing and $C_{DES} = 0.65$ is a constant.

The definition of the DES-limiter $\tilde{d}$ in Eq. (7) allows to use the turbulence model as hybrid model. In the wall distant region, where $d > C_{DES}\Delta$, the model operates as Subgrid Scale (SGS) model for LES whereas it transforms into a RANS model on approaching the wall, where $d < C_{DES}\Delta$ [21]. This hybrid approach is termed Detached Eddy Simulation (DES).

The original DES formulation in Eqs. (6, 7) comes along with a major drawback: Within the RANS regime, a coarse grid resolution parallel to the wall is required whereas a fine resolution is necessary in the LES regime. If this requirement is violated in the RANS regime by a too small cell spacing, the model switches into the LES mode too close to the wall. Concurrently, the wall-normal cell spacing is not yet fine enough to resolve turbulent fluctuations. As a result, unphysical results can arise such as modeled stress depletion and grid induced separation [21]. Therefore, Spalart et al. [18] addressed this drawback by modifying the definition of the DES-limiter using an additional blending function f_d

$$\tilde{d} = d - f_d \max\{0; d - C_{DES}\Delta\} . \tag{8}$$

The blending function f_d in Eq. (8) equals zero within the boundary layer and rapidly approaches one in the free stream area. The exact equation is given in [18]. In contrast to the original DES formulation in Eqs. (6,7), the modified DES-limiter $\tilde{d}$ in Eq. (8) does not depend on the local grid size Δ but on the flow field solution [21]. This modified model is termed Delayed DES (DDES). For a more detailed description, the reader is referred e.g. to Fröhlich and von Terzi [21].

Finally, a constant turbulent Prandtl number Pr_t was prescribed in order to close the equation system. In an experimental review, Kays [22] stated that Pr_t essentially takes a constant value of 0.85 within a turbulent boundary layer. Based on this conclusion, $Pr_t = 0.85$ was used in the here presented simulations.

All simulations were conducted using a finite volume approach. The Pressure-Implicit-Split-Operator (PISO) was applied as pressure corrector with two outer loops within every timestep. Convective and diffusive fluxes were discretized with a second-order central scheme and the time dependency was approximated using a second-order backward scheme. Besides, the numerical mesh consists of a structured O-grid with $16.5 \cdot 10^6$ hexahedral cells. The boundary layer at the wall is fully resolved with a maximum non-dimensional height of the first cell layer $y_1^+ = u_\tau y_1/\nu < 1.5$. Here, y_1 and $u_\tau = \sqrt{\nu \frac{\partial u}{\partial y}\big|_{y=0}}$ denote the height of the first cell layer at the wall and the wall shear stress, respectively. Further mesh details are summarized in Tab. 2.

In the course of the simulations, the following boundary conditions were applied for the fluid medium air: At the inlets, a fixed mean velocity was prescribed and superimposed by random velocity fluctuations with a RMS-value of 15% of the mean flow. The walls represent a no-slip boundary in combination with an effective turbulent viscosity of $\tilde{\nu} = 0$. At the outlet, a fixed mean pressure of 10^5Pa was prescribed. Isothermal boundary conditions were applied for the wall temperature $T_W = 293$ K as well as for the inflow temperature $T_{in} = 333$ K.

The velocity, pressure, and temperature fields were initialized with values from a steady-state simulation. Then, the transient simulation time was defined based on the flow through time (FFT) of the convergent tube segment L/u_z. First, the calculations run for 35 FFTs prior to averaging and then for another 10 FFTs in order to obtain the mean quantities. The timestep Δt was adapted automatically during run-time in order to maintain a Courant number of $u\Delta t/\Delta = 0.9$.

Table 2: Summary of mesh details including the total amount of cells, the height of the first cell layer at the wall and the center cell size [15]

Geometry	Cells	$y_1\ [10^{-5}\text{m}]$	$(\Delta_x, \Delta_y, \Delta_z)_{center}\ [10^{-4}\text{m}]$
0 deg	$16.5 \cdot 10^6$	1.7	(6.6, 6.6, 11)
0.42 deg	$16.5 \cdot 10^6$	1.7-3	(4.7-6.6, 4.7-6.6, 11)
0.61 deg	$16.5 \cdot 10^6$	1.7-2.4	(3.8-6.6, 3.8-6.6, 11)
0.72 deg	$16.5 \cdot 10^6$	1.7-2.1	(3.3-6.6, 3.3-6.6, 11)
hyperbolic	$16.5 \cdot 10^6$	1.7-2.1	(3.3-6.6, 3.3-6.6, 11)

In the following, all results represent mean values. Figure 2 depicts a comparison of circumferential and axial velocities between numerical and experimental outcomes for a constant-diameter tube ($\beta = 0$ deg). The experimental data originates from Biegger

[23], who measured flow velocities using the stereo particle image velocimetry (PIV) technique. In Fig. 2, the radial coordinate r and the velocity components u_φ and u_z are normalized by the inlet tube radius R_0 and the corresponding axial bulk velocity $\overline{u}_{z0}$, respectively. Hereafter, this type of scaling is named *global* normalization since it depends on values from the axial location $z = 0$. In general, the results in Fig. 2 prove overall good accuracy of the numerical prediction. Only some smaller deviations occur for u_z in the tube center and for the maximum value of u_φ. Furthermore, Seibold et al. [24] and Seibold and Weigand [15] investigated the turbulence energy spectrum and reported that the simulation reproduces the correct Kolmogorov slop of $-5/3$ for the resolved scales. Therefore, the numerical setup achieves reasonable turbulence behavior and the grid resolution proved to be fine enough to resolve the large scales. Moreover, Biegger [23] carried out a more detailed validation for the numerical DDES setup. For this purpose, the author conducted an extensive mesh study for a channel flow without swirl and compared the outcomes to a Direct Numerical Simulation (DNS) from literature. Based on this mesh study, Biegger [23] simulated swirling flows with the same numerical setup and obtained overall good agreement with experimental data in terms of mean velocity components and turbulent kinetic energy.

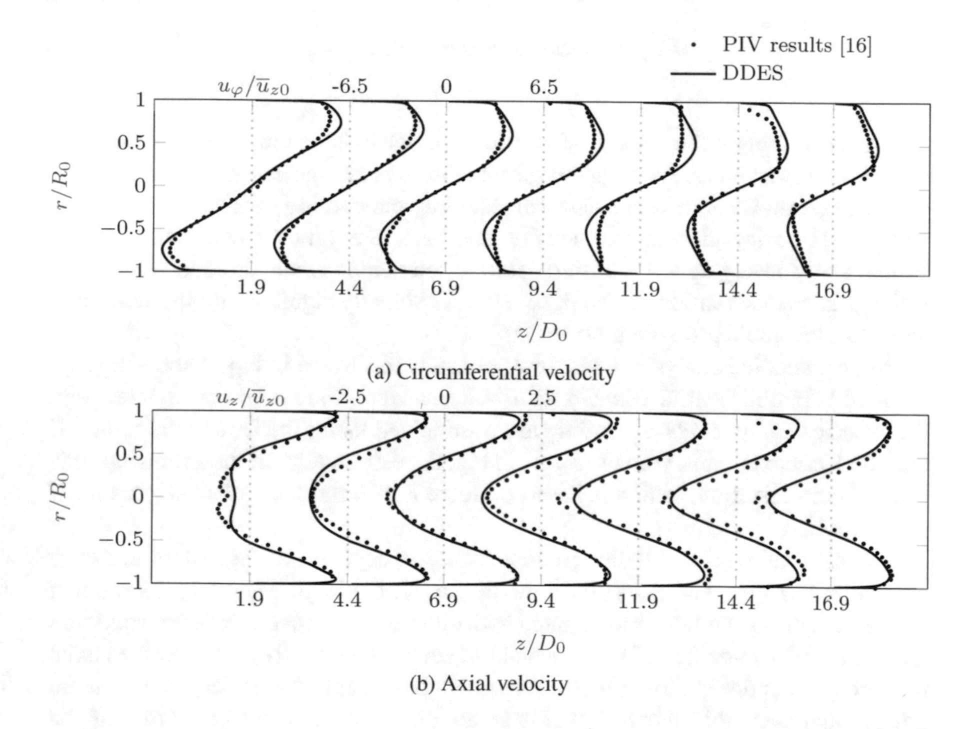

(a) Circumferential velocity

(b) Axial velocity

Fig. 2: Comparison of experimental and numerical results for the geometry $\beta = 0$ deg. The numerical data was originally published in [15, 24]

4 Results

4.1 Flow field

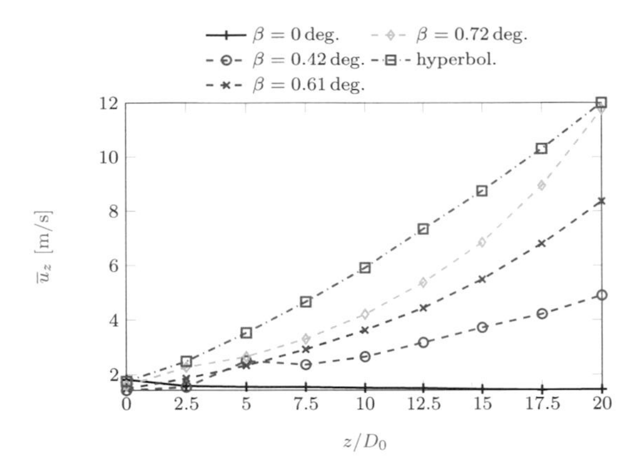

Fig. 3: Local axial bulk velocity $\overline{u}_z$

Figure 3 depicts the values of the local axial bulk velocity $\overline{u}_z$ for the here investigated swirl tubes. As might be expected in advance, the constant-diameter tube ($\beta = 0$ deg.) shows almost constant values. Only the cooling of the gas at the cold wall causes a slight decrease in density that leads to a minor deceleration. On the contrary, all converging tubes feature a flow acceleration in axial direction in order to satisfy mass conservation. Thus, the axial bulk velocity significantly increases and leads to a favorable pressure gradient [15].

A more detailed analysis of the flow pattern is illustrated in Fig. 4 showing both axial and circumferential velocity distributions. Here, the radial coordinate r and the velocity components u_φ and u_z are normalized using the *local* tube radius R and the *local* axial bulk velocity $\overline{u}_z$ from Fig. 3, respectively. In the following, this type of normalization, which is based on local bulk values, is referred to as *local* normalization.

The circumferential velocity component u_φ is depicted in Fig. 4a with locally normalized values. These results show that u_φ is the largest velocity component of the system for the here investigated swirl number. However, the corresponding circumferential velocities of the constant diameter tube ($\beta = 0$ deg.) decline in axial direction due to dissipation effects. In case of convergent geometries, an additional effect counteracts this decrease that is an angular flow acceleration caused by the conservation of circumferential momentum. This acceleration is included in the local normalization using $\overline{u}_z$ in Fig. 4a.

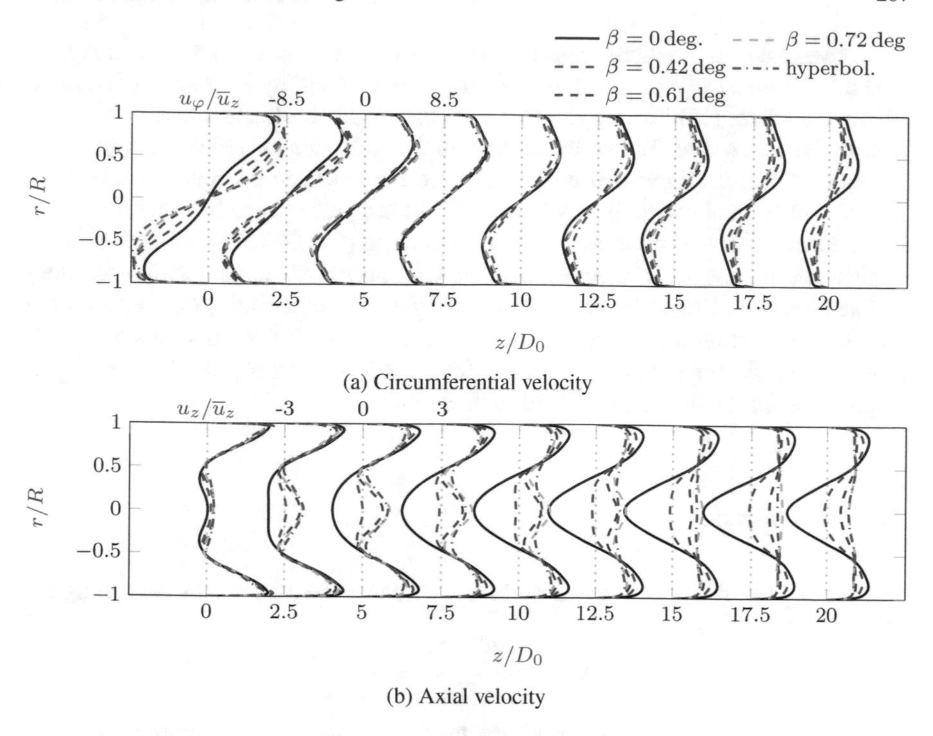

(a) Circumferential velocity

(b) Axial velocity

Fig. 4: Locally normalized velocity field

Further, the normalized velocity profiles of different geometries obviously coincide in the middle section of the tube. In the upstream part, the flow pattern slightly varies since the circumferential velocity has not yet fully evolved. Additionally, some smaller deviations are apparent towards the tube outlet that are caused by an increased impact of dissipation in convergent geometries. Nevertheless, the characteristic flow pattern remains unaffected for the here investigated cases [15, 24].

Moreover, a change in shape is evident when comparing u_φ of a single geometry at different axial positions. In general, the circumferential velocity in a swirl tube matches a Rankine vortex. Such a vortex consists of a solid body rotation in the center, a potential vortex in the outer region and a boundary layer at the wall. In the upstream part of the tube, the solid body rotation dominates the flow pattern, which is indicated by a linear increase of u_φ. Further downstream, the solid body vortex shrinks in favor of the potential vortex.

Investigating the axial velocity u_z in Fig. 4b the constant-diameter tube shows a pronounced backflow in the center that extends to the outlet. This phenomenon corresponds to the axisymmetric vortex breakdown. The velocity magnitude of the reversed flow increases in downstream direction while its area shrinks.

Convergent cross-sections impact considerably on the flow pattern of u_z in Fig. 4b. The axial flow acceleration counteracts the flow reversal by imposing a favorable pressure gradient [15]. In case of $\beta = 0.42$ deg, a slight acceleration reaches stagnant flow in the tube center. The vortex tubes with $\beta \geq 0.61$ deg as well as the hyperbolic geometry cause a sufficient strong acceleration to reach purely positive axial flow. Hence, convergent swirl chambers are able to suppress the axisymmetric vortex breakdown if the flow acceleration is strong enough [15, 24]. Concurrently, purely positive axial flow at the tube outlet indicates a supercritical flow state according to Escudier et al. [9]. Thus, the upstream propagation of disturbances is prevented and the flow within the convergent tubes is made insensitive to disturbances from downstream. As a result, a more robust flow state is obtained, which is of crucial importance for the design process of such systems.

4.2 Heat transfer

The cooling capability of cyclone cooling systems can be assessed by evaluating the Nusselt number Nu

$$Nu = \frac{-\frac{\partial T}{\partial n}|_w D}{T_w - T_{ref}} = \frac{hD}{k} . \tag{9}$$

Here, T, h and k denote the temperature, the heat transfer coefficient and the fluid's thermal conductivity, respectively. The indices w and ref indicate wall and reference values.

The reference temperature T_{ref} in Eq. (9) is of crucial importance for the evaluation of the Nusselt number. Nusselt numbers can only be compared to other investigations with different thermal boundary conditions if the adiabatic wall temperature is used for T_{ref}. In general, this value is difficult to obtain. However, it can be approximated in swirling flows by assuming an adiabatic compression of the fluid from the centerline (index c) to the wall [25]

$$T_{ref} = T_c \left(\frac{p_w}{p_c} \right)^{\frac{\kappa-1}{\kappa}} . \tag{10}$$

Here, κ represents the isentropic exponent of the fluid medium air.

Figure 5 depicts the numerically obtained heat transfer results as circumferentially averaged Nusselt numbers Nu. All of these values consider the local tube diameter D in the definition of Eq. (9) as well as the local reference temperature T_{ref} according to Eq. (10). In addition, experimental results are included in the diagram for comparison. These were measured using the transient liquid crystal technique in combination with a reference temperature that was also based on the adiabatic compression according to Eq. (10). The complete experimental procedure was reported in [23, 27].

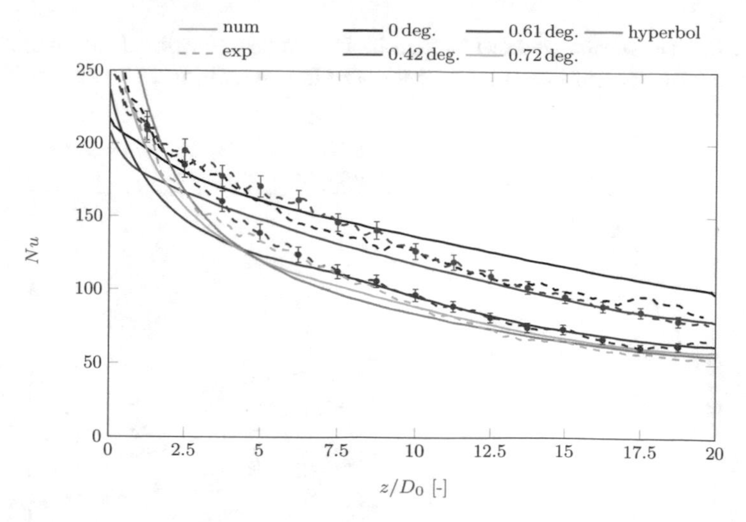

Fig. 5: Axial distribution of Nu from numerical (solid lines) and experimental investigations (dashed lines).

The results in Fig. 5 show overall good agreement between experimental and numerical data. Close to the tangential inlets, all curves feature high values that decay monotonically towards the tube outlet. This decay is caused by the decreasing swirl intensity.

When comparing different geometries, obvious differences are apparent in the downstream part of the tubes. There, the results show a distinct degression when increasing the convergence angle β. However, caution is required when interpreting these results. The aforementioned drop does not indicate a degression of the heat transfer coefficient h but is largely caused by the varying tube diameter D in Eq. (9). Consequently, the varying tube surface impacts considerably on the Nusselt numbers in Fig. 5.

Furthermore, the circumferentially averaged Nusselt numbers can be scaled by a Nusselt correlation from Gnielinski for a fully developed turbulent pipe flow without swirl, which is presented in [26] as:

$$Nu_G = \frac{\frac{\xi}{8} Re Pr}{1 + 12.7\sqrt{\frac{\xi}{8}}\left(Pr^{\frac{2}{3}} - 1\right)} \left[1 + \left(\frac{D}{L}\right)^{\frac{2}{3}}\right] \tag{11}$$

with

$$\xi = \left(1.8 \log_{10} Re - 1.5\right)^{-2} . \tag{12}$$

The scaled results are depicted in Fig. 6 using a *local* normalization. In particular, this means that all Nu-values are normalized by the Gnielinski-correlation from Eqs. (11, 12), which are based on *local* values of the tube diameter D and the Reynolds number Re. The normalized results then indicate a heat transfer enhancement over a pure pipe flow without swirl at the same *local* flow conditions. The results in Fig. 6

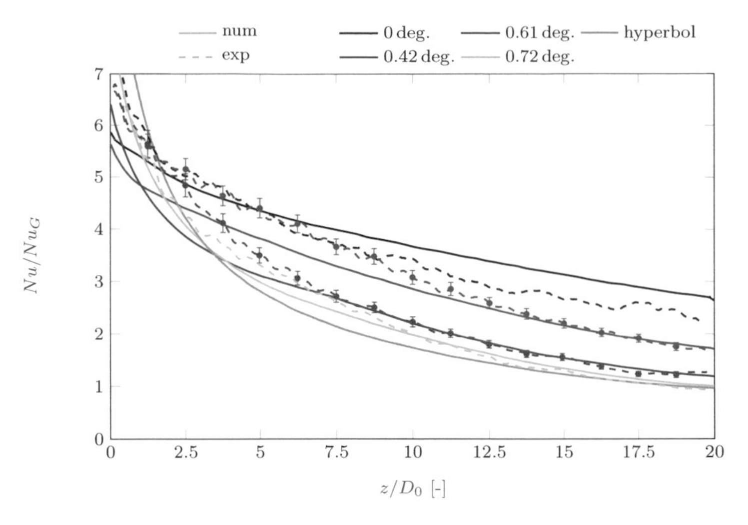

Fig. 6: Axial distribution of *locally* normalized Nu results from numerical (solid lines) and experimental investigations (dashed lines)

show an even more pronounced drop of Nu at the tube outlet in case of convergent geometries. This effect is evoked by the axial flow acceleration that yields higher values for Nu_G at the tube outlet in case of convergent geometries. As a result, the heat transfer enhancement over a non-swirling pipe flow vanishes at the tube outlet for $\beta \geq 0.61$ deg. as well as for the hyperbolic case.

In addition to the local heat transfer, global Nusselt numbers $\overline{Nu}$ can also be evaluated. For this purpose, globally averaged values of $\overline{h}$, $\overline{D}$ and $\overline{k}$ are introduced in Eq. (9). These are calculated as integral means

$$\overline{\phi} = \frac{1}{L} \int_0^L \phi \, \mathrm{d}z \,. \tag{13}$$

Moreover, a proper mean value of the Gnielinski-correlation $\overline{Nu_G}$ is required for normalization of the results. For this purpose, a *local* type of normalization is applied by introducing integral mean values of $\overline{D}$ and $\overline{Re}$ in Eqs. (11, 12). The corresponding results are presented Tab. 3.

The outcomes in Tab. 3 again show overall good agreement between numerics and experiments with a maximum deviation of less than 9%. These global values further show the large potential of cyclone cooling systems by predicting high heat transfer enhancements up to a factor of four compared to a pure pipe flow without swirl.

Table 3: Globally averaged heat transfer

β	$\overline{Nu}$		$\overline{Nu}/Nu_G$	
	num	exp	num	exp
0 deg.	140.8	135.0	3.81	3.65
0.42 deg.	122.6	134.5	2.98	3.27
0.61 deg.	103.8	111.6	2.38	2.55
0.72 deg.	102.7	104.5	2.25	2.29
hyperbolic	105.7	-	2.20	-

5 Usage of computational resources

This section contains an overview of the used computer resources, which are summarized in Tab. 4. Furthermore, the speed-up of the conducted simulations is depicted in Fig. 7 for a calculation with $16.5 \cdot 10^6$ cells. The diagram shows both an ideal and a real speed-up. The latter was determined by comparing each simulation to a calculation with 20 cores (1 node). With a small number of cores used in parallel, the curve rises almost linearly and then flattens out. Based on these results, a parallelization of 660 cores (33 nodes) is selected, which corresponds to 25,000 cells per core. This degree of parallelization takes advantage of massive parallelization but does not waste computational resources.

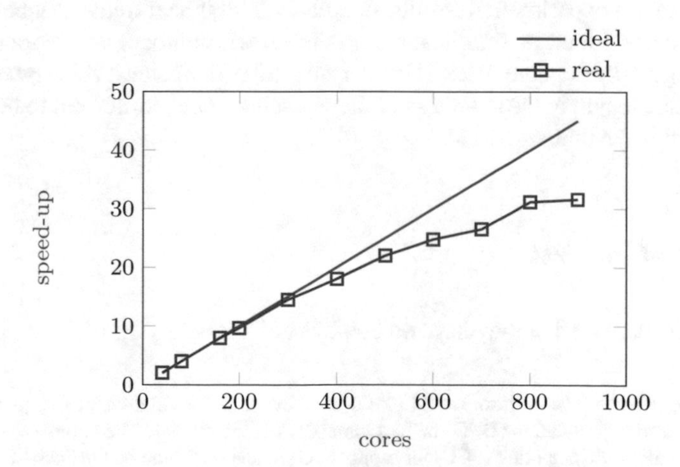

Fig. 7: Speed-up of numerical simulations

Table 4: Usage of resources

platform	processors	grid cells	wall time	software
ForHLR II	660	16.5×10^6	18-36 days	OpenFOAM 6

6 Summary and conclusion

Convergent swirl tubes for cyclone cooling applications were investigated numerically for a Reynolds number of $10,000$ and a geometrical swirl number of 5.3. The numerical setup of the constant-diameter tube was validated with good accuracy. The same geometry was further compared to three tubs that featured linearly decreasing diameter distributions with convergence angles of 0.42 deg, 0.61 deg and 0.72 deg, respectively. Additionally, one geometry with a hyperbolic diameter decrease was analyzed. The results were investigated in terms of flow field and heat transfer.

The constant-diameter tube showed a pronounced flow reversal in the tube center that corresponds to the axisymmetric vortex breakdown. Converging swirl chambers enforced a circumferential and axial flow acceleration. Despite this influence, the circumferential velocity retained its characteristic profile. The axial flow acceleration counteracted the flow reversal in the center and reached unidirectional axial velocities if the acceleration was strong enough. Consequently, convergent tube geometries proved to be capable of suppressing the vortex breakdown phenomenon. Most importantly, these convergent tubes reached a more robust flow that was insensitive to disturbances from the tube outlet. Furthermore, the heat transfer showed overall good agreement with experimental results and reached high heat transfer enhancements up to a factor of four. A significant drop of Nusselt numbers was evident in case of convergent tube geometries. The varying tube diameter, which was used as characteristic length in the definition of the Nusselt number, turned out to be the main factor for this decline.

Conflict of interest

The authors declare that they have no conflict of interest.

Acknowledgements The authors would like to acknowledge the funding of this project by the German Research Foundation (DFG) under Grant No. WE 2549/38-1. The authors also thank the Steinbuch Centre for Computing (SCC) for supply of computational time on the ForHLR II platform.

References

1. Kreith, F., Margolis, D. (1959) Heat transfer and friction in turbulent vortex flow. Applied Scientific Research 8:457–473
2. Glezer, B., Moon, H. K., O'Connell, T. (1996) A novel technique for the internal blade cooling. In: Proceedings of the ASME 1996 International Gas Turbine and Aeroengine Congress and Exhibition, Paper No. 96-GT-181, Birmingham, UK
3. Ligrani, P. M., Oliveira M. M., Blaskovich, T. (2003) Comparison of heat transfer augmentation techniques. AIAA Journal 41(3):337–362
4. Chang, T., Dhir, V. K. (1995) Mechanisms of heat transfer enhancement and slow decay of swirl in tubes using tangential injection. International Journal of Heat and Fluid Flow 16:78–87
5. Sarpkaya, T. (1971) Vortex breakdown in swirling conical flows. AIAA Journal 9(9):1792–1799
6. Escudier, M. P., Keller, J. J. (1985) Recirculation in swirling flows: A Manifestation of vortex breakdown. AIAA Journal 23:111–116
7. Squire, H. B. (1960) Analysis of the vortex breakdown phenomenon, part 1. Department Report No. 102, Imperial College of Science and Technology Aeronautics
8. Benjamin, T. B. (1962) Theory of vortex breakdown phenomenon. Journal of Fluid Mechanics 14:593–629
9. Escudier, M. P., Nickson , A. K., Poole, R. J. (2006) Influence of outlet geometry on strongly swirling turbulent flow through a circular tube. Physics of Fluids 18:125103
10. Bruschewski, M., Grundmann, S., Schiffer, H.-P. (2020) Considerations for the design of swirl chambers for the cyclone cooling of turbine blades and for other applications with high swirl intensity. International Journal of Heat and Fluid Flow 86:108670
11. Ling, J. P. C. (2005) Development of heat transfer measurement techniques and cooling strategies for gas turbines. PhD thesis, University of Oxford, UK
12. Yang, C. S., Jeng, D. Z., Yang, Y.-J., Chen, H.-R., Gau, C. (2011) Experimental study of pre-swirl flow effect on the heat transfer process in the entry region of a convergent pipe. Experimental Thermal and Fluid Science 35:73–81
13. Rafiee, S. E., Sadeghiazad, M. M., Mostafavinia, N. (2015) Experimental and numerical investigation on effect of convergent angle and cold orifice diameter on thermal performance of convergent vortex tube. Journal of Thermal Science and Engineering Applications 7(4):041006
14. Rafiee, S. E., Sadeghiazad, M. M. (2017) Efficiency evaluation of vortex tube cyclone separator. Applied Thermal Engineering 114:300–327
15. Seibold, F., Weigand, B. (2021) Numerical analysis of the flow pattern in convergent vortex tubes for cyclone cooling applications. International Journal of Heat and Fluid Flow 90:108806
16. Biegger, C., Sotgiu, C., Weigand, B. (2015) Numerical investigation of flow and heat transfer in a swirl tube. International Journal of Thermal Sciences 96:319–330
17. Gupta, A. K., Lilley, D. G., Syred, N. (1984) Swirl Flows. Energy and Engineering Science Series, Abacus Press
18. Spalart, P. R., Deck, S., Shur, M. L., Squires, K. D., Strelets, M. Kh., Travin, A. (2006) A new version of detached-eddy simulation, resistant to ambigous grid densities. Theoretical and Computational Fluid Dynamics 20:181–195
19. Spalart, P. R., Jou, W.-H., Strelets, M., Allmaras, S. R. (1997) Comments on the feasibility of LES for wings, and on hybrid RANS/LES approach. In: Advances in DNS/LES
20. Spalart, P. R., Allmaras, S. R. (1994) A one-equation turbulence model for aerodynamic flows. La Recherche Aerospatiale 1:5–21
21. Fröhlich, J., von Terzi, T. (2008) Hybrid LES/RANS methos for simulation of turbulent flows. Progress in Aerospace Sciences 44:349–377
22. Kays, W. M. (1994) Turbulent Prandtl number – Where are we?. Journal of Heat Transfer 116(2):284–295
23. Biegger, C. (2017) Flow and heat transfer investigations in swirl tubes for gas turbine blade cooling. PhD thesis, Institute of Aerospace Thermodynamics, University of Stuttgart, Germany
24. Seibold, F., Weigand, B., Marsik, F. and Novotny, P. (2017) Thermodynamic stability condition of swirling flows in convergent vortex tubes. In: Proceedings of the 12th International Gas Turbine Conference, Tokyo, Japan

25. Kobiela, B. (2014) Wärmeübertragung in einer Zyklonkülkammer einer Gasturbinenschaufel. PhD thesis, Institute of Aerospace Thermodynamics, University of Stuttgart, Germany
26. VDI-Wärmeatlas (2013) 11th edn., Springer Berlin Heidelberg
27. Biegger, C., Weigand, B. (2015) Flow and heat transfer measurements in a swirl chamber with different outlet geometries. Experiments in Fluids 56(4)

On the validity of the linear Boussinesq hypothesis for selected internal cooling features of gas turbine blades

Philipp Wellinger and Bernhard Weigand

Abstract In order to further decrease fuel consumption and pollutant emission of modern, highly efficient gas turbine blades, the internal cooling air consumption needs to be reduced. At the same time, the solid temperature must not exceed a critical level to avoid thermal damage of the turbine blades. Different cooling methods such as impingement cooling, pin fins or ribs are used to increase the internal heat transfer and, thus, to decrease the material temperatures. In order to improve local cooling air consumption the flow field and temperature distribution must be known. High fidelity CFD like LES are extremely difficult or even not feasible for such complex geometries and high Reynolds numbers. Therefore, RANS models must be used accepting a higher degree of inaccuracy. In this study, LES of selected cooling features are conducted for a better understanding of the validity of RANS models which are based on the linear Boussinesq hypothesis. The results given below are part of the research project "Improvement of the Numerical Computation of the Conjugate Heat Transfer in Turbine Blades" supported by the Siemens Clean Energy Center and the German Federal Ministry for Economic Affairs and Energy under grant number 03ET7073F.

Philipp Wellinger
Institute of Aerospace Thermodynamics (ITLR)
University of Stuttgart, Pfaffenwaldring 31 70569 Stuttgart,
e-mail: philipp.wellinger@itlr.uni-stuttgart.de

Bernhard Weigand
Institute of Aerospace Thermodynamics (ITLR)
University of Stuttgart, Pfaffenwaldring 31 70569 Stuttgart,
e-mail: bernhard.weigand@itlr.uni-stuttgart.de

W. E. Nagel et al. (eds.), *High Performance Computing in Science and Engineering '21*,
https://doi.org/10.1007/978-3-031-17937-2_16

1 Introduction

Many of the RANS (Reynolds-averaged Navier–Stokes) turbulence models used today are based on the linear Boussinesq hypothesis (LBH) assuming a linear relation between the unknown Reynolds stresses and the mean strain rate tensor. The Reynolds stresses are expressed by

$$-\rho\overline{u_i'u_j'} = 2\mu_t\overline{S_{ij}} - \frac{2}{3}k\rho\delta_{ij} \tag{1}$$

where ρ is the density, u_i' the velocity fluctuation, μ_t the eddy viscosity, $\overline{S_{ij}}$ the mean strain rate tensor, k the turbulent kinetic energy and δ_{ij} denotes the Kronecker delta. A main assumption of the LBH is that the eigenvectors of the Reynolds stress tensor and the strain rate tensor are aligned. Schmitt [1] introduced a validity parameter ρ_{RS} defined as

$$\rho_{RS} = \frac{\left|\overline{R_{ij}} : \overline{S_{ij}}\right|}{\left\|\overline{R_{ij}}\right\|\left\|\overline{S_{ij}}\right\|} \tag{2}$$

to quantify the misalignment of the eigenvectors. Here, $\overline{R_{ij}} = \overline{u_i'u_j'} - 2/3k\delta_{ij}$ defines the anisotropic stress tensor, $\left|A_{ij} : B_{ij}\right|$ represents the Frobenius inner product and $\left\|A_{ij}\right\| = \sqrt{\left|A_{ij} : A_{ij}\right|}$. The validity parameter ρ_{RS} ranges between 0 and 1. Values of 1 indicate perfect alignment between the eigenvectors of the two tensors. If ρ_{RS} tends to zero a large misalignment occurs and the LBH is violated. Schmitt [1] suggested $\rho_{RS} > 0.86$ as a soft limitation of the validity of the LBH and concluded that the hypothesis fails for even very simple test cases like planar channels. In order to better understand the validity of Eq. (1) for various flow scenarios, ρ_{RS} is evaluated for several more sophisticated test cases. Since an accurate prediction of $\overline{R_{ij}}$ and $\overline{S_{ij}}$ is needed, Large Eddy Simulations (LES) are conducted and validated with experimental or DNS (Direct Numerical Simulations) data.

The complex flow of the internal cooling system of a gas turbine blade is divided up into several test cases to investigate individual flow features separately. Test cases for a planar channel, two different periodic pin fin arrays and a periodic ribbed channel are chosen. It is important that suitable and accurate validation data for the Reynolds stresses and the velocity components are available to validate the LES results. The following test cases satisfy these requirements:

1. planar channel investigated by Abe et al. [2][1] at $Re_b = 24{,}400$
2. periodic pin fin array investigated by Simonin and Barcouda [3] at $Re_b = 34{,}800$
3. periodic ribbed channel investigated by Drain and Martin [4] at $Re_b = 24{,}000$

[1] https://www.rs.tus.ac.jp/t2lab/db/

4. Ames and Dvorak [5] pin fin array at $Re_b = 20{,}000$ with experimental PIV data kindly provided by Siemens Energy[2]

An overview of the four test cases is shown in Fig. 1. The first test case of a planar channel implies two homogeneous directions. It is used as a reference case to validate the methodology. Test cases number two and three are quasi-two-dimensional periodic flows with one homogeneous direction. The fourth test case is a fully three-dimensional periodic flow.

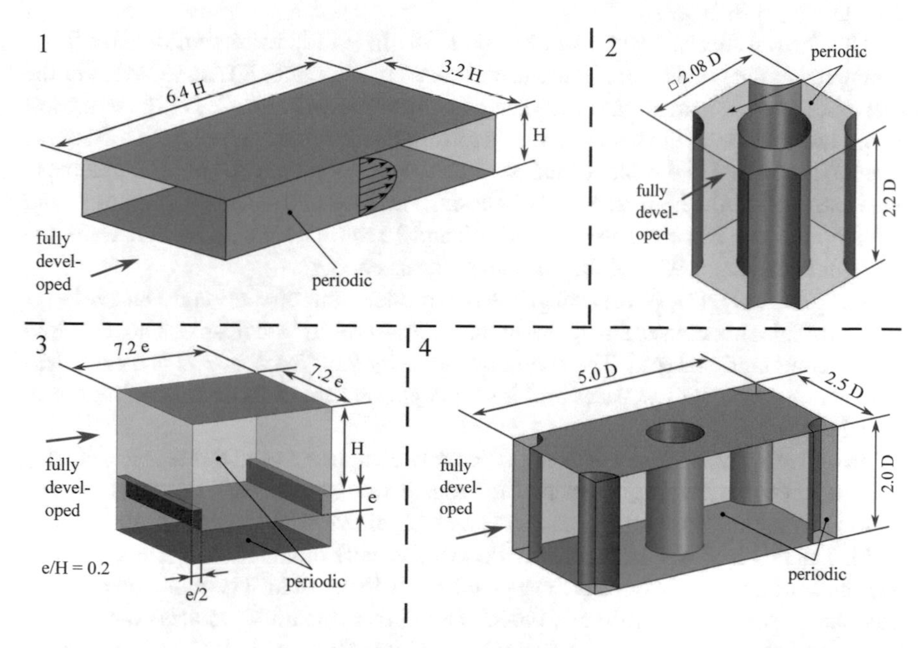

Fig. 1: Geometries and numerical domain of the four studied test cases [2–5].

2 Numerical setup

All simulations were run with the commercial CFD (Computational Fluid Dynamics) solver Simcenter STAR-CCM+ [6]. First, each geometry was calculated using a RANS simulation. The resulting wall normal grid spacing and the velocity field was used to estimate a proper grid resolution and time step size for the LES. Grid size was chosen to obtain wall normal grid spacing of $y_n^+ < 1$ for the first computational cell. The streamwise and spanwise grid spacing in the entire domain was set to $x^+ \approx 50$ and $z^+ \approx 30$, respectively.

[2] conducted at the Siemens Energy Center at the University of Central Florida

The time step size was chosen to ensure $CFL < 1$ in most regions of the flow field. Simcenter STAR-CCM+ offers the opportunity to model convective fluxes with a hybrid method blending between MUSCL 3rd-order upwind and 3rd-order central-differencing reconstruction. A blending factor of 0.02 was chosen to reduce numerical dissipation due to the upwind scheme. Time discretization was performed using a 2nd-order backward differentiation scheme with 5th-order correction. Gradients were computed with a hybrid Gauss-Least Squares method limited with Venkatakrishnans [7] method. All mentioned schemes are described in the Simcenter STAR-CCM+ user guide [6] in detail.

Hexahedral meshes were used for the LES since they were significantly faster compared to the polyhedral meshes usually used in Simcenter STAR-CCM+. For the test cases one and four, meshes were generated with Ansys ICEM CFD [8]. For the test cases two and three, the trimmed mesher available in Simcenter STAR-CCM+ was used to generate a two-dimensional mesh. Subsequently, the two-dimensional mesh was extruded in the homogeneous direction. The overall number of computational cells for the four test cases were 35, 45, 20 and 15 millions. These meshes were fine enough to resolve $\approx 95\%$ of the turbulent kinetic energy.

For this study, the WALE subgrid-scale model from Nicoud and Ducros [9] is applied. This model uses an algebraic formulation for the small scales not resolved by the computational grid. The coefficient used by the WALE model is usually less sensitive to different test cases and it was kept constant at its default value for all simulations.

In order to obtain mean values of the velocity components and turbulence statistics, i.e. the Reynolds stresses, an averaging process was conducted. The mean velocity field and the mean velocity gradients can be obtained from the time averaged flow field. The resolved part of the Reynolds stresses were obtained using the empirical variances and covariances of the instantaneous velocity field. The modeled part was calculated from the subgrid-scale model. In addition, the time-averaged values were spatially averaged in the homogeneous directions (test cases 1-3) to increase the number of evaluation points significantly.

3 Results

3.1 Reference case: planar channel

As a first step, a comparison between the reference data and the LES for the planar channel [2] is shown in Fig. 2. Instead of comparing the Reynolds stress as defined in Eq. (1), only the product of the velocity fluctuations are compared. This term is denoted as the specific Reynolds stresses and is defined by

$$\tau_{ij} = \overline{u_i' u_j'}. \tag{3}$$

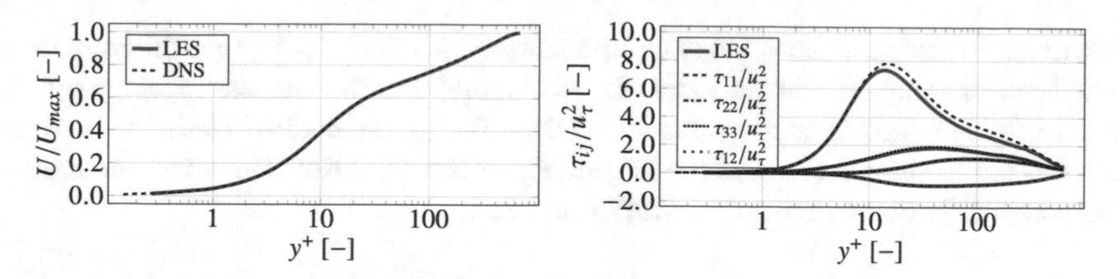

Fig. 2: Comparison of the normalized mean velocity component U/U_{max} and the normalized specific Reynolds stress components τ_{ij}/u_τ^2 for the planar channel obtained by LES with DNS data from [2].

The left diagram shows the mean velocity component normalized by the maximum velocity in the center of the channel U/U_{max} for DNS and LES. The LES predicts almost identical results in the entire domain. Since the DNS uses a much finer grid resolution, y^+ values below 0.3 are available. The diagram on the right compares the specific Reynolds stresses normalized by the square of the friction velocity τ_{ij}/u_τ^2. In this case, a slight underprediction of τ_{11}/u_τ^2 obtained by the LES can be observed. However, a very good overall agreement can be determined. The LES data are applied to compute ρ_{RS}. The comparison with the data given by Schmitt [1] is show in Fig. 3.

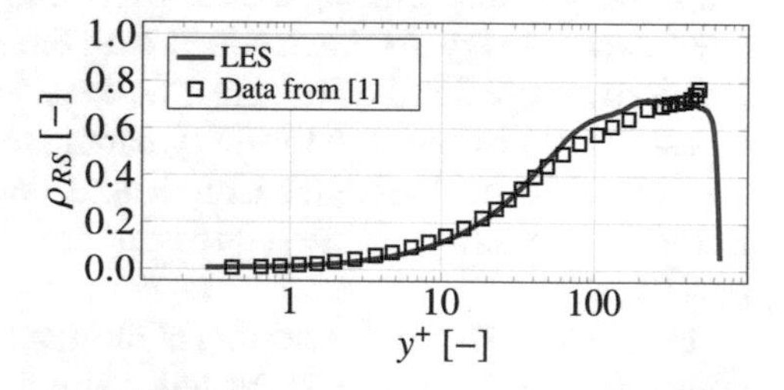

Fig. 3 Comparison of the validity parameter ρ_{RS} for the planar channel computed from LES data with data given by [1], Fig. 2.

Although the Reynolds number of the reference case used by Schmitt [1] is slightly lower, a good agreement between both data can be observed. This indicates a proper implementation of the entire simulation and evaluation process. As already mentioned, the alignment between both tensors is very poor close to the walls and reaches larger values closer to the center of the channel. However, very low values are again obtained in the symmetry plane since there the velocity gradient is zero and the principal specific Reynolds stresses are not.

3.2 Validation of the flow field prediction

The validation of the flow field prediction of test cases 2 - 4 are presented below. The evaluation lines used to compare the predicted flow field with the experimental data are shown in Fig. 4. For the first pin fin array, experimental data of the specific

Reynolds stresses and the velocity components are available along four different lines. The same accounts for the periodic ribbed channel. For the last test case, the pin fin array originally investigated by Ames and Dvorak [5], PIV data are available between the tail of the second row and the beginning of the third row. However, the results were explicitly compared at seven different lines.

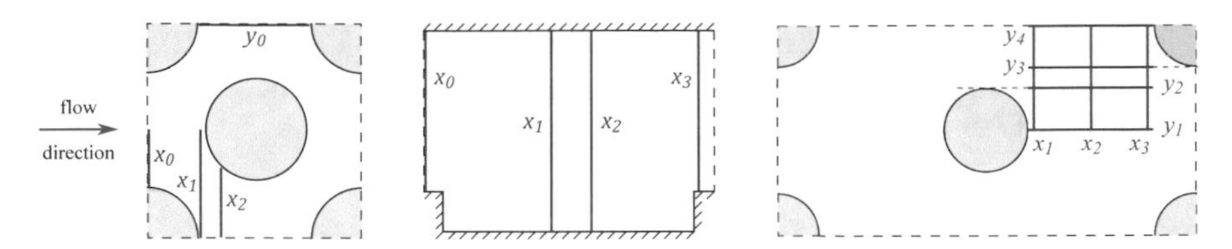

Fig. 4: Position of the experimental data for the periodic pin fin array (left) and the periodic ribbed channel (middle). The evaluation lines of the 2nd pin fin array are located in the symmetry plane (right).

Figures 5 and 6 compare the mean velocity field and the specific Reynolds stresses of the LES with the experimental data of the pin fin array at selected locations. In this case, the bulk velocity U_b at the inlet is used for normalization. The prediction of the velocity field (including the corresponding gradients) and the specific Reynolds stresses are in very good agreement with the experimental data. Since all available locations are predicted quite well, a well resolved flow field of the entire domain can be assumed. Therefore, the validity parameter ρ_{RS} can be accurately computed in the entire flow field using the LES data. A detailed study of this test case including a study of the turbulence structures can be found in the paper published during the project [10].

Figures 7 and 8 show a selection of the numerical results of the normalized velocity components and the specific Reynolds stress components for the ribbed channel in comparison with experimental data. Again, the bulk velocity U_b at the inlet was used for normalization. The agreement between experimental results and numerical data is very good for all positions and components. Based on this data, an accurate prediction of ρ_{RS} is expected in the entire domain.

The last test case was the most complex. Since it does not contain any homogeneous direction the additional space averaging is not applicable. Therefore, longer averaging times were needed. The comparison between the experimental data (PIV) and the LES results normalized by the bulk velocity U_B at several locations is given in Figs. 9 and 10. In this case, larger deviations between both methods can be observed for both velocities and the specific Reynolds stresses. Especially the specific Reynolds stresses tend to be overpredicted. Although the agreement is not as good as for the previous cases, the overall flow behavior is well captured. Therefore, the LES results can be used for a rough prediction of the validity parameter ρ_{RS}.

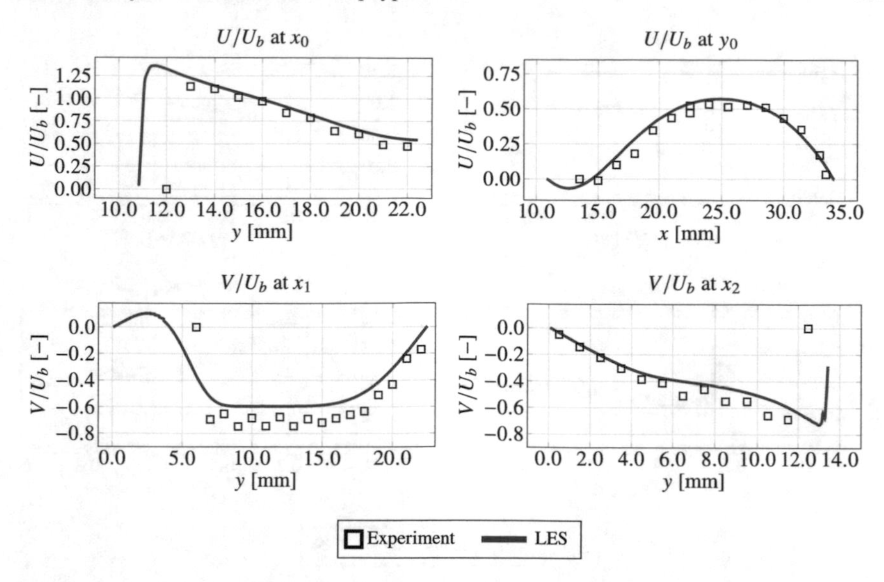

Fig. 5: Comparison of the normalized velocity components U/U_b and V/U_b obtained by LES with experimental data from [3] at various locations of the quasi-two-dimensional pin fin array.

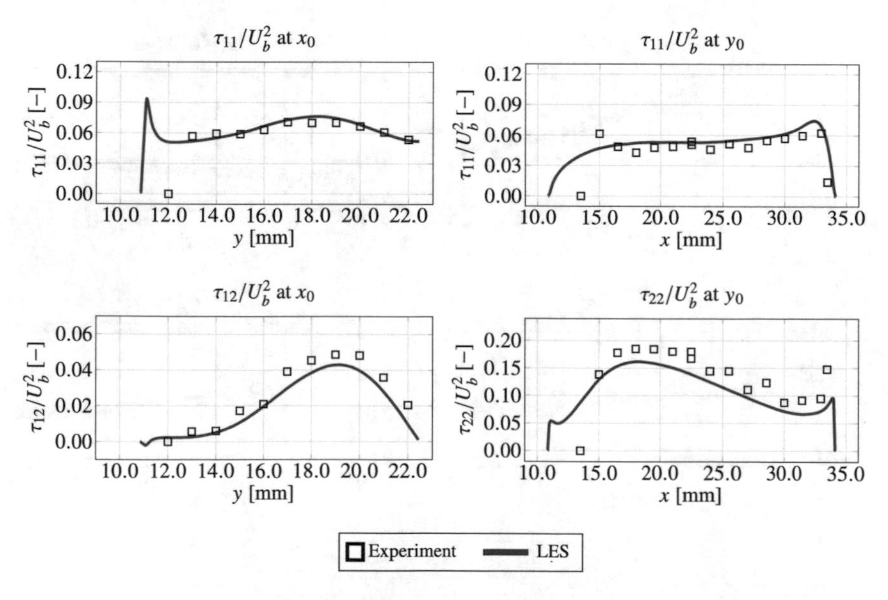

Fig. 6: Comparison of the normalized specific Reynolds stress components τ_{11}/U_b^2, τ_{12}/U_b^2 and τ_{22}/U_b^2 obtained by LES with experimental data from [3] at various locations of the quasi-two-dimensional pin fin array.

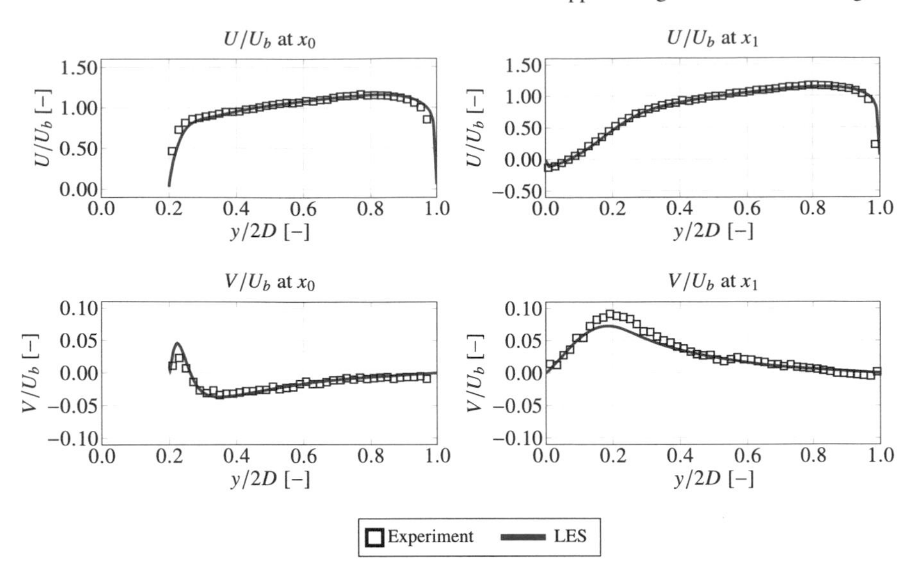

Fig. 7: Comparison of the normalized velocity components U/U_b and V/U_b obtained by LES with experimental data from [4] at various locations of the quasi-two-dimensional ribbed channel.

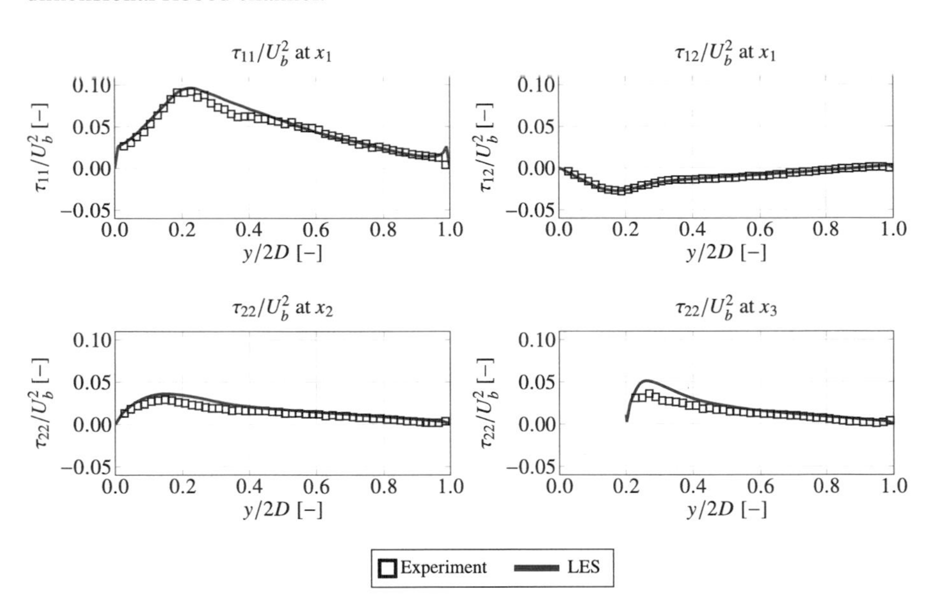

Fig. 8: Comparison of the normalized specific Reynolds stress components τ_{11}/U_b^2, τ_{12}/U_b^2 and τ_{22}/U_b^2 obtained by LES with experimental data from [4] at various locations of the quasi-two-dimensional ribbed channel.

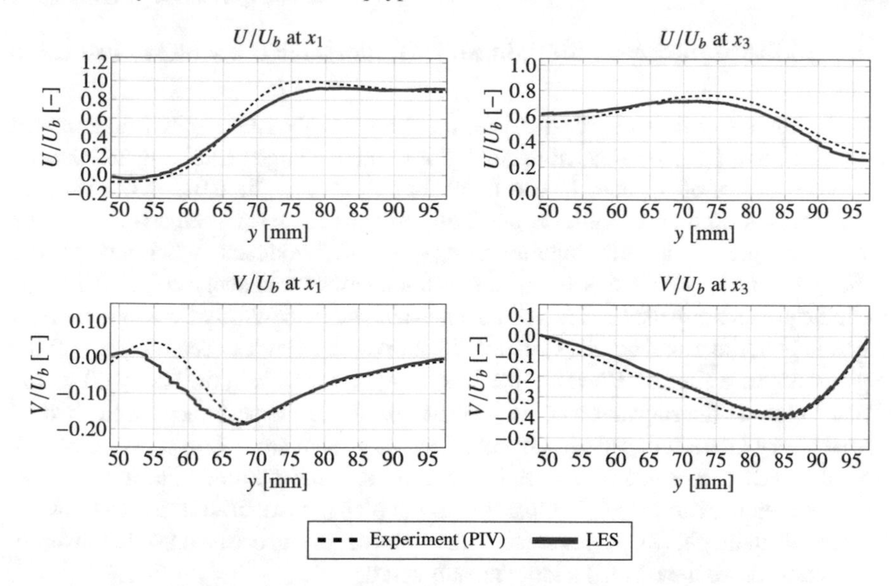

Fig. 9: Comparison of the normalized velocity components U/U_b and V/U_b obtained by LES with experimental data (PIV) provided by Siemens Energy at various locations of the three-dimensional pin fin array.

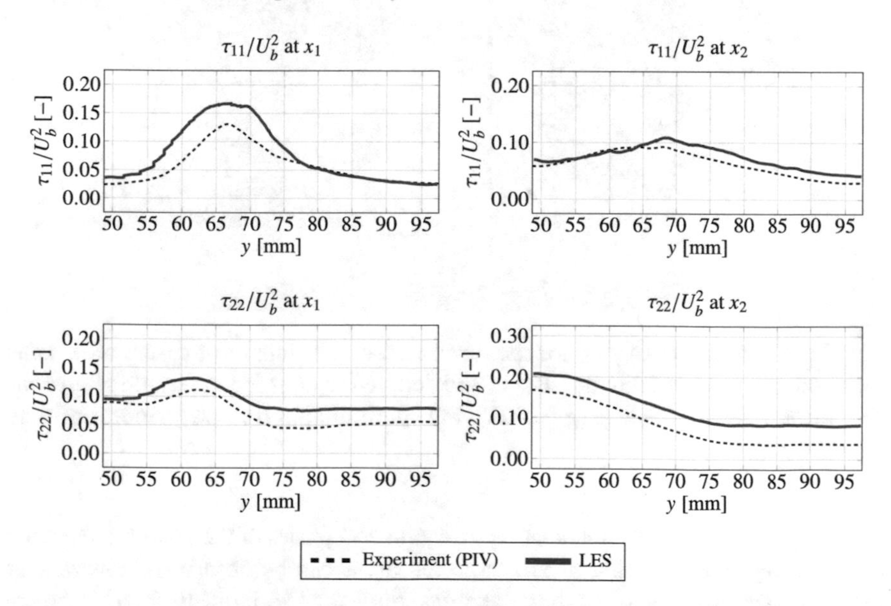

Fig. 10: Comparison of the normalized specific Reynolds stress components τ_{11}/U_b^2 and τ_{22}/U_b^2 obtained by LES with experimental data (PIV) provided by Siemens Energy at various locations of the three-dimensional pin fin array.

3.3 Validity of the linear Boussinesq hypothesis for the selected test cases

The distribution of the validity parameter ρ_{RS} based on the LES results are presented in Figs. 11 and 12. Blue areas indicate regions where a large misalignment between the eigenvectors of $\overline{R_{ij}}$ and $\overline{S_{ij}}$ can be observed. Hence, the LBH is not valid in these regions. Red areas indicate good alignment between the eigenvectors. The black line represent the soft limitation of $\rho_{RS} > 0.86$ introduced by Schmitt [1]. The validity parameter for the quasi-two-dimensional cases are shown in Fig. 11. For the periodic pin fin array the LBH is valid in approximately one-third of the fluid domain. Especially in the wake region (1) and in the narrow section between the pins (2) the eigenvectors are aligned. However, the LBH is not applicable in the near wall region (3) and close to the stagnation point (4). The results for the periodic ribbed channel reveal a very interesting pattern. Overall, only very small areas with $\rho_{RS} > 0.86$ can be identified. The results in the predominantly undisturbed flow field in the upper part between the ribs (1) are comparable to the planar channel. In the area between the ribs at rib height (2), ρ_{RS} increases but mainly remains below 0.86. However, in the section above the ribs (3) a large misaligned occurs.

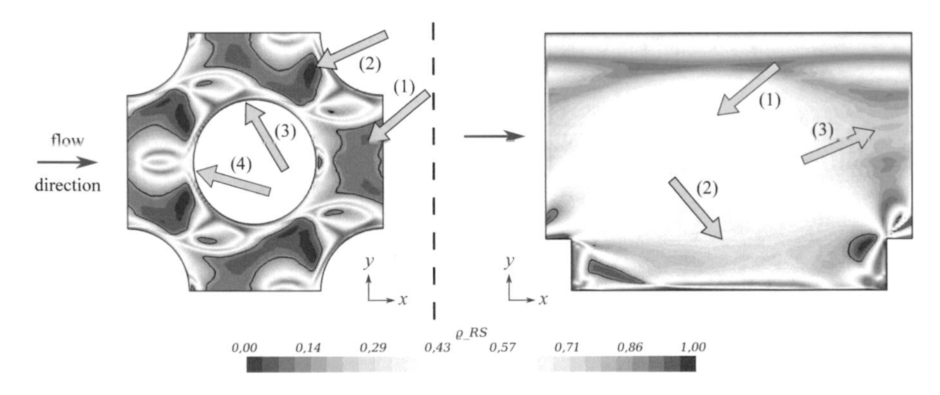

Fig. 11: Contour plot of ρ_{RS} for the quasi-two-dimensional test cases: blue areas represent regions in which the LBH is violated, red regions indicate good alignment. The black lines highlight $\rho_{RS} > 0.86$. left: periodic pin fin array, right: periodic ribbed channel

In Fig. 12 the validity parameter on the symmetry planes of the three-dimensional pin fin array is given. First of all, large red areas can be observed. However, in most regions ρ_{RS} is smaller than the weak limit of 0.86 indicating larger errors introduced by the Boussinesq hypothesis. Similar to the two-dimensional pin fin array, values of ρ_{RS} close to 0.86 can be observed in the wake region behind the pin (1). However, the behavior in the stagnation region (2) differ from each other. For this test case, larger values of ρ_{RS} are noticeable. This could be due the reduced influence of the surrounding pin fins. In addition, blue areas close to solid walls of the pins are less apparent. This might be an effect of the lower Reynolds number, but needs further investigations. Interestingly, no larger, continuous blue areas exist as they occur for the periodic ribbed channel. Only distinct narrow blue lines (3, 4) separating several red regions are noticeable. The physical meaning of this sharp separation in currently unclear and further investigations are needed. Although the flow field is three-dimensional, the results only slightly differ along the pin height in the z-direction (5). The largest gradients occur in front of the pin, where the blue line forms a parabolic shape (4). Additionally, the LBH is not valid in corners in the streamwise direction (6).

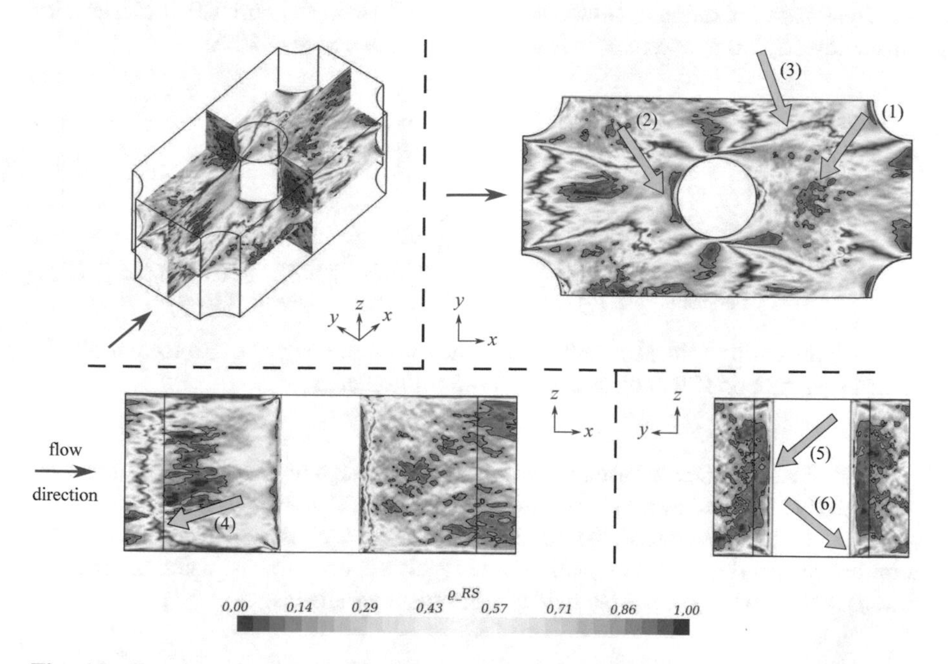

Fig. 12: Contour plot of ρ_{RS} for the three-dimensional pin fin array: blue areas represent regions in which the LBH is violated, red regions indicate good alignment. The black lines highlight $\rho_{RS} > 0.86$.

4 Computational performance

The parallel computing performance of Simcenter STAR-CCM+ on the ForHLR II cluster was investigated. Therefore, the LES of the quasi-two-dimensional periodic pin fin array (test case number two) containing ≈ 45 million cells was run on 40 up to 3600 cores (2, 4, 8, 16, 32, 64, 128 and 180 nodes). Consequently, the number of computational cells per cpu (cpc) vary between ≈ 1.1 million and 12,500. For the speed-up test, only 70 iterations were simulated and the average computational time per iteration was determined. For this purpose, the first 15 and the last 5 iterations were excluded to eliminate potential errors due to loading or saving effects.

Figure 4 shows the parallel speedup against the ideal speedup and the efficiency of Simcenter STAR-CCM+ on ForHLR II. The efficiency is defined as the respective speedup divided by the ideal speedup. A very high efficiency of more than 80% can be determined using up to 64 nodes (35,000 cpc). In addition, a high efficiency just above 50% using 180 nodes (12,500 cpc) can be observed. This highlights the excellent parallelization of Simcenter STAR-CCM+ on the ForHLR II cluster. For this study, 25,000 cpc were used leading to an efficiency of $\approx$ 80%.

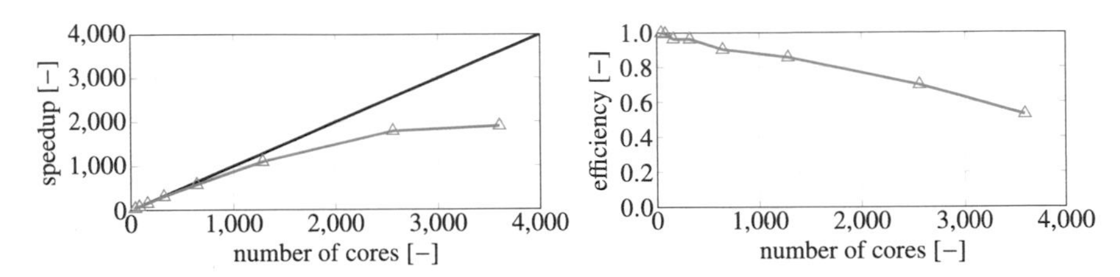

Fig. 13: Speedup vs. ideal speedup (left) and efficiency (right) of Simcenter STAR-CCM+ on the ForHLR II cluster using $\approx$ 45 million cells.

Table 1 summarizes information concerning the usage of computing resources on the ForHLR II. As shown before, Simcenter STAR-CCM+ shows good speedup up to $\approx$ 25,000 cpc. Therefore, the number or cores were chosen in accordance to the conducted speedup test. The wall time for each test case was adjusted to reach an almost symmetrical mean flow field and turbulence statistics.

Table 1: Usage of resources on the ForHLR II

Simulation	# of cells	# of cores	wall time	core hours
planar channel	35 million	1400	8 days	270,000
pin fin Simonin	45 million	1800	12 days	520,000
ribbed channel	20 million	800	10 days	190,000
pin fin Ames	15 million	600	16 days	230,000
				1,210,000

5 Conclusion

The validity of the linear Boussinesq hypothesis has been investigated for four different test cases. For this purpose, Large Eddy Simulations have been performed and the numerical results have been compared to experimental and DNS data. The mean strain rate tensor and the anisotropic stress tensor have been computed from the LES data. The validity parameter ρ_{RS} introduced by Schmitt [1], representing the misalignment between the eigenvectors of those two tensors, has been analyzed. The first test case of a planar channel demonstrates that the applied numerical method correctly reproduces the results given by Schmitt [1] for a similar planar channel. Flow field and Reynolds stresses were well predicted for the second test case of a quasi-two-dimensional pin fin array. Several distinct regions with good alignment between the eigenvectors, e.g. the wake region, have been identified. The LES was also able to predict the experimental results for the third test case of a periodic ribbed channel very well. In the upper part of the channel ρ_{RS} is quite similar to the planar channel. Only small areas with good alignment have been observed. In contrast, a high degree of misalignment has been identified in the region above the ribs. For the three-dimensional pin fin array, larger deviations between the experimental data and the LES have been observed. Nevertheless, the overall flow field is predicted with sufficient accuracy to get a general idea of the distribution of ρ_{RS}. Similar to the quasi-two-dimensional test case, the LBH is valid in the wake region. However, a different behavior close to the stagnation point has been observed.

Furthermore, speedup tests have been conducted. Good scaling up to 12,500 cpc with an efficiency just above 50% has been observed. However, stronger scaling with 25,000 cpc have been used for this study corresponding to an efficiency of $\approx$ 80%.

Acknowledgements The investigations were conducted as part of the Siemens Clean Energy Centre (CEC) joint research program. The authors acknowledge the financial support by Siemens Energy and the German Federal Ministry for Economic Affairs and Energy in the project under grant number 03ET7073F. The authors also thank the Siemens Energy Center, University of Central Florida for the permission to publish the experimental data shown in Figs. 9 and 10. The responsibility for the content lies solely with its authors. This work was performed on the computational resource ForHLR II funded by the Ministry of Science, Research and Arts Baden-Württemberg and DFG ("Deutsche Forschungsgemeinschaft").

References

1. F.G. Schmitt, About Boussinesq's turbulent viscosity hypothesis: historical remarks and a direct evaluation of its validity. Comptes Rendus Mécanique **335**(9-10), 617–627 (2007) doi: 10.1016/j.crme.2007.08.004
2. H. Abe, H. Kawamura, Y. Matsuo, Direct numerical simulation of a fully developed turbulent channel flow with respect to the Reynolds number dependence. J. Fluids Eng. **123**(2), 382–393 (2001) doi: 10.1115/1.1366680
3. O. Simonin, M. Barcouda, Measurements of fully developed turbulent flow across tube bundle. Proc. Third Int. Symp. Applications of Laser Anemometry to Fluid Mech., Lisbon (1986)

4. L.E. Drain, S. Martin, Two-component velocity measurements of turbulent flow in a ribbed-wall flow channel. Int. Conf. on Laser Anemometry-Advances and Application, Manchester, 99–112 (1985)
5. F.E. Ames, L.A. Dvorak, Turbulent Transport in Pin Fin Arrays: Experimental Data and Predictions. J. of Turbomach. **128**(1), 71–81 (2005) doi: 10.1115/1.2098792
6. Siemens PLM Software, STAR-CCM+ Documentation Version 2020.1 (2020)
7. V. Venkatakrishnan, On the accuracy of limiters and convergence to steady state solutions. 31st Aerosp. Sci. Meeting, Reno (1993) doi: 10.2514/6.1993-880
8. ANSYS, Inc., Ansys® ICEM CFD Documentation Release 17.1 (2016)
9. F. Nicoud, F. Ducros, Subgrid-scale stress modelling based on the square of the velocity gradient tensor. Flow, Turbul. and Combust. **62**(3), 183–200 (1999) doi: 10.1023/A:1009995426001
10. P. Wellinger, P. Uhl, B. Weigand, J. Rodriguez, Analysis of turbulence structures and the validity of the linear Boussinesq hypothesis for an infinite tube bundle. Int. J. of Heat and Fluid Flow **91**, 108779 (2021) doi: 10.1016/j.ijheatfluidflow.2021.108779

Development of turbulent inflow methods for the high order HPC framework FLEXI

Daniel Kempf, Min Gao, Andrea Beck, Marcel Blind, Patrick Kopper, Thomas Kuhn, Marius Kurz, Anna Schwarz and Claus-Dieter Munz

Abstract Turbulent inflow methods offer new possibilities for an efficient simulation by reducing the computational domain to the interesting parts. Typical examples are turbulent flow over cavities, around obstacles or in the context of zonal large eddy simulations. Within this work, we present the current state of two turbulent inflow methods implemented in our high order discontinuous Galerkin code FLEXI with special focus laid on HPC applications. We present the recycling-rescaling anisotropic linear forcing (RRALF), a combination of a modified recycling-rescaling approach and an anisotropic linear forcing, and a synthetic eddy method (SEM). For both methods, the simulation of a turbulent boundary layer along a flat plate is used as validation case. For the RRALF method, a zonal large eddy simulation of the rear part of a tripped subsonic turbulent boundary layer over a flat plate is presented. The SEM is validated in the case of a supersonic turbulent boundary layer using data from literature at the inflow. In the course of the cluster upgrade to the HPE Apollo system at HLRS, our framework was examined for performance on the new hardware architecture. Optimizations and adaptations were carried out, for which we will present current performance data.

Daniel Kempf, Min Gao, Andrea Beck, Marcel Blind, Thomas Kuhn, Marius Kurz, Anna Schwarz and Claus-Dieter Munz
Institute of Aerodynamics and Gas Dynamics, University of Stuttgart,
e-mail: {kempf,mg,beck,blind,m.kurz,schwarz,munz}@iag.uni-stuttgart.de

Patrick Kopper
Institute of Aircraft Propulsion Systems, University of Stuttgart,
e-mail: kopper@ila.uni-stuttgart.de

W. E. Nagel et al. (eds.), *High Performance Computing in Science and Engineering '21*,
https://doi.org/10.1007/978-3-031-17937-2_17

1 Introduction

Turbulent flows are relevant for a wide range of engineering applications. Therefore, they are the subject of intensive research. In the field of computational fluid dynamics, turbulent inflow conditions are of great relevance in order to simulate turbulent flows as efficiently as possible by minimizing the computational domain. For the computation of turbulent flows and their challenging multi-scale character, high order methods are well suited due to their inherent low dissipation and dispersion errors. Our open source simulation framework FLEXI[1] has been developed and improved during the last years in the Numeric Research Group at the IAG. A current overview of the framework is given in [1]. The solver discretizes the compressible Navier–Stokes–Fourier equations with an arbitrary high order discontinuous Galerkin spectral element method (DGSEM) in space and uses low storage Runge–Kutta schemes for explicit high order time integration. The focus of FLEXI is on scale-resolving large eddy simulation (LES) of compressible flows, which includes aeroacoustics [2–4], transitional and turbulent flows [5, 6], particle-laden flows [7] and many more.

LES poses significant computational requirements, still preventing its application in industry. The less demanding Reynolds-averaged Navier–Stokes equations (RANS) provides good results in attached flows but suffers from well-known shortcomings with regards to intermittency, transition and separation. A zonal LES, sometimes also called embedded LES, has emerged as a hybrid approach, restricting the expensive application of LES to critical areas or regions of special interest, while advancing the remaining domain with RANS. However, while this approach does reduce the computation effort significantly, it poses the new challenge of requiring a turbulent inflow method at the RANS-LES interface which reconstructs the time-accurate LES inflow conditions from the RANS solution. An overview of existing approaches is given in [8]. In this work, we present two turbulent inflow methods, the recycling-rescaling anisotropic linear forcing (RRALF) by Kuhn et al. [9], a combination of a modified recycling-rescaling approach and an anisotropic linear forcing, and a synthetic eddy method (SEM) by Roidl et al. [10] in combination with an anisotropic linear forcing (ALF) based on Laage de Meux et al. [11]. Both methods are implemented in FLEXI and form the basis for RANS-LES coupling approaches in future applications.

This work is structured as follows. In Sec. 2, we introduce both turbulent inflow methods. In Sec. 3, we present a performance analysis of FLEXI on HAWK, the new cluster at HLRS, and performance optimizations on this system as well as implementation details of the turbulent inflow methods. In Sec. 3, both presented turbulent inflow methods are validated with a simulation of a turbulent boundary layer along a flat plate. In case of the RRALF method, a zonal large eddy simulation of the rear part of a subsonic turbulent boundary layer is presented. The SEM is validated with the case of a supersonic turbulent boundary layer. Sec. 5 concludes the paper with a short outlook on further activities.

---

[1] https://github.com/flexi-framework/flexi

2 Methods

In this section, we introduce both turbulent inflow methods, which will be applied throughout this work. First, in Sec. 2.1 we cover the RRALF method, an approach for a zonal LES. Due to the recycling-rescaling used, the solution can suffer from an artificial low frequency periodicity. In specific cases, like a shock-boundary layer interaction with an oscillating shock, this frequency can influence the shock movement. Therefore, a synthetic eddy method can be favorable in such cases, which is described in Sec. 2.2.

2.1 Recycling-rescaling anisotropic linear forcing

In this work, we present an application of the combined recycling-rescaling anisotropic linear forcing technique by Kuhn et al. [9]. Recycling methods generally produce the most accurate turbulent fields by employing the turbulence mechanisms of the Navier–Stokes equations themselves. They are however restricted to fully developed flows. An extension to zero pressure gradient (ZPG) boundary layer flows was introduced by Lund et al. [12] by recycling and rescaling only the fluctuating part of the solution at the interface. Anisotropic linear forcing introduced by de Laage de Meux et al. [11] on the other hand is a volume forcing approach, introducing a source term in the momentum equation to quickly stimulate the development of turbulence. However, since the forcing is disconnected from the Navier–Stokes equations, it is limited to weak source terms and thus large development regions are required to form realistic turbulence. RRALF combines these two approaches, relying on the Navier–Stokes equations for physical turbulence production but adding a forcing zone in front of the recycling plane to break the limitation to ZPG boundary layers and drive the solution towards the desired boundary layer profile. The rescaling follows [12] by splitting the boundary layer into an inner and an outer region and reconstructing the rescaled profile using

$$u_{i,in} = \left[(\bar{u_i})_{in}^{inner} + (u_i')_{in}^{inner}\right][1 - W(\eta_{in})] + \left[(\bar{u_i})_{in}^{outer} + (u_i')_{in}^{outer}\right] W(\eta_{in}),$$

where u_i is the corresponding component of the instantaneous velocity field, $[\bar{\ }]$ and $[\ ]'$ denoting mean and fluctuating quantities, and $W(\eta_{in})$ is a blending function defined as

$$W(\eta_{in}) = \frac{1}{2}\left[1 + \tanh\left(\frac{a(\eta_{in} - b)}{(1 - 2b)\eta_{in} + b}\right)/\tanh(a)\right],$$

with the weights $a = 4$ and $b = 0.2$. The ALF volume forcing term is taken from [11], which yields

$$\frac{\partial \rho u_i}{\partial t} = R(u_i) + \rho f_i$$

for each of the momentum vector components $i = 1, 2, 3$, where R represents the spatial DG operator and f_i the ALF vector given as $f_i = A_{ij}u_j + B_i$. A_{ij} and B_i are the tensors for the Reynolds stresses and the mean velocity forcing, respectively. Due to the recycling-rescaling, the second order turbulence statistics can overshoot the target data. A modification of the original ALF introduces an explicit damping which corrects the overshoot rapidly.

In the following paragraph, we discuss the algorithmic implementation of the RRALF method in FLEXI. In a preprocessing step, the computational grid of the recycling interface is built based on the properties of the target data. The scaling is implemented as a static geometric scaling of the recycling plane, which is computationally efficient. For every point on the recycling grid, we determine the global element ID and the cell local coordinates of the scaled interface. Additionally, we provide the necessary data for the MPI communication between the processors, which handle the physical boundary and the interface plane. Besides the numerical aspects of the preprocessing step, we also compute the wall-normal scaling function from the target turbulent statistics of the initial flow field. This information is stored to build the communication and the interface as well as to interpolate the turbulent inflow fluctuations during runtime. The second preprocessing step creates an initial solution based on turbulent target statistics. Time-averaged data can be provided as 1D profiles and is interpolated to the flow domain to match the desired Re_θ. Additionally, field data of a precursor simulation can be used. Following Kuhn et al. [9], we add superimposed wall-parallel streaks to improve the development of turbulent structures inside the boundary layer. Further, the forcing zones where ALF is used are defined. During runtime, the main overhead of the RRALF method is created by the communication of data between the interface recycling plane and the boundary as well as the update and application of the ALF forcing in every Runge–Kutta step.

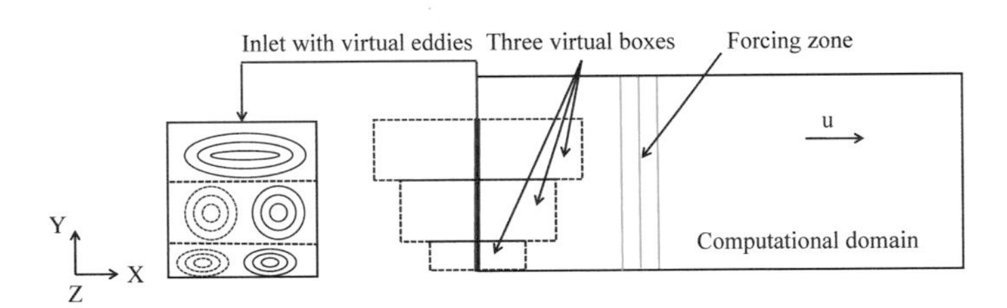

Fig. 1: Schematic of the SEM with an ALF zone and three virtual boxes containing three different types of virtual eddies. The inflow boundary is highlighted red.

2.2 Synthetic eddy method

Another method to produce the desired turbulent fluctuations at inflow boundaries is the synthetic eddy method. SEM was proposed by Jarrin et al. [13], based on the idea that turbulence can be seen as a superposition of coherent structures. Following this idea, SEM creates turbulent fluctuations at the inflow by superimposing virtual eddies in a virtual volume box at the inlet such that the prescribed turbulent statistics are recovered. Since the virtual eddy cores at the inlet are distributed randomly, SEM does not suffer from the well-known spurious periodicity of recycling-based schemes [12]. Pamies et al. [14] extended the SEM by subdividing the single virtual volume box containing one type of virtual vortices into multiple zones containing different types of virtual vortices, as is shown in Fig. 1. This extension allowed to account for the different turbulent scales in the wall-normal direction. Roidl et al. [10] extended this method to compressible flows by enforcing the strong Reynolds analogy. To reduce the required recovery length, we followed the idea of Kuhn et al. [9] that ALF can be employed to improve the ability of the recycling-rescaling method to recover the target turbulent statistics. To this end, we also adopted the ALF method mentioned above to reduce the required recovery length of SEM.

3 Performance

The purpose of this section is twofold. In Section 3.1, the general performance of FLEXI on the new HAWK system is evaluated and code optimizations for the new system are discussed. In a second step, the parallelization strategy for SEM is presented in Section 3.2.

3.1 FLEXI on HAWK

The system change at HLRS from the former Cray XC40 (Hazel Hen) to the current HPE Apollo (HAWK) system included a shift from Intel® Xeon® CPUs to AMD EPYC™ as well as a changed node-to-node interconnect, where an InfiniBand architecture with a 9D-Hypercube topology replaced the Aries interconnect of Hazel Hen. These drastic changes in hardware architecture and especially the increase of available cores per node (from 24 to 128) motivated a detailed investigation of the performance of our open source code FLEXI on the new system. To analyze the performance, we use a free stream problem within a Cartesian box as test case, for which we solve the Navier–Stokes–Fourier equations. To account for different problem sizes, we further vary the number of elements in each direction for the different cases. The polynomial degree was set to $N = 5$, which results in a sixth-order scheme and is a typical choice for production runs. We investigated the number of nodes in the range from 1 to 512. The code was compiled with the GNU compiler version 9.2.0 with the libraries mpt 2.23, hdf5 1.10.5 and aocl 2.1.0. The gradients of the solution are computed with the BR1 lifting method [15], and the split formulation by Pirozzoli [16] is employed to control aliasing errors. For each run, we computed 100 time steps. The influence of fluctuations on the performance of the system is taken into account by running each configuration at least three times to collect statistics. The performance data was obtained with the system's configuration as of June 2021.

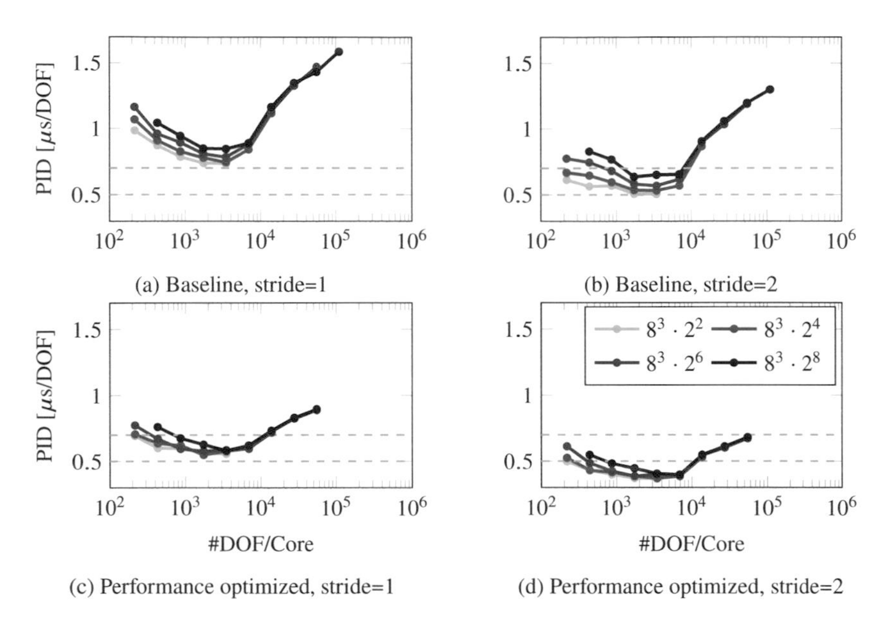

Fig. 2: Results of the scaling analysis for different meshes and loads. Four different cases are investigated: the baseline version of FLEXI as described in Krais et al. [1] and the optimized version for HAWK with a stride of 1 and 2 for each version. The legend lists the number of elements of the investigated meshes.

Fig. 2 presents the strong scaling behavior by plotting the performance index (PID) over the number of degrees of freedom (DOF) per rank for all considered cases and configurations. The PID is defined as

$$\mathrm{PID} = \frac{\text{wall-clock-time} \cdot \#\text{cores}}{\#\text{DOF} \cdot \#\text{time steps} \cdot \#\text{RK-stages}}\,, \tag{1}$$

and represents the time it takes to advance a single degree of freedom to the next stage in the Runge–Kutta time-stepping algorithm.

Fig. 2a shows the results of the baseline version of FLEXI on the HAWK cluster. The qualitative behavior is similar for all plots in Fig. 2, and also matches the behavior observed on the former system as described in Krais et al. [1]. A small number of nodes for a given problem size results in a high load. Here, an increasing amount of data has to be regularly moved to and retrieved from the main memory, since the data does not fit into the fast CPU cache anymore. With increasing load, the memory bandwidth per core of the AMD EPYC™ CPUs becomes the limiting factor and the performance index increases dramatically. This is due to the specific CPU architecture. Each socket consists of 8 CCDs (Core Chiplet Die), which comprise 2 CCXs (Core Complex) each. Both CCXs share a common interface to the I/O and consist of 4 cores each. Thus, this hardware architecture leads to a comparably small memory bandwidth per core, which can deteriorate the performance of memory intensive operations. Therefore, the reduced code performance for high loads is more pronounced on HAWK than on the former Hazel Hen system. For an increasing number of nodes for a given problem size, i.e. lower load, the PID increases again. This is due to decreasing local work on each CPU, which cannot hide the communication latency effectively and the latency becomes dominant. For all cases, the optimal PID is observed in between 10^3 and 10^4 DOF per core, where the communication latency and caching effects are balanced optimally. A detailed discussion of the fundamental scaling behavior of the DGSEM code FLEXI can be found in Krais et al. [1].

To analyze the impact of the architecture and especially the limited memory bandwidth on the performance of FLEXI, the performance analysis was first carried out on all available cores (stride=1, see Fig. 2a) and in a second step, while using only every second available core (stride=2, see Fig. 2b). This artificially increases the available memory bandwidth per active core. Further, the reduced usage of CPUs within each node can have benefits for the internode communication and might raise the power limits for the active cores. The results in Fig. 2 indicate that the case with stride=2 gains about 30 % in performance compared to the stride=1 case. It is important to stress that the overall improvement of the PID for the stride=2 configuration would have to exceed the factor of 2 to compensate for the unused cores, which was not observed in any of the investigated cases. The qualitative behavior is similar for both cases and the significant decrease in performance towards high loads is still noticeable for the stride=2 case.

The Hypercube topology of the interconnect allows only for defined numbers of nodes (64, 128, 256, 512, ...) for large jobs, which renders the aforementioned scaling behavior unfavorable regarding flexibility and efficiency. Therefore, we strived to decrease the memory footprint of FLEXI and improved the code optimizations performed by the compiler. To this end, the lifting procedure was redesigned to compute only the gradients of the variables which are actually required to compute the parabolic fluxes of the Navier–Stokes–Fourier equations. While this causes the data to be not contiguous in memory anymore, it proved to reduce the memory footprint of the lifting procedure by about a fifth and improved the overall performance of FLEXI. By profiling the code, we found two major performance-reducing issues. Two frequently called functions, namely the Riemann solver as well as the solver for the two-point volume split flux, were not inlined by the compiler. To this end, we introduced a two step compilation process to employ profile-guided optimization (PGO). By using PGO, the aforementioned functions get inlined by the compiler and the overall cache usage is improved. In Fig. 2c and Fig. 2d the results with the optimized code version using both stride=1 and stride=2 are depicted. A comparison of Fig. 2a with Fig. 2c demonstrates a performance improvement by about 25 % in

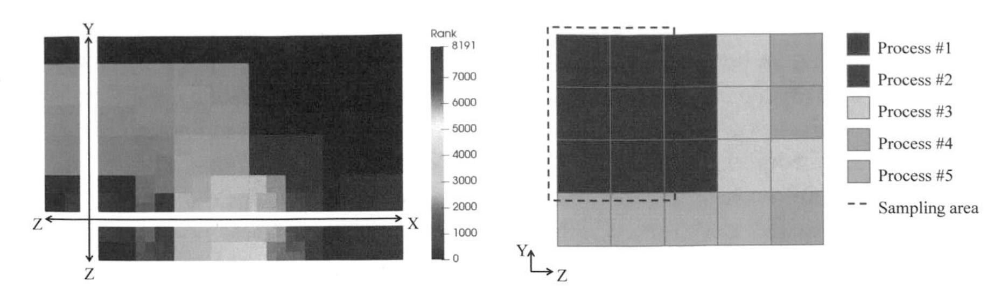

Fig. 3: (left) Distribution of 8192 processing ranks (64 nodes) in the computational domain. Only the ranks covering the inlet, i.e. the Y-Z-plane, up to $1.2\delta_{99,in}$ have to be considered for the SEM and form the sub-communicator. (right) Schematic view of the domain decomposition at the inflow boundary. For the first process (red), only the vortices with their core inside the sampling area have to be evaluated. The sampling area is larger than the occupied area at the inlet by the maximum length scale of the virtual vortices.

case of an optimal load and even up to 40 % for high loads. This indicates the more efficient usage of the CPU cache and the available memory bandwidth, especially towards high loads, where the significant performance losses are mitigated. Since the performance of the optimized code is thus less sensitive to the specific load per core, FLEXI becomes more flexible and efficient, especially regarding the specific Hypercube topology of the interconnect on HAWK. The comparison of Fig. 2c with Fig. 2d again depicts a higher performance for the case with stride=2. However, the observed improvement is not sufficient to compensate for the idling cores.

3.2 Parallel implementation of SEM

As discussed in Sec. 2.2, the SEM superposes virtual eddies in a virtual domain at the inflow boundary, which are then used to derive the turbulent inflow state at each point of the inflow plane. Since the inflow state is computed in a non-local fashion, a consistent parallelization strategy is crucial for the method to be applicable for large-scale HPC applications. Therefore, the following section focuses on the parallel implementation of the reformulated SEM introduced in Sec. 2.2.

In a first step, a new MPI sub-communicator is introduced, which comprises all processes with elements requiring velocity fluctuations at the turbulent inflow boundary. All the information exchange within those processes computing the turbulent fluctuations can thus be handled within this communicator. An exemplary rank distribution at the inflow for practical applications is illustrated in Fig. 3 left. According to the methodology of the SEM, the virtual eddies affect the velocity only locally, while the range of influence is determined by the length scale of each virtual eddy core. Therefore, the involved processors do not have to consider the

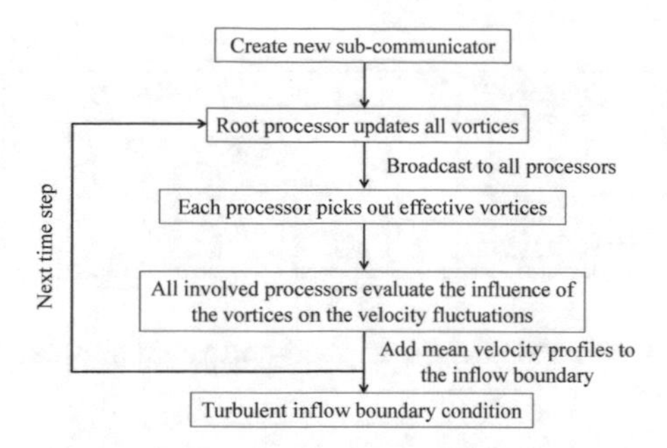

Fig. 4: Flow chart of the MPI implementation of the reformulated SEM.

influence of all virtual vortices. Instead, only the effective vortices for each point have to be evaluated. To this end, each process introduces a rectangular "sampling region" containing a halo region. This region, which is depicted in Fig. 3 right, is larger than the area occupied by the process by the maximum length scale of the vortices and thus contains all vortices which influence the inflow state of this processor. Clearly, the current restriction to rectangular sampling regions is not optimal. Especially if a processor occupies a non-rectangular inflow area, vortices might be evaluated unnecessarily, as shown in Fig. 3 right. In this example, the red processor evaluates all the vortices within its sampling region. This might also include vortices, which do not influence the red process but only the purple process. One way to improve the sampling accuracy in the future could be to introduce an element-local sampling procedure. However, the downside is the increase of the required memory and communication latency. While such improvements could be easily integrated in the future, the proposed approach already gives acceptable computational efficiency at reasonable implementation effort.

In order to synchronize the distribution of the virtual eddy cores in each time step, the root processor updates the locations of the virtual vortices. Afterwards, the new distribution of vortices is broadcasted to all other processors inside the sub-communicator. Finally, each process can evaluate its effective vortices. Figure 4 gives an overview of the implementation of the MPI strategy.

4 Results

In the following section, the introduced turbulent inflow methods are validated via a turbulent flat plate simulation. First, we present a subsonic turbulent boundary layer simulation at $Ma = 0.3$ with a tripped simulation setup in Sec. 4.1. This is followed by a zonal LES of the rear part of the identical tripped boundary layer case using

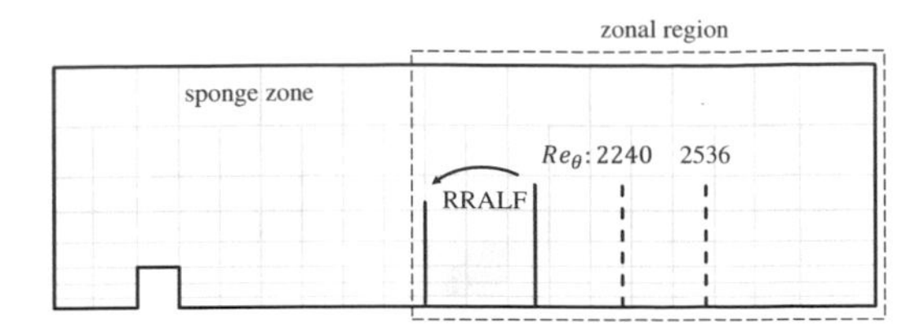

Fig. 5: Sketch of the mesh used for the tripped and zonal simulation of the turbulent flat plate.

the RRALF method as turbulent inflow. The second configuration is a supersonic boundary layer at $Ma = 2.0$, which uses the SEM as turbulent inflow condition and is presented in Sec. 4.2.

4.1 Subsonic turbulent flat plate

Two types of simulations of a weakly compressible turbulent flat plate were carried out with the same numerics and resolution. First, a tripped turbulent boundary layer simulation was conducted. These simulation results were used as target data for a zonal LES of the rear half of the flat plate. The tripped boundary layer used the incompressible, turbulent velocity profile provided by Eitel-Amor et al. [17] at $Re_\theta = 750$ and a Mach number of $Ma = 0.3$ as inflow data. The quadratic trip is placed one boundary layer thickness behind the inflow and has the size of $y^+ = 50$ in wall units referring to the inflow profile. The domain size is $205\ \delta_{99,in} \times 12\ \delta_{99,in} \times 16\ \delta_{99,in}$, $\delta_{99,in}$ being the boundary layer thickness at the inflow of the tripped case. The domain resolves the boundary layer up to about $Re_\theta = 2,800$.

The mesh displayed in Fig. 5 comprises 459,150 elements, which results in about 235 million DOF for a polynomial degree of $N = 7$. The mesh resolution is, improving towards higher Re_θ, $x^+ = 25 \rightarrow 20.5$, $y^+ = 1 \rightarrow 0.8$ and $z^+ = 12 \rightarrow 9.8$. The simulation was carried out for 531 convective time units $T^* = \delta_{99,z}/U_\infty$ and was averaged over the last 115 T^* to obtain the first and second order turbulence statistics. Here, U_∞ denotes the freestream velocity and $\delta_{99,z}$ the boundary layer thickness at the inflow of the zonal domain at $Re_\theta = 1,800$ and is from now on the reference for both cases. The simulation was carried out on HAWK with 256 nodes which results in a load of 7,174 DOF/core. In this particular case, a PID of 0.95 μs was achieved with the optimized version of FLEXI, which corresponds to a reduction of 36% compared to a PID of 1.48 μs for the non-optimized version. Due to additional overhead, e.g. by the collection of turbulent statistics, the PID was slightly increased compared to the results in Sec. 3. In total, the simulation required 445,000 CPUh.

The zonal LES had the same mesh resolution and resolved the boundary layer from $Re_\theta = 1,800 - 2,800$. Hereby, the mesh was reduced to about 112 DOF. The region where ALF was applied starts at the inflow and extends over 6 $\delta_{99,z}$ in streamwise

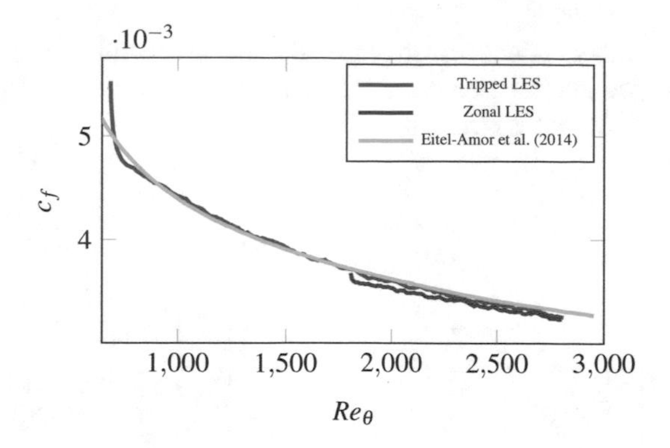

Fig. 6: The friction coefficient c_f of the tripped and the zonal turbulent flat plate with the numerical reference from Eitel-Amor et al. [17].

direction, whereby the first wall normal cell was skipped. The recycling plane was placed at 8 $\delta_{99,z}$ in streamwise direction. The simulation was carried out for 277 T^* and averaged over 69 T^* to obtain the turbulence statistics. The simulation used 227,000 CPUh with 128 nodes, which results in a load of 6,871 DOF/core. Due to an additional overhead, introduced by the turbulent inflow method, the optimized PID increased to 1.41 μs, which corresponds to a reduction of 25% referring to the non-optimized version with a PID=1.88 μs. In Fig. 6, the friction coefficient of the tripped and zonal LES as well as the reference from Eitel-Amor et al. [17] are illustrated. The friction coefficient of the tripped simulation clearly indicates the influence of the trip at low Re_θ. With increasing Re_θ, the friction coefficient trends towards the reference but with a slightly steeper decline. For the zonal LES, the turbulent inflow results in a significant drop in the friction coefficient at the inflow. Due to the choice of the recycling method, the instantaneous fluctuations were recycled without further amplitude scaling. However, in this flow regime the changes of the amplitude of the Reynolds stresses over the recycling distance are noticeable and lead to a relaxation of the flow field at the inflow. Further downstream, the solution of the zonal LES tends towards the numerical reference.

Fig. 7 depicts the first and second order turbulence statistics at $Re_\theta = 2,240$ and $Re_\theta = 2,536$ within the zonal region, for which reference data from Eitel-Amor et al. [17] are available. The time-averaged streamwise velocity in Fig. 7 (left) displays an overall good agreement between the two simulations and the reference data for both cases. At $Re_\theta = 2,240$ and $Re_\theta = 2,536$, the tripped LES matches the reference very well in the viscous sublayer, the buffer layer and in the log-law region. In the outer layer are minor deviations are noticeable due to a remaining influence of the trip. The zonal LES slightly overestimates the reference from literature and the tripped LES from the log-law region. At $Re_\theta = 2,536$, the zonal LES matches the two references very well up to the log-law region. In the outer layer, the results of the zonal LES lies between the two references. The second order turbulent statistics in form of the normal stresses are illustrated in Fig. 7 (right) for both Reynolds numbers. For both Reynolds numbers under investigation, there is an overall very good agreement between the tripped and the zonal LES. However, both simulations tend to slightly underestimate the Reynolds stresses.

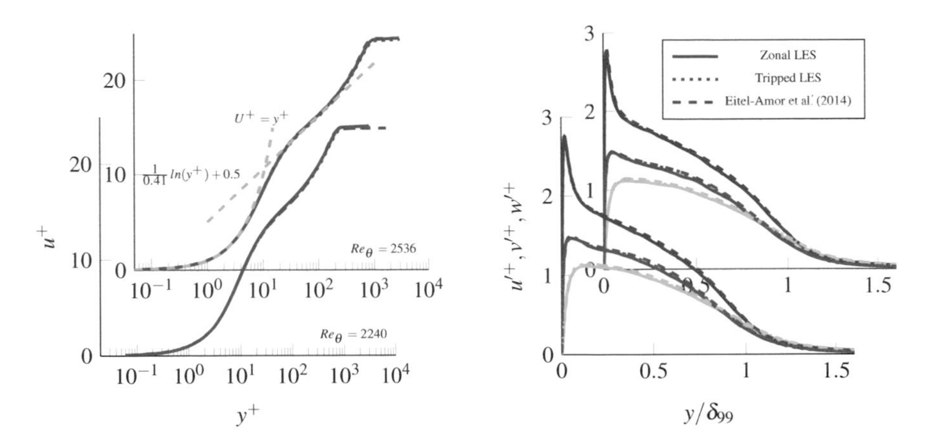

Fig. 7: Comparison of the mean velocity $u^+ = u/u_\tau$ (left) and the Reynolds fluctuations (right) at $Re_\theta = 2,240$ and $Re_\theta = 2,536$ for the tripped and the zonal turbulent flat plate with a numerical reference from Eitel-Amor et al. [17]. The Reynolds fluctuations $u'^+ = \sqrt{\overline{u'u'}}/u_\tau$, $v'^+ = \sqrt{\overline{v'v'}}/u_\tau$ and $w'^+ = \sqrt{\overline{w'w'}}/u_\tau$ are presented by red lines (—), green lines (—) and blue lines (—), respectively.

4.2 Supersonic turbulent flat plate

In the following, we present the LES results of a compressible turbulent boundary layer at a Mach number of $Ma = 2.0$, which is validated with the DNS data provided by Wenzel et al. [18]. The SEM introduced in Sec. 2.2 is used as turbulent inlet condition in conjunction with ALF. The computational domain covers a box with the size of $34.78\ \delta_{99,in} \times 20.55\ \delta_{99,in} \times 3.95\ \delta_{99,in}$. The present simulation (LES02) was conducted with a polynomial degree of $N = 6$ with $102 \times 104 \times 27$ elements in streamwise, wall-normal and spanwise direction, respectively. This leads to a total amount of about 98 million DOF. The grid resolution at the inlet was chosen in wall units as $\Delta y^+_{min} = 1.6$, $\Delta x^+ = 30$ and $\Delta z^+ = 16$. Periodic boundary conditions were used in the spanwise direction and an adiabatic no-slip wall boundary condition was used for the flat plate. At the upper boundary, the DNS data was set as Dirichlet boundary condition to avoid reflections induced by the inconsistency between the inflow and the upper boundaries. Finally, a sponge layer was placed before the outlet and a supersonic outflow boundary condition was used to avoid spurious reflections. The setup is displayed in Fig. 8. The simulation was performed with 8, 192 CPU

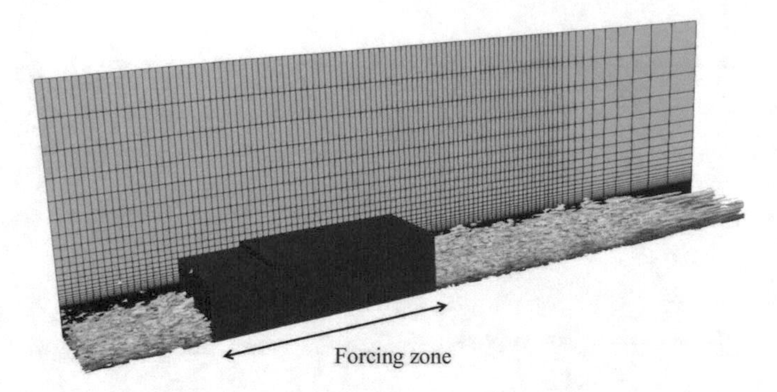

Fig. 8: Q-criterion of the turbulent structures colored by the velocity magnitude. The red region represents the zone where the ALF is enforced. The background mesh is also presented up to 10.5 δ_{99} in the wall-normal direction.

cores, which leads to a load of about $11,992$ DOF/core. We obtained a PID of about 1.81 μs for this LES02 case, which is about 33.3% larger compared to the results in Sec. 3. In order to quantify the performance loss due to the reformulated SEM, we calculated the same numerical case without using the additional ALF (LES01). This resulted in a PID of around 1.68 μs, which is 24.4% larger than the baseline code. In total, the computations used about 100,000 CPUh. In consideration of the communication overhead due to the collection of the turbulent statistics for analysis, it is assumed that the current MPI strategy works well but could be improved even further.

The flow reached a quasi-steady state after about 77 convective time units $T^* = \delta_{99,in}/U_\infty$. In Fig. 8, an instantaneous snapshot of the turbulent structures in the flow domain as well as the forcing zone, where the ALF was enforced, are illustrated. In order to ensure that the forcing term remains weak, the forcing zone extends from 5 $\delta_{99,in}$ to 15 $\delta_{99,in}$ in streamwise direction. After the quasi-steady state was reached, the flow was advanced by another 350 T^* to obtain the turbulent mean flow statistics. In Fig. 9, the mean velocity and the Reynolds stresses of the LES01 and the LES02 are compared to the reference DNS. In order to account for the compressibility effect, the Van Driest transformation of the mean velocity $u^+_{vD} = \int_0^{u^+} \sqrt{\rho/\rho_w}\, du^+$ as well as Morkovin's density scaling of the Reynolds fluctuations $R_{\phi\phi} = \sqrt{\rho/\rho_w}\, \phi'^+$, $\phi = u, v, w$ are applied, where ρ denotes the density and $[\ ]_w$ the statistics at the wall. The results for the LES02 case show better agreement with the reference DNS data than the LES01 case, which did not employ an ALF zone. This highlights the benefit of the proposed method, which combines the SEM with ALF.

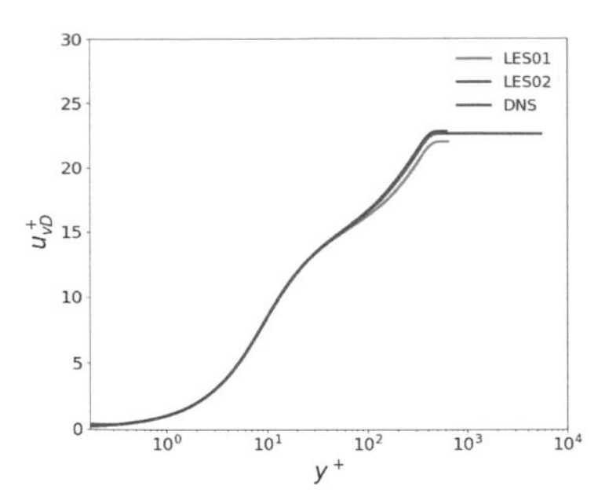

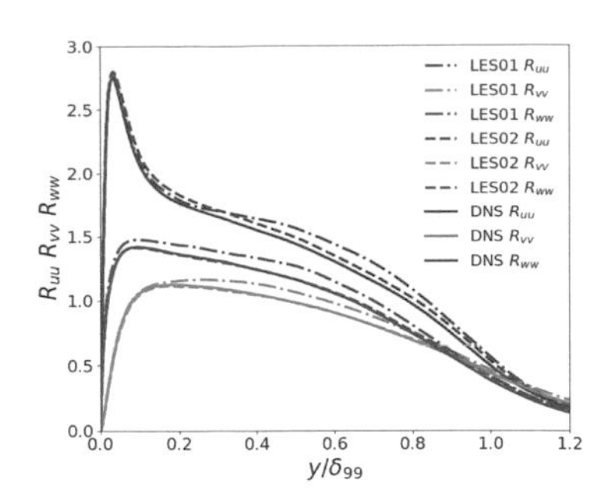

Fig. 9: Comparison of the averaged flow statistics at 15 δ_{99} between SEM (LES01), SEM with additional ALF (LES02), and the reference DNS. Van Driest transformed velocity profiles (left) and Morkovin's density scaled Reynolds stress profiles (right).

5 Summary and outlook

In this work, we have presented the current capabilities of FLEXI with regards to turbulent inflow methods as well as recent performance optimizations motivated by the new hardware architecture of the HPE Apollo HAWK at HLRS. To adapt FLEXI to the new CPU architecture, the optimizations were designed to reduce the memory footprint and to achieve more efficient cache usage. Furthermore, the function call overhead of two essential subroutines was reduced. We achieved a performance gain of 25% at an optimal load and up to 40% at high loads, which was intended to give FLEXI the necessary flexibility for an optimal use of the hypercube topology of HAWK.

We presented two turbulent inflow methods, the recycling-rescaling anisotropic linear forcing (RRALF), a combination of a modified recycling-rescaling approach and an anisotropic linear forcing, and a synthetic eddy method (SEM). Both methods were validated for a turbulent boundary layer along a flat plate. In case of the RRALF method, a zonal large eddy simulation of the rear part of a tripped, subsonic turbulent boundary layer was presented. By applying the RRALF method, we were able to reproduce the target data from the tripped boundary layer within the zonal region. The SEM was validated against a supersonic turbulent boundary layer using data from literature at the inflow. Additionally, we showed that the SEM method benefits from an additional anisotropic linear forcing behind the inflow to reach the target data faster. In future work, the RRALF method will be applied for a zonal RANS-LES coupling and the SEM method will be used to investigate shock-boundary layer interaction with moving walls.

Acknowledgements We thank the Deutsche Forschungsgemeinschaft (DFG, German Research Foundation) for supporting this work by funding - EXC2075 – 390740016 under Germany's Excellence Strategy and the DFG project Rebound - 420603919. The authors also gratefully acknowledge the Deutsche Forschungsgemeinschaft DFG (German Research Foundation) for funding this work in the framework of the research unit FOR 2895. We acknowledge the support by the Stuttgart Center for Simulation Science (SimTech) and the DFG International Research Training Group GRK 2160. Min Gao recognizes the support of the China Scholarship Council (CSC). We all truly appreciate the ongoing kind support by HLRS in Stuttgart.

References

1. N. Krais, A. Beck, T. Bolemann, H. Frank, D. Flad, G. Gassner, F. Hindenlang, M. Hoffmann, T. Kuhn, M. Sonntag, and C.-D. Munz. Flexi: A high order discontinuous Galerkin framework for hyperbolic–parabolic conservation laws. Computers and Mathematics with Applications, 2020.
2. T. Kuhn, J. Dürrwächter, F. Meyer, A. Beck, C. Rohde, and C.-D. Munz. Uncertainty quantification for direct aeroacoustic simulations of cavity flows. Journal of Theoretical and Computational Acoustics, 27(01):1850044, 2019.
3. H.M. Frank and C.-D. Munz. Direct aeroacoustic simulation of acoustic feedback phenomena on a side-view mirror. Journal of Sound and Vibration, 371:132–149, 2016.
4. A. Beck, J. Dürrwächter, T. Kuhn, F. Meyer, C.-D. Munz, and C. Rohde. hp-multilevel Monte Carlo methods for uncertainty quantification of compressible Navier–Stokes equations. SIAM Journal on Scientific Computing, 42(4):B1067–B1091, 2020.
5. A.D. Beck, T. Bolemann, D. Flad, N. Krais, J. Zeifang, and C.-D. Munz. Application and development of the high order discontinuous Galerkin spectral element method for compressible multiscale flows. In High Performance Computing in Science and Engineering'18, pages 291–307. Springer, 2019.
6. A. Beck, T. Bolemann, D. Flad, H. Frank, G. Gassner, F. Hindenlang, and C.-D. Munz. High-order discontinuous Galerkin spectral element methods for transitional and turbulent flow simulations. International Journal for Numerical Methods in Fluids, 76(8):522–548, 2014.
7. A. Beck, P. Ortwein, P. Kopper, N. Krais, D. Kempf, and C. Koch. Towards high-fidelity erosion prediction: On time-accurate particle tracking in turbomachinery. International Journal of Heat and Fluid Flow, 79:108457, 2019.
8. M.L. Shur, P.R. Spalart, M.K. Strelets, and A.K. Travin. Synthetic turbulence generators for RANS-LES interfaces in zonal simulations of aerodynamic and aeroacoustic problems. Flow, turbulence and combustion, 93(1):63–92, 2014.
9. T. Kuhn, D. Kempf, A. Beck, and C.-D. Munz. A novel turbulent inflow method for zonal large eddy simulations with a discontinuous Galerkin solver. Submitted to Computers and Fluids, 2020.
10. B. Roidl, M. Meinke, and W. Schröder. A reformulated synthetic turbulence generation method for a zonal RANS–LES method and its application to zero-pressure gradient boundary layers. International Journal of Heat and Fluid Flow, 44:28–40, 2013.
11. B. De Laage de Meux, B. Audebert, R. Manceau, and R. Perrin. Anisotropic linear forcing for synthetic turbulence generation in large eddy simulation and hybrid RANS/LES modeling. Physics of Fluids, 27(3):035115, 2015.
12. T.S. Lund, X. Wu, and K.D. Squires. Generation of turbulent inflow data for spatially- developing boundary layer simulations. Journal of Computational Physics, 140(2):233–258, 1998.
13. N. Jarrin, S. Benhamadouche, D. Laurence, and R. Prosser. A synthetic-eddy-method for generating inflow conditions for large-eddy simulations. International Journal of Heat and Fluid Flow, 27(4):585–593, 2006.
14. M. Pamiès, P.-E. Weiss, E. Garnier, S. Deck, and P. Sagaut. Generation of synthetic turbulent inflow data for large eddy simulation of spatially evolving wall-bounded flows. Physics of fluids, 21(4):045103, 2009.
15. F. Bassi and S. Rebay. A high-order accurate discontinuous finite element method for the numerical solution of the compressible Navier–Stokes equations. J. Comput. Phys., 131:267–279, 1997.

16. S. Pirozzoli. Numerical methods for high-speed flows. Annual Review of Fluid Mechanics, 43:163–194, 2011.
17. G. Eitel-Amor, R. Örlü, and P. Schlatter. Simulation and validation of a spatially evolving turbulent boundary layer up to Re_θ= 8300. International Journal of Heat and Fluid Flow, 47:57–69, 2014.
18. C. Wenzel, B. Selent, M. Kloker, and U. Rist. DNS of compressible turbulent boundary layers and assessment of data/scaling-law quality. Journal of Fluid Mechanics, 842:428–468, 2018.

A narrow band-based dynamic load balancing scheme for the level-set ghost-fluid method

Daniel Appel, Steven Jöns, Jens Keim, Christoph Müller, Jonas Zeifang and Claus-Dieter Munz

Abstract We present a dynamic load balancing scheme for compressible two-phase flows simulations using a high-order level-set ghost-fluid method. The load imbalance arises from introducing an element masking that applies the costly interface-tracking algorithm only to the grid cells near the phase interface. The load balancing scheme is based on a static domain decomposition by the Hilbert space-filling curve and employs an efficient heuristic for the dynamic repartitioning. The current workload distribution is determined through element-local wall time measurements, exploiting the masking approach for an efficient code instrumentation. The dynamic repartitioning effectively carries over the single-core performance gain through the element masking to massively parallelized simulations. We investigate the strong scaling behavior for up to 16384 cores, revealing near optimal parallel efficiency and a performance gain of factor five on average compared to previous, unbalanced simulations without element masking. The load balancing scheme is applied to a well-studied two- and three-dimensional shock-drop interaction in the Rayleigh–Taylor piercing regime, providing an overall runtime reduction of approximately 65%.

1 Introduction

Due to its inherent complexity, the modeling of compressible multi-phase flow remains an active research topic. The two main simulation approaches for these types of flows are sharp-interface and diffuse-interface methods. In the diffuse-interface method the phase interface is resolved explicitly. In contrast, sharp-interface methods treat the two phases as immiscible fluids separated by an interface. Thus, a physically sound

---

Daniel Appel, Steven Jöns, Jens Keim, Christoph Müller, Jonas Zeifang, and Claus-Dieter Munz
Institute of Aerodynamics and Gasdynamics, Pfaffenwaldring 21, 70569 Stuttgart, Germany,
e-mail: daniel.appel@iag.uni-stuttgart.de

Jonas Zeifang
Faculty of Sciences, Universiteit Hasselt, Agoralaan Gebouw D, BE-3590 Diepenbeek, Belgium

W. E. Nagel et al. (eds.), *High Performance Computing in Science and Engineering '21*,
https://doi.org/10.1007/978-3-031-17937-2_18

coupling of the fluids has to be ensured and an additional interface-tracking algorithm is required. The sharp-interface method was first introduced by Fedkiw et al. [11] and has since been improved by many authors. Merkle and Rohde [24] presented a ghost-fluid method based on the solution of a multi-phase Riemann problem for the evaluation of the ghost states at the interface. The idea was extended by Fechter et al. [7–9] to approximate Riemann solvers and applied in a level-set ghost-fluid framework. Further improvements of the high-order level-set ghost-fluid method were presented by Jöns et. al [19]. The simulation of multi-phase flows with complex topological changes, present e.g. in shock-droplet and droplet-droplet interactions, is a common high performance computing application due to its requirement for a high spatial resolution. However, the used level-set ghost-fluid framework exhibits a dynamic variation of the computational workload, which is caused by the locally applied interface-tracking algorithm and the evaluation of the multi-phase Riemann problem at the phase interface. To ensure an efficient resource utilization on large-scale computer systems, dynamic load balancing strategies redistribute the workload evenly among the processors at runtime. In this work, we present a load balancing approach for the level-set ghost-fluid method. The approach is based on a static domain decomposition through a Hilbert space-filling curve and employs the efficient H2-heuristic of Miguet and Pierson [25] for the dynamic repartitioning. Similar approaches were successfully applied to particle-based methods [29], direct-coupled multi-physics simulations of acoustics [27], atmospheric modeling [22] and multiscale flow simulations using Chimera grid techniques [3]. To evaluate the present workload distribution, we perform element-local wall time measurements. The code instrumentation exploits the masking concept to minimize the runtime overhead through these measurements to a negligible extent. The remainder of the paper is organized as follows: First, the governing equations of the level-set ghost-fluid framework are presented in section 2. Section 3 briefly introduces the numerical framework to solve these equation systems. This is followed by a description of the dynamic load balancing approach in section 4. Turning to simulation results, we first investigate the strong scaling behavior of the load balancing scheme in section 5.1 and apply our framework to the simulation of a two-dimensional and a three-dimensional shock-droplet interaction in section 5.2. We conclude the report with a succinct summary.

2 Governing equations

In this work, we consider two pure, immiscible phases. Our domain Ω can generally be divided into a liquid region Ω^L and a vapor region Ω^V, which are separated by an interface Γ. The fluid flow in Ω^L and Ω^V are each described by the compressible Euler equations:

$$\begin{pmatrix} \rho \\ \rho \mathbf{v} \\ \rho e \end{pmatrix}_t + \nabla \cdot \begin{pmatrix} \rho \mathbf{v} \\ \rho \mathbf{v} \otimes \mathbf{v} + \mathbf{I} p \\ \mathbf{v}(\rho e + p) \end{pmatrix} = 0, \tag{1}$$

where ρ denotes the density, $\mathbf{v} = (u, v, w)^T$ the velocity vector, p the pressure and e the specific total energy. The total energy per volume of the fluid ρe is composed of the internal energy per volume $\rho \epsilon$ and the kinetic energy:

$$\rho e = \rho \epsilon + \frac{1}{2} \rho \mathbf{v} \cdot \mathbf{v}. \tag{2}$$

The pressure and specific internal energy of each phase are linked via an appropriate equations of state (EOS). Within our framework, algebraic as well as multiparameter EOS can be used. The tabulation technique of Föll et al. [12] ensures an efficient evaluation.

Since we consider two-phase flows, the description of the phase interface is essential. In our work we follow [33], in which the position of the interface is implicitly given by the root of a level-set function $\phi(\mathbf{x})$. From the level-set field, geometrical properties such as the interface normal vector $\mathbf{n}^{\mathbf{LS}}$ and the interface curvature κ can be calculated [19]. The level-set transport is governed by

$$\phi_t + \mathbf{v}^{\mathbf{LS}} \cdot \nabla \phi = 0, \tag{3}$$

where $\mathbf{v}^{\mathbf{LS}}$ is the level-set velocity-field. It is obtained at the interface and then extrapolated into the remaining domain following [1], by solving the Hamilton–Jacobi equations

$$\frac{\partial v_i^{\mathrm{LS}}}{\partial \tau} + \mathrm{sign}(\phi) \mathbf{n}^{\mathbf{LS}} \cdot \nabla v_i^{\mathrm{LS}} = 0, \tag{4}$$

with the direction-wise components v_i^{LS} of the velocity field $\mathbf{v}^{\mathbf{LS}}$ and the pseudo time τ.

In general, the level-set function should fulfill the signed distance property. However, Eq. (3) does not preserve this property. Hence, the level-set function has to be reinitialized regularly. According to [33], we use the Hamilton–Jacobi equation

$$\phi_t + \mathrm{sign}(\phi) \left(|\nabla \phi| - 1 \right) = 0, \tag{5}$$

to retain the signed distance property.

3 Numerical method

Our numerical method can be separated into a fluid solver and an interface capturing scheme. Both use the same computational mesh of non-overlapping hexahedral elements Ω_e. In the following, we will briefly describe both building blocks. A detailed discussion can be found in [8, 9, 19, 26].

3.1 Fluid solver

The compressible Euler equations are discretized by the discontinuous Galerkin spectral element method (DGSEM) [20]. In the high-order DGSEM, the solution is a polynomial of degree N, defined at Gauss–Legendre points in each element. The coupling between the neighboring elements is achieved via a classical numerical flux function, e.g., HLLC or Rusanov. Since the DGSEM is a high-order method, discontinuities, e.g., shocks and phase boundaries, will ultimately lead to the appearance of the unwanted Gibbs phenomena. To stabilize the solution in these areas, we combine the DGSEM with a finite-volume (FV) sub-cell method [2, 16, 21, 30, 32]. Areas in which sub-cells are required are identified by a modal indicator [30] or the position of the level-set root, for shocks or the phase-interface, respectively. In these elements, the solution representation is switched conservatively from a polynomial of degree N to $N+1$ equidistantly spaced FV sub-cells. To reduce the loss of accuracy in the sub-cells, the underlying FV scheme is extended to a second-order TVD scheme in combination with a minmod limiter.

3.2 Interface capturing

The level-set transport equation (3) is discretized with a DGSEM method for hyperbolic equations with non-conservative products [6, 19], by the use of the framework for path-conservative schemes of [5]. Herein, the solution is also described by a polynomial of degree N. The coupling between the elements is handled via a so-called path-conservative jump term, which we approximate with a path-conservative Rusanov Riemann solver following [6].

Theoretically, the level-set function is a smooth signed-distance function. However, in practical applications discontinuities may occur, e.g. in the case of topological changes. To handle these, we use the FV sub-cell approach for the level-set transport equation as well. The two additional operations required for the interface tracking, reinitialization and velocity extrapolation, rely on Hamilton–Jacobi equations, which are each solved with a fifth-order WENO scheme [18] in combination with a third-order low-storage Runge–Kutta method with three stages.

3.3 The level-set ghost-fluid method

The two building blocks described above are the fundamentals of our sharp interface method. The remaining step is a physically sound coupling of the two fluid regions. We follow the methodology presented in [9, 24], with a ghost-fluid method based on the solution of two-phase Riemann problem. Given this solution, the numerical flux at the interface as well as the velocity required for the extrapolation procedure can be calculated for each phase.

Our numerical framework is able to predict complex two-phase flows. Phase transition problems were considered in [10, 17]. Cases without phase transition, e.g., colliding droplets, merging droplets and shock-drop interactions were shown in [19, 26, 36].

4 Load balancing

4.1 Characterization of load imbalance

One of the major challenges for efficient high performance computing (HPC) is the load balance [22, 34], i.e. the even distribution of the application workload across the processor units. The need for *dynamic* load balancing arises if the workload changes during runtime, as encountered in adaptive spatial grids [14] or local time stepping techniques [4]. Aside from the numerical scheme, workload variations may also originate from the considered physics. Examples range from particle-laden flows [29], where the computational costs correlate with the local particle concentration, to atmospheric modeling [22] and computational aeroacoustics [27], both of which couple two physical models to capture the multiscale nature of problem.

Similarly, the outlined framework for multiphase flows introduces an uneven workload distribution by applying the interface-tracking algorithm only in the vicinity of the phase interface and solving the two-phase Riemann problem at the interface itself. More specific, the solution of Eqs. (3)-(5) is only necessary in a narrow band encompassing the interface [13]. Outside this narrow band, the level-set function is set to the band's fixed radius and the velocity field to zero.

The computational domain $\mathcal{D}$ can thus be decomposed into three intersecting sets of grid cells: the bulk elements $\mathcal{D}_{\text{bulk}} \equiv \mathcal{D}$ that discretize only the Euler equations, the narrow band elements $\mathcal{D}_{\text{NB}} \subset \mathcal{D}_{\text{bulk}}$ in which additionally the aforementioned level-set equations are solved, and the elements containing the phase interface itself, $\mathcal{D}_{\Gamma} \subset \mathcal{D}_{\text{NB}}$, subject to the calculation of the interface curvature and to the solution of the two-phase Riemann problem. The computational costs associated with these three subsets consequently rise in the listed order. From an implementational perspective, the distinction of $\mathcal{D}_{\text{bulk}}$, $\mathcal{D}_{\text{NB}}$ and $\mathcal{D}_{\Gamma}$ is accomplished by introducing element masks that filter the subset of elements relevant to the considered operations.

4.2 Domain decomposition and repartitioning

In general, the present framework operates on unstructured, curved grids. Consequently, also the Cartesian grids in the investigated test cases below are internally represented as unstructured meshes. For the given mesh representation, Hindenlang [15] compared domain decomposition strategies for parallel simulations based on space-filling curves (SFC) and graph partitioning. He demonstrated that graph partitioning outperforms space-filling curves in terms of the ratio of inter-domain surfaces to the total number of elements, which indicates a lower communication amount per data operation. However, as the differences vanish with an increasing number of subdomains (cores) and a space-filling curve greatly simplifies data structures and parallel I/O, an SFC-based decomposition strategy was favored. In particular, the Hilbert curve provided more compact subdomains than the Morton curve, and thus led to a better domain composition for massively parallelized simulations.

With respect to dynamic load balancing, SFC-based strategies show further advantages over graph partitioning approaches. Space filling curves generally provide a fast mapping from the n-dimensional to the one-dimensional space while spatial locality is preserved. This locality allows to reduce the distribution of N_{elem} grid cells among P processors to the one-dimensional partitioning problem of decomposing a task-chain of N_{elem} elements with positive weights w_i, $i = 1, \ldots, N_{\text{elem}}$ into K consecutive segments such that the *bottleneck value*, i.e. the maximum load among all processors, is minimized. This problem is also referred to as Chains-on-Chains-Partitioning (CCP) in the literature [31]. Herein, the weight w_i represents the computational cost of the grid cell i and the bottleneck value is defined as

$$B = \max_{1 \leq k \leq K} \{L_k\} \tag{6}$$

$$\text{with the processor load } L_k = \sum_{j=s_{k-1}+1}^{s_k} w_j \, , \tag{7}$$

where the symbol s_k denotes the index of the last task assigned to processor k.

Thus, SFC-based dynamic load balancing means to periodically recompute the sequence of separator indices $s_0 = 0 \leq s_1 \leq \cdots \leq s_K = N_{\text{elem}}$ during runtime, by the use of the updated chain of task weights w_i. The associated partitioning algorithms typically supply medium-quality domain decompositions at low computational costs. Graph partitioning, by contrast, often provides decompositions of higher quality, but also at higher computational costs, which is not suitable for the applications with frequent changes in the workload distribution [34].

Besides, SFC-based partitioning relies only on geometric mesh information. Graph-based methods, on the other hand, abstract grid cells and shared cell faces as vertices V and edges E of an induced graph $\mathcal{G} = (V, E)$. They thus require more detailed information on the costs of the data and communication operations, respectively, to accurately define $\mathcal{G}$ and deduce an optimal partitioning.

Due to complex interface dynamics in compressible two-phase flows, the shape and the position of the narrow-band region change significantly throughout the simulation. This implies both pronounced spatial and temporal workload variations. The necessitated frequent repartitioning at runtime, following the considerations above, thus favors a SFC-based load balancing approach. Besides, this choice eases the implementation over graph-based methods and allows to exploit the established static domain decomposition framework for mesh parallelization of Hindenlang [15]. We employ the well-known, efficient H2-heuristic of Miguet and Pierson [25] with a minor modification to minimize the deviation of the last processor's load L_P from the ideal bottleneck value

$$B^* = \frac{1}{K} \sum_{i}^{N_{\text{elem}}} w_i \,. \tag{8}$$

4.3 Code instrumentation and algorithmic details

To evaluate the processor load (7), the current computational cost of each grid cell has to be determined. This can naturally be achieved through cell-local wall time measurements. However, as they entail an additional runtime overhead and require extensive implementational efforts, this approach is considered to be unsuited for most scientific applications [27].

In contrast, Ortwein et al. [28] showed significant performance improvements for a hybrid particle-mesh method with element-local wall time measurements compared to a constant weighting factor which relates the workload of a grid cell and the particle operations:

$$w_i = 1 + \nu \, n_{\text{particles},i} \,. \tag{9}$$

This ansatz assumes the total cost of one element to be composed of a constant part through the grid cell operation itself, and a variable part, which scales linearly with the number of particles $n_{\text{particles},i}$ located on that element. As the proportionality constant ν is in fact runtime-dependent and specific to the simulation scenario, the authors of the cited work favored the element-local wall time measurements.

In the present work, we pursue a third, less intrusive approach, which exploits the introduced distinction of the three element subsets outlined in the preceding section. We only measure the total wall time of the level-set and ghost-fluid operations listed in section 3.3 and distribute it evenly among the associated elements. This averaging ansatz presumes that each of the masked elements contributes equally to the total wall time of the measured subroutine. Nevertheless, it accounts for the fact that the cost ratio of the three element subsets relative to each other is again setup-specific. This ratio depends e.g. on the chosen numerical flux function and two-phase Riemann solver, the polynomial degree used for the fluid and the level-set equations, respectively [26], and the frequency of the level-set reinitialization and the velocity extrapolation, which are not necessarily executed every iteration. Unlike in

Ortwein et al. [28], the runtime overhead of our approach for the time measurements has proven to be negligible, as only very few calls to the code instrumentation functions are necessary per time step.

Given the workload distribution $\{w_i\}$, the processor loads L_k define the current imbalance

$$I_{\max} = \max_k (I_k)\,, \tag{10}$$

$$\text{with the processor imbalance } I_k = \frac{L_k}{B^*} - 1 \tag{11}$$

and $I_{\max} = 0$ indicates a perfect load distribution. The imbalance is evaluated every nth time step, where $n \approx \mathcal{O}(10^2)$ is adjusted to the temporal workload variations in the considered simulation setup. Upon exceeding the threshold $I_{\text{threshold}}$, the imbalance triggers the actual repartitioning and the subsequent simulation restart. It should be noted that the runtime overhead caused by the imbalance evaluation grows significantly with increasing parallelism due to the involved collective communication operation. The choice of evaluation period n hence needs to account for the wall time of the evaluation operation itself, on the one hand, and, on the other hand, the performance gain through a repartitioning as well as the dynamics of the workload distribution.

The dynamic load balancing algorithm can thus be summarized as follows:

1. measure the element wall times $\{w_i\}_{i=1}^{N_{\text{elem}}}$ through outlined averaging procedure
2. compute the processor workloads $\{L_k\}_{k=1}^{K}$
3. evaluate the current imbalance $I_{\max}$
4. if $I_{\max} > I_{\text{threshold}}$, execute the repartitioning and restart the simulation

Fig. 1 depicts the domain decomposition before and after the repartitioning for a two-dimensional droplet resting in a quiescent gas ($\mathbf{v} = \mathbf{0}$ on Ω), by the use of a representative grid resolution and solver setup.

5 Numerical results

5.1 Strong scaling behavior

In this section, we investigate the strong scaling behavior of the developed load balancing algorithm. For reference, we include performance data of a previous code version without the introduced element masking. All runs were performed on the HPE Apollo System *Hawk* at HLRS, which is equipped with 5632 compute nodes of two AMD EPYC 7742 CPUs (64 cores each) and 256GB memory. The internode communication deploys the InfiniBand HDR200 interconnect with a bandwidth of 200GBit/s.

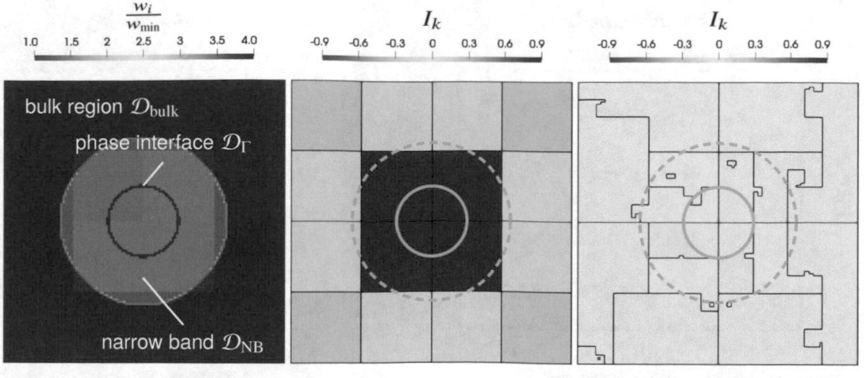

(a) Dimensionless element wall times (b) Processor imbalances before repartitioning: $I_{\max} = 0.921$ (c) Processor imbalances after repartitioning: $I_{\max} = 0.023$

Fig. 1: Domain decomposition for $K = 16$ processors for a two-dimensional resting droplet in a quiescent gas, using a representative solver setup and grid resolution (128 DOFs per diameter D).

The investigated test case considers a three-dimensional resting droplet in a quiescent gas, centered in a cubic domain of edge length $5D$ and a resolution of 50 DOFs/D.

The resolution and the solver setup are chosen to be representative of three-dimensional compressible droplet dynamics simulations. Therefore, we restrict ourselves to the strong scaling behavior, i.e. we keep the problem size constant while we successively increase the number of cores. The stationary physics allow us to isolate the performance gain and the scalability through the repartitioning. The dynamic repartitioning is triggered after five time steps, followed by another 65 time steps to evaluate the performance.

The main analyzed metric is the performance index PID, which expresses the wall time required by a single core to advance one DOF for one stage of the RK time integration scheme:

$$\text{PID} = \frac{\text{total runtime} \times \#\text{cores K}}{\#\text{DOFs} \times \#\text{time steps} \times \#\text{RK-stages}} . \tag{12}$$

We also report the parallel efficiency to quantify the relative performance degradation for an increasing number of cores. The parallel efficiency η relates the obtained performance index to the value obtained for a single compute node, that is in the present case

$$\eta = \frac{\text{PID}_{K=128}}{\text{PID}} . \tag{13}$$

Fig. 2a depicts the obtained PID over a range of 1 to 128 compute nodes ($K = 128, 256, \ldots, 16384$). The introduction of the element masking itself leads to a performance gain in terms of the PID of approximately five for a single compute node.

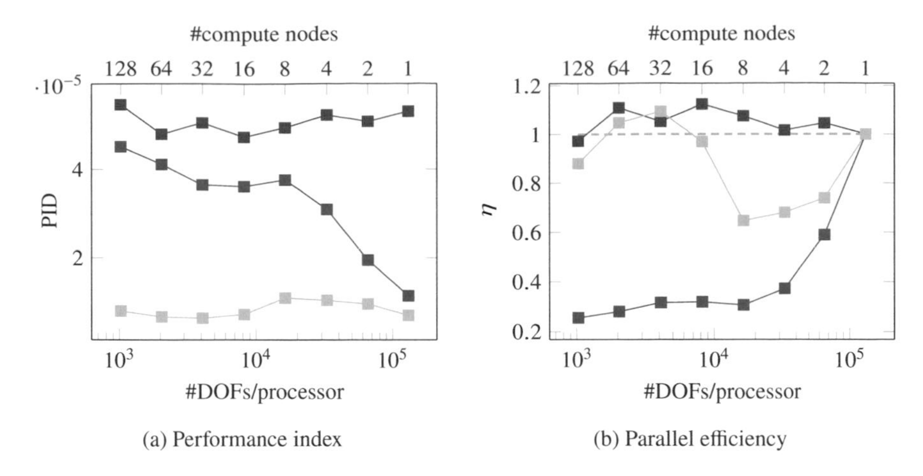

(a) Performance index

(b) Parallel efficiency

Fig. 2: Strong scaling behavior for a three-dimensional resting droplet in quiescent gas, with one compute node comprising 128 physical cores:
without element masking, unbalanced masking, balanced masking

As the domain granularity, i.e., the number of nodes, increases, this gain diminishes and the PID approaches the reference value obtained without the masking. This trend is to be expected, since a larger percentage of subdomains contains exclusively narrow band elements, and thus does not benefit from the reduced workload due to the element masking. In the present setup, for example, the narrow band region covers only 1% of the elements. Through the dynamic repartitioning, the performance gain is preserved over the entire range of cores.

Fig. 2b evaluates the parallel efficiency of the observed PID curves. In general, the implementation benefits from the excellent scaling behavior of the FLEXI framework [21] which it is built upon. The strong scaling behavior is governed by two opposing effects: For decreasing processor workloads (#DOFs/processor), the memory requirement per core is reduced such that a larger proportion of the used data can be stored in the fast cache of the CPU, which increases the performance. Concurrently, the communication effort raises relative to the core-local work and eventually outweighs the caching effect. This gives rise to an optimal processor workload, which is reached at 4048 DOFs/processor for the load balancing implementation. The significant drop in the parallel efficiency to $\eta = 0.25$ for the element masking alone underlines the need for a dynamic load balancing. In total, a compute time of approximately 2900CPUh was spent on the depicted scaling results.

It should be noted that the runtime overhead through the wall time measurements is negligible, in particular for three-dimensional setups. The additional runtime for evaluating the imbalance, recomputing the partitioning and executing the simulation restart is overcompensated by the reduced wall time, that is gained in return, after 40 time steps on average for the present test case.

5.2 Shock-droplet interaction

In contrast to the rather generic setup with a static workload distribution in the preceding section, we now apply the developed load balancing method to more complex and realistic scenarios featuring temporal workload variations.

We first revisit the two-dimensional shock-droplet interaction in Jöns et al. [19], that was previously studied by Winter et al. [35]. The authors of those works investigated the deformation of an initially cylindrical water column after the interaction with a shock in air. In the first stage of the deformation, the water column is flattened. This flattening is independent of the Weber number

$$\mathrm{We} = \frac{\rho^V (v^V)^2 D}{\sigma}, \tag{14}$$

which expresses the relative importance of the inertial to the surface tension forces. The second stage starts with the onset of interfacial instabilities at the air-water interface, where different droplet breakup modes can be distinguished depending on the Weber number. At relatively small surface tension forces (We small), breakup occurs in the shear-induced entrainment (SIE) regime, which is characterized by the formation of a thin sheet at the top and the bottom droplet equator. Large surface tension forces (We large), in contrast, suppress the growth of these sheets and have a smoothing effect on the interface geometry. This so-called Rayleigh–Taylor piercing (RTP) regime results in a bag-shaped droplet and is considered below due to the negligible effect of the viscous forces.

We adopt the setup of Jöns et al. [19] and initialize a shock wave of Ma = 1.47 upstream of the water droplet with a Weber number of We = 12. The droplet is discretized with 170 DOFs/D and we impose a symmetry boundary condition along the symmetry axis. The numerical domain contains 10^6 DOFs and we deploy 4 compute nodes, which yields a nominal processor workload of 2048 DOFs and results in a total compute time of 1200CPUh. For further details on the numerical setup, the material parameters and the initial conditions as well as a more comprehensive discussion of the involved physics, the reader is referred to the cited publications. In Fig. 3, we provide the results at the dimensionless time $t^* = 1.58$ to illustrate the impact of the dynamic repartitioning. Initially, the domain is partitioned homogeneously as no a priori knowledge of the workload distribution is available. The depicted partitioning then reflects the current workload distribution, with the narrow band elements being computationally two orders of magnitude more expensive than the bulk elements. The performance gain through the dynamic load balancing for this realistic simulation is expressed more adequately in terms of the overall runtime saving; it has been evaluated for a coarser mesh with 42 DOFs/D and amounts to 64%.

Ultimately, we extend the given two-dimensional setup to three dimensions. The resolution is reduced to 42 DOFs/D, which results in a numerical domain of 8×10^6 DOFs. We deploy 32 compute nodes to attain the same nominal processor workload as before. The simulation finished after a total compute time of 8200CPUh. Fig. 4

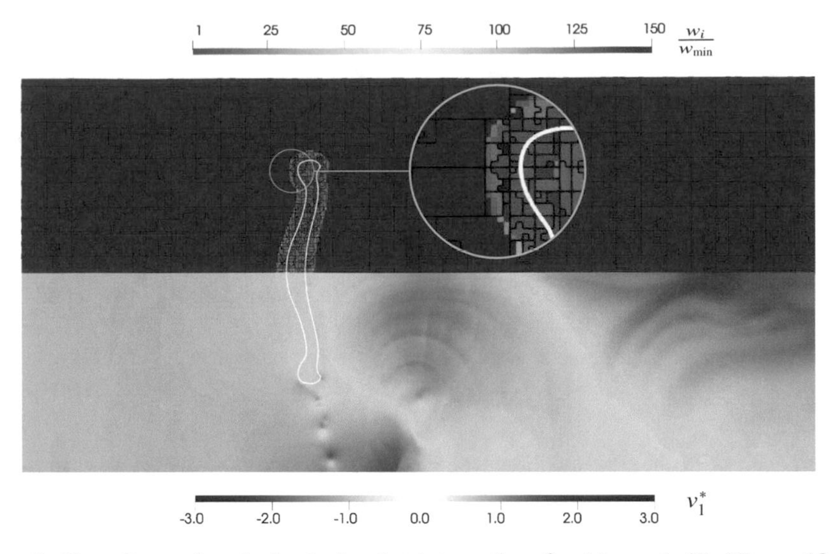

Fig. 3: Two-dimensional shock-droplet interaction for Ma = 1.47, We = 12 at $t^* = 1.58$: dimensionless element wall times (top) and dimensionless axial velocity (bottom).

depicts the droplet contour together with the non-dimensional velocities in axial and radial direction, v_1^* and v_2^*, respectively. The droplet exhibits a small concave deformation on the upstream side and has a convex shape on the downstream side, which indicates the onset of the characteristic bag growth [35]. An axially stretched recirculation zone develops in the wake of the drop, however, with a less pronounced detachment than in the cited work. The deviation may be attributed to the unmatched and insufficient resolution as well as the neglected interface viscosity. The overall runtime reduction by 69% emphasizes the effectiveness of the proposed dynamic load balancing strategy.

6 Conclusion

We have presented a dynamic load balancing scheme for a high-order level-set ghost-fluid method used for compressible two-phase flow simulations. The load imbalance arises from introducing an element masking that applies the costly interface-tracking routines only to the necessary grid cells. The load balancing scheme is based on a static domain decomposition through the Hilbert space-filling curve and employs an efficient heuristic for the dynamic repartitioning. The current workload distribution is determined through element-local wall time measurements, which exploit the masking approach for a minimally intrusive code instrumentation with negligible runtime overhead. The developed load balancing scheme transfers the single-core

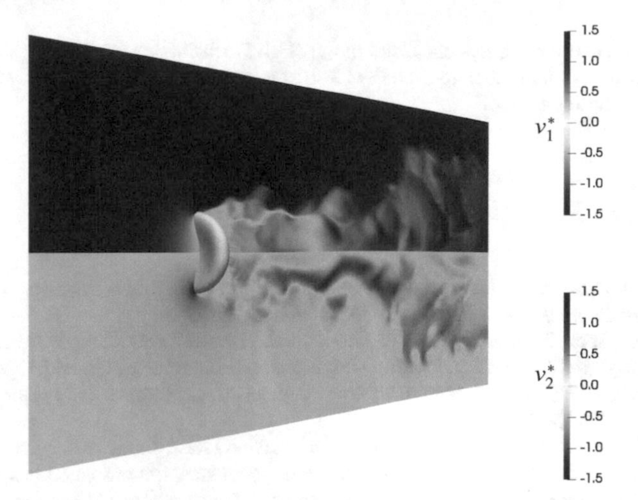

Fig. 4: Three-dimensional shock-droplet interaction for Ma = 1.47, We = 7339 at $t^* = 1.58$: dimensionless axial velocity (top) and dimensionless radial velocity (bottom).

performance gain from the masking approach to massively parallelized simulations. We provided strong scaling results, which reveal a parallel efficiency beyond 65 % and superlinear scaling at the optimal processor workload. Compared to unbalanced simulations without element masking, the performance gain amounts to approximately five throughout the investigated range up to 16384 cores. We have demonstrated the capability of the dynamic repartitioning scheme for two- and three-dimensional shock-droplet interactions in the Rayleigh–Taylor piercing regime with an overall runtime reduction by 64% and 69%, respectively.

The runtime overhead of the load balancing scheme is mainly attributed to the parallel I/O associated with the simulation restart, which requires $\mathcal{O}(10^1 \ldots 10^2)$ time steps to be compensated through the induced performance gain. This amortization rate could be reduced – i.e. the repartitioning could be called more frequently – by recalculating only the static data like mesh metrics upon restart, while the dynamic data (state variables) are migrated via point-to-point communication. Furthermore, the current imbalance is evaluated periodically, which causes a growing runtime overhead with increasing parallelism due to the involved collective communication. Since the value of the evaluation period relies on an educated guess, future work may adapt the idea of Menon et al. [23] to determine it automatically at runtime based on application characteristics.

Acknowledgements We gratefully acknowledge the Deutsche Forschungsgemeinschaft (DFG) with their support of D. Appel, C. Müller and J.Zeifang through the project GRK 2160/1 "Droplet Interaction Technologies", S. Jöns through the project SFB-TRR 75, Project number 84292822 -"Droplet Dynamics Under Extreme Ambient Conditions", J. Keim and C.-D. Munz under Germany's Excellence Strategy - EXC 2075 - 390740016. The simulations were performed on the national supercomputer HPE Apollo System *Hawk* at the High Performance Computing Center Stuttgart (HLRS) under the grant number *hpcmphas/44084*, of which we used approximately 2.7M CPUh in the current accounting period.

References

1. Tariq D. Aslam. A partial differential equation approach to multidimensional extrapolation. *Journal of Computational Physics*, 193(1):349–355, 2004.
2. Andrea D. Beck, Thomas Bolemann, David Flad, Hannes Frank, Gregor J. Gassner, Florian Hindenlang, and Claus-Dieter Munz. High-order discontinuous Galerkin spectral element methods for transitional and turbulent flow simulations. *International Journal for Numerical Methods in Fluids*, 76(8):522–548, 2014.
3. Andrea D. Beck, Thomas Bolemann, David Flad, Nico Krais, Jonas Zeifang, and Claus-Dieter Munz. Application and development of the high order discontinuous galerkin spectral element method for compressible multiscale flows. In *High Performance Computing in Science and Engineering '18*, pages 291–307. Springer International Publishing, 2019.
4. Carsten Burstedde, Omar Ghattas, Michael Gurnis, Tobin Isaac, Georg Stadler, Tim Warburton, and Lucas Wilcox. Extreme-scale AMR. In *SC'10: Proceedings of the 2010 ACM/IEEE International Conference for High Performance Computing, Networking, Storage and Analysis*, pages 1–12. IEEE, 2010.
5. Manuel Castro, José Gallardo, and Carlos Parés. High order finite volume schemes based on reconstruction of states for solving hyperbolic systems with nonconservative products. applications to shallow-water systems. *Mathematics of Computation*, 75(255):1103–1134, 2006.
6. Michael Dumbser and Raphaël Loubère. A simple robust and accurate a posteriori sub-cell finite volume limiter for the discontinuous Galerkin method on unstructured meshes. *Journal of Computational Physics*, 319:163–199, 2016.
7. Stefan Fechter, Felix Jaegle, and Veronika Schleper. Exact and approximate Riemann solvers at phase boundaries. *Computers & Fluids*, 75:112–126, 2013.
8. Stefan Fechter and Claus-Dieter Munz. A discontinuous Galerkin-based sharp-interface method to simulate three-dimensional compressible two-phase flow. *International Journal for Numerical Methods in Fluids*, 78(7):413–435, 2015.
9. Stefan Fechter, Claus-Dieter Munz, Christian Rohde, and Christoph Zeiler. A sharp interface method for compressible liquid-vapor flow with phase transition and surface tension. *Journal of Computational Physics*, 336:347–374, 2017.
10. Stefan Fechter, Claus-Dieter Munz, Christian Rohde, and Christoph Zeiler. Approximate Riemann solver for compressible liquid vapor flow with phase transition and surface tension. *Computers & Fluids*, 169:169–185, 2018.
11. Ronald P. Fedkiw, Tariq Aslam, Barry Merriman, and Stanley Osher. A non-oscillatory Eulerian approach to interfaces in multimaterial flows (the ghost fluid method). *Journal of Computational Physics*, 152(2):457–492, 1999.
12. Fabian Föll, Timon Hitz, Christoph Müller, Claus-Dieter Munz, and Michael Dumbser. On the use of tabulated equations of state for multi-phase simulations in the homogeneous equilibrium limit. *Shock Waves*, 2019.
13. Pablo Gómez, Julio Hernández, and Joaquín López. On the reinitialization procedure in a narrow-band locally refined level set method for interfacial flows. *International Journal for Numerical Methods in Engineering*, 63(10):1478–1512, 2005.
14. Daniel F. Harlacher, Harald Klimach, Sabine Roller, Christian Siebert, and Felix Wolf. Dynamic load balancing for unstructured meshes on space-filling curves. In *2012 IEEE 26th International Parallel and Distributed Processing Symposium Workshops & PhD Forum*. IEEE, 2012.
15. Florian Hindenlang. *Mesh Curving Techniques for High Order Parallel Simulations on Unstructured Meshes*. PhD thesis, Universität Stuttgart, 2014.

16. Florian Hindenlang, Gregor J. Gassner, Christoph Altmann, Andrea D. Beck, Marc Staudenmaier, and Claus-Dieter Munz. Explicit discontinuous Galerkin methods for unsteady problems. *Computers & Fluids*, 61(0):86–93, 2012.
17. Timon Hitz, Steven Jöns, Matthias Heinen, Jadran Vrabec, and Claus-Dieter Munz. Comparison of macro- and microscopic solutions of the Riemann problem II. Two-phase shock tube. *Journal of Computational Physics*, 429:110027, 2021.
18. Guang-Shan Jiang and Danping Peng. Weighted ENO schemes for Hamilton–Jacobi equations. *SIAM Journal on Scientific Computing*, 21(6):2126–2143, 2000.
19. Steven Jöns, Christoph Müller, Jonas Zeifang, and Claus-Dieter Munz. Recent advances and complex applications of the compressible ghost-fluid method. In Muñoz-Ruiz, María Luz, Carlos Parés, and Giovanni Russo, editors, *Recent Advances in Numerical Methods for Hyperbolic PDE Systems. SEMA SIMAI Springer Series*, volume 28, pages 155–176. Springer, Cham, 2021.
20. David A. Kopriva. *Implementing spectral methods for partial differential equations: Algorithms for scientists and engineers*. Springer Publishing Company Incorporated, 1st edition, 2009.
21. Nico Krais, Andrea D. Beck, Thomas Bolemann, Hannes Frank, David Flad, Gregor Gassner, Florian Hindenlang, Malte Hoffmann, Thomas Kuhn, Matthias Sonntag, and Claus-Dieter Munz. FLEXI: A high order discontinuous Galerkin framework for hyperbolic-parabolic conservation laws. *Computers & Mathematics with Applications*, 81:186–219, 2021.
22. Matthias Lieber and Wolfgang E. Nagel. Highly scalable SFC-based dynamic load balancing and its application to atmospheric modeling. *Future Generation Computer Systems*, 82:575–590, 2018.
23. Harshitha Menon, Nikhil Jain, Gengbin Zheng, and Laxmikant Kalé. Automated load balancing invocation based on application characteristics. In *2012 IEEE International Conference on Cluster Computing*, pages 373–381. IEEE, 2012.
24. Christian Merkle and Christian Rohde. The sharp-interface approach for fluids with phase change: Riemann problems and ghost fluid techniques. *ESAIM: Mathematical Modelling and Numerical Analysis*, 41(06):1089–1123, 2007.
25. Serge Miguet and Jean-Marc Pierson. Heuristics for 1d rectilinear partitioning as a low cost and high quality answer to dynamic load balancing. In *International Conference on High-Performance Computing and Networking*, pages 550–564. Springer, 1997.
26. Christoph Müller, Timon Hitz, Steven Jöns, Jonas Zeifang, Simone Chiocchetti, and Claus-Dieter Munz. Improvement of the level-set ghost-fluid method for the compressible Euler equations. In Grazia Lamanna, Simona Tonini, Gianpietro Elvio Cossali, and Bernhard Weigand, editors, *Droplet Interaction and Spray Processes*. Springer, Heidelberg, Berlin, 2020.
27. Ansgar Niemöller, Michael Schlottke-Lakemper, Matthias Meinke, and Wolfgang Schröder. Dynamic load balancing for direct-coupled multiphysics simulations. *Computers & Fluids*, 199:104437, 2020.
28. P Ortwein, T Binder, S Copplestone, A Mirza, P Nizenkov, M Pfeiffer, C.-D. Munz, and S Fasoulas. A load balance strategy for hybrid particle-mesh methods. *arXiv preprint arXiv:1811.05152*, 2018.
29. Philip Ortwein, Tilman Binder, Stephen Copplestone, Asim Mirza, Paul Nizenkov, Marcel Pfeiffer, Torsten Stindl, Stefanos Fasoulas, and Claus-Dieter Munz. Parallel performance of a discontinuous Galerkin spectral element method based PIC-DSMC solver. In Wolfgang E. Nagel, Dietmar H. Kröner, and Michael M. Resch, editors, *High Performance Computing in Science and Engineering '14*, pages 671–681, Cham, 2015. Springer International Publishing.
30. Per-Olof Persson and Jaime Peraire. Sub-cell shock capturing for discontinuous Galerkin methods. In *44th AIAA Aerospace Sciences Meeting and Exhibit*. American Institute of Aeronautics and Astronautics, 2006.
31. Ali Pınar and Cevdet Aykanat. Fast optimal load balancing algorithms for 1d partitioning. *Journal of Parallel and Distributed Computing*, 64(8):974–996, 2004.
32. Matthias Sonntag and Claus-Dieter Munz. Efficient parallelization of a shock capturing for discontinuous Galerkin methods using finite volume sub-cells. *Journal of Scientific Computing*, 70(3):1262–1289, 2016.

33. Mark Sussman, Peter Smereka, and Stanley Osher. A level set approach for computing solutions to incompressible two-phase flow. *Journal of Computational Physics*, 114(1):146–159, 1994.
34. James D Teresco, Karen D Devine, and Joseph E Flaherty. Partitioning and dynamic load balancing for the numerical solution of partial differential equations. In *Numerical Solution of Partial Differential Equations on Parallel Computers*, pages 55–88. Springer, 2006.
35. Josef M. Winter, Jakob W.J. Kaiser, Stefan Adami, and Nikolaus A. Adams. Numerical investigation of 3d drop-breakup mechanisms using a sharp interface level-set method. In *11th International Symposium on Turbulence and Shear Flow Phenomena, TSFP 2019*, 2019.
36. Jonas Zeifang. *A Discontinuous Galerkin Method for Droplet Dynamics in Weakly Compressible Flows*. PhD thesis, Universität Stuttgart, 2020.

Numerical simulation of flake orientation during droplet impact on substrates in spray painting processes

Qiaoyan Ye and Martin Dreher and Bo Shen

Abstract A numerical study of flake orientation by droplet impact on dry and wetted solid surfaces for spray painting processes has been carried out. A dynamic contact angle model was applied for the calculation of viscous droplet impact on the dry surface after experimental validation. A user-defined 6-DOF model that concerns to the rigid body motion was implemented in a commercial CFD program to calculate the flake movement inside the droplet. The simulated flake orientations show interesting results that are helpful to understand and to improve painting processes.

1 Introduction

Effect pigments (flat, 5 µm to 40 µm broad flakes) are widely used in industrial spray painting processes for imparting colour and coating appearance, enabling e.g. metallic effect coatings that are very popular in the automotive industry. From practical observations, it is already well known that the initial pigment/flake orientation in the paint layer, primarily at the early stage of the film formation and before the subsequent solvent evaporation and baking, influences the final metallic effect significantly. Therefore, it is strongly suspected that processes during viscous droplet impact have a decisive influence on flake orientation. For a strong metallic effect, usually a flake orientation parallel to the substrate is desired.

Regarding the droplet impact dynamics, there have been already many research works [1–8]. In experimental studies, mainly large droplets (e.g. $D > 500\,\mu m$) with low viscosity under hydrophilic/hydrophobic surface conditions were investigated.

Qiaoyan Ye and Martin Dreher
Fraunhofer Institute for Manufacturing Engineering and Automation, Stuttgart, Germany,
e-mail: Qiaoyan.Ye@ipa.fraunhofer.de

Bo Shen
University of Applied Sciences Esslingen, Esslingen, Germany,
e-mail: Bo.Shen@hs-esslingen.de

W. E. Nagel et al. (eds.), *High Performance Computing in Science and Engineering '21*,
https://doi.org/10.1007/978-3-031-17937-2_19

Different outcomes of droplet impact on dry/wet substrates were analysed [4]. The air entrapment by droplet impact was studied [3]. The maximal droplet-spreading diameters were correlated with non-dimensional numbers, i.e., Weber and Reynolds numbers, using published experimental results [5]. The contact angle models that have to be applied in the numerical simulation were studied and discussed in detail [7, 8].

There are not so many studies that focus on the flake movement inside the droplet and in the film. For small droplets (50 μm to 300 μm), especially for opaque liquids, like in spray painting processes, it is very difficult to obtain high quality time-resolved imaging of the flake orientation during droplet impingement experimentally. Although a few predictions of flake orientation were carried out based on some mathematical analysis [9] and the flow field of two-dimensional droplet impact calculation [10], the corresponding results should be further verified.

The objective of the present paper is to carry out a detailed numerical study of flake orientation, focusing on droplet impact processes on dry and wetted solid walls. Thereby, a contact angle model that is necessary in the numerical simulation was developed for the paint droplet impact process. In addition, a user-defined 6DOF (6-degrees-of-freedom) solver was implemented in a CFD-program to perform the rigid body (flake) motion calculation within the impacting droplet. The developed models were applied in a parameter study, to further clarify the existing dependencies on application and fluid parameters more quantitatively.

2 Basic numerical methods

The droplet impact and spreading on a surface is an example of an interfacial flow problem that can be calculated using the Volume of Fluid (VOF) method. Hereby, two or more immiscible fluids can be modelled by solving a single set of momentum and mass equations and tracking the volume fraction of each of the fluids throughout the domain. The numerical simulations in this work were carried out with the commercial CFD code ANSYSFluent based on the finite-volume approach. Time dependent VOF calculations were performed using an explicit scheme. A geometric reconstruction scheme for the volume fraction discretization was used, ensuring a sharp and low-diffusion interface discretization. To accommodate surface tension effects, the CSF-model with wall adhesion modelling was used. The PRESTO scheme was applied for the pressure discretization.

It is well known that the VOF-method is quite sensitive to grid resolution. A study of grid independence was performed by comparing measured and calculated spread and height factors of the droplet impact. A reasonable cell size was obtained at about $D/\Delta x = 150$ and applied around the droplet and liquid film on the substrate. Here D denotes the droplet diameter and Δx the grid size. In the far-field, coarser hexahedral meshes were used to reduce the total number of cells and therefore computational cost. For the simulation of droplet impact on the wet solid surface, a relative large computational domain had to be used. In order to reduce further computational cost, a dynamic mesh adaption was applied, which was necessary especially for large droplets. Consequently, the domains in the present study contain 20 to 150 million cells. Further detailed important boundary conditions and models are described as follows.

2.1 Simulation of viscous droplet impact on dry solid surfaces

A computational domain of $2D \times 2D \times 1.5D$ was created for the simulation of droplet impact on the dry solid surface. Thereby, only a quarter of the spherical droplet was calculated and symmetry boundary conditions were used to reduce the total amount of cells. A flow field aware variable time step with the CFL (Courant–Friedrichs–Lewy) number < 1 was set. The resulting time step sizes ranged from 5×10^{-9} s to 10^{-7} s. Based on common spray-painting conditions, droplet impact velocities between $0.5\,\mathrm{m{\cdot}s^{-1}}$ to $10\,\mathrm{m{\cdot}s^{-1}}$ were applied. The initial droplet position was chosen to be a few micrometers above the wall surface, so that the surrounding gas field can be calculated, which is necessary in the study of droplet impact dynamics. The pressure inside the droplet, induced by the surface tension of the liquid, was calculated and properly initialized. The remaining domain is set to ambient pressure. A dry smooth wall with no-slip boundary condition was used.

It is well known that the description of moving contact line or contact angle in the numerical simulation of droplet impact on dry surfaces is not trivial and quite challenging. Nevertheless, dynamic contact angles are widely used in VOF-simulations to accommodate adhesion behaviour. However, the applicability of the corresponding models is quite problem dependent, such as the variety of liquid and substrate properties, as well as the numerous operating parameters. The hysteresis of contact line and contact angle makes the model development more difficult. Further information can be referred to [7, 8].

We have carried out the experimental observation of viscous droplet impact on dry solid surfaces, focusing on the contour evolution during the spreading and receding processes. The experimental results provided useful information to develop the model for dynamic contact angle that was applied as boundary condition in the simulation. Equation (1) describes the model in a mathematical manner,

$$\theta_D = \begin{cases} \theta_{\mathrm{A}} & \text{for } v_{\mathrm{cl}} > 0 \\ \theta_{\mathrm{int}} = 90^\circ & \text{for } \overline{v}_{\mathrm{cl}} < |\delta| \cap v_{\mathrm{cl}} < \delta \\ \theta_{\nabla\phi} & \text{for } \theta_{\nabla\phi} < \theta_{\mathrm{A}} \\ \theta_{\mathrm{E}} & \text{for } \theta_{\nabla\phi} \leq \theta_{\mathrm{E}} \end{cases} \tag{1}$$

where $\delta = -0.1$. In the spreading phase, where the contact line velocity (v_{cl}) is significantly positive, a constant advancing contact angle (θ_{A}) is deployed. At maximum wetting spread, when the contact line movement stops and the droplet usually starts to recede, an intermediate contact angle (θ_{int}) is set. This helps to suppress contact line movement in accordance to the experimental observation. If the angle between the fluid interface and the wall, calculated by the gradient of the VOF-field ($\nabla\phi$), gets smaller than the advancing contact angle, the interphase-angle is set as contact angle ($\theta_{\nabla\phi}$). It is secured, that this angle decreases monotonically to the static or equilibrium contact angle (θ_E). On the wall cells and far away from the contact line, the equilibrium contact angle (θ_E) is used, which ensures the wetting behaviour on the wall is more reasonable.

2.2 Simulation of viscous droplet impact on wet solid surfaces

A thin film with a height $H = 30\,\mu m$ to $60\,\mu m$ and an infinitive width was applied in the present simulation. At first, droplet impact on the wet solid surface was calculated in a two-dimensional axisymmetric coordinates, in order to track the gas-liquid interface with a large amount of parameters and to define the impact events with some dimensionless numbers, such as Re-, Oh- and We-number. Then, a computational domain of $8D \times 8D \times 3D$ was created for the simulation of droplet impact with flake on the wet solid surface. Clearly, the number of computational cells will be tremendously increased by homogeneous grid because of the large domain required. Therefore, local grid adaption and dynamic mesh adaption were carefully performed, ensuring the necessary grid resolution. For large droplets, e.g., $300\,\mu m$, the region with inhomogeneous grid could be enlarged because of the wave, which results in also an increasing the computational cells. As shown in Fig. 1, for instance, the mesh size around the interface between two phases was set to be $1\,\mu m$ obtained by using $D/\Delta x = 150$ for a $150\,\mu m$ droplet. Far away from the interface, a coarse mesh with size of $16\,\mu m$ was used, finally resulting a four-lever mesh adaption around the interface.

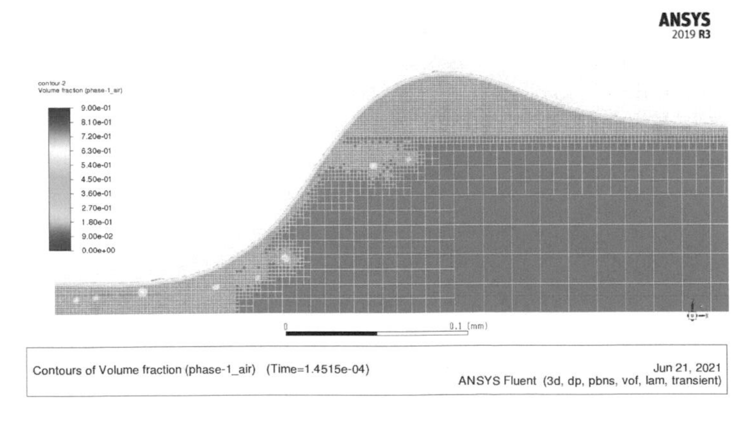

Fig. 1: Mesh size distribution for the droplet impact with D=150 μm by using dynamic mesh adaption model.

2.3 Simulation of flake orientation

For the simulation of flake movement, the effect pigment (flake) was incorporated into the domain mesh utilizing the overset meshing option as a dynamic mesh method. Hereby, a separate component mesh that may be adapted to its special geometry, is necessary. As shown in Fig. 2, Cartesian grids with local mesh-refinement in the relevant regions were created for the overall simulation domain. The component mesh for the flake (Fig. 2b) and the background grid are overlaid as show in Fig. 2a and coupled by an interpolated transition area, in which the flow solutions that are separately calculated in both grids are exchanged. The resolution of the component grid corresponds to that of the background grid. Comparing to other methods, such as remeshing method, the overset meshing yields VOF-friendly meshing setup and a good performance in the flake movement calculation because of the elimination of distorted cells and the necessity of remeshing. In the present study, only one

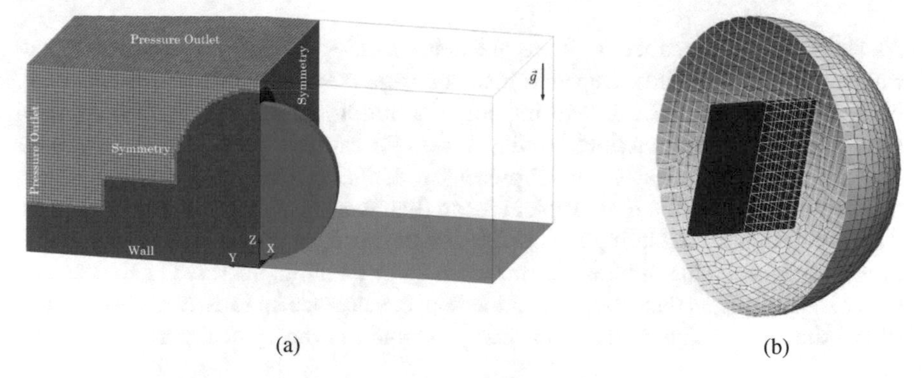

Fig. 2: a) 3D mesh setup with boundary conditions and initial droplet position, b) flake/component mesh.

rectangular shaped flake in the computational domain is applied. The size of the considered flake was chosen to be $1 \times 16 \times 16\,\mu m^3$ that is a reasonable estimate concerning typical effect pigment size distributions. Yet, this results in corresponding moments of inertia of $1.6 \times 10^{-23}\,kg{\cdot}m^2$, which are too marginal for the currently available 6DOF motion solver by ANSYS. In consequence, a custom motion equation solver has been implemented via a user-defined-function (UDF). This 6DOF solver has been developed under the assumption of rigid body motion. Therefore, the flake motion may be simplified to the movement of its centre of mass and the rotation around it. Consequently, the momentum conservation (2) is calculated in the global inertial coordinate system, and the angular momentum (3) is determined in body coordinates:

$$\dot{\mathbf{v}} = \frac{1}{m}\Sigma\mathbf{F} \tag{2}$$

$$\dot{\omega} = \mathbf{I}^{-1}(\ \Sigma\mathbf{M} - \omega \times \mathbf{I}\omega\ ) \tag{3}$$

Here, m denotes the mass, $\mathbf{I}$ the inertia tensor and $\mathbf{F}$, $\mathbf{M}$ the flow induced forces and moments, respectively. The points over the velocity ($\mathbf{v}$) and the angular velocity (ω) represent time derivatives. In Equation (2), the sum includes pressure, viscosity and gravitational forces. An Adams–Moulton algorithm of 4th order with a rather complex variable time step formulation has been derived and implemented to time integrate the above given ODEs. This enables simulations of flake movement or orientation stable with a quite reasonable time step size, such as $dt = 10^{-8}$ s to 10^{-7} s that is solely adjusted by the flow solver.

3 Results and discussion

3.1 Evolution of interface of gas-liquid by droplet impact on wet solid surface without flake

As stated in the introduction, there are many studies about outcomes of droplet impact on wet substrates. Depending on the impact velocity and liquid viscosity, basically, the outcome of droplet impact on a thin liquid film can be divided into spreading/deposition, cratering/receding, crown formation and crown with splashing. The latter will not be studied in the present investigation because it is very seldom in spray coating applications. Typical mean diameter of spray droplet ranges from 30 µm to 50 µm. Considering the correlation between droplet impact velocity and droplet diameter using different atomizers in spray painting processes [11], droplets with diameter of less than 100 µm and the impact velocities up to 10 m·s^{-1} were used in the simulation. The model paint was used and the rheological parameters that

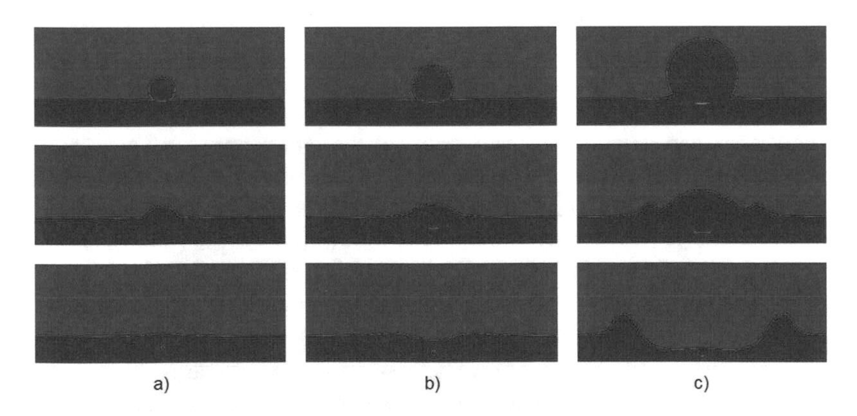

Fig. 3: Phenomena of drop impact on a wet solid surface (red: liquid, blue: air). Film $H = 30$ µm. a) $D = 30$ µm, $U = 4$ m·s^{-1}, b) $D = 45$ µm, $U = 4$ m·s^{-1}, c) $D = 80$ µm, $U = 8$ m·s^{-1}.

describe the shear thinning viscosity of the model paint were summarized in our previous work [12]. Figure 3 shows typical outcomes of the paint droplet impact on a thin liquid film. By small droplet and with a low impact speed the liquid of the droplet deposits subsequently in the liquid layer (Fig. 3a). With increasing impact energy of the droplet, namely either with a larger diameter or with a higher velocity, a small crater can be formed in the target liquid layer (see Fig. 3b). If the impact energy is high enough, at the circumference of the crater the so-called crater rim or crown rises above the original surface of the target liquid layer (Fig. 3c).

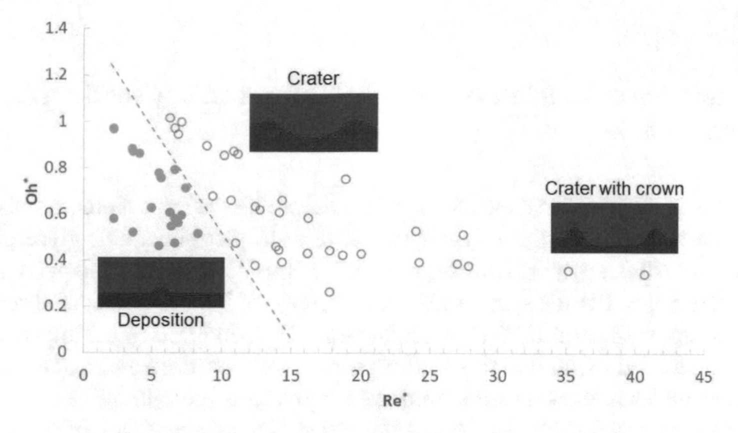

Fig. 4: Droplet impact formations at different Reynolds and Ohnesorge numbers, the ratio between film thickness and drop diameter, $H/D = 0.375$ to 2.16.

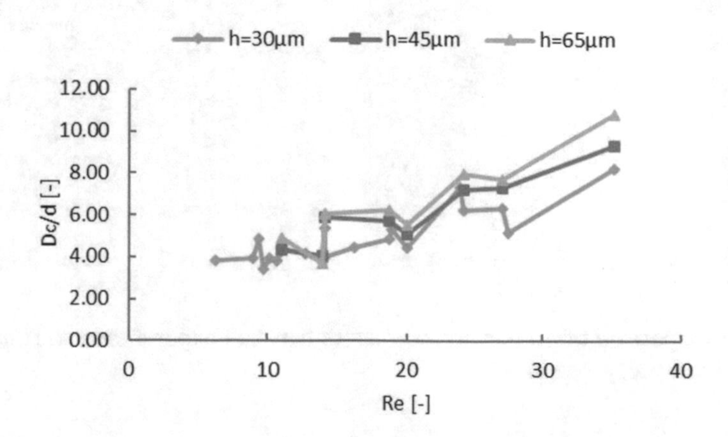

Fig. 5: Crater sizes as a function of the Reynolds number.

The dependence of the outcomes on the Reynolds and Ohnesorge numbers ($\mathrm{Re} = \rho u D/\eta, \mathrm{Oh} = \eta/\sqrt{(\rho\sigma D)}$) are depicted in Fig. 4. Based on the simulation results, the dimensionless crater sizes, namely the ratio between the crater diameter D_c and the droplet diameter D are plotted as a function of the Reynolds number in Fig. 5, indicating a strong correlation with the Reynold number. The corresponding quantitative correlations for curves in both Fig. 4 and 5 were obtained in [12]. The simulation results in this section deliver useful information for further investigations, such as flake orientation during droplet impact on wet solid surfaces, entrainment of air bubbles in the impact process as well as film leveling.

3.2 Validation of simulation of droplet impact on dry solid surface without flake

Validations of the numerical models were carried out for the impact process of viscous droplet on dry solid surfaces. Thereby, using a droplet generator, glycerol/water droplets ($D = 400\,\mu\mathrm{m}$, $\eta = 20\,\mathrm{mPa{\cdot}s}$, $\sigma = 0.063\,\mathrm{N{\cdot}m^{-1}}$, $\theta_\mathrm{E} = 55°$) and paint droplets ($D = 300\,\mu\mathrm{m}$, $\eta = 17\,\mathrm{mPa{\cdot}s}$, $\sigma = 0.025\,\mathrm{N{\cdot}m^{-1}}$, $\theta_\mathrm{E} = 53°$) were created. Subsequently, the droplet spread-factors d/D_0 and height ratios h/D_0 induced by the impact process that was recorded using a high-speed camera were measured and calculated. The advancing angles in the simulation were set for glycerol/water droplet $\theta_A = 120°$ and paint droplet $\theta_A = 95°$, respectively. Figure 6 shows a comparison of these values between the experimental data and the simulation results. A quite good agreement can be observed, which indicates that the proposed dynamic contact angle model delivers good performance and is suitable for the following further application.

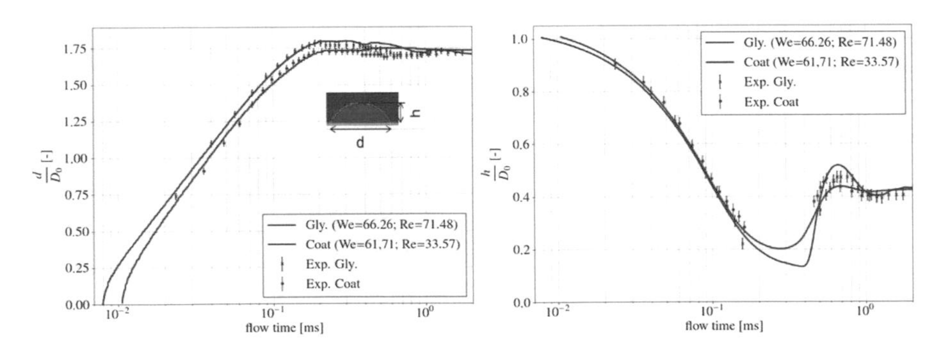

Fig. 6: Comparison of droplet spread-factors between experiment and simulation during the impact process.

Table 1: Parameters used in the simulation for Newtonian drops.

Parameter	Range	Parameter	Range
Droplet diameter D (μm)	50, 100 and 300	Flake size (μm^3)	$1 \times 16 \times 16$
Droplet velocity U ($m \cdot s^{-1}$)	0.5 to 10	Flake density ($kg \cdot m^{-3}$)	3200
Liquid viscosity η (Pa·s)	0.01, 0.02 and 0.04	Ratio of flake to droplet l/D	0.053 and 0.32
Surface tension σ ($N \cdot m^{-1}$)	0.025 and 0.063	Reynolds-number, $Re = \sigma UD/\eta$	5 to 100
Liquid density ρ ($kg \cdot m^{-3}$)	1020 and 1200	Weber-number, $We = \sigma U^2 D/\sigma$	3 to 1200
Static contact angle (°)	50 to 60	Ohnesorge-number, $Oh = \eta/\sqrt{\sigma D \rho}$	0.1 to 2.2
Height of film (μm)	60 and 30		

3.3 Flake orientation by droplet impact on dry solid surfaces

Considering the shear thinning and thixotropic non-Newtonian behaviour of most paint liquids, fluids with a constant viscosity between 10 mPa·s to 40 mPa·s were chosen that represent a good approximation of the viscosity of typical paint materials at high shear rate by the atomization as well as the impact processes. Moreover, a surface tension of $\sigma = 0.025\,N \cdot m^{-1}$ was used in the simulations for paint liquids. Depending on different atomizer and droplet size, the droplet impact velocity could be from $0.5\,m \cdot s^{-1}$ to $50\,m \cdot s^{-1}$ [11]. Nevertheless, velocity $< 10\,m \cdot s^{-1}$ was used, since a representative mean diameter of paint droplet is usually smaller than 50 μm. Table 1 summarizes some parameters used in this study. The position and orientation of effect pigments within a droplet may be randomly distributed. For the present simulation, the flake has initially been located at the vertical centre at one half of the droplet radius and oriented vertically to the solid surface. This initial orientation angle is considered as the worst-case whilst a horizontal orientation as optimal in the resulting film. In order words, the ideal flake is aligned parallel to the solid surface after the curing process. Figure 7 shows the contour evolution and the flake orientation during a droplet impact process. The colour bar gives also the development of the contact angle based on Eq. (1) during the droplet impact process. Figure 7a) depicts the initial state. The inertia driven spreading phase, as shown in Fig 7b) and c), turns the flake quite substantially. In Fig. 7d) the maximum spreading is reached and surface tension effects become dominant. During this stage, despite droplet contour movement, the mentioned contact line hysteresis is observed, which is identical to our experimental findings. Here, the flake is found to increase its orientation angle again. This is partly reversed during recoiling and whilst the final equilibrium state is reached, as shown in Fig 7 e), f), g) and h). Air entrapment by droplet impact and air bubble (blue blobs) development on the wall, namely at the beginning a thin air layer/small bubbles on the wall and late large bubbles with time, can also be observed in Fig. 7, which is another topic and will not be discussed in the present report.

The effects of the droplet liquid properties and impact parameters on the flake orientation are obtained and shown in Fig. 8. Glycerol/water and paint drops with $\eta = 20\,\text{mPa·s}$ were applied. For small droplets, such as $D = 50\,\mu\text{m}$, the quasi-static state is reached at $t = 0.3\,\text{ms}$. Since the size ratio between flake and droplet diameter is relative large, $l/D = 0.32$ for $50\,\mu\text{m}$ droplet, the movement of the flake is quite limited. For $300\,\mu\text{m}$ droplets, a larger change in orientation angle can be observed. It takes a relative long time to reach the static state for large We- or Re-number. The final flake angle in the quasi-static state depends, presumably, mainly on the Re-number. It seems that the flake orientation improves with increasing Re-number, which should be further studied by increasing the parameter variation.

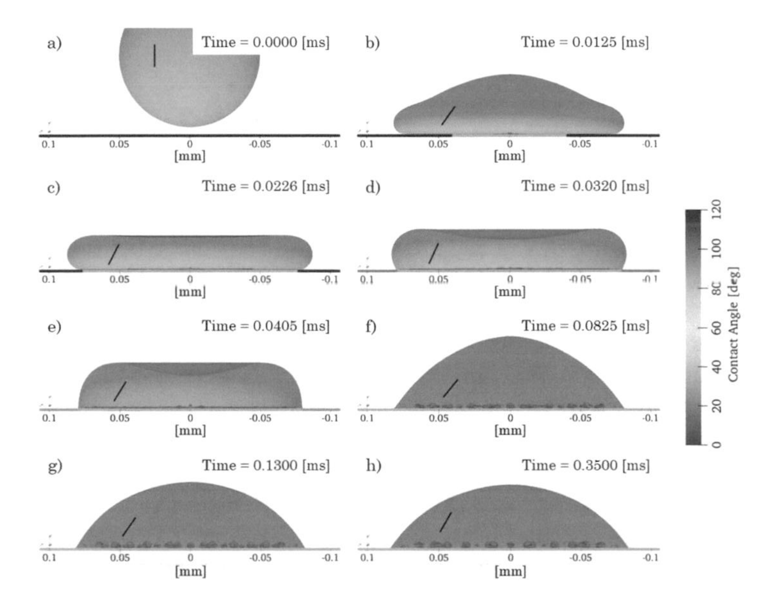

Fig. 7: Evolution of the droplet contour and flake orientation during dry wall droplet impact ($D = 100\,\mu\text{m}$, $\eta = 20\,\text{mPa·s}$, $\sigma = 0.063\,\text{N·m}^{-1}$, $\theta_E = 55°$, $U = 6\,\text{m·s}^{-1}$, $We = 68.6$, $Re = 36$).

The interaction between flake and flow field may be observed more detailed in Figure 9. At $t = 0.17\,\text{ms}$ (Fig. 9c), there is no macroscopic contour movement left, which is referred to as quasi-static state. Although the velocity magnitude is quite small, there are eddies arising around the pigment (Fig. 9d, c), resulting in some disorientation of the flake. This effect requires further investigation. Also, the entrapment of air bubbles on the solid surface, which are later noticed also in the paint film, can be seen in Fig. 7 and Fig. 9.

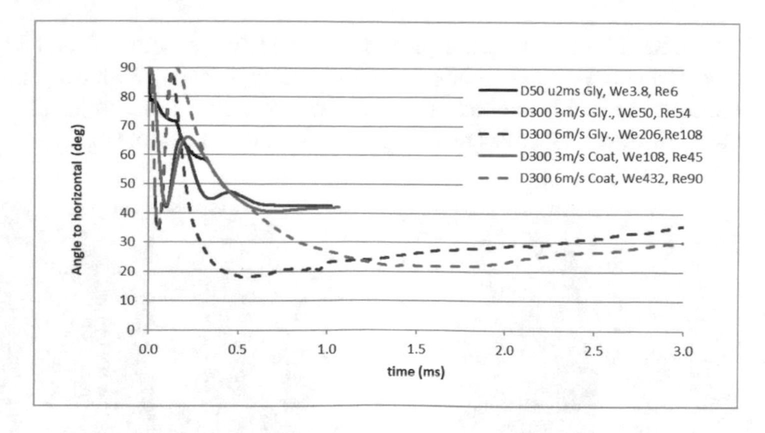

Fig. 8: Comparison of flake orientation with different viscous droplets.

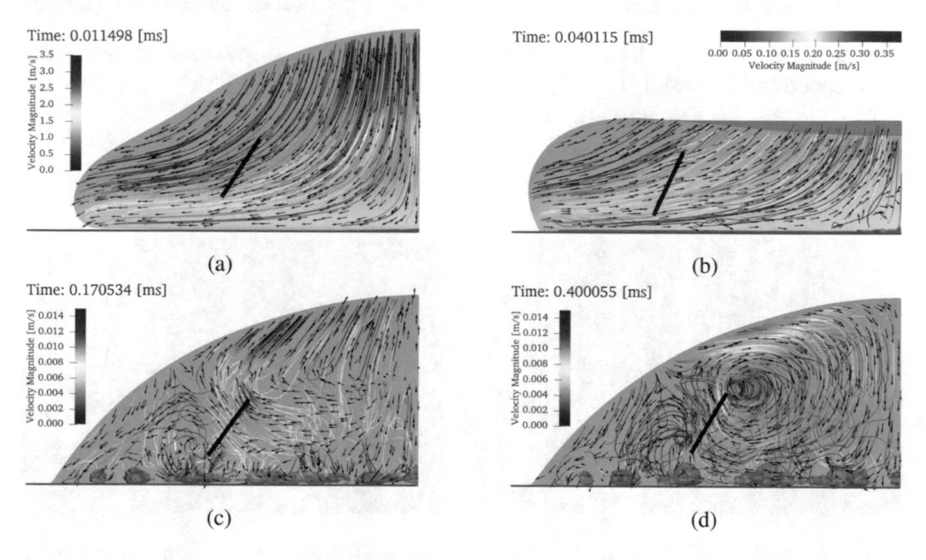

Fig. 9: Flake within coat droplet (D = 100 μm, θ_E = 53°, U = 6 m·s^{-1}, We = 153.3, Re = 30.7); overlaid with velocity vector (m·s^{-1}) and streamline: a) advancing process b) receding process, c) and d) quasi-static state.

3.4 Flake orientation by droplet impact on wet solid surfaces

Flake orientation by droplet impact on the wet surface is more concerned than on the dry surface, since the optical properties is mainly influenced by flakes located close to the film surface. Since the simulation is quite time consuming for large droplet because of relative large computational domain in this case, detailed parameter studies were carried out with droplet with diameter of 50 μm.

Typical spreading and cratering processes are shown in Fig. 10 and Fig. 11, respectively. The crater occurs by increasing the impact velocity. With deepening the crater, as shown in Fig. 11, the flake angle decrease and even tends to parallel to the solid surface. However, by crater receding the flake angle increases again.

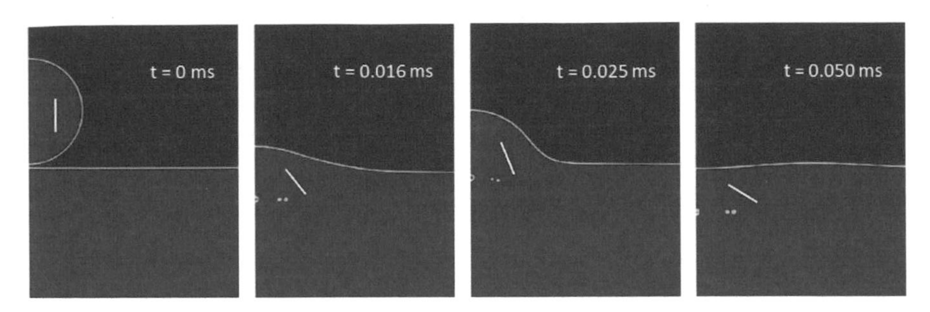

Fig. 10: Evolution of droplet-, film-contour, and flake orientation (black ruler corresponds to 50 μm), (D = 50 μm, η = 20 mPa·s, σ = 0.063 N·m^{-1}, U = 2 m·s^{-1}, We = 3.8, Re = 6, Oh = 0.33).

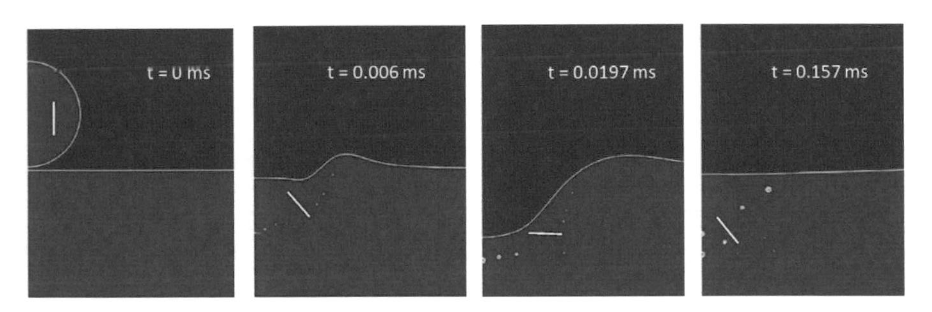

Fig. 11: Evolution of droplet-, film-contour, and flake orientation (black ruler corresponds to 50 μm), (D = 50 μm, η = 20 mPa·s, σ = 0.063 N·m^{-1}, U = 10 m·s^{-1}, We = 95, Re = 30, Oh = 0.33).

Some comparison of flake orientations using different viscous droplets (glycerine and paint) is depicted in Fig. 12. At quasi-static state, we can observe that for droplet with lower viscosity (20 mPa·s) the flake angle with lower impact velocity is smaller (32°) than that (50°) with high impact velocity. The paint droplet with constant viscosity (η = 40 mPa·s) that undergoes only the spreading process shows the largest flake angle (58°). The surface tension shows weak effect on the final flake orientation at the quasi-static state, however, the time to reach the static state is longer for the paint droplet.

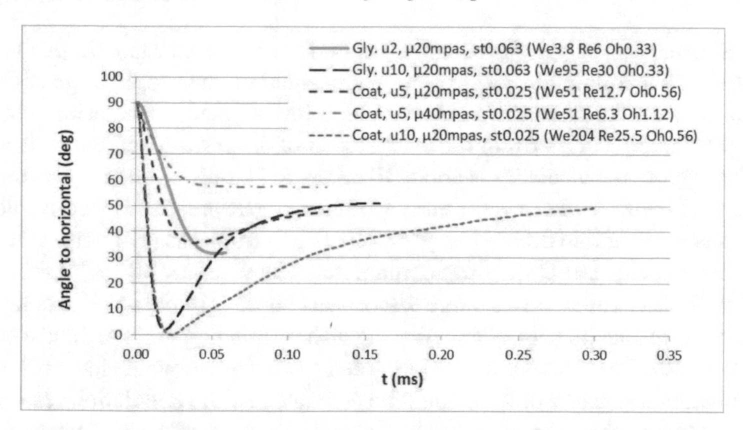

Fig. 12: Comparison of flake orientations by droplet impact on the wet solid surface using different viscous droplets, $H = 60\,\mu\text{m}$, $D = 50\,\mu\text{m}$, We, Re and Oh are dimensionless numbers and are defined in Table 1.

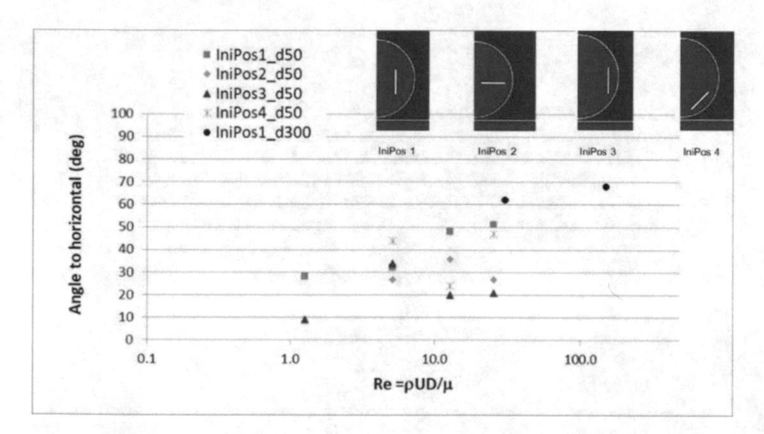

Fig. 13: Effect of initial position of the flake on the final flake orientation, $H = 60\,\mu\text{m}$, $D = 50\,\mu\text{m}$ and $300\,\mu\text{m}$, $\sigma = 0.025\,\text{N·m}^{-1}$, $\rho = 1020\,\text{kg·m}^{-3}$, $\eta = 20\,\text{mPa·s}$, $U = 0.5\,\text{m·s}^{-1}$ to $10\,\text{m·s}^{-1}$.

The effect of the initial position of the flake inside droplet on the final flake orientation was also studied. Simulations with four representative initial flake positions shown in Fig. 13, were carried out. Initial position 2 and 3 show the lower angle to horizontal. In general, the final angles are quite scatted with increasing Reynolds number, indicating a worse flake orientation by a higher droplet impact velocity for a droplet with diameter around $D = 50\,\mu\text{m}$. For a large droplet, the flake distribution in the droplet may be more complicated. Noticeable gas-liquid interface wave occurs because of crown formation / splashing by large Reynolds and Weber numbers, which results in a complicated flake movement and an extreme time consuming for the corresponding numerical simulation.

The aforementioned parameter study was performed at a constant droplet viscosity. The effect of shear thinning viscosity of paint liquid on the droplet impact [12] and consequently on the flake movement are also investigated. Simulation results are summarized in Fig. 14 a - d). In the main region of droplet impact, which is also the cratering region, the viscosity is about 10 mPa·s to 20 mPa·s. The maximum liquid viscosity is 80 mPa·s where is far away from the impact centre. For convenience, a characteristic shear rate defined as $\dot{\gamma}^* = 2U/D$ is introduced [12] here, where U is the impact velocity and D the droplet diameter. The $\dot{\gamma}^*$ is 2×10^5 (s^{-1}) for the case shown in Fig. 14. The corresponding viscosity is approx. 14 mPa·s. The comparison of simulation results (Fig. 14 d and e)) using a shear thinning viscosity and a constant viscosity of 10 mPa·s shows a quite similar flake orientation, which indicates that the flake orientation in a real droplet could be estimated by the simulation results using the constant viscosity corresponding to the paint viscosity at the characteristic shear rate.

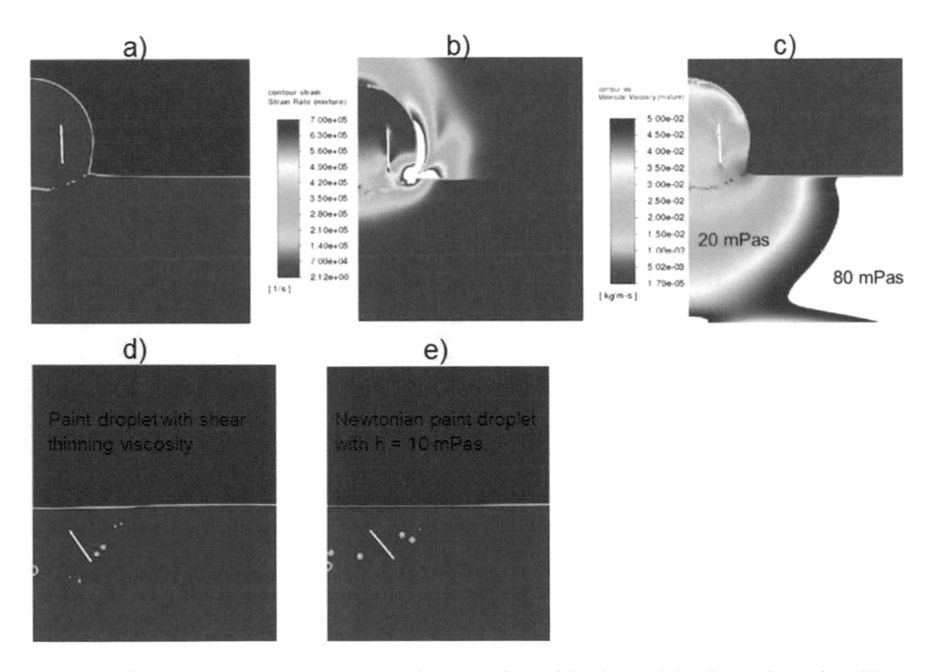

Fig. 14: Simulation results using paint droplet with shear thinning viscosity (U = 5 $m \cdot s^{-1}$, D = 50 μm): a) VOF contours (red: air, blue: liquid); b) Strain rate (s^{-1}) distribution (values that are large than 7×10^5 are blanked out); c) viscosity distribution (values that are large than 50 mPa·s are blanked out); d) Flake orientation at static state; e) Flake orientation at static state using constant droplet viscosity of 10 mPa·s.

4 Computational performance

The numerical flow simulations were performed using the commercial flow solver ANSYSFluent v19.5 on the HPE APOLLO (HAWK) of the High Performance Computing Center (HLRS) in Stuttgart. This supercomputer is equipped with 5632 compute nodes (2021), where each node has 256 GB memory and maximal 128 cores/node can be suggested to use. Figure 15 shows the performance study for simulations of flake orientation by droplet impact on the dry solid surface with a grid of 24 million cells for the large case and 12 million cells for the small case. Although dynamic load-balancing is used during simulation with AnsysFluent, clearly, by the large case, the speed-up decreases tremendously, when the number of nodes is larger than 16. A reasonable number of 8 to 10 nodes were therefore applied.

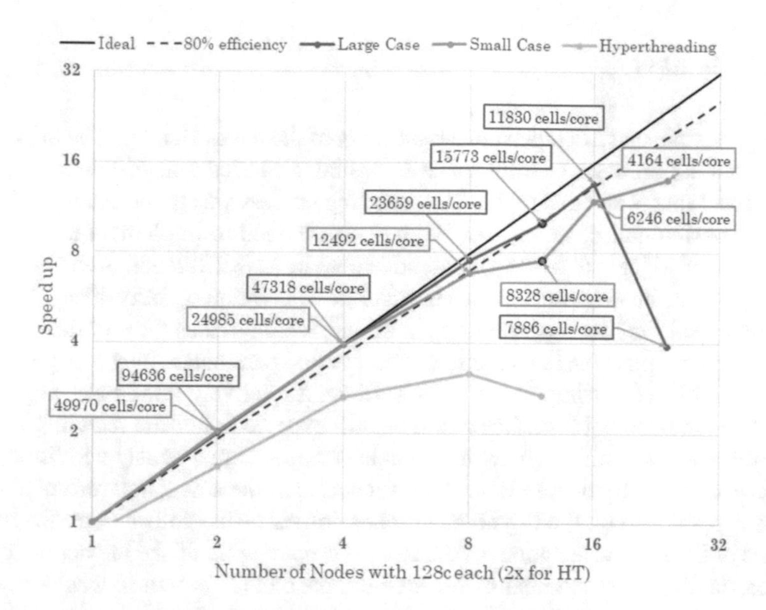

Fig. 15: Scaling with number of nodes.

For the simulation of flake orientation by droplet impact on the wet surface, more memory per core is required because of the overset method coupled with dynamic mesh adaption. We reduced the number of the used cores per node extensively. The corresponding performance is depicted in Fig. 16. Obviously, the calculation with 32 cores/node and 256 cores used in total shows a better performance. It seems that a higher number of cores per node will not benefit our current application.

A simulation case with 130 million cells was calculated up to 0.1 $m \cdot s^{-1}$ to 0.5 $m \cdot s^{-1}$ of physical time with time step size $< 10^{-7}$ s. Using 8 nodes and 256 cores, it consumed about 150 to 760 hours. The latter corresponds to the case with a large droplet.

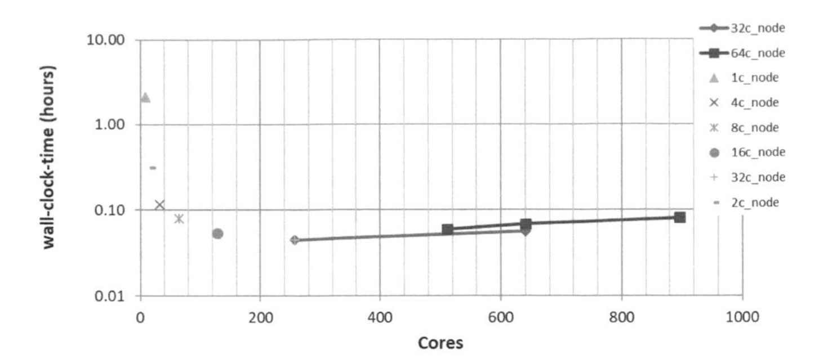

Fig. 16: Mean wall-clock-time of one time step for the droplet impact on the wet solid surface with 130 Mio. computational cells.

5 Conclusions

It is the first time that a detailed numerical study of the flake orientation by the viscous droplet impact on dry/wet solid surfaces has been carried out. A dynamic contact angle model that is suitable to the spray painting process was proposed and validated by the experimental observations. We have succeeded to implement a rigid body motion solver for calculating pigment movement in ANSYSFluent. Some numerical treatments have been applied, ensuring accurate calculation of derivatives and stable solution by solving the flake motion equations, which makes the simulation of the flake movement possible in a practical sense. A first parameter study was performed. In the case of droplet impact on a dry surface, a relatively large flake angle with respect to the horizontal surface could be still expected after the receding process at quasi-static state, although the flake angle changes tremendously with increasing Reynolds number. By droplet impact on wet surfaces, the simulation results of crater size in dependence on the Reynolds number deliver useful information for further analysis of the flake orientation. In the case of droplet with a flake impact on the wet surface, detailed parameter study for medium droplet ($D = 50\,\mu\text{m}$) was performed. Effects of initial flake position, droplet viscosity and impact velocity on the final flake orientation were investigated. For a given low viscous droplet and at the quasi-static state, it seems that the flake angle increases with increasing impact velocity because of the effect of crater receding. It is clearly observed that higher viscosities of droplet and film reduce the flake movement, resulting in smaller differences between the initial and the final flake angles. It is found that the height of film has weak effect on the flake orientation.

Further detailed parameter study should be performed with non-Newtonian liquids, especially for large droplets, such as $D = 80\,\mu\text{m}$ to $200\,\mu\text{m}$, for which time consumed numerical simulations have to be applied.

Acknowledgements The present investigations were supported by the German Federal Economical Affairs and Energy through the Arbeitsgemeinschaft industrieller Forschungsvereinigungen (AIF). The simulations were performed in the High Performance Calculation Centre Stuttgart (German federal project: DropImp). These supports are gratefully acknowledged by the authors.

References

1. Chandra, S. and Avedisian, C. T.: On the collision of a droplet with a solid surface. Proc. R. Soc. London, Ser. A 432, 13 (1991)
2. Thoroddsen, S. T. and Sakakibara, J.: Evolution of the fingering pattern of an impacting drop. Phys. Fluids 10(6):1359—1374 (1998)
3. Thoroddsen,S. T., Takehara, K. and Etoh, T. G.: Bubble entrapment through topological change. Phys. Fluids 22(051701):1—4 (2010)
4. Weiss Daniel A., Yarin Alexander L.: Single drop impact onto liquid films: neck distortion, jetting, tiny bubble entrainment, and crown formation. Journal of Fluid Mechanics, vol.385: 229–254 (1999)
5. Arogeti, M., Sher, E. and Bar-Kohany, T.: A single spherical drop impact on a flat, dry surface – a unified correlation. Atomization and Sprays, 27(9):759–770 (2017)
6. Kim, E. and Baek, J.: Numerical study of the parameters governing the impact dynamics of yield-stress fluid droplets on a solid surface. Journal of Non-Newtonian Fluid Mechnics **173-174**,62–71(2012)
7. Šikalo, Š., Wilhelm, H.-D., Roisman, I. V., Jakirlić, S. and Tropea, C.: Dynamic contact angle of spreading droplets: Experiments and simulations. Physics of Fluids 17, 062103 (2005); https://doi.org/10.1063/1.1928828
8. Linder, N., Criscione, A., Roisman, Ilia V., Marschall, H., Tropea, C.: 3D computation of an incipient motion of a sessile drop on a rigid surface with contact angle hysteresis. Theor. Comput. Fluid Dyn. 2015, DOI 10.1007/s00162-015-0362-9
9. Kirchner, E.: Flow-induced orientation of flakes in metallic coatings – II. The orientation mechanism, Progress in Organic Coatings, **124**, 104–109 (2018)
10. Schlüsener, T.: Untersuchungen zum Einfluss der thermo- und hydrodynamischen Vorgänge bei der Lackapplikation und -trocknung auf die Farbtonausbildung wasserbasierter Metallic-Lacke, PhD thesis in German, University Darmstadt, (2000)
11. Ye, Q. and Domnick, J.: Analysis of droplet impingement of different atomizers used in spray coating processes, J. Coat. Technol. Res., 14 (2) 467—476 (2017)
12. Shen, B., Ye, Q., Tiedje, O., Domnick, J.: On the impact of viscous droplets on wet solid surfaces for spray painting processes, accepted by ICLASS 2021, 15th Triennial International Conference on Liquid Atomization and Spray Systems, Edinburgh, UK, 29 Aug. – 2 Sept. (2021)

A low-pass filter for linear forcing in the open-source code OpenFOAM – Implementation and numerical performance

Jordan A. Denev, Thorsten Zirwes, Feichi Zhang and Henning Bockhorn

Abstract Direct Numerical Simulations or Large-Eddy Simulations often require that turbulence is forced and maintained throughout the solution time in the complete computational domain. In numerical codes written in physical space—like the open source library OpenFOAM—a modeling technique called linear forcing is often used for this purpose. It consists of adding a body-force term to the Navier–Stokes equations which is linearly proportional to the velocity. When compared to codes written in spectral space, in physical space this technique can only capture integral length scales half the size of those in spectral codes and is therefore inferior in terms of numerical efficiency. As shown by Palmore and Desjardins 2018 ([1], *Physical Review Fluids 3, 034605*) this drawback can be overcome through low-pass filtering of the velocity field used for forcing. However, the filter proposed in [1], although one-dimensional, is implicit and increases considerably the CPU-time. In order to overcome this, the present work proposes a new three-dimensional, explicit, low-pass Laplace-filter which is numerically efficient, shows very good scaling features and is easy to implement and parallelize. With this, the integral length-scale of turbulence increases more than two times thus resolving successfully a larger scale range of turbulence. A second improvement proposed in this work concerns the particular form of the linear forcing term: it applies the forcing individually to each velocity component. It is found that the new forcing term prevents unphysical, numerically triggered growth of only one velocity component and hence stabilizes the numerical process in OpenFOAM.

Jordan A. Denev and Thorsten Zirwes
Steinbuch Centre for Computing, Karlsruhe Institute of Technology, Hermann-von-Helmholtz-Platz 1, 76344 Eggenstein-Leopoldshafen, Germany, e-mail: jordan.denev@kit.edu

Thorsten Zirwes, Feichi Zhang, and Henning Bockhorn
Engler-Bunte-Institute/Combustion Technology, Karlsruhe Institute of Technology, Engler-Bunte-Ring.7, 76131 Karlsruhe, Germany

W. E. Nagel et al. (eds.), *High Performance Computing in Science and Engineering '21*,
https://doi.org/10.1007/978-3-031-17937-2_20

1 Introduction

Turbulence is the most common fluid flow type in engineering. Its ability to strongly enhance mixing of species, to intensify heat transfer and chemical reactions allows decreasing the size of gas turbines, of car engines as well as of other equipment in the process industry. Through properly controlling the features of turbulence—like its intensity and length scales—one can control the burning regimes and hence the burning intensity of premixed flames as it has been summarized in the Borghi regime diagram and its modifications [2–4]. Another application which strongly depends on the turbulence properties is connected to the distribution and clustering of spray particles and their interaction with eddies of the continuous phase as described in [5].

Appreciating the importance of turbulence for engineering applications, more accurate numerical methods for resolving the time and spatial scales of turbulence on very fine numerical grids—like Direct Numerical Simulation (DNS) or Large-Eddy Simulation (LES)—have been developed. These methods require that corresponding initial and boundary conditions which reflect the development of turbulent eddies in space and time are imposed. The generation of proper boundary conditions for the simulation of these methods is not a trivial task and more details can be found in [6–8]. Although with such boundary conditions the desired properties of turbulence are well defined at the inflow boundaries, the energy cascade and the action of viscous forces could still change the properties of turbulence on the way downstream until it reaches the close vicinity of the flame front.

Therefore, a second class of numerical methods for DNS has been developed which generate and maintain the proper turbulence features throughout the complete numerical domain. This is achieved through imposing an additional body force on the gas mixture thus maintaining the desired turbulence properties everywhere in the domain, including areas of special interest like the very vicinity of a flame front. The method is known as "forcing of turbulence" [9–12] and the present work introduces features that lead to improvements of its numerical performance.

Historically, forcing of turbulence has been first introduced for codes operating in wave-number space [13, 14]. Here, the forcing is applied to low-wave number modes [10] which resembles the energy input of large-scale turbulent vortices. However, as recognized by Rosales and Meneveau [10], the forcing of low-wave numbers is not easy to implement in codes which are written only in terms of physical space. But extending the forcing to such codes is desirable as then simulation setups are no more limited to periodic boundary conditions thus enabling the study of, e.g., freely propagating premixed turbulent flames.

An important step allowing the forcing to be extended to codes in physical space has been given by the work of Lundgren [9] who introduced a method which nowadays is known as "linear forcing". Although he still used a pseudospectral computation, his idea was to apply a forcing function which is directly proportional to the velocity.

The usability of the idea of Lundgren [9] for codes in physical space has been recognized and studied in detail by [10]. In their work [10], they introduce the linear forcing for codes in physical space by adding a local force that is proportional to the velocity in each computational point. In their work the authors showed that "linear forcing gives the same results as in spectral implementations" and that "the extent of Kolmogorov $-5/3$ range is similar to that achieved using the standard band-limited forcing schemes".

One general problem of physical forcing observed in [10] is that the integral length scale (defined as $\ell = u_{rms}^3/\varepsilon$ with u_{rms} being the rms of velocity and ε being the dissipation rate of the kinetic energy of turbulence) is smaller than in spectral space. As the smallest scales resolved in numerical simulations are determined from the resolution of the numerical grid, this means that the scale range which can be resolved in physical space is smaller than that resolved by codes in spectral space on the same numerical grid.

Carroll and Blanquart [15] found that the implementation of [10] exhibits an oscillatory behavior at the beginning of the computations. As the oscillations would require relative long computations until steady-state is reached, [15] suggest a slight modification of the forcing term which they show to reduce this oscillatory nature without altering the physics of turbulence. The code of Carroll and Blanquart uses a staggered numerical grid and we found that for the OpenFOAM implementation which uses a collocated grid, one of the components could raise non-proportionally high (on the expense of the other two), if the velocity components have a mean value which is different from zero and the forcing is following the implementation suggested by [15]. Therefore, in the present work we extend the idea of [15] and use forcing that is applied to each velocity component individually. This way the collocated grid arrangement is found to work stable and the forcing can be controlled for each component separately thus allowing, at least in theory, to obtain even non-isotropic turbulence. However, application of this technique to obtain non-isotropic turbulence would require a further investigation which goes beyond the scope of the present work.

Palmore and Desjardins [1] investigated the problem with the reduced integral scale of the turbulent eddies in physical forcing. They suggested a remedy to this problem by applying a low-pass filter to the velocity field before calculating the forcing term. It has been shown that the range of resolved scales can be doubled by applying a low-pass filter. In [1], they apply successively a one-dimensional (1D) filter in each spatial direction, x, y and z, which requires solving three times a complex, tridiagonal matrix. The application of this filter leads to a considerable increase in the CPU-time of the computations: between 48 and 652 percent, depending on the order of the filter which in turn determines its sharpness.

The present work extends the ideas of Palmore and Desjardin [1] by introducing a new, explicit, three-dimensional (3D) Laplacian filter in physical space. Similar to [1], this new filter also increases by a factor of two the integral length-scale of the turbulence, but, as shown in chapter 4, it is computationally considerably cheaper than the filter proposed in [1].

The paper starts with a mathematical description of the physical forcing (chapter 2) where the new low-pass Laplace-filter is introduced to the reader. The code for this filter—in the framework of the open source software OpenFOAM—is presented in the text. The idea to use a separate (individual) filtering for each velocity component is also introduced and described mathematically in chapter 2. In chapter 3, first a DNS-simulation of [10] is repeated and used for validation of the case without filtering. Then, on the same setup the effect of filtering on the integral length-scale is simulated and assessed. The numerical performance of the introduced Laplace-filter and of the idea for individual filtering of each velocity component is thoroughly examined in chapter 4. A summary of the work completed and an outlook to future work is given in chapter 5.

2 Mathematical description

In the present study, the incompressible Navier–Stokes equations are solved in the framework of Direct Numerical Simulations (DNS) of the turbulent flow:

$$\frac{\partial u_i}{\partial x_i} = 0, \tag{1}$$

$$\frac{\partial u_i}{\partial t} + \frac{\partial u_i u_j}{\partial x_j} = -\frac{1}{\rho_0}\frac{\partial p}{\partial x_i} + \frac{\partial(\nu 2 S_{ij})}{\partial x_j} + f_i, \tag{2}$$

where u is the fluid velocity, ν the kinematic viscosity, p the pressure and f the forcing term. The density ρ_0 is constant in case of incompressible flows. The strain rate tensor S is given by:

$$S_{ij} = \frac{1}{2}\left(\frac{\partial u_i}{\partial x_j} + \frac{\partial u_j}{\partial x_i}\right). \tag{3}$$

Following the idea of Lundgren [9], the forcing term f is presented by:

$$f_i = C u_i, \tag{4}$$

where C is a constant. For a stationary isotropic turbulence, Lundgren [9] considered the balance of the turbulent energy equation and could show that the value of the constant C corresponds to:

$$C = \varepsilon/(3U^2) \tag{5}$$

where U^2 is the mean square of one component of the velocity: $U^2 = \langle u_i\, u_i\rangle/3 = u_{rms}^2$ and ε is the mean (in spatial sense) dissipation rate of turbulence. Here, u_i are the velocity fluctuations and summation over repeating indices is assumed. The angular brackets denote spatial averaging.

Rosales and Meneveau [10] realized that the forcing-term from equation 5 is suitable for forcing in physical space and showed the applicability of this idea in their work. They noticed that the constant C from equation 5 has a meaning of a turnover time scale and kept the value of this constant unchanged during the computations.

Carroll and Blanquart [15] improved the convergence towards steady-state by rearranging the forcing term into the form:

$$f_i = C\left(\frac{k_0}{k}\right)u_i, \tag{6}$$

where k_0 is the desired kinetic energy and k is the instantaneously calculated kinetic energy. Equation 6 reduced the amplitude of the oscillations compared to equation 5 and the computations with the former reached quicker steady-state conditions. After reaching these conditions, the two equations deliver identical (in statistical sense) results.

Palmore and Desjardins [1] made a physical interpretation of equation 6. Considering the equation for the energy production, they could show that the pre-factor k_0/k dynamically adjusts the forcing in such a way that a constant dissipation rate is achieved. In the same sense, by modifying the constant C to depend on the current k or ε values, Basenne et al. [16] introduced physical forcing that maintains either constant turbulent kinetic energy or constant turbulent dissipation (also constant enstrophy for incompressible flows), or also a combination of the two showing further shortening of the initial transient period.

Palmore and Desjardins [1] applied a low-pass filter for the velocity field before using it as a source-term in the Navier–Stokes equations. Through this, they could double the integral length-scale ℓ of the resulting turbulent velocity field. For the forcing type represented by equation 6, the use of a filtered velocity field is written as:

$$f_i = C\left(\frac{k_0}{k}\right)\tilde{u}_i. \tag{7}$$

Here, $\tilde{\cdot}$ denotes the low-pass filtering operation. As already mentioned, in the work of [1] the filtering was applied to each spatial direction individually.

In the present work we introduce two new features to linear forcing in physical space which improve the numerical performance of this method.

The first new feature proposed in this work consists of introducing of a real three-dimensional filter for the filtering operation in equation 7. This filter is an explicit Laplace-filter. Since this filter is applied to the velocity field and smoothes the velocities in analogy to physical diffusion, its form is determined from the implementation of the incompressible flow solvers of OpenFOAM:

$$\tilde{\mathbf{u}} = \mathbf{u} + \nabla\cdot(B\nabla\mathbf{u}) + \nabla\cdot\left(B\left((\nabla\mathbf{u})^T - \frac{1}{3}\nabla\cdot\mathbf{u}\right)\right), \tag{8}$$

with $\mathbf{u}$ being the vector of the instantaneous velocity components. We use this form of the filter in the simulations in chapter 3. Before applying the filter, the velocity components have been centered with respect to their space-average value.

For an incompressible fluid, in analogy to the viscous term from equation 2, the above equation can be simplified to:

$$\tilde{\mathbf{u}} = \mathbf{u} + \nabla\cdot\left(B\left(\nabla\mathbf{u} + (\nabla\mathbf{u})^T\right)\right). \tag{9}$$

The filter constant B is not bound to the physical viscosity. Its value depends on the computational grid:

$$B = \frac{(Vol)^{\frac{2}{3}}}{W_{\text{coeff}}}, \tag{10}$$

where Vol is the volume of the numerical cell and W_{coeff} is a user-specified constant with the meaning of a width (diffusion) coefficient. Through variation of the value of W_{coeff}, respectively of B, the desired length-scale of turbulence can be achieved. Details about the particular value of the constant B and the way it has been obtained in the present work are given in section 3.

A Laplace-filter is already available in OpenFOAM for LES models, but the diffusive term there is formulated only for scalar quantities, not vector quantities like the velocity here. So, the filter has been modified accordingly and the coding is provided to the reader in the following; one major difference to the available OpenFOAM-filter is that now we made the filter entirely explicit thus allowing its value to be immediately obtained and directly used in the forcing term. In OpenFOAM's syntax, the implementation of the Laplace-filter for the velocity field is written as:

```
1  Foam::tmp<Foam::volVectorField> laplaceFilter
2  (
3      const Foam::volVectorField& U,
4      scalar widthCoeff,
5      const fvMcsh& mcsh
6  )
7  {
8      scalar averageCellVolume = average(mesh.V()).value();
9
10     dimensionedScalar coeff(dimArea,
11       Foam::pow(averageCellVolume, 2.0/3.0) / widthCoeff);
12
13     return U + fvc::laplacian(coeff, U) +
14       fvc::div(coeff*dev(T(fvc::grad(U))));
15 }
```

In the next sections it will be shown that the explicit and localized nature of the Laplace-filter leads to considerable savings of the CPU-time usage when compared to the filters originally used by [1]. Similarly to the original work of [1], the present Laplace-filter also leads to doubling of the integral velocity length scale of the resulting turbulent field and hence to an increase of resolved length scales similar to the one observed by spectral codes.

The second new feature introduced in the present work consists of modifying equations 6 or 7 so that each velocity component can be forced individually. For velocity in direction of axis i this results in:

$$f_i = C\left(\frac{u^2_{i,rms,0}}{u^2_{i,rms}}\right)u_i \quad (a) \qquad \text{or} \qquad f_i = C\left(\frac{u^2_{i,rms,0}}{u^2_{i,rms}}\right)\tilde{u}_i \quad (b). \tag{11}$$

Here, $u^2_{i,rms,0}$ is the targeted value of the squared velocity fluctuations of component u_i and $u^2_{i,rms}$ is the instantaneously calculated value of these fluctuations; both values are averaged in space. In case of application of a filter, the filtered velocity field $\tilde{u}_i$ according to equation 11b is applied instead.

We found that the above form of the forcing equation 11a, unlike equation 6, prevents a divergence from a non-proportional growth of one velocity component on the expense of the other two in case when the average value of this component $\langle u_i \rangle$ is not zero. Such a non-zero average value is typical for setups with a prescribed flow direction as in the case of premixed flames with inflow velocity equal to the flame velocity. Therefore, the proposed modification enables also the handling of cases with non-periodic, i.e. inflow-outflow boundary conditions. Although not explicitly demonstrated in the present work, but by simply setting $u^2_{1,rms,0} \neq u^2_{2,rms,0} \neq u^2_{3,rms,0}$, equation 11 enables a forcing which, at least in theory, would produce also a non-isotropic turbulence.

3 DNS results with the proposed filter for linear forcing

In order to assess the modeling and the numerical performances of the suggested filter (equation 8), a Direct Numerical Simulation following case 3 from the work of Rosales and Meneveau [10] is carried out. This case has been chosen because it has the largest forcing constant and therefore produces the highest kinetic energy of turbulence. In the following, some details of the DNS are presented. The domain dimensions are 2π m in all directions and periodic boundary conditions apply on all boundaries. The kinematic viscosity is equal to $\nu = 0.004491,\text{m}^2/\text{s}$. The value of the forcing constant for both cases (with and without filtering) is set to $C = 0.2$ and the forcing scheme is according to equation 11 with a value of $u^2_{i,rms,0}$ equal to 0.44. While the DNS of [10] was simulated on a numerical grid of 128^3 cells, here the calculations have been made also on a two times finer grid with 256^3 cells. One reason for this is the need to assess the numerical performance of the introduced filter on finer grids which are typical for DNS. A second reason is to compensate the numerical error of the second-order scheme applied in the present work as compared to the sixth-order numerical scheme used originally by [10]. The results presented in this chapter are obtained on the finest grid with 256^3 numerical cells.

The initial turbulent velocity field was generated using a recently implemented OpenFOAM utility called `createBoxTurb` (see description in [17] for more details) according to the spectrum of [18]. This is different from the initial spectra used in [10], however, according to [10] the spectra of the initial velocity field has no effect on the final value of the integral length scale. To make a verification of the current results before the application of the filter, the energy spectrum from the current work is compared with that from [10] (Figure 2, case 3c). The comparison is performed after reaching a final stationary state and is presented in Figure 1. The figure shows that the spectrum from the present work, when reaching a stationary state, practically overlaps with that of [10].

The value of the filter constant B for all grids was set equal to 0.0277 which corresponds to $W_{coeff} = 0.087$ for the 128^3 grid and to $W_{coeff} = 0.02175$ for the 256^3 grid (see equation 10). The above value of B was first iteratively adjusted (by trial and error) on a grid with 64^3 grid points ($W_{coeff} = 0.348$). As shown further in this section, the target for the adjustment was to reach—at the statistically stationary state (times larger than $100s$)—a doubling of the integral length-scale of the turbulent field ℓ in comparison with the original study of [10]. Each of the above test runs had a duration of 100 minutes by using only a single core.

Figure 2 allows to get a first impression of the influence of the filter on the velocity field presented by the magnitude of the velocity vector. It can be clearly seen that the structures appearing on the left-hand side of the Figure are on average visually larger than those without the filter; also the amount of structures at the left-hand side is less than in the case without the filter.

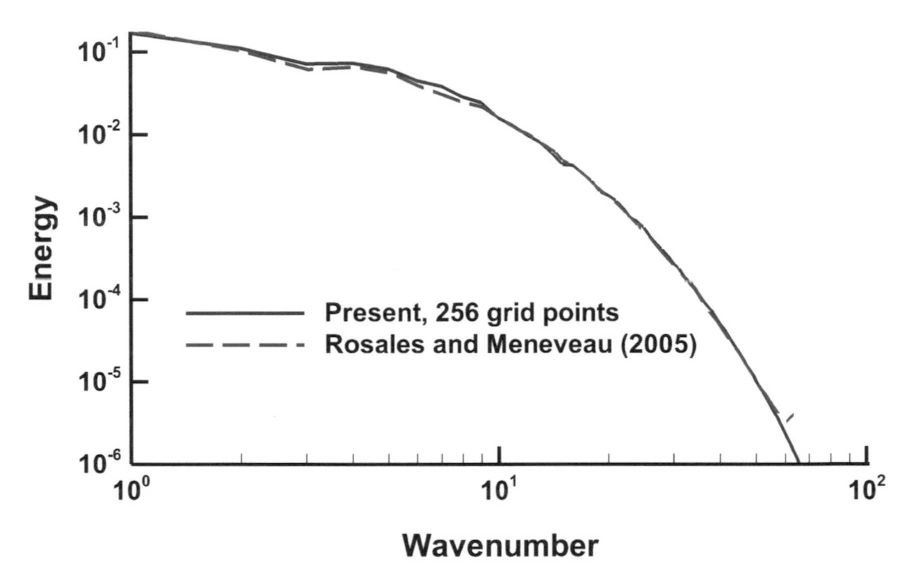

Fig. 1: Energy spectrum of the velocity field without filtering. Comparison of present stationary results at $t = 300$ s with stationary results of [10].

In the following, the effect of the filter on the statistical quantities of the obtained turbulence is examined. First of all, it has been observed that the filter leads to a decrease of the value of the energy dissipation ε. This can be seen in Figure 4 where the spatially averaged value of ε is plotted against the number of time-steps; for convenience, the first 5000 time-steps with wide-spread values for ε are omitted. It can be seen that the resulting value for the energy dissipation rate is approximately two times smaller with the filter (0.115) than without the filter (0.254).

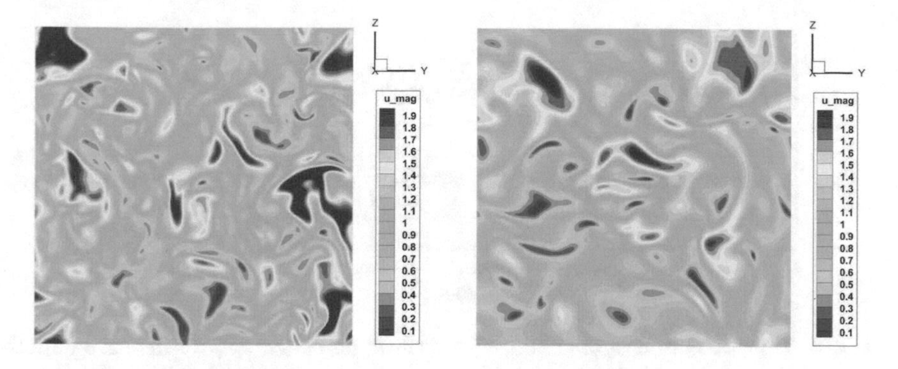

Fig. 2: Turbulent structures of the velocity magnitude $|\mathbf{u}|$ at $x = 2.1m$ and $t = 300$ s appearing in the case without the filter (left) and with it (right).

As it is shown in Figure 3, the forcing scheme according to equations 11a and 11b leads statistically to the same amount of kinetic energy of turbulence. Bearing in mind the two times higher values for the energy dissipation when the filter is applied, the relation $\ell = u_{rms}^3/\varepsilon$ reveals a major outcome from the Laplace-filtering, namely that the integral length scale ℓ also increases by a factor of two. Therefore, through the application of the filter, the main drawback of forcing in physical space, i.e. the smaller size of the integral length scale, - can be successfully overcome and thus the range of resolved scales is considerably increased. Following the ideas of [1], the smallest eddies are always fixed by the grid size and therefore the two times larger integral length scale means that the range of scales captured by the same numerical grid is also doubled. The corresponding increase of the integral length scale is shown in Figure 5.

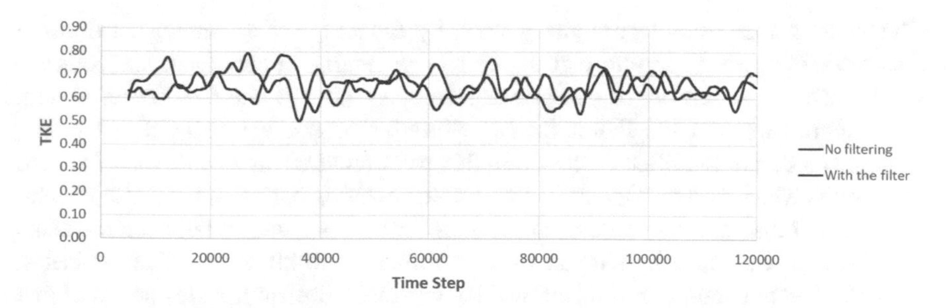

Fig. 3: Time evolution of the turbulence kinetic energy TKE (m^2/s^2) for the case with the filter and when no filtering is applied.

Showing the advantage which filtering brings for the range of scales resolved, the question about the numerical performance of the computations with the filter becomes important. This performance is evaluated in section 4.

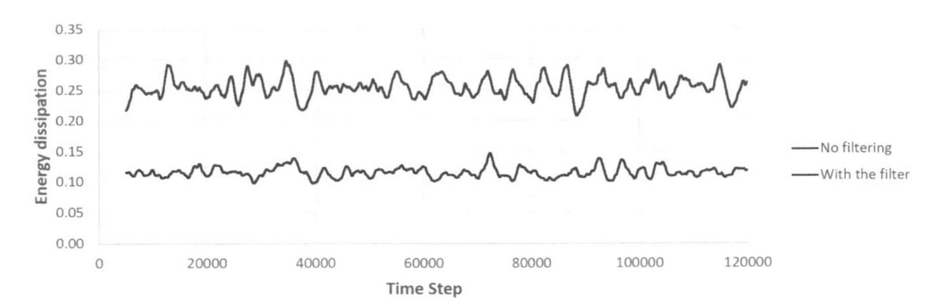

Fig. 4: Time evolution of the energy dissipation rate ε (m^2/s^3) for the case with the filter and when no filtering is applied.

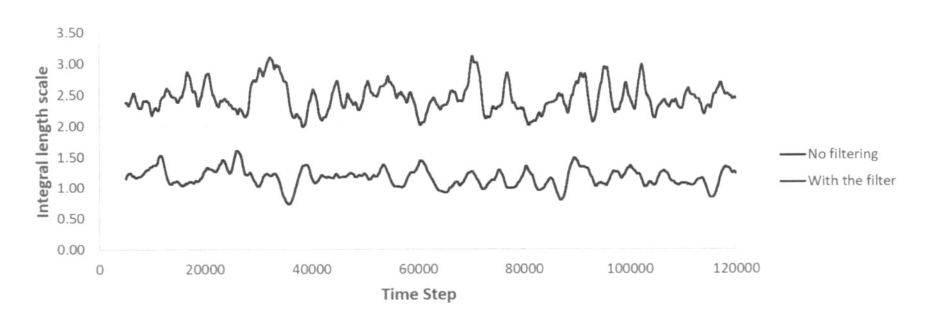

Fig. 5: Integral length scale ℓ (m) compared for the case with the filter and when no filtering is applied.

4 Parallel performance

The method for parallel computing used by OpenFOAM is known as domain decomposition, where the entire grid with a large number of computational cells are partitioned into a number of sub-domains and assigned to one CPU core. Ideally each sub-domain or CPU shares the same number of grid elements. directly. The implementation of the filter described in this work requires the evaluation of spatial derivatives which utilizes OpenFOAM's implementation based on the finite volume method. For this, pair-to-pair communication in MPI is required between the tasks that share internal domain boundaries. In addition, all-to-all communication has to be performed to obtain global quantities like the total turbulent kinetic energy, global dissipation rate and velocity fluctuation.

The scale-up performance depends mainly upon the number of allocated cells to each single CPU core n_{cell}. A good parallel scalability is expected for sufficiently large n_{cell}. In case of small n_{cell}, the time occupied by communications or data exchanges between the processors cores may constitute a large portion of the total computing time and reduce parallel efficiency. Table 1 lists the measured computing performance by running the proposed solver in parallel for a test case with $N_{cell} = 256^3$

grid cells on the HoreKa cluster at SCC/KIT. The number of processor cores N_{core} in the x-axis is selected to be multiples of 76 (n_{core} = 76, 152, 304, 608, 1216 and 2432), because each computing node from Horeka contains 76 cores. In this case, the consumed wall clock time for running the solver for 2000 iteration time steps (Δt) has been recorded and the measurement has been repeated for 3 times in each case by using different n_{core}. The averaged consumed clock times per time step $t_{\Delta t}$ from these 3 simulations, shown in the second column of Tab. 1, are used to evaluate the speed-up factor and the efficiency factor with respect to parallel scalability

$$S_n = \frac{t_{\Delta t,ref}}{t_{\Delta t,n}}, \qquad E_n = \frac{n_{core,ref}}{n_{core}} \cdot \frac{t_{\Delta t,ref}}{t_{\Delta t,n}}. \tag{12}$$

S_n and E_n are normalized to the measured computing time with $n_{core,ref}$ = 76 (corresponding to use of one single node from Horeka). P given in the 5th column of Tab.1 represents an estimation of consumed clock time per time step and per cell on one single CPU core:

$$P = \frac{t_{\Delta t} \cdot n_{core}}{N_{cell}}. \tag{13}$$

Table 1: Scale-up performance of running the implemented forced-turbulence code with OpenFOAM-v2006 on the HoreKa cluster at SCC/KIT on a computational mesh with 256^3 cells.

n_{core} [-]	$t_{\Delta t,n}$ [s]	S_n [-]	E_n [-]	P [μs/Δt/cell]	n_{cell} [-]
76	4.420	1	1	20.0	220.8×10^3
152	1.932	2.29	1.14	17.5	110.4×10^3
304	0.904	4.88	1.22	16.4	55.2×10^3
608	0.457	9.67	1.21	16.6	27.6×10^3
1216	0.292	15.12	0.95	21.1	13.8×10^3
2432	0.276	16.0	0.50	40.0	6.9×10^3

The measured parallel scalability is further illustrated in Fig. 6 with the speed-up factor on the left and the efficiency factor on the right. A superlinear scaling can be detected by increasing n_{core} from 76 to 608 for the current setup, which may be explained by a beneficial usage of the cache memory instead of assessing the RAM by using more processor cores. In addition, the solution procedures of the governing equations occupy a major breakdown of the total computing time compared with the consumed time for data-exchanges, communication and synchronization between the processors. Further increase of n_{core} from 1216 to 2432 results however in a considerable decrease of E below unity (see the 4th column in Tab. 1 and Fig. 6 on the right), which may be attributed to a strongly increased share of communication compared with the computational effort for solving the mathematical equations. Therefore, the optimal number of cells per CPU core, shown in the last column of Tab. 1, should be selected as large as more than 10,000 for good parallel scalability.

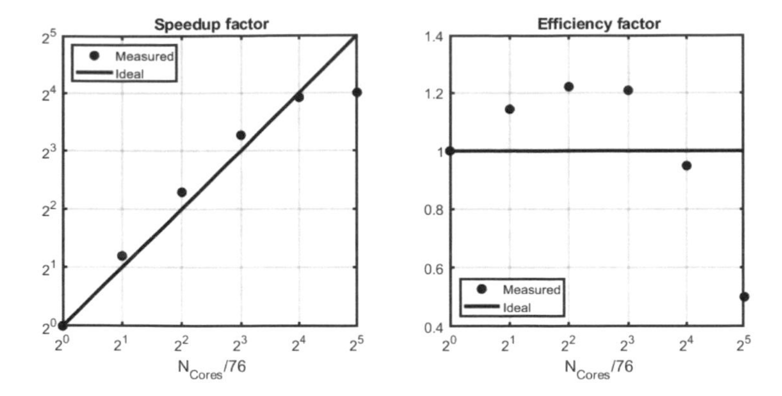

Fig. 6: Strong scaling performance of running DNS of forced turbulence on a computational grid with 256^3 cells on the HoreKa cluster at KIT/SCC Karlsruhe.

5 Conclusions

Numerical methods like DNS and LES are established tools for the detailed investigation of turbulent flows. These methods are known to be very computationally intensive and therefore the increase of their numerical efficiency is a subject of active research. In order to sustain a stable isotropic turbulence that does not decay in time, a class of methods exists that is called linear forcing. The present work introduces two numerical improvements of linear forcing applicable for codes operating in physical space.

The first one introduces a new explicit Laplace-filter for linear forcing in physical space by equation 8. The new filter is a low-pass filter that is three-dimensional in space. It is implemented in OpenFOAM and is listed in chapter 2. On a given computational mesh and with a fixed domain size, the application of the filter is shown to increase two times the integral length scale ℓ resolved by the numerical simulation. This is done with only 9 % increase of the CPU-time. Compared to a recent work of [1], where the filtering operation requires at least 48% longer CPU-time, there is a considerable improvement in the numerical efficiency with the present filter.

The second improvement is introduced through equation 11. The novelty is that the forcing is applied to each velocity component separately thus allowing its level to be adjusted individually in the course of the computations. This prevents the appearing of non-physical solutions that lead to much stronger forcing of one velocity component on the expense of the other two when forcing is applied according to equation 6 suggested by [15]. The overhead for this improvement is below 1% of the overall CPU-time of the simulations.

In chapter 4 it has been shown that the strong scaling with the new filtering method performs as expected for the incompressible solvers in OpenFOAM. A recommendation is given to use more than 10,000 cells per core to achieve good parallel scalability.

Acknowledgements The authors gratefully acknowledge the financial support by the Helmholtz Association of German Research Centers (HGF), within the research field MTET (Materials and Technologies for the Energy Transition), subtopic "Anthropogenic Carbon Cycle" (38.05.01). This work utilized computing resources provided by the High Performance Computing Center Stuttgart (HLRS) at the University of Stuttgart and on the ForHLR II and HoreKa Supercomputers at the Steinbuch Centre for Computing (SCC) at the Karlsruhe Institute of Technology.

References

1. J. A. Palmore and O. Desjardins, "Linear forcing in numerical simulations of isotropic turbulence: Physical space implementations and convergence properties," *Physical Review Fluids*, vol. 3, pp. 034 605(1)–034 605(18), 2018.
2. R. Borghi, "On the Structure and Morphology of Turbulent Premixed Flames," in *Recent Advances in Aeronautical Science*. Plenum Press, 1985.
3. J. Warnatz, U. Maas and R.W. Dibble, *Combustion.*, 4th ed. 378 p., Springer, Berlin, 2006.
4. T. Poinsot and D. Veynante, *Theoretical and numerical combustion.*, 2nd ed. 522 p., Edwards, 2005.
5. N. S. H. Lian and Y. Hardalupas, "Evaluation of the topological characteristics of the turbulent flow in a 'box of turbulence' through 2D time-resolved particle image velocimetry," *Experiments in Fluids*, vol. 58:118, p. 17, 2017.
6. F. Galeazzo, F. Zhang, T. Zirwes, P. Habisreuther, H. Bockhorn, N. Zarzalis, and D. Trimis, "Implementation of an Efficient Synthetic Inflow Turbulence-Generator in the Open-Source Code OpenFOAM for 3D LES/DNS Applications," in *High Performance Computing in Science and Engineering '20*. Springer, 2020, (accepted).
7. M. Klein, A. Sadiki, and J. Janicka, "A digital filter based generation of inflow data for spatially developing direct numerical or large eddy simulations," *Journal of computational Physics*, vol. 186, no. 2, pp. 652–665, 2003.
8. N. Kornev and E. Hassel, "Synthesis of homogeneous anisotropic divergence-free turbulent fields with prescribed second-order statistics by vortex dipoles," *Physics of Fluids*, vol. 19(6), pp. 068 101(1)–068 101(4), 2007.
9. T. Lundgren, "Linearly forced isotropic turbulence," in *Annual Research Briefs*. Center for Turbulence Research, Stanford University, Stanford, CA, 2003, pp. 461–473.
10. C. Rosales and C. Meneveau, "Linear forcing in numerical simulations of isotropic turbulence: Physical space implementations and convergence properties," *Physics of Fluids*, vol. 17, p. 095106, 2005.
11. J. A. Denev and H. Bockhorn, "Dns of lean premixed flames," in *High Performance Computing in Science and Engineering '13*, W. E. Nagel, D. H. Kröner, and M. M. Resch, Eds. Cham: Springer International Publishing, 2013, pp. 245–257.
12. J. A. Denev, I. Naydenova, and H. Bockhorn, "Lean premixed flames: A direct numerical simulation study of the effect of lewis number at large scale turbulence," in *High Performance Computing in Science and Engineering '14*, W. E. Nagel, D. H. Kröner, and M. M. Resch, Eds. Cham: Springer International Publishing, 2015, pp. 237–250.
13. V. Eswaran and S. B. Pope, "An examination of forcing in direct numerical simulations of turbulence," *Comput. Fluids*, vol. 16, p. 257, 1988.
14. S. M. Sullivan and R. Kerr, "Deterministic forcing of homogeneous, isotropic turbulence," *Phys. Fluids*, vol. 6, p. 1612, 1994.
15. L. Carroll and G. Blanquart, "A proposed modification to Lundgren's physical space velocity forcing method for isotropic turbulence," *Physics of Fluids*, vol. 25, pp. 105 114(1)–105 114(9), 2013.
16. M. Bassenne, J. Urzay, G. I. Park, and P. Moin, "Constant-energetics physical-space forcing methods for improved convergence to homogeneous-isotropic turbulence with application to particle-laden flows," *Physics of Fluids*, vol. 28, pp. 035 114(1)–035 114(16), 2016.

17. T. Saad, D. Cline, R. Stoll, and J.C. Sutherland, "Scalable tools for generating synthetic isotropic turbulence with arbitrary spectra," *AIAA Journal*, vol. 55(1), pp. 327–331, 2017.
18. G. Comte-Bellot and S. Corrsin, "Simple Eulerian Time Correlation of Full- and Narrow-Band Velocity Signals in Grid-Generated, Isotropic Turbulence," *Journal of Fluid Mechanics*, vol. 48(2), pp. 273–337, 1971.

Numerical simulation of vortex induced pressure fluctuations in the runner of a Francis turbine at deep part load conditions

Jonas Wack, Marco Zorn and Stefan Riedelbauch

Abstract For hydropower applications far off-design operating points like deep part load are more and more investigated, as these turbines can play a key role for the compensation of fluctuations in the electrical grid. In this study the single-phase simulation results of a Francis turbine at model scale are investigated for three mesh resolutions with the commercial CFD software ANSYS CFX. For the investigated deep part load operating point the typical inter-blade vortices can be observed. Further vortex structures are on the one hand traveling upstream close to the suction side and are on the other hand a result of a flow detachment at the runner trailing edge. The evaluation of the mesh resolution shows that a mesh refinement especially in the region of the inter-blade vortices results in a better prediction of the pressure minimum of these vortices.
The strong scaling test indicates an acceptable scaling up to 1536 cores for the mesh with 56 million cells. For the mesh with 82 million cells the scaling is acceptable even up to 2048 cores. A comparison of the MPI methods Open MPI and HMPT MPI showed that the latter is 16.5% slower.

1 Introduction

Due to the higher share of volatile renewable energies like solar and wind, other technologies are necessary that can compensate fluctuations in the electrical grid. For this purpose hydropower can play a key role. However, this requires more start-stop sequences and operation outside of the designed operating range [1]. These modified operating conditions result in higher structural loads and thus can have a negative impact on the lifetime of the turbine. Consequently, for future turbine designs reliable load evaluations at off-design conditions are required in the design process.

J. Wack, M. Zorn and S. Riedelbauch
Institute of Fluid Mechanics and Hydraulic Machinery, Pfaffenwaldring 10, D-70569 Stuttgart, e-mail: jonas.wack@ihs.uni-stuttgart.de

W. E. Nagel et al. (eds.), *High Performance Computing in Science and Engineering '21*,
https://doi.org/10.1007/978-3-031-17937-2_21

To integrate off-design operating points into the design process is challenging. For example at deep part load conditions stochastic pressure fluctuations occur [2] that cannot be captured by a standard RANS approach [3]. For that reason, either algebraic stress models [4] or hybrid RANS-LES models [3, 5] are typically applied for these operating conditions. However, especially the latter are computationally expensive as they need fine meshes and small time steps. In the design process a low computational effort is desired, as a lot of different designs are simulated and compared. Contrary to this, it is the goal of this project to provide highly resolved CFD results. Even though these simulations are not feasible in the design process, they can serve as a reference to derive more simplified simulation setups (e.g. smaller mesh and reduction of the simulation domain) that give the best compromise between computational effort and accuracy.

2 Numerical setup

CFD simulations of a Francis turbine model (24 guide vanes and 13 runner blades) of mid-range specific speed ($n_q = 40-60$) were performed with the commercial software ANSYS CFX. The investigations were carried out for a deep part load operating point with the following characteristics: discharge $Q/Q_{BEP} \approx 0.33$, discharge factor $Q_{ed} = 0.07$, speed factor $n_{ed} = 0.37$ and 8° guide vane opening. Within this study, single-phase simulations are carried out and consequently cavitation effects are neglected. Taking cavitation into account is an interesting research objective and will be published in the future.

The simulation domain consists of the following components: spiral case (SC), stay and guide vanes (SVGV), runner (RU), draft tube (DT) and downstream tank. In this study the focus is on the impact of mesh refinement on the resolution of vortex structures in the runner channels, as different studies already highlighted the necessity of fine grids to properly resolve the inter-blade vortices at deep part load conditions [6, 7]. For these analyses, three different meshes are investigated that are listed in table 1. All meshes have in common that the majority of cells are used for resolving the vortex structures in the runner. The coarsest mesh has around 25 million cells (25M) and uses wall functions in the boundary layer as the averaged y^+-value is around 40. This mesh is already finer compared to standard industrial applications in the field of hydraulic machinery. Mesh 56M is refined in the main flow, with the focus on a better resolution of the vortex structures in the runner, while the boundary layer resolution and consequently the y^+-value remains constant. This allows to investigate the impact of a better mesh resolution of the vortices. Contrary to that, mesh 82M has the same mesh resolution in the main flow as mesh 56M but the mesh in the boundary layer is refined to have an averaged y^+-value that is slightly below one. With this methodology it is possible to distinguish between mesh effects that result from a refinement of the main flow and those that originate from a refinement of the boundary layer.

Table 1: Mesh size of the subdomains for the used grids in million elements.

Mesh	SC	SVGV	RU	DT	Tank
25M	0.5	7.9	10.8	5.2	0.6
56M	1.6	11.2	31.1	10.1	1.7
82M	1.9	16.9	48.0	13.8	1.7

A comparison of simulations with the SBES [8] and the SAS [9] turbulence model (both hybrid RANS-LES models) did not show significant differences [10] and consequently only the more recently developed SBES model has been selected for more thorough investigations. This model is blending between the SST model in the RANS region and the WALE model in the LES region. The idea of hybrid RANS-LES models is to resolve large eddies in the main flow, while the computational effort can be reduced by using a RANS approach in the wall boundary layers [11]. Due to the selected turbulence model it is necessary to perform the simulations with a small time step that fulfills LES criteria. For mesh 25M the time step is approximately 0.4° of runner revolution while for the meshes 56M and 82M it is around 0.2°. With these settings the RMS averaged Courant number is below one for all simulations. Due to the small time step sizes the unsteady simulations converge quite fast within one time step and consequently only four coefficient loops are performed.

For spatial discretization a bounded central differencing scheme is selected and for temporal discretization a second order backward Euler scheme is applied. At the spiral case inlet a constant mass flow is set and the static pressure is set at the outlet of the tank. It can be expected that runner seal leakage has a higher impact on the velocity distribution for low discharges as the leakage flow is almost independent of the operating point [12]. For that reason seal leakage is considered in the numerical model by defining a sink at the runner inlet (outlet boundary condition) and a source at the runner outlet (inlet boundary condition). The seal leakage flow is assumed to be 0.33 % of the discharge for the investigated deep part load operating point. Between the rotating runner and stationary components a transient rotor stator interface is applied. To ensure that the interface between runner and draft tube is far away from the inter-blade vortices, it is located at the end of the draft tube cone. Consequently, walls of the draft tube that are within the runner domain are prescribed as counter-rotating walls.

Within this study all averaged results comprise 50 runner revolutions. Before the averaging has started, additional 20 runner revolutions have been simulated within the initialization process.

3 Experimental setup

A model test has been performed for the Francis turbine with multiple sensors in the stationary and rotating frame. Theses sensors include both, pressure sensors as well as strain gauge sensors on the runner blades. To investigate cavitation effects measurements have been carried out for three different pressure levels: $h_s = -10\,\mathrm{m}$, 5.6 m and 6.9 m. Even for the highest pressure level ($h_s = -10\,\mathrm{m}$) some cavitation regions could be observed, which is a result of the strong off-design operating point. More information on the experimental setup can be found in [10, 13].

4 Results

First, a comparison of the integral quantities head and torque is carried out. The deviations compared to the experiment with $h_s = -10\,\mathrm{m}$ are listed in table 2. For this comparison the experimental results for the highest pressure level are used as it has the smallest cavitation volume and for that reason is closest to the single-phase treatment of the simulations.

Table 2: Head and torque deviation compared to experiment with $h_s = -10\,\mathrm{m}$.

Mesh	25M	56M	82M
Head	-4.5%	-4.9%	-4.4%
Torque	+7.1%	+3.2%	+2.6%

For all meshes the head is underestimated by around 4.5%. A more significant difference between the meshes can be found for the torque. All simulation results overestimate torque, but the coarsest mesh has the highest deviation. Especially the refinement in the main flow from mesh 25M to 56M results in a significant improvement of almost 4%. Refining the mesh of the boundary layer (mesh 82M) leads to a further improvement of 0.6% of torque deviation.

The remaining deviations in head and torque can result from several effects. First, the neglection of cavitation causes some uncertainty. Furthermore, in the measurement the angle of the guide vane opening has somc uncertainty. While a small deviation in guide vane opening has only a relatively small effect for operating points with high power, it is much more significant for operating points with small guide vane opening like deep part load conditions. Additionally, the complex flow structures in the runner channels might even need a further mesh refinement.

4.1 Vortex movement

For the investigated deep part load operating point the highest pressure fluctuations are caused by vortices. Three main vortex structures can be observed that are visualized in Fig. 1 for mesh 82M with an isosurface of the velocity invariant ($Q = 5 \cdot 10^5\,\mathrm{s}^{-2}$). The inter-blade vortex ranges from the hub, at a location close to the leading edge, to the runner outlet close to the shroud, where it decays into small vortices. Over a wide range of the runner channel, the inter-blade vortex is located closer to the suction side (not visible in Fig. 1). Only close to the runner outlet it is located in the middle between pressure and suction side.

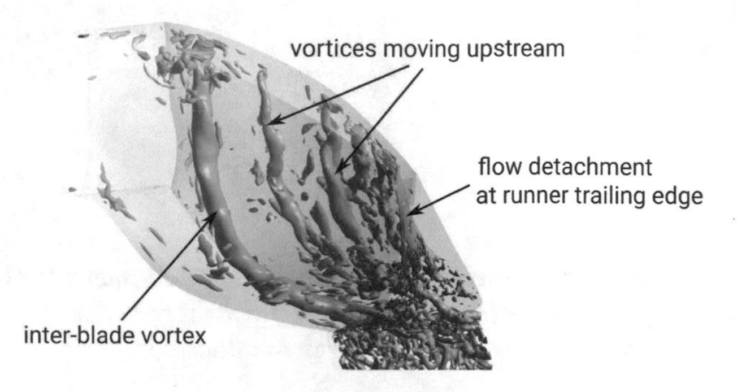

Fig. 1: Vortex structures in one runner channel.

Furthermore, the investigated operating point is characterized by a huge backflow region in the center of the draft tube cone. Due to this reversed flow, it comes to a detached flow at the trailing edge on the pressure side close to the hub. Vortices are generated from the flow detachment that are moving upstream close to the suction side in areas close to the hub.

To evaluate the effect of mesh refinement, it is investigated if the pressure minimum in the vortex core of the inter-blade vortices is well resolved. For that reason an isosurface of the pressure (threshold: vapor pressure for $h_s = 5.6\,\mathrm{m}$) is visualized in Fig. 2. While for the coarsest mesh the instantaneous pressure (top) is not falling below vapor pressure in the region of the inter-blade vortex, the results for mesh 56M and 82M show a low pressure region at the inter-blade vortex. This can be explained by the fact that these meshes are refined in the main flow compared to mesh 25M and consequently can better resolve the pressure minimum in the vortex core.

Similar results can be found for the time-averaged pressure (Fig. 2 bottom). However, for that quantity it can be observed that the isosurface is a connected region from hub to the runner outlet for mesh 82M, while the isosurface is only present in the rear part of the runner channel for mesh 56M. This shows that with the better wall resolution the inter-blade vortex develops slightly stronger. Nevertheless, it can be stated that the mesh refinement in the main flow has a bigger impact on the pressure minimum in the inter-blade vortices compared to the wall resolution.

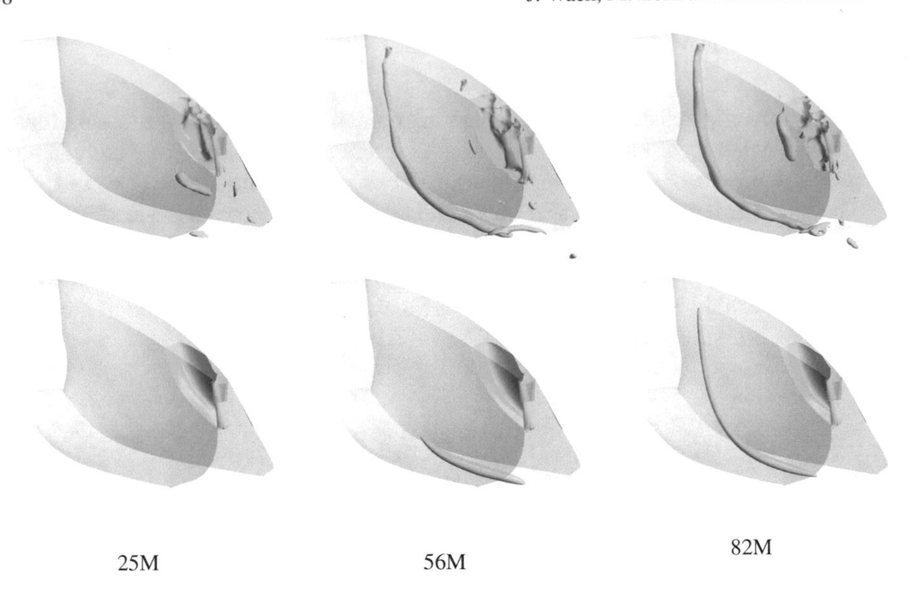

Fig. 2: Visualization of low pressure regions in one runner channel by an isosurface of pressure with threshold vapor pressure for h_s = 5.6 m. Top: Instantaneous snapshot. Bottom: Time-averaged values for 50 runner revolutions.

In addition to the inter-blade vortex, a low pressure region can be found in Fig. 2 at the trailing edge near the hub. This region develops due to the backflow in the draft tube cone that causes the flow detachment at the trailing edge.

The main fatigue damage potential of vortices results from the vortex movement. While a vortex that remains at the same location causes a constant load, it is the movement that results in permanent changes of the pressure field on the runner blades. Consequently, it is of interest to have a closer look on the vortex movement in the runner channel. This movement is visualized in Fig. 3 by the golden isosurface, which represents the envelope of vortex movement. The idea of this envelope is to visualize the region that falls below vapor pressure (for pressure level h_s = 5.6 m) in at least one time step using the minimum pressure of the simulated 50 runner revolutions. For better clarity this envelope is clipped at a surface that is located close to the runner trailing edge. The part of this surface, where the pressure falls below vapor pressure, is shown in Fig. 3 and colored with the minimum pressure within the 50 runner revolutions. In addition, an isosurface of the instantaneous pressure (threshold vapor pressure, isosurface is not clipped) is visualized in purple. This isosurface allows to visualize the vortex size and compare it to the envelope, which enables to get an impression about the vortex movement.

A comparison of the simulation results shows that for mesh 25M the envelope is much smaller in the inter-blade vortex region compared to the finer meshes. This is in agreement to the previous results and can be explained by the poorer resolution of the pressure minimum in the vortex core. The significantly higher pressure in the inter-blade vortex core for mesh 25M can also be seen in Fig. 3 in the minimum pressure on the surface close to the runner outlet. There, very intense pressure minima occur for meshes 56M and 82M, while this turns out to be significantly weaker for mesh 25M.

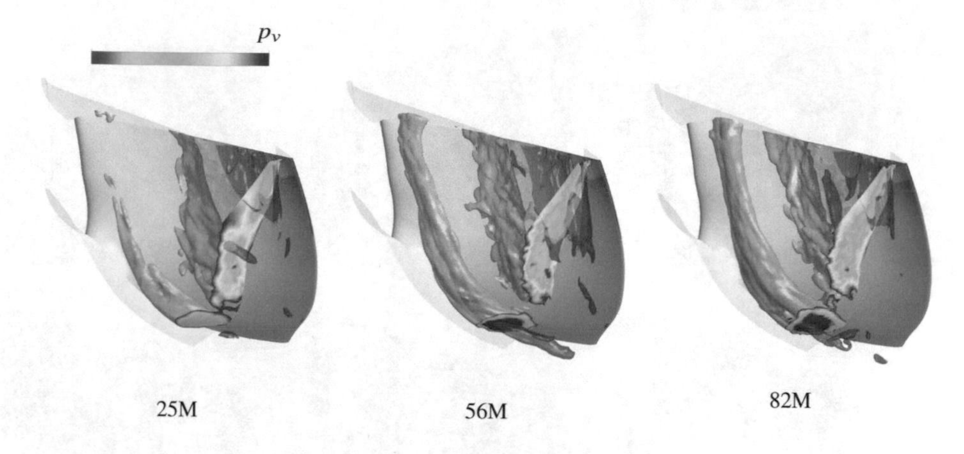

Fig. 3: Vortex movement in a runner channel. The movement is visualized by an isosurface (gold) that captures the region where the pressure is falling below vapor pressure p_v for at least one time step within the simulated 50 runner revolutions. Furthermore an isosurface of the pressure (threshold set to vapor pressure) for a randomly chosen time step is displayed in purple. The envelope of vortex movement is clipped at a plane in spanwise direction. On the part of this plane where the minimum pressure is falling below vapor pressure the minimum pressure is visualized and the rest of the plane is clipped.

Generally, it seems that the vortex movement in the hub region is smaller compared to the shroud region. The second envelope region results from the other vortices and the detachment region at the runner trailing edge. It gives an impression how far vortices with significant intensity travel upstream. The results indicate that close to the hub this region extends further upstream for meshes 56M and 82M, which again is related to a better vortex resolution.

4.2 Pressure fluctuations on the runner blade

The impact of the vortices on the pressure fluctuations on one runner blade can be found in Fig. 4. It shows the standard deviation on suction (top) and pressure side (bottom). Additionally, an isosurface of the time-averaged velocity invariant

($\overline{Q} = 5 \cdot 10^5\ \mathrm{s}^{-2}$) is displayed to get an impression about the location of the vortices. The vortices that are moving upstream close to the suction side are not visible by the time-averaged velocity invariant because of their strong movement.

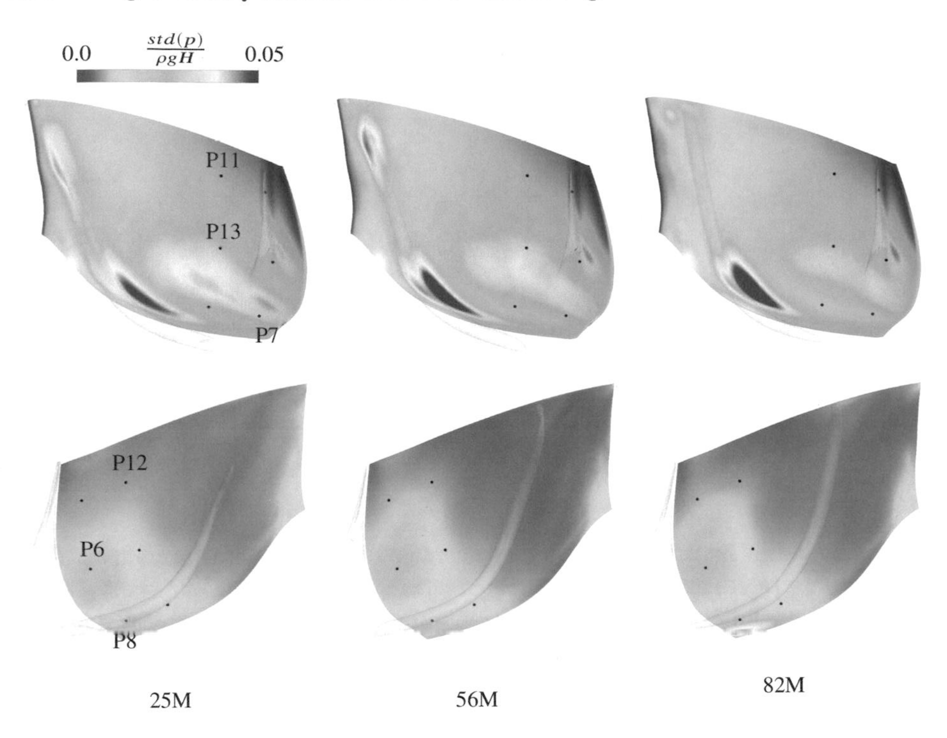

Fig. 4: Standard deviation of pressure on the suction side (top) and pressure side (bottom) of a runner blade. The standard deviation is calculated from 50 runner revolutions.

Over a wide range of the runner blade the standard deviation and consequently the pressure fluctuations are significantly higher on the suction side compared to the pressure side. This can be explained by the fact that the vortices are located closer to the suction side. While the inter-blade vortex causes for all meshes the maximum standard deviation close to the shroud in the middle of the blade suction side, over a wide range of the pressure side no impact of the inter-blade vortex on the pressure fluctuations can be detected. Only close to the trailing edge, where this vortex is located in the middle between pressure and suction side, an increased standard deviation can also be observed on the pressure side.

A comparison of the different simulations shows qualitatively similar results. Nevertheless, differences can be found. While mesh 25M and 56M have a region of high pressure fluctuations close to the leading edge near the hub on the suction side, for mesh 82M significantly smaller pressure fluctuations occur in that region. This might result from a slightly different location of the inter-blade vortex in this area that is shifted slightly downstream for mesh 82M compared to the other meshes. Furthermore, a region of high standard deviation can be found on the suction side in the area around measurement location P7 for mesh 25M. This region is located further away from the shroud for the other meshes and might result from a different interaction between the inter-blade vortex with the other vortices that is caused by the mesh refinement in the main flow. On the pressure side all simulations show increased pressure fluctuations at the leading edge close to the shroud that are very likely a result of the rotor stator interaction. Further regions of higher standard deviation can only be found close to the trailing edge for all simulation results. For mesh 82M a region of significantly higher pressure fluctuations compared to the other meshes occurs close to the shroud at the trailing edge.

While different simulation results can be compared all over the runner blade (see Fig. 4), a comparison to experimental results is only possible at specific measurement locations. In Fig. 5 the results of an FFT are presented for six sensor locations. Sensors with an even number (top) are located on the pressure side and sensors with an odd number (bottom) are on the suction side. The respective sensor location is marked in Fig. 4. To get an impression about the impact of cavitation, experimental results are displayed for $h_s = -10$ m and 5.6 m.

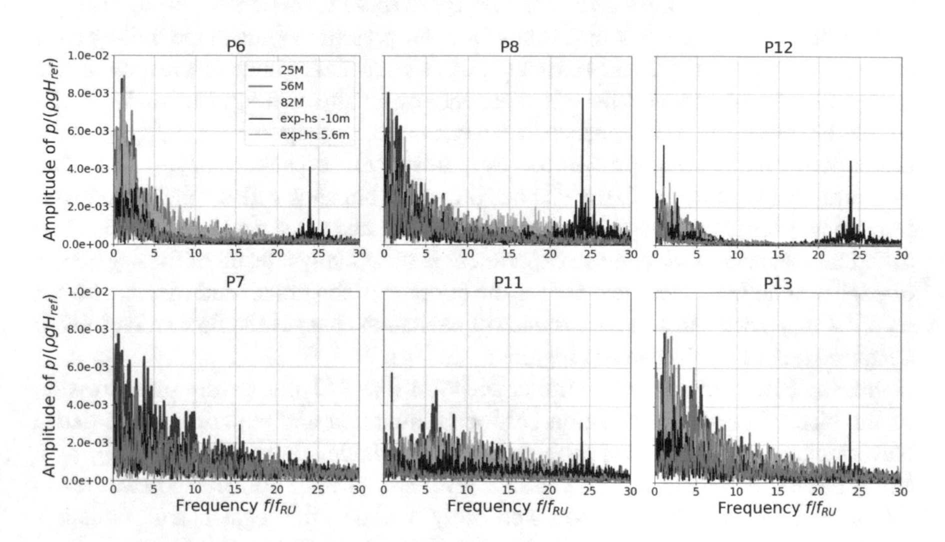

Fig. 5: FFT results for different locations in on the pressure (top) and suction side (bottom) of one runner blade. The positions are marked in Fig. 4.

Sensor P6 is located around midspan close to the trailing edge on the pressure side. For $f/f_{RU} < 5$ and especially $f/f_{RU} < 3$ high amplitudes are present at this sensor location in the simulation and measurement results. However, all simulations show a significant overestimation of the amplitude. A comparison of the experimental results at different pressure level shows that the amplitude of the pressure fluctuations decreases with a reduction of the pressure level. For that reason the overestimation of the simulation results is probably a result of the neglection of cavitation.

At sensor P7 noticeable amplitudes of pressure fluctuations occur over a wide spectrum of frequencies. This broadband spectrum is a result of vortices that are located close to the sensor position. The amplitude of the simulation results decreases with increasing mesh size. For mesh 82M the amplitude is in good agreement to the experimental results at both pressure levels. The difference in the amplitude of pressure fluctuations is also visible in Fig. 4 and is very likely a result of shifted vortex locations.

A variety of different frequencies with noticeable amplitude can also be found for sensor P8 ($f/f_{RU} < 10$). At this location this broadband spectrum is caused by the inter-blade vortex that is located close to that sensor (see Fig. 4). The simulation results show a trend of overestimating the amplitude. Again, mesh 82M is closest to the experimental results. However, the differences are not as big as for sensor P7. For rotor stator interaction ($f/f_{RU} = 24$) the measurement with $h_s = -10\,\mathrm{m}$ shows a huge peak that is neither present in the simulation results nor at the measurement at $h_s = 5.6\,\mathrm{m}$. It is not clear at this point why this peak is so pronounced but it might be a result of measurement issues and should be taken with caution.

Sensor P11 is located just upstream of the low pressure region at the trailing edge close to the hub (see Fig. 2 and 4). It can be observed that the simulation results behave totally different compared to the measurement - especially with $h_s = 5.6\,\mathrm{m}$. While the simulation results show a broadband frequency spectrum up to around $f/f_{RU} < 20$, the experiment with $h_s = 5.6\,\mathrm{m}$ has only a noticeable peak at $f/f_{RU} = 1$. The measurement with $h_s = -10\,\mathrm{m}$ also has this peak but shows also some broadband frequency spectrum that has a lower amplitude compared to the simulation results. The difference between simulation an experiment is most likely a result of the neglection of cavitation in the simulation. Due to the absence of the broadband frequencies for $h_s = 5.6\,\mathrm{m}$, it seems that the occurrence of cavitation changes the flow and results in a suppression of vortices in this region.

Sensor P12 is located at a similar location like P11 but on the pressure side of the blade. There, the deviation between simulation and experiment is smaller. Nevertheless, all simulations show relevant amplitudes up to around $f/f_{RU} = 5$, while the measurement at low pressure level shows only noticeable peaks up to $f/f_{RU} = 1.5$. Again, this deviation is probably caused by the neglection of cavitation, which is already indicated by preliminary two-phase simulation results.

Finally, the FFT results for sensor P13, which is located around midspan, also show a broadband frequency spectrum with overestimated amplitude by the simulations. Due to the location close to the low pressure region at the trailing edge near the hub the neglection of cavitation is the probable cause of the deviation. A comparison of the different pressure levels of the experimental results indicates that the size of the cavitation region has an impact on the pressure fluctuations. While for $h_s = 5.6\,\mathrm{m}$ the highest amplitude occurs around $f/f_{RU} = 5$, the maximum amplitude is shifted to lower frequencies for $h_s = -10\,\mathrm{m}$.

Preliminary results of two-phase simulations support that the occurrence of cavitation results in a damping of the amplitude of pressure fluctuations. Furthermore, frequency shifts, as observed at sensor location P13, can be captured by considering cavitation in the simulations. Detailed comparisons of single- and two-phase simulation results will be performed in the future.

5 Computational resources

The results of the numerical simulations have been performed on the HPE Apollo (Hawk) at the High Performance Computing Center Stuttgart. The architecture of the HPE Apollo consists of 5632 compute nodes with AMD EPYC Rome processors with 128 cores and 256 GB memory. On Hawk an infiniband HDR based interconnect with a 9-dimensional enhanced hypercube topology is used.

A strong scaling test is carried out for the meshes 56M and 82M on the file system ws10. For mesh 56M the speedup curve that is normalized to the simulation with 256 cores and time per time step are shown in Fig. 6 for the ANSYS CFX versions 19.5 and 21.1. A normalization of the speedup to 128 cores is not possible as this case needs more than 256 GB of memory.

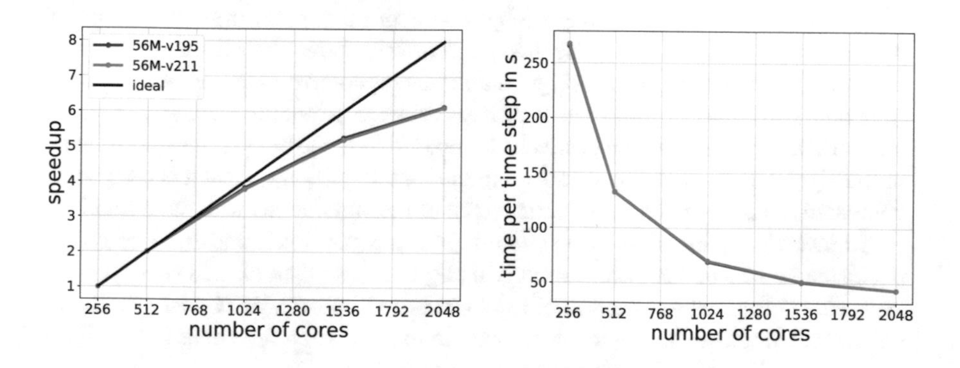

Fig. 6: Speedup and time per time step for mesh 56M.

For the two versions of ANSYS CFX the speedup is almost identical. As the time per time step is also not affected by the version of the solver, it can be stated that parallel performance has not been improved from version 19.5 to 21.1. The code shows an acceptable scaling up to approximately 1536 cores, where parallel performance is above 87%. This corresponds to around 36000 cells per core.

As ANSYS CFX version 21.1 did not show any improvement in parallel performance for mesh 56M, the speedup test for mesh 82M is only performed for version 19.5. The results are presented in Fig. 7. For the speedup the results are normalized to the simulation with 512 cores, which again is selected due to memory requirements. The results indicate a not ideal but reasonable scaling up to 1536-2048 cores, where parallel performance is 87% or 84%, respectively.

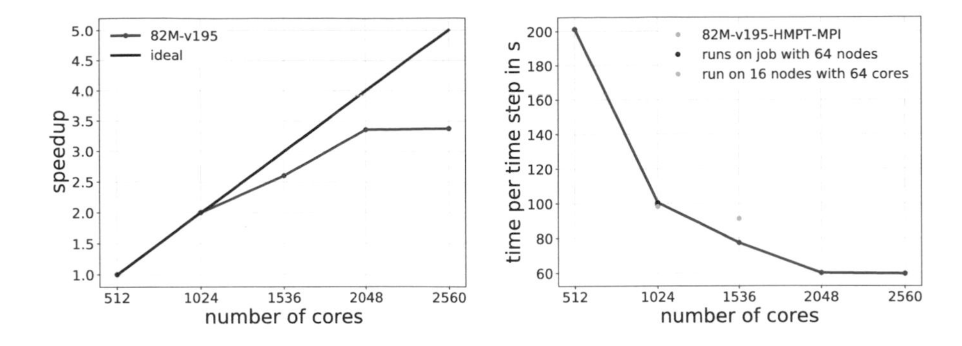

Fig. 7: Speedup and time per time step for mesh 82M.

As all jobs that use less than 64 nodes are performed in a special partition, for one case it is tested how the ideal use of the hypercube topology affects the simulation time. This is achieved by running a job on 64 nodes that performs eight simulations on 1024 cores each. Analyzing the time per time step shows that all simulations of the 64-node-job are slightly but not significantly faster compared to the 8-node-job that runs on the special partition. The fastest and slowest simulation of the 64-node-job are 0.6% faster or slower, respectively, compared to the averaged time per time step of the eight simulations. A comparison to the 8-node-job shows that it is 0.8% slower compared to the averaged time per time step of the simulations with the 64-node-job.

To investigate the effect of only using 64 cores per node, one 1024-core run is performed on 16 nodes instead of only using 8 nodes when all 128 cores per node are used. This gives a reduction of time per time step of 2.3%. Consequently, it is not advisable to use this approach as the gain in simulation time does not justify the usage of significantly higher computational resources.

Finally, the impact of the MPI method is investigated. Currently it is possible to use Open MPI or HMPT MPI on Hawk with ANSYS CFX. The comparison of these methods for a job on 1536 cores shows that the simulation with HMPT MPI is 16.5% slower. Consequently, Open MPI is used for all simulations on Hawk.

Within the reporting period around 18 million core-h have been used. From that approximately 11 million core-h were needed for the results of the single-phase simulations presented in this study. The other core-h were used for two-phase simulations that are not finished yet. The two-phase results will be published in the future.

6 Conclusion and outlook

In this study a deep part load operating point of a Francis turbine at model scale has been investigated. Three different meshes have been used for the single-phase simulations. The results show that head is almost not affected by the mesh resolution, while for the torque the mesh refinement results in a reduction of the deviation to the experiment. Especially the refinement of the main flow has an impact on torque as well as the vortex movement. A finer resolution of the boundary layer also has some effect that, however, is lower compared to the refinement of the main flow.

For the different meshes a slightly changed location of the vortices is possible. This is accompanied by shifted locations of high pressure fluctuations as they are mainly caused by the vortices. A comparison to experimental results indicates that at some pressure sensors the amplitude of the pressure fluctuations are overestimated due to the neglection of cavitation. Furthermore, cavitation can result in a shifted frequency spectrum that cannot be captured by single-phase simulations.

The strong scaling test did not show a difference in the parallel performance of the ANSYS CFX versions 19.5 and 21.1. For simulations with the mesh 56M the parallel performance is acceptable up to 1536 cores. With the finest mesh (82M) even 2048 cores are feasible. In terms of the MPI method the Open MPI is preferable as it is 16.5% faster compared to the HMPT MPI method.

The pressure field in the runner of the presented single-phase simulations will now be used as input for structural mechanical investigations. With the results a reduced simulation setup will be developed that can be used in the design process of future turbines. Furthermore, two-phase simulations are carried out to investigate the impact of cavitation on the pressure fluctuations more detailed.

Acknowledgements The simulations were performed on the national supercomputer HPE Apollo Hawk at the High Performance Computing Center Stuttgart (HLRS) under the grant number 44047. Furthermore, the authors acknowledge the financial support by the Federal Ministry for Economic Affairs and Energy of Germany in the project FrancisPLUS (project number 03EE4004A).

Supported by:

References

1. Seidel, U., Mende, C., Hübner, B., Weber, W., Otto, A.: Dynamic loads in Francis runners and their impact on fatigue life. In: IOP Conference Series: Earth and Environmental Science, vol. 22 p 032054 (2014)
2. Dörfler, P., Sick, M., Coutu, A.: Flow-induced pulsation and vibration in hydroelectric machinery: Engineer's guidebook for planning, design and troubleshooting. Springer (2012)
3. Wack, J., Riedelbauch, S., Yamamoto, K., Avellan, F.: Two-phase flow simulations of the inter-blade vortices in a Francis turbine. In: Proceedings of the 9th International Conference on Multiphase Flow, Florence, Italy (2016)
4. Conrad, P., Weber, W., Jung, A.: Deep Part Load Flow Analysis in a Francis Model Turbine by means of two-phase unsteady flow simulations. In: Proceedings of the Hyperbole Conference, Porto, Portugal (2017)
5. Yamamoto, K.: Hydrodynamics of Francis turbine operation at deep part load condition. Ph.D. thesis École Polytechnique Fédérale de Lausanne (2017)
6. Stein, P., Sick, M., Dörfler, P., White, P., Braune, A.: Numerical simulation of the cavitating draft tube vortex in a Francis turbine. In: Proceedings of the 23rd IAHR Symposium on Hydraulic Machinery and Systems, Yokohama, Japan p 228 (2006)
7. Wack., J., Riedelbauch, S.: Numerical simulations of the cavitation phenomena in a Francis turbine at deep part load conditions. In: Journal of Physics: Conference Series, vol. 656 p 012074 (2015)
8. Menter, F.R.: Stress-Blended Eddy Simulation (SBES) - A new Paradigm in hybrid RANS-LES Modeling. In: Proceedings of the 6th HRLM Symposium, Strasbourg, France (2016)
9. Menter, F.R., Egorov, Y.: The Scale-Adaptive Simulation Method for Unsteady Turbulent Flow Predictions. Part 1: Theory and Model Description. In: Flow, Turbulence and Combustion, vol. 85, pp. 113-138 (2010)
10. Wack, J., Beck, J., Conrad, P., von Locquenghien, F., Jester-Zürker, R., Riedelbauch, S.: A Turbulence Model Assessment for Deep Part Load Conditions of a Francis Turbine. In: Proceedings of the 30th IAHR Symposium on Hydraulic Machinery and Systems, Lausanne, Switzerland (2021)
11. Menter, F.R.: Best practice: Scale-resolving simulations in ANSYS CFD - Version 2.0. ANSYS Germany GmbH (2015)
12. Čelič, D., Ondráčka, H.: The influence of disc friction losses and labyrinth losses on efficiency of high head Francis turbine. In: Journal of Physics: Conference Series, vol. 579 p 012007 (2015).
13. von Locquenghien, F., Faigle, P., Aschenbrenner, T.: Model test with sensor equipped Francis runner for Part Load Operation. In: Proceedings of the 30th IAHR Symposium on Hydraulic Machinery and Systems, Lausanne, Switzerland (2021)

Validation of ACD and ACL propeller simulation using blade element method based on airfoil characteristics

Michael Schollenberger, Mário Firnhaber Beckers and Thorsten Lutz

Abstract Investigations concerning the aerodynamic interactions of propellers at new configurations such as distributed propulsion, wingtip mounted propellers, etc., are characterized by a large, multidimensional parameter space. Simplified models like the steady-state Actuator Disk (ACD) or the unsteady Actuator Line (ACL) are therefore common to reduce the calculation time. Since the quality of the models depends on the accuracy of the input data, it is essential that these are represented in a physically correct manner. In this study a validation and comparison of blade element method (BEM) based ACD/ACL models with fully resolved simulations and experimental data known from literature is performed. It is demonstrated that both the effects of the propeller on the wing and vice versa are accurately captured with BEM-based propeller models.

1 Introduction

The CFD simulation of propeller flows is of increasing relevance due to research into new configurations such as distributed propulsion, wingtip mounted propellers, etc., which benefit from positive aerodynamic interactions between the propellers and the other components of the aircraft. Investigations concerning the aerodynamic interactions involving propellers are characterized by a large, multidimensional parameter space. In order to reduce the calculation time simplified models like the steady-state Actuator Disk (ACD) or the unsteady Actuator Line (ACL) are thereby necessary. Stokkermans has impressively demonstrated the capability of ACD and ACL to simulate the complex effects of the propeller slipstream on the wing when using prescribed propeller loadings as input [1]. However, for the design of novel

---

Michael Schollenberger, Mário Firnhaber Beckers and Thorsten Lutz
University of Stuttgart, Faculty of Aerospace Engineering and Geodesy, Institute of Aerodynamics and Gas Dynamics (IAG), Pfaffenwaldring 21, 70569 Stuttgart,
e-mail: schollenberger@iag.uni-stuttgart.de

W. E. Nagel et al. (eds.), *High Performance Computing in Science and Engineering '21*,
https://doi.org/10.1007/978-3-031-17937-2_22

configurations it is crucial to also capture the effect of the wing on the propeller loads, both on the averaged propeller coefficients and on the phase dependence over one propeller revolution. In addition to the use of prescribed loadings, in the DLR TAU-code [2] the blade forces for ACD and ACL can be calculated via blade element method (BEM) [3, 4]. In the present study the airfoil characteristics of the propeller blades in the terms of lift and drag polars (c_l, c_d) are first extracted from a steady-state single blade simulation and are thereafter used for ACD, ACL calculations and validated by comparison with fully resolved simulations and experimental data. Three configurations were simulated in this study, see Fig. 1:

- A single propeller blade: to generate the airfoil characteristics for the ACD, ACL models
- An isolated propeller with an axisymmetric nacelle: to compare and validate the slipstream data
- An installed tractor propeller mounted at the wingtip of an semi-span wing: to compare and validate the effect of the slipstream on the wing as well as the wing effect on the propeller

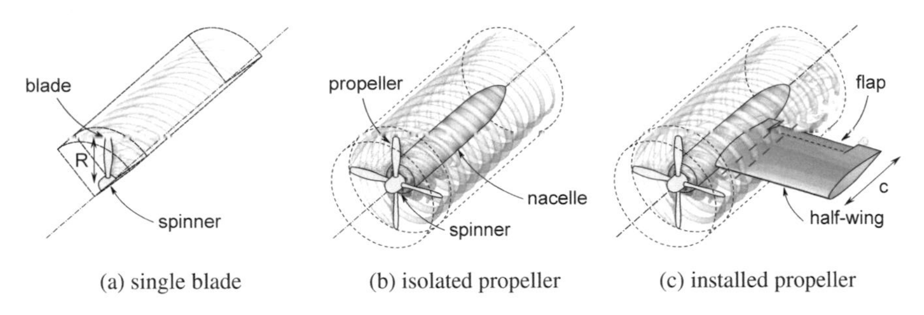

(a) single blade (b) isolated propeller (c) installed propeller

Fig. 1: Geometries of the simulated configurations.

The geometries are identical to those of Stokkermans et al. [1] and Sinnige et al. [5]. The single propeller blade, with a diameter of 0.237 m and a blade pitch angle of $\beta_{75} = 23.9°$, was simulated for a range of advance ratios between $J = 0.6 - 1.2$. The four-bladed propeller at the isolated and installed case uses the same blade geometry and was simulated with an advance ratio of $J = 0.8$ at a freestream velocity of 40 m/s ($Ma = 0.12$) and a freestream angle of attack of $\alpha = 0°$. The wing of the installed case has a semi-span of $b = 0,327\ m$ with a symmetrical NACA 64_2A015 airfoil, a chord length of 0.240 m and a plain flap with a deflection of $\varphi = +10°$ to generate lift.

2 Simulation methods and numerical setup

2.1 Propeller Simulation

Three different approaches to model the propeller impact within the CFD simulation are used: Actuator Disk (ACD), Actuator Line (ACL) and fully resolved propeller blades (full).

2.1.1 Fully resolved propeller (full)

The fully resolved propeller simulation in TAU is enabled by the Chimera technique, see Stürmer [7]. Thereby, a propeller grid rotates in front of the fixed background grid with a cylindrical hole at the position of the propeller. A sufficient overlap of the two grids ensures a smooth transition of the flow values. In this study a viscid modeling of the propeller blades is achieved by resolving the blade boundary layer to capture the propeller friction losses.

2.1.2 Actuator Disk (ACD)

In the standard TAU version, an Actuator Disk method is available, where the blade forces averaged over one propeller revolution and introduced stationary into the flow field on a circular boundary condition (BC), see Raichle et al. [3]. The local effective angle of attack α and the local inflow velocity are directly determined from the flow at each point of the disk. The steady-state calculation enormously reduces the computation time compared to the fully resolved simulation. However, the ACD has limitations due to the simplification made, the slipstream is stationary and blade tip vortices are not taken into account.

2.1.3 Actuator Line (ACL)

An unsteady Actuator Line (ACL) method was implemented into the TAU code by the presenting authors [4]. Here the blade forces are also introduced into the flow field on the propeller disk BC, but along discrete lines rotating in time. The calculation process of the ACL consists of four steps: 1. the calculation of the time-dependent ACL positions and the inflow conditions, 2. the calculation of the sectional blade loads via BEM, 3. the distribution of the forces and 4. the insertion into the flow field. The determination of the local effective angle of attack is more difficult than with the ACD, since it is not possible to take the flow values directly at the line. The circulation bound to the blade element causes an upwash upstream and a downwash downstream from the ACL, both of similar magnitude. By averaging the flow values over an angular range up- and downstream from each ACL point, this effect can therefore be eliminated and the effective angle of attack can be determined. The angular ranges are defined by two angles. In order to prevent discontinuities by reducing the force applied to one cell, the forces are distributed two-dimensionally in the radial and chordwise directions of the blade. In the TAU ACL three different probability density functions

(PDF) were implemented for the distribution, see [4]. Besides the usual isotropic Gauss distribution, an anisotropic Gauss distribution following Churchfield et al. [8] is available, where the forces are distributed in chordwise direction as a function of the local blade chord length. In addition, an anisotropic Gumbel distribution can be used, where the force distribution is more similar to that of a typical airfoil. Compared to the ACD, the unsteady ACL simulation increases the computation time, but compared to the fully resolved computation, significantly fewer cells are required for the propeller meshing, which reduces the computational effort.

2.2 Single blade simulation

For BEM calculation with the ACD and ACL models the airfoil characteristics must be available as input data in the form of lift and drag polars. To obtain these, different methods are available [9]. Here, 3D CFD simulations are carried out with a single propeller blade via a quarter propeller model, in order to reduce the number of cells. Periodic BC enable a steady-state rotational in- and outflow through the computational domain. The tangential velocity is generated by the rotation of the grid and the axial velocity by the farfield BC. To determine a propeller polar the inflow conditions have to be varied, either by variation of the propeller rotation Ω at constant inflow velocity v_∞ or by variation of the inflow velocity at constant rotation. After the simulations, the coefficients c_l and c_d have to be determined at different sections of the propeller blade for each inflow case. They are obtained from the effective angle of attack and by integration of the difference in pressure coefficients c_p between the upper and the lower side of the respective blade section and the skin friction coefficients c_f. The local induced angle of attack α_i can be extracted from the CFD data using the reduced axial velocity technique (RAV), see Johansen et al. [10]. For this purpose, two planes are used, shaped as a circular ring sector in front of and behind the blade section, on which the flow data are averaged, see Fig. 4b. The effective angle of attack is than estimated by linear interpolation between the planes.

2.3 Grids and numerical setup

The hybrid grids used in this study are shown in figure 2. Structured cells are utilized in the cylindrical slipstream area up to the end the nacelle with a radius of 1.1 R. The boundary layer around the spinner, the nacelle and the wing is resolved with structured cells and $y^+ < 1$. The structured domain is embedded in a refined unstructured domain with a radius of 10 R and a length of 30 R. The latter domain is itself embedded in a coarser farfield area with a distance of up to 50 R. By using a farfield BC all influences by wind tunnel walls are neglected. In order to take advantage of the symmetry of the installed propeller configuration a symmetry plane BC is applied here. For the single blade simulation, a one-quarter section of the propeller grid is used with rotationally periodic BC on the sides.

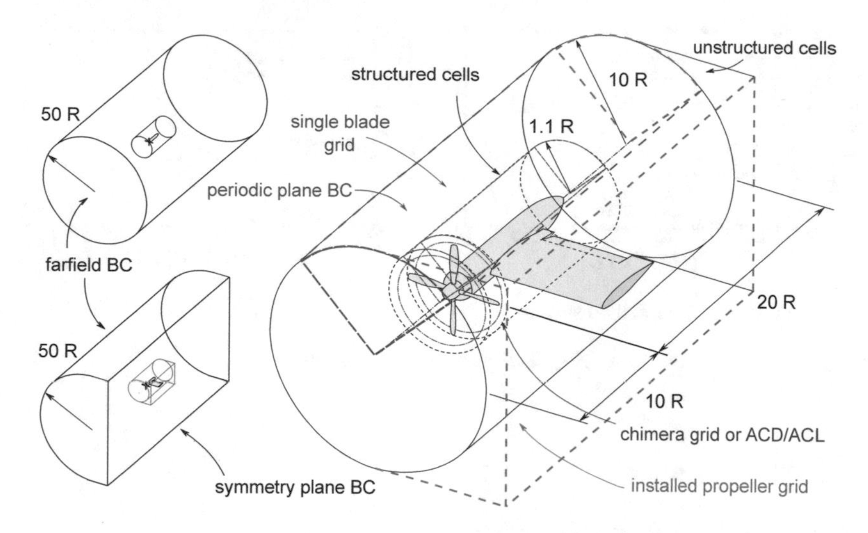

Fig. 2: Computational Grids.

The research was preceded by two grid convergence studies: one for discretization of the single propeller blade and one for discretization of the propeller with the ACD/ACL methods. The results are shown in Fig. 3. The number of cells in chordwise and radial direction as well as the number of cells inside the boundary layer were varied for the single propeller blade. The selected grid has a deviation from the Richardson-extrapolated value of $\Delta c_t = 0.26\%$ and $\Delta c_p = 0.19\%$. The airfoil is discretized by 116 points, the blade by 62 points in radial direction, excluding the hub, and the boundary layer by 24 structured cells and $y^+ < 1$. The same propeller blade grid was also used for the fully resolved simulation, copied to obtain four blades. For the ACD/ACL study, the discretization of the propeller disk BC was varied in radial n_r and tangential n_t direction, the selected grid has a deviation of $\Delta c_{t,p} < 0,5\%$. The single blade grid has about $\approx$ 1.7 millions cells. With ACD/ACL, the isolated propeller grid has $\approx$ 8.7 million cells and the installed propeller grid $\approx$ 12 million. With the fully resolved propeller $\approx$ 5 million more cells are necessary. For all calculations the second order central scheme was used for spatial discretization, the implicit Backward Euler for time discretization and the one-equation Spalart–Allmaras model to generate the turbulence. No laminar-turbulent transition was considered. The calculation time of the time-resolved simulations covered at least five propeller revolutions. A physical timestep corresponding to 4° rotation of the propeller was chosen. For the last revolution, over which the results were later averaged, the physical timestep was reduced to a value corresponding to 1° rotation.

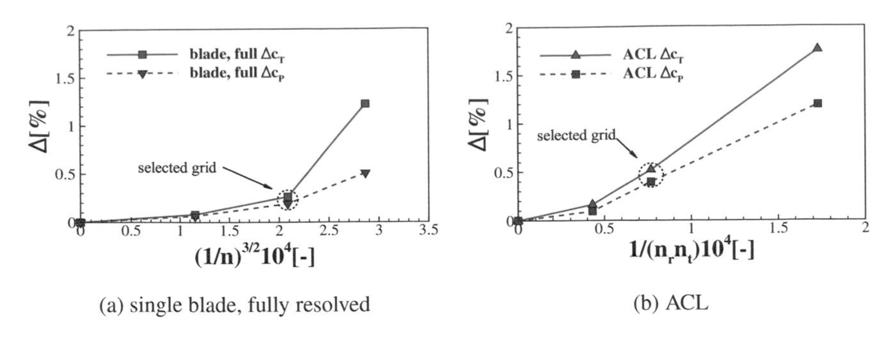

(a) single blade, fully resolved (b) ACL

Fig. 3: Spatial discretization.

3 Results

3.1 Airfoil characteristics

For the determination of the propeller blade polars, seven advance ratios were simulated with the single blade. Two methods were thereby applied: first, the rotational speed was varied at a constant inflow velocity, and second, the inflow velocity was varied at a fixed rotational speed, both producing almost the same thrust and power coefficients. For the following polar calculation the varying velocities are used, because it was assumed that this corresponds more to the velocity change due to the wing influence. Simulation data from Stokkermans [6] is shown as a comparison, which slightly differs with increasing distance from the advance ratio $J = 0.8$, which was used as the (Reynolds number) design point of the blade grid. However, the general characteristic is matched.

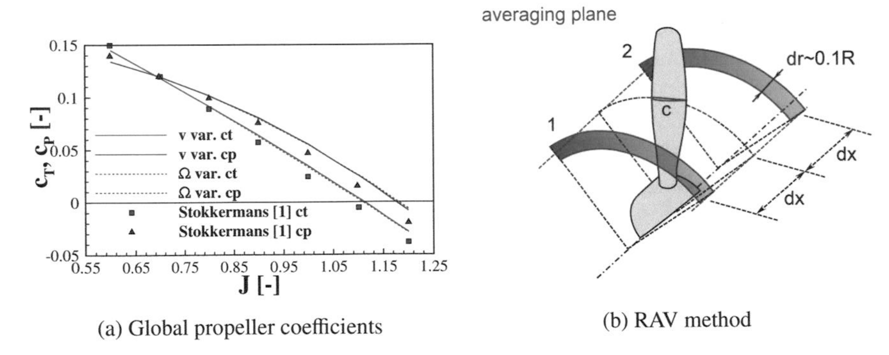

(a) Global propeller coefficients (b) RAV method

Fig. 4: Determination of the propeller airfoil characteristics. Data from Stokkermans [6].

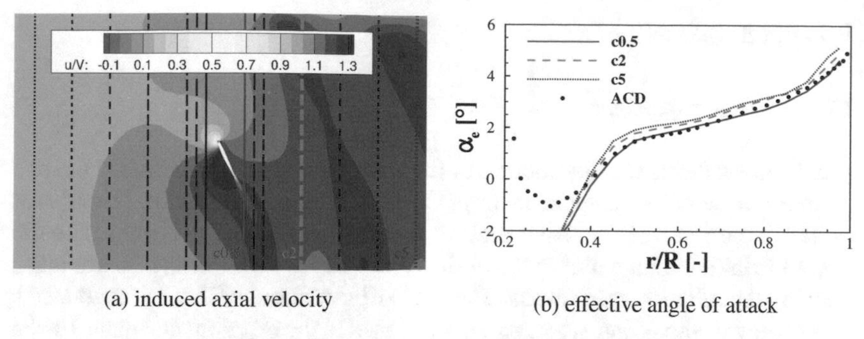

(a) induced axial velocity (b) effective angle of attack

Fig. 5: Propeller blade angle of attack.

Figure 5a shows the axial velocity induced by the propeller blade and different positions of the RAV averaging planes ($0.5c, 0.75c, 1c, 2c, 3c, 4c, 5c$). As can be seen in figure Figure 5b, the distance between the plane and the propeller disk has an influence on the angle of attack determined with the RAV, which varies by approx. $\alpha = 0.5°$. For comparison, the effective angle of attack determined with the ACD is also shown. Figure 6 shows the forces along the radius obtained with polars based on the different RAV distances as well as the single blade results for $J = 0.8$. The thrust and tangential force distributions present the influence of the polars that are used as input. For the polar with RAV distance $5c$, with the largest angle of attack difference between ACD and RAV, the forces are recognizably underestimated. The ACD forces with RAV distance $0.5c$ capture the blade data most accurately, although an underestimation remains visible in the medium blade area. The results demonstrate that it is crucial for the method of extracting the polars to use an effective angle of attack as closely as possible to that of the propeller model.

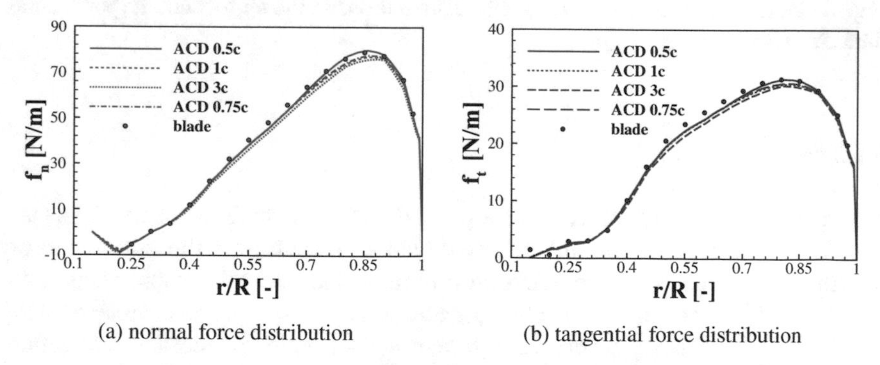

(a) normal force distribution (b) tangential force distribution

Fig. 6: Radial force distributions at the propeller disk.

3.2 Isolated propeller

3.2.1 Propeller slipstream

Figure 7 shows the radial distribution of the total pressure coefficient and the axial and tangential velocity components from the ACD, ACL, fully resolved simulations as well as experimental data from Stokkermans [1] at different positions. The time averaged axial and tangential velocity distributions for ACD and ACL are a little lower than the fully resolved results. The blade tips vortex at $r/R = 1$ is predicted by both unsteady methods, but is less pronounced than in the experimental data. The less pronounced blade tip vortices, caused by numerical diffusion, were also described by Stokkermans [1]. In the root region, the tangential velocity is slightly underestimated by the polar data, which is also reflected in lower tangential forces. However, in general, the slipstream values are captured well by the BEM-based models.

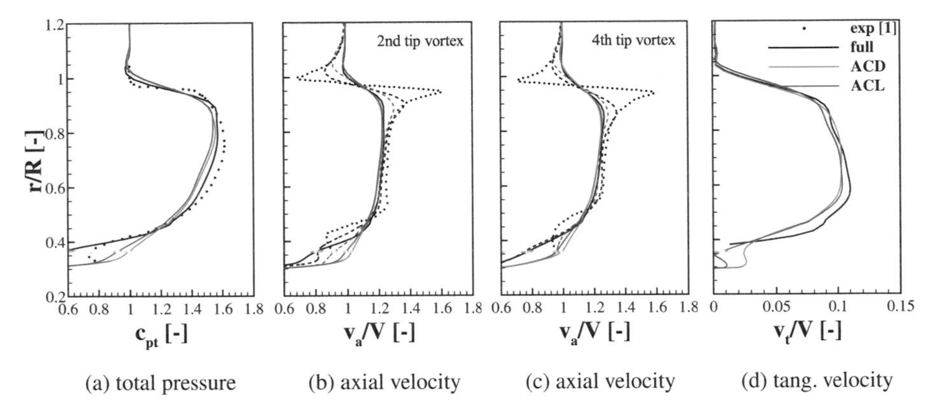

Fig. 7: Radial total pressure and velocity distributions in the slipstream. Experimental data from Stokkermans [1].

3.2.2 Propeller blade flow

Figure 8 shows the ACL forces on the propeller disk BC with the three PDFs (see Sec. 2.1.3) compared to the fully resolved blades. In contrast to the isotropic force distribution, which is concentrated on a narrow line, the anisotropic Gauss and Gumbel distributions follow the blade geometry. The closest match is obtained with the Gumbel distribution, where the asymmetric distribution in chordwise direction is visible. Figure 9 shows the pressure distribution around a propeller blade in an orthogonal section at $r/R = 0.75$ for all distributions as well as the fully resolved simulation. For better comparability, the blade section is also plotted in the ACL results. Due to the ACL force insertion on the propeller disk BC, the pressure distribution around the blade is in a way projected onto the disk. The too dense concentration of the isotropic PDF is also manifested in the pressure distribution.

The asymmetric pressure distribution in the chordwise direction is only reflected by the Gumbel distribution. However, all ACL approaches show a shifted pressure distribution. An improvement could probably be achieved by a distribution around the pressure point instead of the geometrical center of the blade.

(a) Iso. Gauss (b) Aniso. Gauss (c) Gumbel (d) full

Fig. 8: Force distribution of ACL and fully resolved propeller.

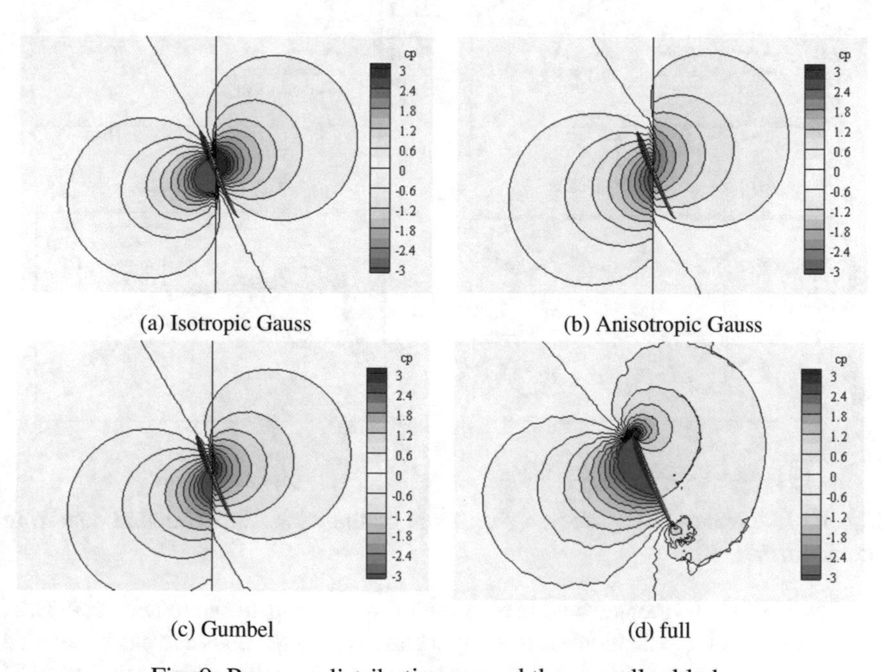

(a) Isotropic Gauss (b) Anisotropic Gauss

(c) Gumbel (d) full

Fig. 9: Pressure distribution around the propeller blade.

3.3 Installed propeller

3.3.1 Propeller wing interaction

Figure 10a shows the influence of the propeller slipstream on the averaged pressure distribution of the wing near the outboard flap edge. The fully resolved simulation captures the experimental data [1] accurately. On the upper side, ACD and ACL predict a slightly lower pressure. Figure 10b shows the pressure distribution in the propeller blade tip area for the ACL and the fully resolved methods (single timestep, time-accurate) as well as for comparison reason the distribution for the ACD method (steady-state). Compared to the ACD distribution, the time-accurate distributions present pressure fluctuations caused by the blade tip vortices. The position and the amplitude of these fluctuations are similar between the ACL and the fully resolved solutions.

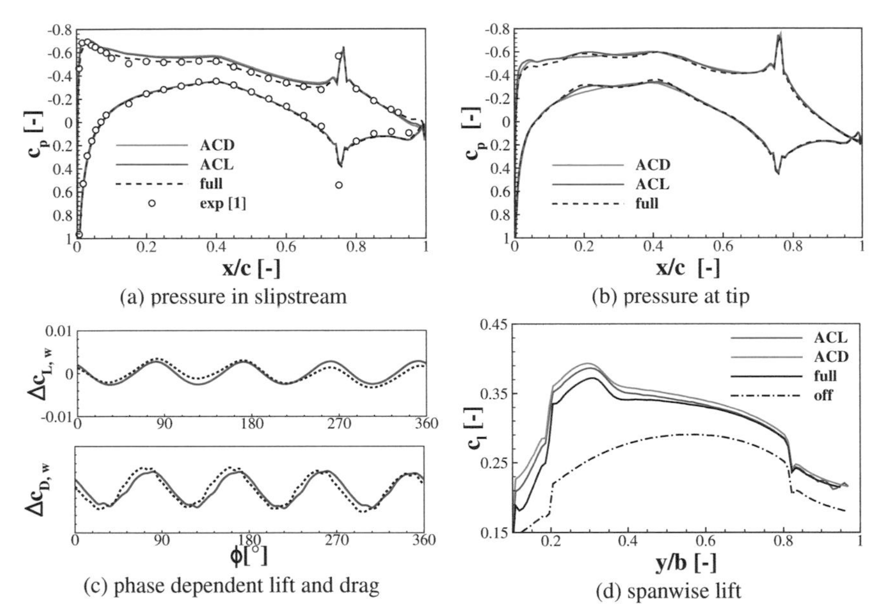

Fig. 10: Influence of the propeller slipstream on the wing. Experimental data from Stokkermans [1].

Figure 10d shows the spanwise lift distribution of the propeller simulation compared to the isolated wing. The increase in lift is captured by all methods and can be divided into regions inside and outside the area covered by the propeller slipstream. The ACD generally overestimates the lift increase, which is also indicated in the simulations by Stokkermans [1] using prescribed forces. The ACL overestimates the lift in the region covered by the propeller slipstream, but the distribution outside of it agrees well with the fully resolved one. The effect of the slipstream on the wing is in principle captured correctly by the BEM-based models. However, compared to results with prescribed forces [1] a slightly larger deviation occurs.

3.3.2 Wing propeller interaction

The advantage of BEM-based ACD and ACL simulations over prescribed forces is that, in addition to the influence of the propeller on the wing, the change of the propeller forces due to the wing's presence are also captured, which is demonstrated in Figure 11. Figure 11a shows the variation of the blade lift coefficient over one revolution at a blade position of $r/R = 0.75$ for all three methods.

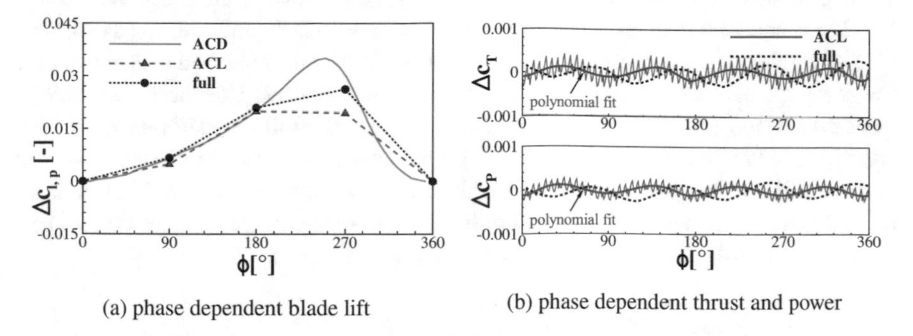

(a) phase dependent blade lift

(b) phase dependent thrust and power

Fig. 11: Influence of the wing slipstream on the propeller.

The influence of the wing leads to non-constant blade loads. When passing through the reduced velocity field in front of the wing, an increase in lift occurs on the propeller blade. With the ACD, for which a continuous evaluation is shown, a maximum appears at about 250°. The ACL and fully resolved methods are evaluated only at discrete positions of the blades. In principle, the lift variations correspond well across all methods. However, the ACL shows a shift of the maximum to an earlier propeller position in comparison. While the ACD captures the position dependence of the propeller forces, they are introduced into the flow field in a purely steady-state manner. With the ACL and fully resolved methods the individual blade forces vary accurately in time. Figure 11b shows the unsteady propeller thrust and power variation for both methods. The amplitudes match for both methods, whereby the periodicity of the ACL is superimposed by higher-frequency fluctuations. The reason for this is assumed to be that the distribution of the discrete ACL line loads on the neighboring cells varies slightly depending on the position of the line relative to the grid. The small shift in the phase dependence by less than an eighth of a rotation can again be seen. This shift can be attributed to the ACL-internal calculation of the local effective angle of attack (see Sec. 2.1.3). The phase shift depends on the distance of the averaging ranges and the blade and decreases with a smaller distance. The strongest influence of the wing is thus not captured when the blade passes in front of the wing, but instead when the upstream averaging range passes in front of it. In the results shown, the averaging range is from 25 – 40°. At smaller distances, the influence of the blade dominates and the effective angles of attack are not calculated correctly. A reduction of the phase shift could be achieved with a modified effective angle of attack determination method, which is planned for the future: instead of averaging over constant lines, the ranges could be coupled to the chord length for each ACL point, whereby a larger distance at the root would then be possible along with a smaller distance at the tip.

4 Scaling test

To investigate the efficiency of the TAU code, a preliminary scaling test on the HPC Hawk was conducted using the isolated propeller case with the fully resolved propeller rotated by the chimera-technique. This propeller simulation method was chosen for the scaling test, because it is considered the most complex. Unsteady simulations were conducted for various number of domains with 3000 inner iterations, which corresponds to a one third rotation of the propeller with a physical timestep of 4° and 100 inner iterations per time step. The grid was partitioned with the TAU internal private partitioner with a varying numbers of subgrids, ranging from 64 to 1280 domains. The results of the scaling test are shown in Figure 12.

nodes	cores	points/domain
1	64	100961
1	128	50481
2	256	25240
3	384	16827
4	512	12620
5	640	10096
6	768	8413
7	896	7212
8	1024	6310
9	1152	5609
10	1280	5048

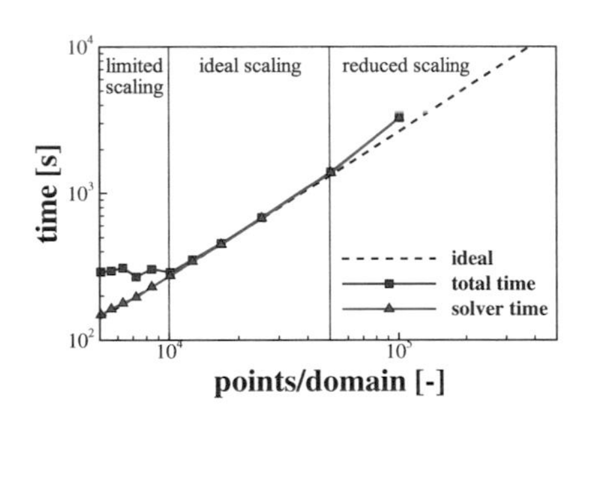

Fig. 12: Test cases and scaling behavior of the DLR TAU-code on Hawk.

The monitored values were the pure solver time without initialization and output routines and the total computation time. The TAU-code, using the private partitioner and the chimera-technique, shows a reasonably efficient scaling behavior on the HPC Hawk with decreasing ratio of points/domain. A performance optimum exists in the range of 10,000 to 50,000 grid points per domain, where an almost ideal scaling behavior is reached. For a lower ratio the total computation time seems to be limited by the input/output time even though the pure solver time continues to decrease. A further increase of the number of cores was therefore rejected, since the total computation time would not decrease any further. The simulations of the study with the different grids (described in Sec. 2.3) were performed accordingly with a points/domain ratio in the ideal scaling range.

5 Conclusion

In the study, simulations using Blade Element method (BEM) based Actuator Disk (ACD) and Actuator Line (ACL) propeller modeling approaches are compared with fully resolved simulations and experimental data known from literature, for an isolated and an installed propeller. The characteristics of the propeller blades as input data for the BEM are determined by steady-state 3D simulations of a single blade with different inflow velocities. The effective angles of attack are determined with the Reduced Axial Velocity method (RAV). It is shown that the effective angle of attack distribution determined by the RAV method varies with the distance of the averaging planes from the propeller disk. Therefore, the effective angles of attack distribution should be chosen as close as possible to that of the method used for propeller simulation (ACD, ACL). With appropriate polars, force, pressure and velocity distributions obtained from ACD and ACL calculations agree well with fully resolved and experimental data. Using the ACL method, the force and pressure distribution of a line with a Gumbel distribution agree better with the fully resolved blade than with an isotropic or anisotropic Gauss distribution. The slipstream calculated with BEM-based ACD and ACL achieves a similar effect on the wing as the fully resolved one. Compared with predefined forces, a slightly larger deviation must be accepted, however. Due to the BEM the phase-dependent influence of the wing on the propeller flow is also captured with ACD and ACL. The study demonstrates that BEM-based propeller models are suitable for design studies, to consider the interactions between propeller and wing in both directions.

Acknowledgements This study was carried out within the LuFo V 3 project ELFLEAN, which is funded by the Federal Ministry for Economic Affairs and Energy. The authors gratefully thank Leo Veldhuis from TU-Delft for providing the propeller geometry for the single blade and fully resolved simulations to validate the BEM-based models.

Supported by:

on the basis of a decision
by the German Bundestag

References

1. Stokkermans, T.C., v. Arnhem, N., Sinnige, T., Veldhuis, L.L.: Validation and Comparison of RANS Propeller Modeling Methods for Tip-Mounted Applications. AIAA SciTech Forum, (2018). doi:10.2514/6.2018-0542
2. Schwamborn, D. and Gerhold, T. and Heinrich, R.: The DLR TAU-code: Recent applications in research and industry, ECCOMAS CFD 2006 CONFERENCE,(2006), https://elib.dlr.de/22421/
3. Raichle A.: Flusskonservative Diskretisierung des Wirkscheibenmodells als Unstetigkeitsfläche. PhD thesis, DLR, (2017).
4. Schollenberger M. , Lutz T. and Krämer E.: Boundary Condition Based Actuator Line Model to Simulate the Aerodynamic Interactions at Wingtip Mounted Propellers. New Results in Numerical and Experimental Fluid Mechanics XII, (2020)
5. Sinnige, T., de Vries, R., Della Corte, B., Avallone, F., Ragni, D., Eitelberg, G., Veldhuis, L.L: Unsteady Pylon Loading Caused by Propeller-Slipstream Impingement for Tip-Mounted Propellers, Journal of Aircraft 55:4, (2018)
6. Stokkermans, T.C.: Aerodynamics of Propellers in Interaction Dominated Flowfields: An Application to Novel Aerospace Vehicles, (2020). https://doi.org/10.4233/uuid:46178824-bb80-4247-83f1-dc8a9ca7d8e3
7. Stuermer, A.: Unsteady CFD Simulations of Propeller Installation Effects. 42nd AIAA/ASME/SAE/ASEE Joint Propulsion Conference & Exhibit, (2006). doi:10.2514/6.2006-4969
8. Churchfield, M.J., Schreck, S.J., Martinez, L.A., Meneveau, C., Spalart, P.R.: An Advanced Actuator Line Method for Wind Energy Applications and Beyond. 35th Wind Energy Symposium, AIAA SciTech Forum, (2017). doi:10.2514/6.2017-1998
9. Schollenberger M. and Lutz T.: Comparison of Different Methods for the Extraction of Airfoil Characteristics of Propeller Blades as Input for Propeller Models in CFD. New Results in Numerical and Experimental Fluid Mechanics XIII, (2021).
10. Johansen J. and Sørensen N.N.: Aerofoil Characteristics from 3D CFD Rotor Computations, Wind Energy, 7,4, (2004). https://doi.org/10.1002/we.127

Transport and Climate

The simulations in the category "Transport and Climate" have consumed a total of approximately 50 million core-hours over the past granting period. Four projects have utilized the system Hawk (HLRS), and nine were hosted on the now retired system ForHLR II (SCC). The majority (67%) of computational resources in this category was spent on ForHLR II.

In the project "GLOMIR+" by Schneider et al. the platform ForHLR II was leveraged for the analysis of spectra measured via satellite-bound infrared atmospheric sounding interferometry. The particular focus of the post-processing campaign is to identify atmospheric trace gases such as H2O, HDO, CH4, N2O and HNO3. The results of these retrievals constitute highly sought-after data for a variety of other studies in earth system science.

The authors of the second project "MIPAS" (Kiefer et al.) have performed similar analysis of Michelson interferometer data taken aboard the satellite Envisat over one decade. Here the distributions of more than 30 species of trace gases are analyzed, constituting one of the largest available databases for the composition of the middle atmosphere. The post-processing operations carried out successfully on Hawk are not only computationally intensive, but they also feature an exceptional load on I/O.

The third project by Bauer et al. ("WRFSCALE") uses the large-eddy technique for the description of atmospheric turbulence in the context of the Weather and Research Forecasting (WRF) model. The project is aimed at understanding the interaction between land-surface processes and the dynamics of the atmosphere, which is studied for a mid-western US region covered by the Land Atmosphere Feedback Experiment (LAFE). Their setup exhibits good parallel scaling on Hawk making use of the full number of available compute cores. Overall, the project "WRFSCALE" convincingly showcases how the nested large-eddy approach down to a grid resolution of 20 meters can contribute to further our understanding of atmospheric processes.

Markus Uhlmann

Institute for Hydromechanics,
Karlsruhe Institute of Technology,
Kaiserstr. 12,
76131 Karlsruhe,
Germany,
e-mail: markus.uhlmann@kit.edu

The HPC project GLOMIR+ (GLObal MUSICA IASI Retrievals - plus)

Matthias Schneider, Benjamin Ertl, Christopher Diekmann and Farahnaz Khosrawi

1 Introduction

The HPC project GLOMIR+ (GLObal MUSICA IASI Retrievals - plus) has the objective to generate a global, multi-year dataset for various atmospheric trace gases based on thermal nadir spectra measured by the sensor IASI (Infrared Atmospheric Sounding Interferometer) aboard the EUMETSAT's polar orbiting satellites Metop-A, -B, and -C. The Metop/IASI mission combines high spectral resolution and low measurement noise with high horizontal and temporal coverage (12 km ground pixel, global, twice daily). Figure 1 shows the measurement geometry of IASI. The polar orbit in combination with a wide across track scanning angle ensures a global coverage twice per day. The IASI measurements started in 2007 and in the context of an already approved successor mission (launch of IASI-NG on Metop-SG satellites is currently scheduled for early 2024) observations are guaranteed until the late 2030s. This unprecedented climate and weather research potential is due to the strong support by the European meteorological organizations. However, it comes along with the challenge of the large amount of spectra that have to be processed. The currently three orbiting IASI instruments provide about 3.8 million measured spectra each 24 h. For their analysis, the application of high performance computing is indispensable. For the global retrievals the MUSICA PROFFIT-nadir retrieval code is used. It has been developed recently within the ERC (European Research Council) project MUSICA (MUlti-platform remote Sensing of Isotopologues for investigating the Cycle of

Matthias Schneider, Benjamin Ertl, Christopher Diekmann and Farahnaz Khosrawi
Institute of Meteorology and Climate Research – Atmospheric Trace Gases and Remote Sensing (IMK-ASF), Karlsruhe Institute of Technology (KIT), Hermann-von-Helmholtz Platz 1, Eggenstein-Leopoldshafen, 76344, Germany,
e-mail: matthias.schneider@kit.edu, benjamin.ertl@kit.edu, christopher.diekmann@kit.edu, farahnaz.khosrawi@kit.edu

Benjamin Ertl
Steinbuch Centre for Computing (SCC), Karlsruhe Institute of Technology (KIT), Hermann-von-Helmholtz Platz 1, Eggenstein-Leopoldshafen, 76344, Germany, e-mail: benjamin.ertl@kit.edu

W. E. Nagel et al. (eds.), *High Performance Computing in Science and Engineering '21*,
https://doi.org/10.1007/978-3-031-17937-2_23

Atmospheric water, [13]. The target of the MUSICA processing is the atmospheric composition of H_2O, HDO/H_2O ratio, CH_4, N_2O and HNO_3 (the HDO/H_2O ratio retrieval is truly innovative). The processing has been continuously improved during MUSICA follow-up projects and the current MUSICA IASI data product and the corresponding processing chain is presented in [15] and shown as a flowchart in Figure 2. The processing relies on six different components:

1. Preprocessing stage: the merging of the EUMETSAT IASI Level 1c data (L1c, the calibrated IASI spectra) and Level 2 products (L2, the first guess profiles of atmospheric humidity and temperature).
2. PROFFIT-nadir retrieval: this algorithm inverts the spectral measurements and generates the atmospheric composition profiles [11–13].
3. Output generation: tool for generating storage efficient MUSICA IASI netcdf output files [23] in agreement to the CF (Climate and Forecast, `http://cfconventions.org/`) metadata conventions. The high quality of the products is demonstrated by several validation studies, e.g. [1, 10, 24].
4. A posteriori data reusage: algorithms ensuring tailored data reusage, e.g. for the generation of {H2O,δD} pairs [3] or for the synergetic use of MUSICA IASI and TROPOMI methane products [16].
5. Retrieval simulator: the retrieval simulator is needed for high quality observation-model comparison studies, e.g. [14].
6. Re-gridded L3 products: the level 3 data generation tool is still not available, but highly asked for by the scientific community. It will enable the generation and distributing of the data products on a regular horizontal and time step grid tailored to the needs of the individual data user.

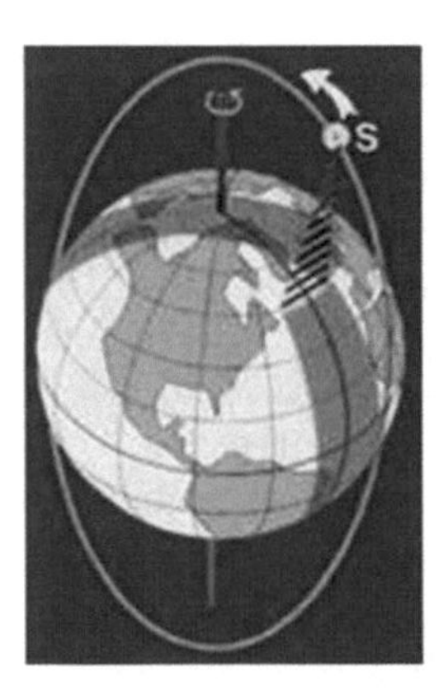

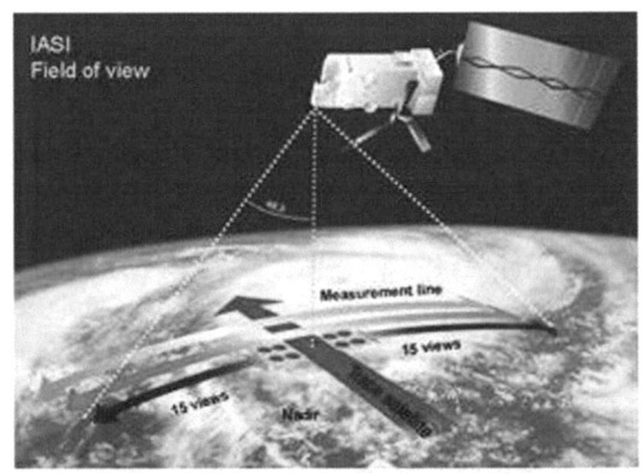

Fig. 1: Depiction of a polar orbiting satellite (left, source: `http://tornado.sfsu.edu/`) and indication of the IASI observation geometry (right, source: `https://www.eumetsat.int`).

In Sec. 2 we give details on the executed computations, the parallelism and scaling of our computations, and the used HPC resources. The second part of this paper presents the scientific results we achieved based on the HPC calculations (Sec. 3).

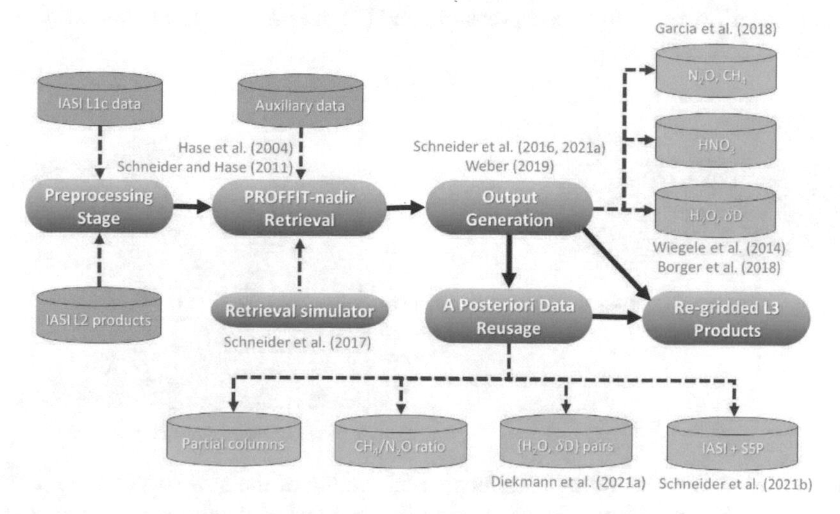

Fig. 2: Flow chart with the different components (blue symbols) of the MUSICA IASI processing chain.

2 Data processing

2.1 Computations, parallelism, and scaling

Using a single processing unit the retrieval of one observation takes about 30 seconds. We make retrievals for all spectra measured under cloud-free conditions, which are typically 25000 observations per orbit. Currently three different IASI instruments are in operation and provide spectra for 42 orbits per 24 h. This results in about 600000 observations per day meaning a processing time of 1.8 million seconds (208 days) when using a single processing unit. High performance computing is thus indispensable for processing the satellite data. In the interest of optimum utilization of the HPC resources, our calculations are distributed across the different processors, so that they can operate in parallel (data parallelism). For this purpose, we use the Python Multiprocessing package and the Python Luigi package. The Multiprocessing package supports spawning processes using an API similar to the threading module. It allows the full leverage of processors on a machine (`https://docs.python.org/3.6/library/multiprocessing.html`). The Python Luigi package (2.7, 3.6, 3.7 tested) helps to build complex pipelines of batch jobs. It handles dependency resolution, workflow management, visualization, handling failures, command line integration, and much more (`https://luigi.readthedocs.io/en/stable/index.html`). In the following, we give details on the parallelism and scaling of our three computational most intensive applications: (1) the PROFFIT-nadir retrieval algorithm, (2) the CF data generation and output compression, and (3) the processing in context of the a posteriori data reusage.

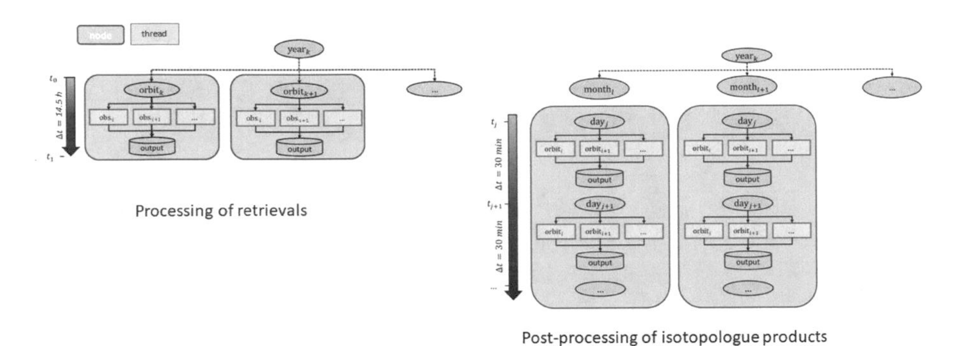

Fig. 3: Charts demonstrating parallelism and scaling of the MUSICA IASI retrieval processing (left panel) and the processing for a posteriori data reusage (right panel).

1. The main MUSICA IASI retrieval Python program uses a thread pool to distribute the input data (individual spectra observations and corresponding auxiliary data like temperature profiles) to the available number of worker threads, which can execute the Fortran retrieval code in parallel (see left chart on Fig. 3). The thread pool schedules as many worker threads as possible (40 threads on a single HoreKa computing node, i.e. 2 threads per processor core). The available computing performance is fully explored until the retrievals of all the observations of a single orbit are completed. The retrieval of one observation takes about 30 seconds. One orbit contains typically 25 observations. Because 40 observations can be processed in parallel, we can process a whole orbit in typically less than 15 h on a single node.
2. The compression and generation of CF conform netcdf MUSICA IASI data files are also scaling with the number of processors that can be used on a given compute node [23]. For this purpose, our Python compression code utilizes the Python Luigi package. The centralized Luigi scheduler coordinates the operations that can be executed in parallel. The tasks are distributed to the available workers and it is ensured that two instances of the same task are not running simultaneously. This procedure ensures that all available processors are utilized to capacity.
3. The processing in context of the a posteriori isotopologue data reusage distributes the input data on the available computing resources as depicted on the right side of Fig. 3. For each observation, the post-processing (Python program) reads in a subset of the retrieval output variables and takes a fraction of seconds for

processing. Therefore, the memory of one computing node offers the capacity for post-processing all orbit files for one day simultaneously by distributing them to the available processors. Using this parallelization method, the processing of one day takes about 40 min. Thus, one single computing node can be used to do the a posteriori processing of one month of data ($\approx$ 30 Mio. observations) in about 20 h.

2.2 Summary of used computation resources

The computationally most intensive tasks were as follows:

1. **Execution of the retrievals using the algorithm PROFFIT-nadir**: we processed IASI (Infrared Atmospheric Sounding Interferometer) spectra measured aboard the satellites Metop-A, -B, and -C between June 2019 and December 2020. The retrievals for 2019, the generation of the first half of 2020 have been performed on ForHLR II. The second half of 2020 has been processed on HoreKa.
2. **Generation of the data output files (including data compression):** we made an output data compression and generated the CF (Climate and Forecast metadata standard) conform netcdf files for the whole period of MUSICA IASI retrieval data (for more than six years between October 2014 and December 2020). For the October 2014 to June 2019 period, this processing has been made on ForHLR II. These data have been published with DOIs [15, 17, 18]. The output data generation and compression for the period between July 2019 and December 2020 has been made on HoreKa.
3. **Isotopologue data post-processing:** we made the MUSICA IASI isotopologue post-processing for the October 2014 to June 2019 period on ForHLR II. These data have been published with DOIs [3,5]. The post-processing for the July 2019 to December 2020 period has been realized on HoreKa.
4. **Forward operator and LAGRANTO backward trajectory calculations:** for the West African Monsoon period in 2016 and 2018 we have made forward operator calculation on ForHLR II. These calculations allow simulating the MUSICA IASI observations of the atmospheric state as modelled by COSMO-iso. For the same period and area we performed LAGRANTO back trajectory calculations using the fields as modeled by COSMO-iso. These calculations have also been made on ForHLR II. The COSMO-iso model data were provided by the project partner (ETH Zürich) of the MOTIV (MOisture Transport pathways and Isotopologues in water Vapour) project. In Table 1 we give an overview of the computing resources used for the different tasks and of the period when the processing has been made.

Table 1: Summary of consumed computing resources. We calculate here with 20 cores on one node and two CPUs per core (i.e. 40 CPUs per node)

Task	Period	Consumed CPUh
(1) Execution of the retrievals	03/2020–02/2021	10 millions
(2) Generation of the data output	10/2020 and 02/2021	0.75 million
(3) Isotopologue data post-processing	10/2020–12/2020	0.75 million
(4) Forward operator and trajectory	06/2020–12/2020	0.5 million
		Total: 12 million

3 Scientific results

The GLOMIR+ HPC activities contributed to different international research projects. This results in a large number of scientific publications with the written acknowledgment "Retrieval calculation for this work were performed on the supercomputer ForHLR funded by the Ministry of Science, Research and the Arts Baden-Württemberg and by the German Federal Ministry of Education and Research". In the following, we give some brief examples of these contributions (order by scientific field), and list the related scientific publications.

3.1 Data set dissemination and presentation

During the GLOMIR+ project and the related precursor HPC projects we have been able to perform MUSICA IASI retrievals for the whole period between October 2014 and December 2020. In the beginning of 2021 we published the data in the research data repository RADAR4KIT (`https://radar.kit.edu/`, [4, 17, 18] and describe the data products in detail in two publications in the Copernicus journal "Earth System Science Data" (`https://www.earth-system-science-data.net/`, [3, 15]. The full standard data product [17] contains the trace gas profiles of H_2O, δD (the standardised HDO/H_2O ratio), CH_4, N_2O and HNO_3 as well as atmospheric and surface temperatures. The full standard data files also contain auxiliary data like uncertainty covariances, information of constraint and representativeness (e.g. averaging kernels), which is essential for a comprehensive re-use of the data. The file with the extended data product [18] contains additional information about the response of the data on different error sources (e.g. unrecognized clouds, uncertainties in surface emissivity), but only considers 74 exemplary observations for a polar mid-latitudinal and tropical site. The MUSICA IASI full retrieval products (standard and extended output files) are described in detail in the Earth System Science Data publication of [15]. The

{H2O,δD} pair product [5] contains H_2O and δD data generated by a posteriori processing method. Such a posteriori processing is essential for a successful scientific usage of the data. Details on the data product and the a posteriori processing are given in the Earth System Science Data publication of [3]. Table 4 gives an overview on the related publications and Fig. 4 gives an example of the MUSICA IASI data showing a continuous time series (Oct. 2014 – Dec. 2019) of the different trace gas products over Karlsruhe (49.0°N, 8.5°E).

Table 2: Publications in the field of MUSICA IASI data set dissemination and presentation

Description of publication	Reference	DOI
Presentation of full data product	Schneider et al. (2021a)	10.5194/essd-2021-75
Presentation of {$H_2O,\delta D$} pair data product		10.5194/essd-2021-87
Dissemination of full standard output data product	Schneider et al. (2021c)	10.35097/408
Dissemination of full extended output data product	Schneider et al. (2021d)	10.35097/412
Dissemination of {$H_2O,\delta D$} pair data product	Diekmann et al. (2021c)	10.35097/415

3.2 Research of the atmospheric water cycle

The data generated during GLOMIR+ have been essential for achieving the scientific goals of the two international projects MOTIV (MOisture Transport and Isotopologues in water Vapour) and TEDDY (TEsting isotopologues as Diabatic heating proxy for atmospheric Data analYses). Table 3 gives an overview on the related publications. In [2] the MUSICA IASI {H2O,δD} pair data are used together with COSMOiso model simulations to disentangle the different moisture transport pathways to the subtropical North Atlantic middle troposphere. We have been able to demonstrate that the {H2O,δD} pair data allow identifying the different transport pathways. This gives new possibilities for investigating the related processes and its possible response to global warming. Similarly, [4] has used the measured and simulated {H2O, δD} pair data together with Lagrangian trajectory calculations for research on the West African monsoon. We have found distinct {H2O,δD} pair distributions

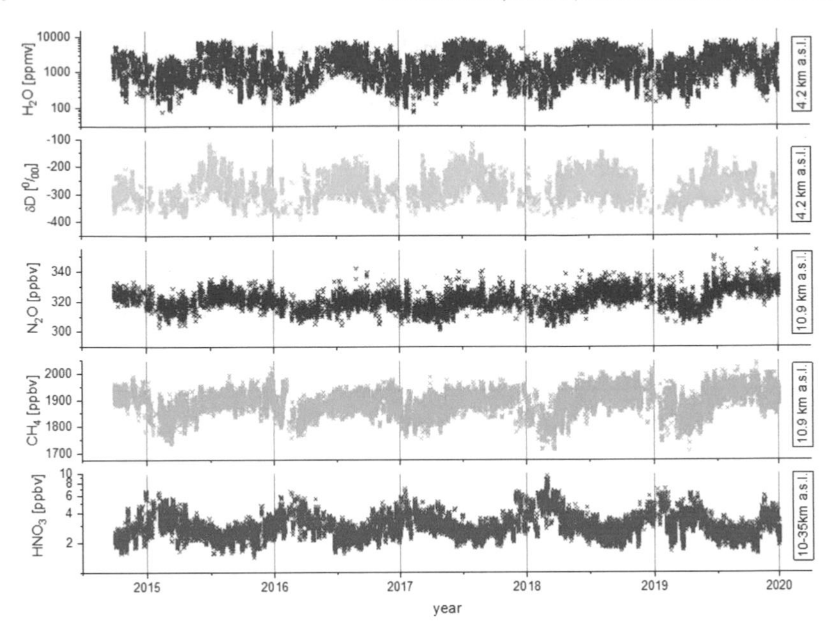

Fig. 4: October 2014 to December 2019 time series of MUSICA IASI data products retrieved over Karlsruhe (49.0°N, 8.5°E): H_2O, δD (the standardised HDO/H_2O ratio), CH_4, N_2O and HNO_3.

for air masses that experienced a convective process. These results indicate that the isotopologue data can be used for quantifying the importance of convective transport for free tropospheric humidity. In [7] we present a new method for clustering pair distributions that allows for constraining with respect to the correlation of the pairs (CoExDBSCAN: Correlation Extended DBSCAN). We apply this method for pattern recognitions with the $\{H2O,\delta D\}$ pair data. [8] shows that CoExDBSCAN can very efficiently identify temporal sequences that are characterised by a specific correlation of data pairs. In [9] we use CoExDBSCAN for clustering the temporal sequences of $\{H2O,\delta D\}$ pairs along an atmospheric Lagrangian trajectory. This allows for an automated recognition of moisture processes along a Lagrangian trajectory.

In [20] we show that assimilating the MUSICA IASI water vapour isotopologue data in addition to traditional atmospheric observations (humidity, temperature, wind, etc.) can significantly improve the knowledge about the atmospheric state, which in turn allow for improved weather predictions (see example of Fig. 5). The MUSICA IASI water vapour isotopologue data have also been of central importance for the PhD thesis of Christopher J. Diekmann [6]. He further developed the MUSICA IASI water vapour isotopologue processing and interpreted the MUSICA IASI $\{H2O, \delta D\}$ pair data with the help of COSMOiso model simulations and also by using complementary observations from other satellite sensors. Fig. 6 gives an example of the latter. There the IMERGE (Integrated Multi satellitE Retrievals for GPM, `https://gpm.nasa.gov/data/imerg`) satellite rain data are used for

Table 3: Publications in the field of atmospheric water cycle research.

Description of publication	Reference	DOI (or URL)
Investigating the North Atlantic moisture transport pathways	Dahinden et al. (2021)	10.5194/acp-2021-269
Water isotope signals and the West African Monsoon	Diekmann et al. (2021b)	10.1029/2021JD034895
Method for clustering $\{H_2O, \delta D\}$ data (CoExDBSCAN)	Ertl et al. (2020)	10.5220/00101312010540115
Time point clustering with CoExDBSCAN	Ertl et al. (2021b)	caiac.pubpub.org/pub/a3py333z
Time point clustering of $\{H_2O, \delta D\}$ pair data	Ertl et al. (2021a)	10.1007/978-3-030-77961-0_23
Assessment of isotopologue data assimilation	Toride et al. (2021)	10.1029/2020GL091698
PhD thesis	Diekmann (2021)	doi:10.5445/IR/1000134744

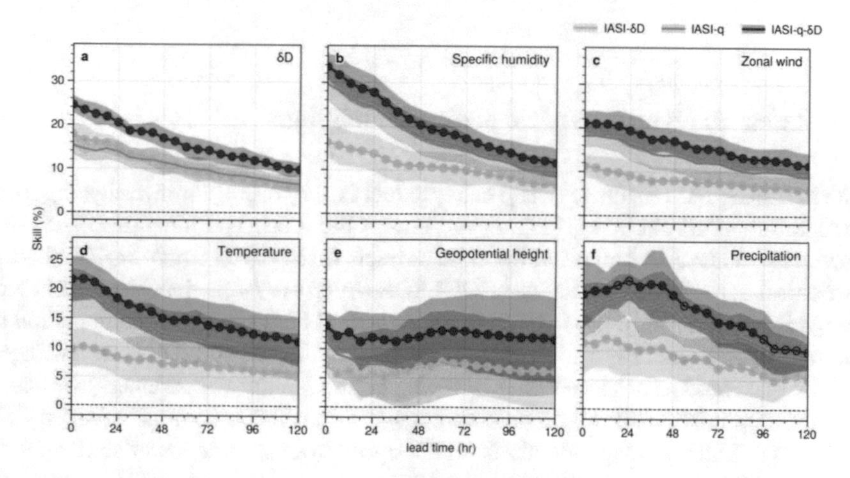

Fig. 5: Improvements of forecast skills for δD, humidity, wind, temperature, geopotential height, and precipitation (a-f) by assimilating MUSICA IASI data in addition to the traditional atmospheric observations (green: additional assimilation of H_2O; yellow: of δD; violet: of $\{H_2O, \delta D\}$ pairs). Figure has been taken from [20].

classifying the IASI observations in affected by rain and not affected by rain. For air that has been affected by rain a distinct {H2O,δD} pair distribution can be clearly observed. Currently, we also discuss a MUSICA IASI water isotopologue data usage in support of the project EUREC4A (`http://eurec4a.eu/`). EUREC4A aims at an advanced understanding of the interplay between cloud processes and the large-scale atmospheric state and circulation important for weather and climate prediction.

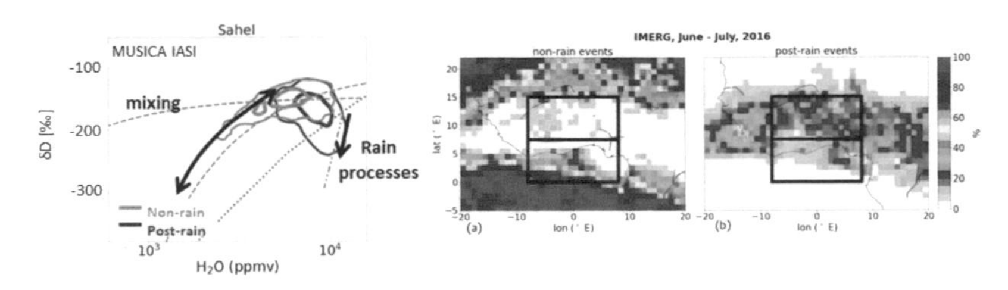

Fig. 6: Left: H2O, δD pair distributions as observed over the Sahel zone by the MUSICA IASI data for locations with a previous rain event (post-rain, red contours) and no rain during 48 h before the observation (non-rain, blue contours). Right: Location and frequency of occurrence of non-rain (a) and rain events (b). For the rain event identification the IMERG satellite data products are used. Figures have been from [6].

3.3 Research on atmospheric methane emissions

The MUSICA IASI methane (CH_4) data product is very useful for investigating local methane sources. Table 4 gives an overview of the related publications. With this product we have also supported the ESA projects FRM4GHG (`https://frm4ghg.aeronomie.be/`) and COCCON-PROCEEDS (`https://www.imk-asf.kit.edu/english/3225.php`). In [16] we present a method for a synergetic combination of the MUSICA IASI CH_4 profile data products with the CH_4 total column averaged (XCH4) data product of TROPOMI (TROPOspheric Monitoring Instrument). We show that the combination allows for detecting lower tropospheric column averaged CH_4 amounts (TXCH4) independently from the upper tropospheric/lower stratospheric amounts (this is not possible by one of the data sets alone). In [22] we use this synergetic combination method for investigating surface near CH_4 anomalies due to waste disposal site emissions nearby Madrid (see Fig. 7).

Table 4: Publications in the field of atmospheric methane emissions.

Description of publication	Reference	DOI
Synergetic combination of MUSICA IASI and TROPOMI CH_4	Schneider et al. (2021b)	10.5194/amt-2021-31
Quantification of local CH_4 emissions from ground and space	Tu et al (2021b)	10.5194/acp-2021-437

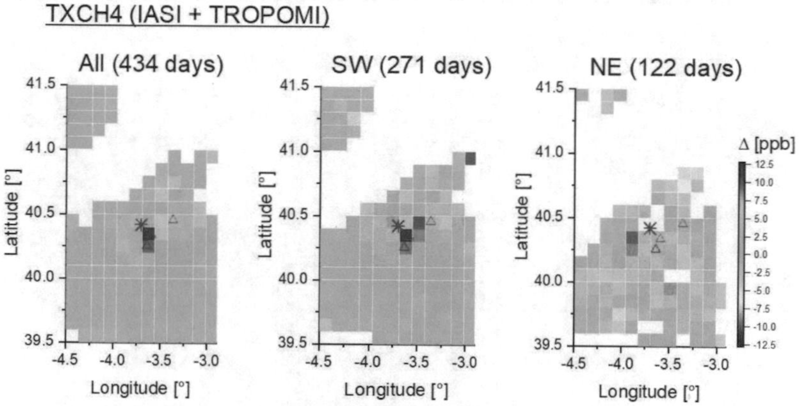

Fig. 7: Methane anomalies as observed in the tropospheric column averaged product (generated by the synergetic combination of TROPOMI and MUSICA IASI CH_4 data) ordered by predominant wind directions: all winds (left); south-west winds (centre); north-east winds (right). Similar to Fig. 8 of [22].

3.4 Validation of satellite data

Comparison of different remote sensing data products is essential for understanding the data quality. We have participated with our MUSICA IASI H_2O and δD data products in two inter-comparison studies (see Table 5). The first study resulted in the publication of [21]. There water vapour total column data (XH2O) obtained from IASI by our MUSICA processor and data generated from TROPOMI measurements have been successfully compared with different ground-based XH2O remote sensing data sets. Fig. 8 gives an example of the good agreement between the MUSICA IASI XH2O data and the XH2O reference measured within the COCCON (Collaborating Carbon Column Observation Network). Furthermore, in the context of the ESA project S5P+I H2O-Iso our MUSICA IASI δD data product has been used to estimate the precision of a newly developed TROPOMI column integrated δD data product [19].

Table 5: Publications in the field of satellite data validation

Description of publication	Reference	DOI (or URL)
Validation of total column water vapour	Tu et al. (2021a)	10.5194/amt-14-1993-2021
Validation of column integrated water vapour isotopologue ratios	Schneider et al. (2021e)	s5pinnovationh2o-iso.le.ac_uk/ wp-content/uploads/2021/02/ S5p-I-VR-Version 1.1.pdf

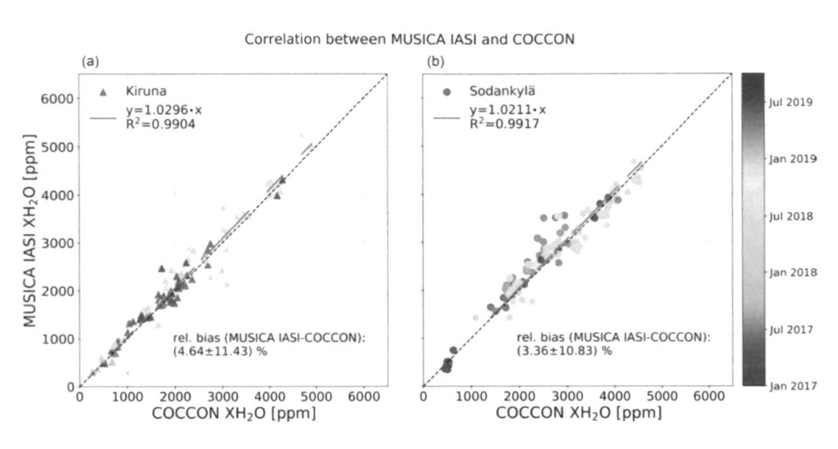

Fig. 8: Scatter plot of the COCCON XH2O compared with coincident MUSICA IASI retrievals at the Kiruna site (a) and the Sodankylä site (b). Figure taken from [21].

References

1. Borger, C., Schneider, M., Ertl, B., Hase, F., García, O. E., Sommer, M., Höpfner, M., Tjemkes, S. A., and Calbet, X.: Evaluation of MUSICA IASI tropospheric water vapour profiles using theoretical error assessments and comparisons to GRUAN Vaisala RS92 measurements, Atmos. Meas. Tech., 11, 4981–5006, doi:10.5194/amt-11-4981-2018, 2018.
2. Dahinden, F., Aemisegger, F., Wernli, H., Schneider, M., Diekmann, C. J., Ertl, B., Knippertz, P., Werner, M., and Pfahl, S.: Disentangling different moisture transport pathways over the eastern subtropical North Atlantic using multi-platform isotope observations and high-resolution numerical modelling, Atmos. Chem. Phys. Discuss., 21, 16319–16347, https://doi.org/10.5194/acp-21-16319-2021, 2021.
3. Diekmann, C. J., Schneider, M., Ertl, B., Hase, F., García, O., Khosrawi, F., Sepúlveda, E., Knippertz, P., and Braesicke, P.: The global and multi-annual MUSICA IASI {H2O, δD} pair dataset, Earth Syst. Sci. Data, 13, 5273–5292, https://doi.org/10.5194/essd-13-5273-2021,2021a.

4. Diekmann, C. J., Schneider, M., Knippertz, P., de Vries, A. J., Pfahl, S., Aemisegger, F., Dahinden, F., Ertl, B., Khosrawi, F., Wernli, H., Braesicke, P.: A Lagrangian perspective on stable water isotopes during the West African Monsoon, J. Geophys. Res.: Atmospheres, 126, e2021JD034895. https://doi.org/10.1029/2021JD034895, 2021b.
5. Diekmann, C. J., Schneider, M., Ertl, B.: MUSICA IASI water isotopologue pair product (a posteriori processing version 2), Institute of Meteorology and Climate Research, Atmospheric Trace Gases and Remote Sensing (IMK-ASF), Karlsruhe Institute of Technology (KIT). DOI: 10.35097/415, 2021c.
6. Diekmann, C. J., Analysis of stable water isotopes in tropospheric moisture during the West African Monsoon, PhD thesis, successfully completed at the Faculty of Physics, Karlsruhe Institute of Technology (KIT), doi:10.5445/IR/1000134744, 2021.
7. Ertl, B., Meyer, J., Schneider, M., and Streit, A.: CoExDBSCAN: Density-based Clustering with Constrained Expansion, Proceedings of the 12th International Joint Conference on Knowledge Discovery, Knowledge Engineering and Knowledge Management - Volume 1: KDIR, pages 104–115, INSTICC, SciTePress, ISBN 978-989-758-474-9, doi:10.5220/0010131201040115, 2020.
8. Ertl, B., Meyer, J., Schneider, M., Diekmann, C. and Streit, A.: A Semi-Supervised Approach for Trajectory Segmentation to Identify Different Moisture Processes in the Atmosphere, Computational Science – ICCS 2021, Springer International Publishing, https://doi.org/10.1007/978-3-030-77961-0_23, 2021a.
9. Ertl, B., Meyer, J., Schneider, M., and Streit, A.: Semi-Supervised Time Point Clustering for Multivariate Time Series, Advances in Artificial Intelligence, accepted for publication, https://caiac.pubpub.org/pub/a3py333z, 2021b.
10. García, O. E., Schneider, M., Ertl, B., Sepúlveda, E., Borger, C., Diekmann, C., Wiegele, A., Hase, F., Barthlott, S., Blumenstock, T., Raffalski, U., Gómez-Peláez, A., Steinbacher, M., Ries, L. and de Frutos, A. M.: The MUSICA IASI CH_4 and N_2O products and their comparison to HIPPO, GAW and NDACC FTIR references, Atmos. Meas. Tech., 11, 4171-4215, doi:10.5194/amt-11-4171-2018, 2018.
11. Hase, F., Hannigan, J.W., Coffey, M. T., Goldman, A., Höpfner, M., Jones, N. B., Rinsland, C. P., and Wood, S.: Intercomparison of retrieval codes used for the analysis of high-resolution, J. Quant. Spectrosc. Ra., 87, 25–52, 2004.
12. Schneider, M. and Hase, F.: Optimal estimation of tropospheric H_2O and δD with IASI/METOP, Atmos. Chem. Phys., 11, 11 207–11 220, https://doi.org/10.5194/acp-11-11207-2011, 2011.
13. Schneider, M., Wiegele, A., Barthlott, S., González, Y., Christner, E., Dyroff, C., García, O. E., Hase, F., Blumenstock, T., Sepúlveda, E., Mengistu Tsidu, G., Takele Kenea, S., Rodríguez, S., and Andrey, J.: Accomplishments of the MUSICA project to provide accurate, long-term, global and high-resolution observations of tropospheric $\{H2O, \delta D\}$ pairs – a review, Atmos. Meas. Tech., 9, 2845-2875, doi:10.5194/amt-9-2845-2016, 2016.
14. Schneider, M., Borger, C., Wiegele, A., Hase, F., García, O. E., Sepúlveda, E., and Werner, M.: MUSICA MetOp/IASI $H2O, \delta D$ pair retrieval simulations for validating tropospheric moisture pathways in atmospheric models, Atmos. Meas. Tech., 10, 507-525, doi:10.5194/amt-10-507-2017, 2017.
15. Schneider, M., Ertl, B., Diekmann, C. J., Khosrawi, F., Weber, A., Hase, F., Höpfner, M., García, O. E., Sepúlveda, E., and Kinnison, D.: Design and description of the MUSICA IASI full retrieval product, Earth Syst. Sci. Data Discuss. [preprint], https://doi.org/10.5194/essd-2021-75, in review, 2021a.
16. Schneider, M., Ertl, B., Diekmann, C. J., Khosrawi, F., Röhling, A. N., Hase, F., Dubravica, D., García, O. E., Sepúlveda, E., Borsdorff, T., Landgraf, J., Lorente, A., Chen, H., Kivi, R., Laemmel, T., Ramonet, M., Crevoisier, C., Pernin, J., Steinbacher, M., Meinhardt, F., Deutscher, N. M., Griffith, D. W. T., Velazco, V. A., and Pollard, D. F.: Synergetic use of IASI and TROPOMI space borne sensors for generating a tropospheric methane profile product, Atmos. Meas. Tech. Discuss. [preprint], https://doi.org/10.5194/amt-2021-31, in review, 2021b.

17. Schneider, M., Ertl, B., Diekmann, C. J.,: MUSICA IASI full retrieval product standard output (processing version 3.2.1), Institute of Meteorology and Climate Reasearch, Atmospheric Trace Gases and Remote Sensing (IMK-ASF), Karlsruhe Institute of Technology (KIT), DOI: 10.35097/408, 2021c.
18. Schneider, M., Ertl, B., Diekmann, C. J.: MUSICA IASI full retrieval product extended output (processing version 3.2.1), Institute of Meteorology and Climate Research, Atmospheric Trace Gases and Remote Sensing (IMK-ASF), Karlsruhe Institute of Technology (KIT), DOI: 10.35097/412, 2021d.
19. Schneider, M., Röhling, A. N., Khosrawi, F., Diekmann, C. J., Trent, T., Bösch, H., and Sodemann, H.: Validation Report (VR), Version 1.1, 2020-11-11, ESA project: S5p+I H2O-ISO, https://s5pinnovationh2o-iso.le.ac.uk/wp-content/uploads/2021/02/S5p-I-VR-Version1.1.pdf, 2021e.
20. Toride, K., Yoshimura, K., Tada, M., Diekmann, C., Ertl., B., Khosrawi, F., and Schneider, M.: Potential of mid-tropospheric water vapor isotopes to improve large-scale circulation and weather predictability, Geophys. Res. Lett., 48, e2020GL091 698, https://doi.org/10.1029/2020GL091698, 2021.
21. Tu, Q., Hase, F., Blumenstock, T., Schneider, M., Schneider, A., Kivi, R., Heikkinen, P., Ertl, B., Diekmann, C., Khosrawi, F., Sommer, M., Borsdorff, T., and Raffalski, U.: Intercomparison of arctic XH2O observations from three ground-based Fourier transform infrared networks and application for satellite validation, Atmos. Meas. Tech., 14, 1993–2011, https://doi.org/10.5194/amt-14-1993-2021, 2021a.
22. Tu, Q., Hase, F., Schneider, M., García, O., Blumenstock, T., Borsdorff, T., Frey, M., Khosrawi, F., Lorente, A., Alberti, C., Bustos, J. J., Butz, A., Carreño, V., Cuevas, E., Curcoll, R., Diekmann, C. J., Dubravica, D., Ertl, B., Estruch, C., León-Luis, S. F., Marrero, C., Morgui, J.-A., Ramos, R., Scharun, C., Schneider, C., Sepúlveda, E., Toledano, C., and Torres, C.: Quantification of CH_4 emissions from waste disposal sites near the city of Madrid using ground- and space-based observations of COCCON, TROPOMI and IASI, Atmos. Chem. Phys. Discuss. [preprint], https://doi.org/10.5194/acp-2021-437, in review, 2021b.
23. Weber, A.: Storage-efficient analysis of spatio-temporal data with application to climate research, Master Thesis, doi:10.5281/zenodo.3360021, https://zenodo.org/record/3360021#.Xl4pZEoo-bg, 2019.
24. Wiegele, A., Schneider, M., Hase, F., Barthlott, S., García, O. E., Sepúlveda, E., González, Y., Blumenstock, T., Raffalski, U., Gisi, M., and Kohlhepp, R.: The MUSICA MetOp/IASI H_2O and δD products: characterisation and long-term comparison to NDACC/FTIR data, Atmos. Meas. Tech., 7, 2719-2732, doi:10.5194/amt-7-2719-2014, 2014.

Global long-term MIPAS data processing

Michael Kiefer, Bernd Funke, Maya García-Comas, Udo Grabowski, Andrea Linden, Axel Murk and Gerald E. Nedoluha

Abstract The aim of this project is to perform a level 2 (L2) processing of the version 8 global infrared spectra data set (V8 L1b data) of the Earth's atmosphere, measured in limb-viewing geometry by the space-borne instrument MIPAS (Michelson Interferometer for Passive Atmospheric Sounding) operated by the European Space Agency (ESA) from 2002–2012. MIPAS was a Fourier transform mid-infrared limb scanning high resolution spectrometer which allowed for simultaneous measurements of more than 30 atmospheric trace species related to atmospheric chemistry and global change. At the Institute for Meteorology and Climate Research (IMK) MIPAS spectra are used for retrieval of vertically resolved profiles of abundances of trace species of the atmosphere. The trace gas distributions are used for the assessment of e.g. stratospheric ozone chemistry, stratospheric cloud physics and heterogeneous chemistry, stratospheric exchange processed with troposphere and mesosphere, intercontinental transport of pollutants in the upper troposphere, effects of solar proton events on stratospheric chemistry, mesospheric dynamics, atmospheric coupling, thermospheric temperature, and validation of climate-chemistry models. In the reporting period 2020/2021 MIPAS data processing was performed on the XC40 (Hazel Hen) and on the HPE Apollo (Hawk) supercomputers. The latter machine was mainly used to perform computationally expensive non-local thermodynamic equilibrium (NLTE) calculations. In the test phase/configuration of the HPE Apollo computer small obstacles had to be overcome, however, our approach which has

M. Kiefer, U. Grabowski and A. Linden
Institut für Meteorologie und Klimaforschung, Karlsruhe Institute of Technology, Karlsruhe, Germany, e-mail: michael.kiefer@kit.edu

M. García-Comas and B. Funke
Instituto de Astrofísica de Andalucía, Granada, Spain

A. Murk
Institute of Applied Physics, University of Bern, Bern, Switzerland

G. Nedoluha
Remote Sensing Division, Naval Research Laboratory, Washington DC, USA

W. E. Nagel et al. (eds.), *High Performance Computing in Science and Engineering '21*,
https://doi.org/10.1007/978-3-031-17937-2_24

been proven to successfully and efficiently work on the XC 40, also initially worked for the HPE Apollo and showed good performance. A configuration change with respect to the /tmp-filesystem of the compute nodes, which was heavily used by our processing software for efficiency reasons, unfortunately altered this state. We were forced to use the temporary workspace filesystem instead, which implies a performance degradation. Most of the processing work presented here was done in close collaboration between the Instituto de Astrofísica de Andalucía (IAA) in Granada, Spain, and KIT-IMK. The focus of the current work is the processing of atmospheric trace gases which require computationally expensive NLTE calculations, i.e., H_2O, CO, NO, NO_2. The progress in the quality of the V8 water vapour data product is demonstrated. First presentations of V8 L2 data of CO, NO, and NO_2 are shown. The vertical profiles of version V8 data processed at IMK are available to external data users on `http://www.imk-asf.kit.edu/english/308.php`.

1 The MIPAS/Envisat mission

The Michelson Interferometer for Passive Atmospheric Sounding (MIPAS) was part of the core-payload of the sun-synchronous polar orbiting satellite Envisat, operated by the European Space Agency (ESA). Envisat orbits the Earth 14.4 times per day. MIPAS measured in the mid-infrared spectral region 4.15–14.6 μm with a design spectral resolution of 0.035 cm^{-1}. It measured thermal emission interferograms of the Earth's limb, centered around a latitude/longitude point (called limb scan or geolocation), whereby variation of the limb tangent altitude provided altitude-resolved information [1, 2]. Mid-infrared limb observations are essentially independent of the illumination status of the atmosphere. Hence, e.g., the atmosphere at polar winter conditions can be observed without problems.

The MIPAS mission improved, and still is improving, the understanding of the composition and dynamics of the Earth's atmosphere by measurement of 4D distributions of more than 30 trace species relevant to atmospheric chemistry and climate change. Operation of satellite and instrument and level-1b (L1b) data processing have been performed by ESA.

MIPAS was operational in its original, full spectral resolution (FR) specification from June 2002 to March 2004. Due to an instrument problem the measurements could not be resumed before the beginning of 2005. Since then a reduced spectral resolution (RR) of 40% of the FR value was used, but at the same time a finer vertical scan grid was implemented.

Just weeks after celebrating its tenth year in orbit, communication with the Envisat satellite was suddenly lost on 8th April 2012, which led to the end of the mission being declared by ESA.

2 L1b data delivery and L2 data generation

2.1 L1b data processing and delivery by ESA

The L1b data processing, i.e. essentially Fourier transformation of the measured interferograms, phase correction, and calibration, were performed by the European Space Agency. A description of this processing step can be found in [3]. During the MIPAS mission the acquired knowledge about deficiencies in the L1b data led to the release of several data versions with respective improvements. Among these improvements are better values for the line of sight (LOS), i.e. the knowledge on the exact altitude at which the instrument points.

A major problem of older data versions was caused by an instrumental drift in time due to the ageing of the MIPAS detectors, in conjunction with their non-linear response curve. This instrumental drift in turn leads to a drift in the atmospheric temperature and constituents calculated from the spectra. Therefore for the V7 L1b data product a new calibration scheme was devised to compensate for this effect of the detector ageing (see [4, 5]). However, it soon turned out that there was an inconsistency between the FR and RR measurement periods. Hence a further correction was implemented for V8 L1b data. This V8 L1b data is the final release, since the financial and administrative efforts for the MIPAS mission were suspended by ESA in 2019.

It should be noted that initially our HPC-project was proposed to last from April 2017 to December 2019 and it was planned to do the L2 processing with the V7 data set. However, as soon as it became clear that there would be one further L1b data version with clear improvements over V7, the decision was to wait for the new V8 L1b data set. The delivery of the V8 L1b data was expected for mid 2017. However, although the processing of the V8 L1b data had been finished by end of January 2018, there were some investigations and corrective actions necessary by ESA. The data delivery to the end users did not start before January 2019.

2.2 L2 Data generation

Retrieval of abundances of atmospheric constituents from infrared spectra requires the inverse solution of the radiative transfer equation. Measured radiances are compared to radiative transfer calculations and residuals are minimized in a least squares sense by adjustment of the abundances of constituents which are fed into a forward model. The Retrieval Control Program (RCP), together with the forward model, is the core of the processing chain.

The forward model, the Karlsruhe Optimized and Precise Radiative Transfer Algorithm (KOPRA) [6] is a computationally optimized line-by-line model which simulates radiative transfer through the Earth's atmosphere under consideration of all relevant physics: the spectral transitions of all involved molecules, atmospheric refraction, line-coupling and non-local thermodynamic equilibrium (NLTE). Along with the spectral radiances, KOPRA also provides the Jacobian matrices, i.e. the sensitivities of the spectral radiances to changes in the atmospheric state parameters.

Atmospheric state parameters are retrieved by constrained multi-parameter non-linear least-squares fitting of simulated to measured spectral radiances. The inversion is regularized with a Tikhonov-type constraint which minimizes the first order finite differences of adjacent profile values [7, 8].

Instead of performing a simultaneous retrieval of all target parameters from a limb sequence, the retrieval is decomposed in a sequence of analysis steps. First, spectral shift is corrected, then the instrument pointing is corrected along with a retrieval of temperature, using CO_2 transitions. Then the dominant emitters in the infrared spectrum are analyzed one after the other (H_2O, O_3, HNO_3...), each in a dedicated spectral region where the spectrum contains maximum information on the target species but least interference by non-target species. Finally, the minor contributors are analyzed sequentially, whereby pre-determined information on major and minor contributors is used. The analysis is done limb sequence by limb sequence, i.e. limb scans are processed independently from each other by separate calls of the RCP.

3 Computational considerations

The original V8 L1b data consists of binary data sets with one data file containing all the spectra of one orbit (14.4 orbits per day, 70–100 geolocations per orbit, 17–35 IR spectra per geolocation, approx. 40000 radiance values per spectrum). The size of one L1b file is approx. 300 MB. At IMK facilities each file was converted once into a less complex binary file to keep the reading and conversion costs low at the actual data processing, since every single spectrum will be read multiple times. Converted data files are put into archives representing 35 days each (the repeat cycle of the Envisat orbits). The resulting 110 cycle archives, together with small meta data files, are stored at the HPSS. The latter files contain information for the efficient access of the archives. Since the actual processing of the infrared spectra includes a lot of I/O, reading of spectral and auxiliary data and writing of intermediate results, our setup included the use of the /tmp-filesystem of the compute nodes (essentially a RAM-disk). This approach had been chosen to reduce the I/O-load on the temporary workspace ($WORK) filesystem of the XC40 "Hazel Hen" computer. Initially, during the testing phase, this approach also worked for the "Hawk". However, the configuration of the /tmp-filesystem was changed (due to memory leakage problems, it seems) for the production phase and the approach did no longer work.

As outlined in the last report, 4800 cores were selected as the "sweet spot" and used throughout the processing of the L2 data on the XC40 "Hazel Hen", and increasing the number of used cores would lead to an increased number of corrupted result data, mainly due to I/O-problems. However, in the course of the adaptation of the processing software to the "Hawk" machine it was found that the I/O-problems might have been caused by an error in a script. Testing of this issue was stopped short due to the shutdown of "Hawk" in May 2020 caused by the infamous security incident. After "Hawk" was available again it turned out that indeed most probably the faulty script was responsible for some of the I/O-error problems. However, the configuration change with respect to the /tmp-filesystem of the compute nodes, which was heavily used by our processing software for efficiency reasons, unfortunately forced us to use the $WORK filesystem for all processing steps instead, which implies a severe performance degradation. The combined impact of the corrected script (improvement) and the necessity to switch from the compute node's /tmp to $WORK (deterioration) leads to a number of 6000 parallel cores, which can be used safely, without having to accept I/O-error problems any more. A maximum number of 8000 cores has been tried and this seems to be the limit for the current configuration. We are eagerly hoping for the node configuration with respect to /tmp to switch back to the state as it was during the "Hawk" testing phase.

The processing steps consist of:

- preprocessing, always performed at IMK
- collection of preprocessed data into archives, according to the chosen size of jobs on the "Hawk"
- transfer of archives containing preprocessed data from IMK to the $WORK directory
- in parallel: extraction of the necessary V8 L1b data from the HPSS to the $WORK directory
- submission of compute jobs, i.e. retrieval (core processing step) with the Retrieval Control Program (RCP) on "Hawk" with approx. 6000 cores used in parallel
- postprocessing to generate some elementary result diagnostics and corresponding diagnostics plots, again approx. 6000 cores are used in parallel
- collection of results and diagnostics into archives, and transfer of these back to IMK

At the time of this writing (Nov. 2021) 62% of the granted core hours have been used.

4 First results

Until the final shutdown of the Cray XC40 spectral shift, temperature and line-of-sight ([8]), H_2O, and O_3 have been processed from V8 L1b data. This has been done for the entire mission, i.e. for 2002–2012 for all major measurement modes: nominal (NOM), middle and upper atmosphere (MA and UA), noctilucent cloud (NLC) and upper troposphere lower stratosphere (UTLS) modes. 20% of the allocated compute time to the MIPAS_V7 project on the Cray XC40 has been used for this task.

The "Hawk" computer has been used for the processing of H_2O (a modified version, see below), CO, NO, and NO_2, i.e. those species which due to NLTE treatment are most expensive to calculate.

The following list gives a short overview over the status of the data processing:

Water Vapour H_2O data for all major measurement modes (NOM, MA, UA, and UTLS) has been processed for the entire mission already on the XC40. However, there is a modification to the setup, which is currently in the processing. The modification essentially shows better validation results in the upper stratosphere and above.

Nitric Oxide NO data for all major measurement modes (NOM, MA, UA, and UTLS) has been processed for the entire mission.

Nitrogen Dioxide NO_2 data for all major measurement modes (NOM, MA, UA, and UTLS) has been processed for the entire mission.

Carbon Monoxide CO data for all major measurement modes (NOM, MA, UA, and UTLS) has been processed for the entire mission.

4.1 Water vapour

As mentioned in our previous report, water vapor retrievals from NOM and MA/UA/NLC measurement modes need to consider NLTE effects. Non-LTE means that the populations of the emitting vibrational states, from which atmospheric variables are derived, do not follow the Boltzmann distribution, which depends only on the local kinetic temperatures. Since the atmospheric radiation emission is a function of the populations of the vibrational levels, these populations must therefore be calculated in detail, considering all possible processes that populate and de-populate them. The correct abundances can only be retrieved with a complete model in which all levels, bands and mechanisms affecting the emitting levels are included. Indeed, the achieved accuracy of the retrieved abundances strongly depends on the non-LTE model.

For this task, we use our non-LTE model GRANADA (Generic RAdiative traNsfer AnD non-LTE population algorithm, [9]), which is a sophisticated algorithm that calculates one by one the impact of the processes at each atmospheric layer, considering thermal and non-thermal collisions with other atmospheric molecules, chemical productions and losses, absorption and emission of external radiation and exchange of energy among atmospheric layers. In the case of water vapor, the effect of non-LTE on retrievals is noticeable above 50 km [10]. For V8 data we thus include full non-LTE calculations not only in MA/UA/NLC mode retrievals but also in NOM retrievals.

For the H_2O retrievals, GRANADA calculates the state populations from the surface up to 120 km in 1 km-steps not only for the H_2O 6.3-micron emitting levels, but also for a number of other H_2O and O_2 states, that also affect the former. In short, the setup for the calculation of H_2O populations accounts for 23 water vapor vibrational levels, including three water vapor isotopologues, and 44 molecular oxygen electronic-vibrational levels. The H_2O levels are connected by 185 radiative transitions, 10 of them considering full radiative transfer (calculated line-by-line with Kopra by means of the Curtis matrix formalism). Fifteen radiative transitions

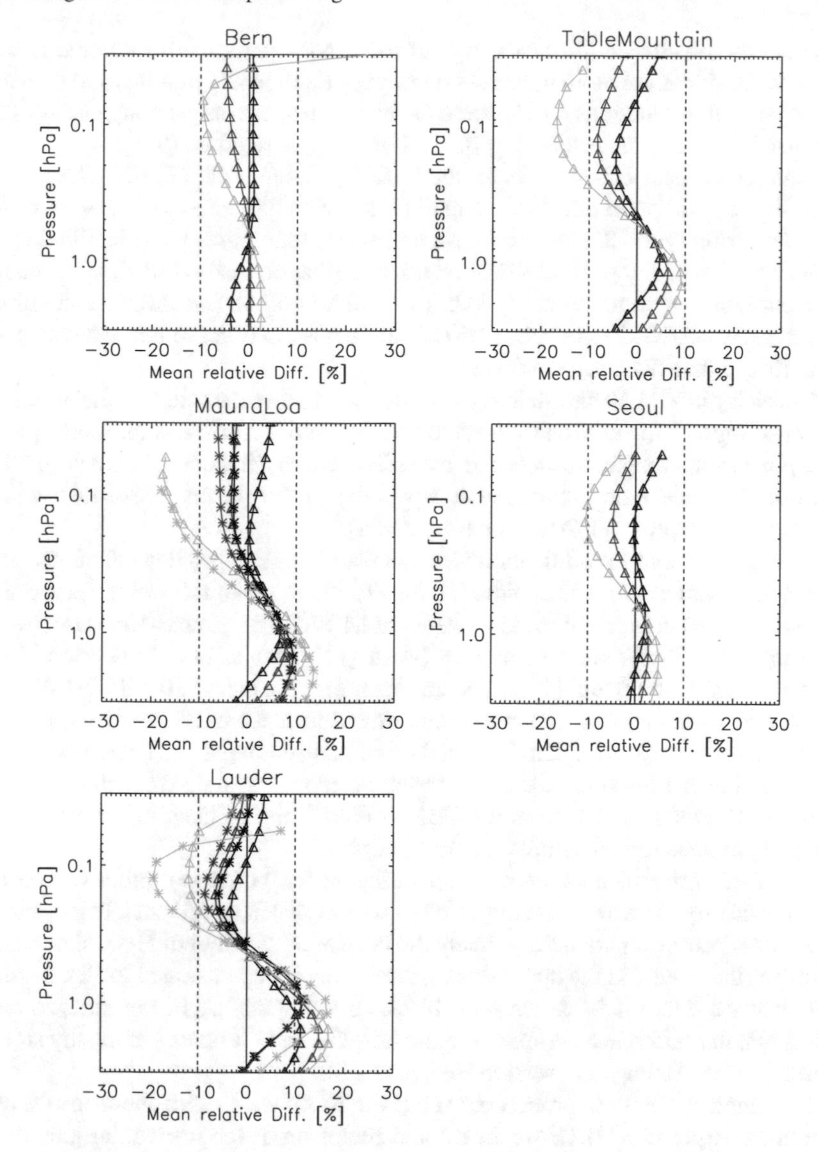

Fig. 1: Comparison of MIPAS H_2O-retrieval results to ground-based microwave instrument data. Green: V5 products, red: current V8 products, black: V8 products with horizontal H_2O-gradient. Symbol for full resolution data is the asterisk; triangle indicates reduced resolution data. The names of the microwave instrument stations are given in the plot title.

connect the electronic O_2 states, two of them considering full radiative transfer. Finally, consideration of 46 processes describing thermal and non-thermal collisions and chemical productions for H_2O and O_2 molecules, not only among each of them but also with other molecules (e.g., CO_2, N_2or O_3), is required [9].

The use of these calculations for the H_2O-O_2 system with GRANADA coupled to RCP is expensive in terms of computer time, which increases by a factor of four in comparison with LTE retrievals. Nevertheless, that seems to be justified by the benefit of the quality of the MIPAS NOM water vapor, leading to H_2O abundance improvements of up to 5% at 45-50 km and 10% at 60-70 km. Above that altitude, i.e., for MA/UA/NLC retrievals, non-LTE are a must in order to retrieve reasonable H_2O (better than 50% below 90 km).

The delay in V8 L1b data delivery was used to improve the setup for the processing of water vapour. In addition to the setup for V5 and V7 processing, more spectral data points and computational expensive radiative transfer calculations in full NLTE are used to achieve an improvement especially in the altitude range of the lower Mesosphere (approx. 50–70 km or 1–0.05 hPa).

The gain of these modifications for the water vapour profiles calculated from NOM observation mode is depicted in Fig. 1, where mean relative differences are shown between coincident measurements of MIPAS and ground-based microwave instruments (GBMW) at several sites (Bern [11], Mauna Loa, Lauder and Table Mountain [12], and Seoul [13]). The criterion for data pairs MIPAS/GBMW to be considered as coincident are a maximum time difference of 24 hours and a spatial difference of at most 1000 km. The method of this comparison is described in [14].

In the lower Mesosphere the V8 based water vapour data (red) differs almost everywhere less than 10% from the GBMW data. This is a clear improvement over V5 data products (green, details see figure caption).

One further modification in the setup for the retrieval of H_2O profiles was devised after the first reprocessing was completed (on the XC40). It turned out that inclusion of one further retrieval parameter, namely the horizontal gradient of H_2O, significantly improves the results in the upper stratosphere. This is demonstrated in Fig. 1, where this new V8 data set is shown with black curves/symbols. In the altitude range 0.1–2 hPa the difference compared to the GBMW H_2O profiles are mostly smaller than for the preceding data version (red, also V8).

In summary the new processing setup for NOM data, with inclusion of computational expensive NLTE treatment and inclusion of the horizontal gradient of H_2O, clearly leads to very satisfactory agreement of the V8 water vapour data with independent comparison data in the upper stratosphere and above.

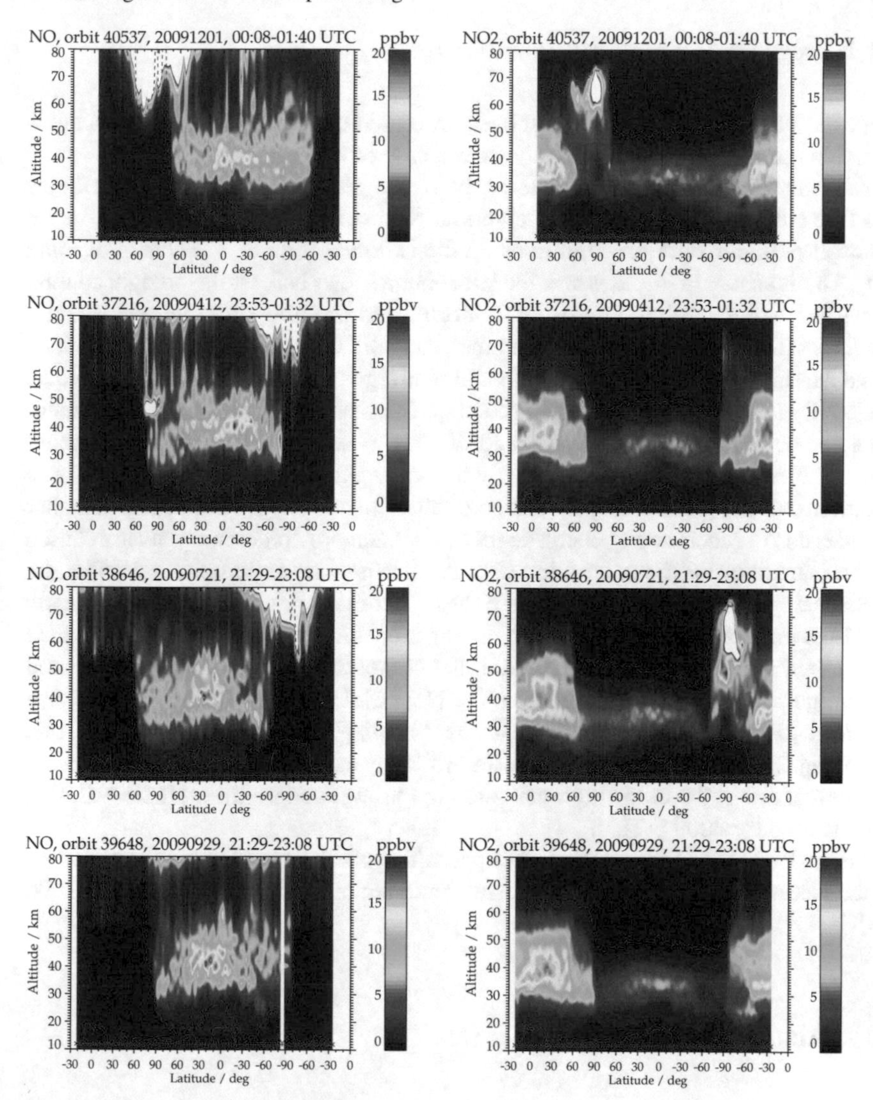

Fig. 2: Result for NO (left column of plots) and NO_2 (right) for four orbits representing northern winter, northern spring, northern summer, and northern fall (from top to bottom).

4.2 Nitrogen oxide and nitrogen dioxide

Since the V8 data products have just been processed there are no thorough analyses available yet. Hence, by now we only show a case of interest.

Nitrogen compounds play a role in the ozone chemistry, e.g. NO contributes as a free radical catalyst to ozone depletion. Normally NO is only available in the atmosphere under sunlight conditions. In the darkness it quickly reacts to become NO_2. This is shown in Fig. 2, where the left column shows NO, while the right column shows NO_2 volume mixing ratios for four orbits representing different atmospheric conditions (northern winter, spring, summer, and fall). Clearly NO is well constrained to the dayside of the orbits (marked by red plus signs near the bottom of the plots), while NO_2 mainly shows up during the night (marked by crosses). However, there are some occurrences of high NO (and of NO_2 as well) in the polar mesosphere (i.e., between 50 and 80 km), especially for hemispheric winter conditions. These enhancements are produced by ionization of nitrogen and oxygen molecules at higher altitudes due to precipitating energetic particles. Such precipitation events are caused by the interaction of the solar wind with the Earth's magnetic field and affect the atmosphere in the polar regions, where these particles are guided by the magnetic field lines towards the Earth. Under polar winter conditions, the produced NO is then transported downwards with the meridional circulation without being photolyzed by sunlight and partly converted into NO_2. [15, 16]. Enhanced NO and NO_2 is also visible at 45–50 km during the 2009 Northern hemispheric polar spring (see second row of Fig. 2) as a consequence of the extraordinarily efficient and confined downward transport of NO in the following weeks after the strong sudden stratospheric warming event in January 2009 [17].

The MIPAS V8 NO product from the upper atmosphere mode measurements will enter the successor of the empirical global reference atmospheric model NRLMSISE-00 [18].

4.3 Carbon monoxide

Since the V8 data products have just been processed there are no thorough analyses available yet. Hence, by now we only present monthly means for a selected year (2009). Figure 3 shows latitude/altitude distributions of monthly averages of CO for all twelve months of 2009 (note that the winter month December is put on top to have the four seasons continuously displayed). The most notable feature is the downward transport of CO over the hemispheric winter poles. With some caution, CO (plus additional chemistry modelling) can be used as tracer for atmospheric dynamics. This is done (together with H_2O and several other species) to retrieve the long term large scale meridional circulation of the middle atmosphere [19].

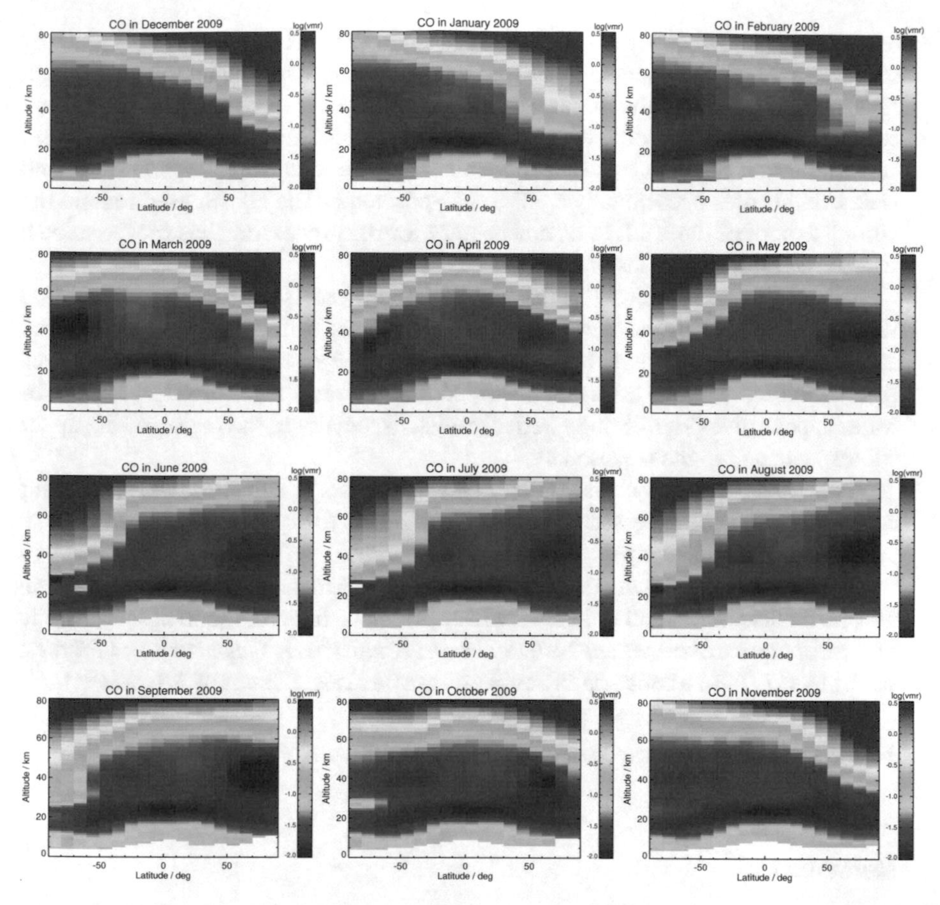

Fig. 3: Monthly means of CO in the middle atmosphere. The rows show, from top to bottom: northern winter (DJF), northern spring (MAM), northern summer (JJA), and northern fall (SON). The depicted values are common logarithms of the CO volume mixing ratio (ppmv).

5 Summary and conclusions

The HPE Apollo "Hawk" has been successfully used to run the IMK/IAA L2-processor with the V8 L1b data of the MIPAS instrument on board the European Environmental Satellite Envisat. Three species (NO, NO_2, CO) have been processed for the entire mission duration (2002–2012)., i.e. for approximately 2.5 million geolocations each. Additionally, a further improved version of H_2O has just entered processing.

The transfer of our processing environment and programs from the XC40 to the HPE Apollo "Hawk" has been performed successfully. After the testing phase, however, there was a change with respect to the /tmp-filesystem of the compute nodes. This change prevents the use of the /tmp-filesystem of the nodes in a way which is essential for the efficiency of our retrieval system. We designed our system in a way to use /tmp (essentially a RAM-disk) to store temporary data and intermediate result files to minimize the traffic for the $WORK filesystem. As a consequence the current setup is far from being efficient with respect to I/O-load on the $WORK. Parallel calculations on 6000 cores has been found to be safe with respect to I/O-problems. Using 8000 cores occasionally leads to I/O-problems. The HPSS, used to hold the complete pool of the V8 L1b data necessary for the processing, has proved to be a reliable and sufficiently fast data repository.

It could already be shown that the new V8 L1b data set is superior to the preceding versions and especially that the effects of an instrumental drift could largely be reduced and that now there is much better consistency between FR and RR temperatures [8]. Additionally, it has been demonstrated that the most recent retrieval setup for water vapour gives significantly better results, especially in the region of the upper stratosphere and in the mesosphere.

A part of the data already is available on a data server where already the preceding data versions can be accessed by interested scientists (`http://www.imk-asf.kit.edu/english/308.php`).

The V8 L2 data set, with description of improvements, appropriate characterisations, error discussion, and validation, will be presented in a cross-journal special issue of *Atmospheric Measurement Techniques* and *Atmospheric Chemistry and Physics*, see `https://www.atmos-meas-tech.net/special_issue1094.html`.

Acknowledgements Sylvia Kellmann (KIT-IMK) provided a part of the figures.

References

1. Fischer, H., Blom, C., Oelhaf, H., Carli, B., Carlotti, M., Delbouille, L., Ehhalt, D., Flaud, J.M., Isaksen, I., López-Puertas, M., McElroy, C.T., Zander, R.: Envisat-MIPAS, an instrument for atmospheric chemistry and climate research. European Space Agency-Report SP-1229, C. Readings and R. A. Harris (eds.), ESA Publications Division, ESTEC, P. O. Box 299, 2200 AG Noordwijk, The Netherlands (2000)
2. Fischer, H., Birk, M., Blom, C., Carli, B., Carlotti, M., von Clarmann, T., Delbouille, L., Dudhia, A., Ehhalt, D., Endemann, M., Flaud, J.M., Gessner, R., Kleinert, A., Koopmann, R., Langen, J., López-Puertas, M., Mosner, P., Nett, H., Oelhaf, H., Perron, G., Remedios, J., Ridolfi, M., Stiller, G., Zander, R.: MIPAS: an instrument for atmospheric and climate research. Atmos. Chem. Phys. **8**(8), 2151–2188 (2008). DOI 10.5194/acp-8-2151-2008
3. Kleinert, A., Aubertin, G., Perron, G., Birk, M., Wagner, G., Hase, F., Nett, H., Poulin, R.: MIPAS Level 1B algorithms overview: operational processing and characterization. Atmos. Chem. Phys. **7**, 1395–1406 (2007)
4. Birk, M., Wagner, G.: Complete in-flight detector non-linearity characterisation of MIPAS/Envisat (2010). Available at: https://earth.esa.int/documents/700255/707720/Technical+note+DLR+on+MIPAS+non_linearity_0810.pdf (last access: 27 May 2021)
5. Kleinert, A., Birk, M., Wagner, G.: Technical note on MIPAS non-linearity correction (2015). Available at: https://earth.esa.int/documents/700255/707720/Kleinert_20151030___TN_KIT_DLR_nonlin_20151030.pdf, (last access: 29 October 2020)
6. Stiller, G.P. (ed.): The Karlsruhe Optimized and Precise Radiative Transfer Algorithm (KOPRA), *Wissenschaftliche Berichte*, vol. FZKA 6487. Forschungszentrum Karlsruhe, Karlsruhe (2000)

7. von Clarmann, T., Glatthor, N., Grabowski, U., Höpfner, M., Kellmann, S., Kiefer, M., Linden, A., Mengistu Tsidu, G., Milz, M., Steck, T., Stiller, G.P., Wang, D.Y., Fischer, H., Funke, B., Gil-López, S., López-Puertas, M.: Retrieval of temperature and tangent altitude pointing from limb emission spectra recorded from space by the Michelson Interferometer for Passive Atmospheric Sounding (MIPAS). J. Geophys. Res. **108**(D23), 4736 (2003). DOI 10.1029/2003JD003602
8. Kiefer, M., von Clarmann, T., Funke, B., García-Comas, M., Glatthor, N., Grabowski, U., Kellmann, S., Kleinert, A., Laeng, A., Linden, A., López-Puertas, M., Marsh, D., Stiller, G.P.: IMK/IAA MIPAS temperature retrieval version 8: nominal measurements. Atmos. Meas. Tech. **14**(6), 4111–4138 (2021). DOI 10.5194/amt-14-4111-2021
9. Funke, B., López-Puertas, M., García-Comas, M., Kaufmann, M., Höpfner, M., Stiller, G.P.: GRANADA: A Generic RAdiative traNsfer AnD non-LTE population algorithm. J. Quant. Spectrosc. Radiat. Transfer **113**(14), 1771–1817 (2012). DOI 10.1016/j.jqsrt.2012.05.001
10. Stiller, G.P., Kiefer, M., Eckert, E., von Clarmann, T., Kellmann, S., García-Comas, M., Funke, B., Leblanc, T., Fetzer, E., Froidevaux, L., Gomez, M., Hall, E., Hurst, D., Jordan, A., Kämpfer, N., Lambert, A., McDermid, I.S., McGee, T., Miloshevich, L., Nedoluha, G., Read, W., Schneider, M., Schwartz, M., Straub, C., Toon, G., Twigg, L.W., Walker, K., Whiteman, D.N.: Validation of MIPAS IMK/IAA temperature, water vapor, and ozone profiles with MOHAVE-2009 campaign measurements. Atmos. Meas. Tech. **5**(2), 289–320 (2012). DOI 10.5194/amt-5-289-2012
11. Deuber, B., Kämpfer, N., Feist, D.G.: A new 22-GHz radiometer for middle atmospheric water vapour profile measurements. IEEE Transactions on Geoscience and Remote Sensing **42**(5), 974 – 984 (2004). DOI 10.1109/TGRS.2004.825581
12. Nedoluha, G.E., Gomez, R.M., Neal, H., Lambert, A., Hurst, D., Boone, C.D., Stiller, G.P.: Validation of long term measurements of water vapor from the midstratosphere to the mesosphere at two Network for the Detection of Atmospheric Composition Change sites. J. Geophys. Res. **118**(2), 934–942 (2013). DOI 10.1029/2012JD018900
13. De Wachter, E., Haefele, A., Kämpfer, N., Ka, S., Lee, J.E., Oh, J.J.: The Seoul water vapor radiometer for the middle atmosphere: Calibration, retrieval, and validation. IEEE Trans. Geosci. Remote Sens. **49**(3), 1052–1062 (2011). DOI 10.1109/TGRS.2010.2072932
14. Nedoluha, G.E., Kiefer, M., Lossow, S., Gomez, R.M., Kämpfer, N., Lainer, M., Forkman, P., Christensen, O.M., Oh, J.J., Hartogh, P., Anderson, J., Bramstedt, K., Dinelli, B.M., Garcia-Comas, M., Hervig, M., Murtagh, D., Raspollini, P., Read, W.G., Rosenlof, K., Stiller, G.P., Walker, K.A.: The SPARC water vapor assessment II: intercomparison of satellite and ground-based microwave measurements. Atmos. Chem. Phys. **17**(23), 14543–14558 (2017). DOI 10.5194/acp-17-14543-2017
15. Funke, B., López-Puertas, M., Holt, L., Randall, C.E., Stiller, G.P., von Clarmann, T.: Hemispheric distributions and interannual variability of NOy produced by energetic particle precipitation in 2002-2012. J. Geophys. Res. Atmos. **119**(23), 13,565–13,582 (2014). DOI 10.1002/2014JD022423
16. Funke, B., López-Puertas, M., Stiller, G.P., von Clarmann, T.: Mesospheric and stratospheric NOy produced by energetic particle precipitation during 2002-2012. J. Geophys. Res. Atmos. **119**(7), 4429–4446 (2014). DOI 10.1002/2013JD021404
17. Funke, B., Ball, W., Bender, S., Gardini, A., Harvey, V.L., Lambert, A., López-Puertas, M., Marsh, D.R., Meraner, K., Nieder, H., Päivärinta, S.M., Pérot, K., Randall, C.E., Reddmann, T., Rozanov, E., Schmidt, H., Seppälä, A., Sinnhuber, M., Sukhodolov, T., Stiller, G.P., Tsvetkova, N.D., Verronen, P.T., Versick, S., von Clarmann, T., Walker, K.A., Yushkov, V.: HEPPA-II model-measurement intercomparison project: EPP indirect effects during the dynamically perturbed NH winter 2008-2009. Atmos. Chem. Phys. **17**(5), 3573–3604 (2017). DOI 10.5194/acp-17-3573-2017
18. Picone, J.M., Hedin, A.E., Drob, D.P., Aikin, A.C.: NRLMSISE-00 empirical model of the atmosphere: Statistical comparisons and scientific issues. J. Geophys. Res. **107**(A12), 1468–1484 (2002). DOI 10.1029/2002JA009430
19. von Clarmann, T., Grabowski, U., Stiller, G.P., Monge-Sanz, B.M., Glatthor, N., Kellmann, S.: The middle atmospheric meridional circulation for 2002-2012 derived from MIPAS observations. Atmos. Chem. Phys. **21**(11), 8823–8843 (2021). DOI 10.5194/acp-21-8823-2021

WRF simulations to investigate processes across scales (WRFSCALE)

Hans-Stefan Bauer, Thomas Schwitalla, Oliver Branch and Rohith Thundathil

Abstract Several scientific aspects ranging from boundary layer research and land modification experiments to data assimilation applications were addressed with the Weather Research and Forecasting (WRF) model from the km-scale down to the turbulence-permitting scale.
Due to the transition to the new Hawk system, most of the work done in the subprojects during the reporting period was related to cleaning up, configuration testing, transfer of data, and publication of the results of earlier simulations. Investigations were extended to a second case of the Land Atmosphere Feedback Experiment (LAFE) in the central United States. The model grid increment was refined from 100 m with 100 vertical levels to 20 m with 200 vertical levels, resulting in more detailed turbulence structures and stronger variability in the boundary layer. These are promising results and match well with lidar observations.

1 Introduction and motivation

Numerical models provide the full 4D representation of the atmosphere and a consistent state with respect to the 3D thermodynamic atmospheric fields, cloud water, cloud ice, and diagnostic variables such as precipitation. Therefore, they are excellent tools to improve our understanding of atmospheric processes across scales and complement to observation-based approaches.

Commonly-used mesoscale models with resolutions on the km scale do not permit the explicit simulation of smaller-scale processes such as large turbulent eddies (100–2000 m), and the whole turbulence spectrum is parameterized. Further improvements

Hans-Stefan Bauer, Thomas Schwitalla, Oliver Branch and Rohith Thundathil
Institute of Physics and Meteorology, Garbenstrasse 30, 70599 Stuttgart, Germany,
e-mail: hans-stefan.bauer@uni-hohenheim.de,
thomas.schwitalla@uni-hohenheim.de, oliver_branch@uni-hohenheim.de,
rohith.thundathil@uni-hohenheim.de

W. E. Nagel et al. (eds.), *High Performance Computing in Science and Engineering '21*,
https://doi.org/10.1007/978-3-031-17937-2_25

are expected if a chain of model runs is performed down to the turbulence-permitting (TP) scale (100 m resolution and below), as further details of land-surface atmosphere (LA) interaction are resolved. Nevertheless, processes between the atmosphere and the land surface are still to date not completely understood and subject of ongoing research.

The WRF model system [4], described with more detail in earlier reports, provides the opportunity to perform LES simulations under realistic conditions because of the wide range of scales it can be applied over. Using a nesting strategy that covers the synoptic weather situation and the mesoscale circulations, in combination with data assimilation, will ensure a forcing of the LES that is as close as possible to the real weather situation.

2 Work done since March 2020

During the report period, apart from some testing, only sub-project one "LES simulations to better understand boundary layer evolution and high-impact weather (LES-PROC)" performed simulations.

The sub-projects "Seasonal land surface modification simulations over the United Arab Emirates (UAE-1)" and "Assimilation of Lidar water vapor measurements (VAP-DA)" only performed some domain tests during the preparation of the follow-up proposal for the project. Causes were the delays in the introduction of the new super computer Hawk and its file system and the necessity to publish the results of the simulations performed in the previous report period [2, 5].

The sub-project "Turbulence-permitting particulate matter forecast system (Open Forecast)" performed the last simulation in February 2020 in the previous report period. The corresponding report was submitted in 2020. The related EU project ended and the results were published in [3]. Therefore, this annual report focuses on sub-project LES-PROC and the needed scaling tests done to optimize the operation of WRF on the new Hawk system.

2.1 Cases and model setup

The **L**and **A**tmosphere **F**eedback **E**xperiment (LAFE) took place in August 2017 at the Southern Great Plains site of the Atmospheric Radiation Measurement Program (ARM) in Oklahoma. Many different instruments were brought to the site and the measurement strategy was optimized in a way to derive as much information as possible. More details about the campaign and the applied measurement strategy can be found in [7].

We so far focused our simulations on two LAFE cases on the 23rd of August 2017 and 08th of August 2017. Both were clear sky days with operation of the lidar systems. On the 23rd of August, the operation was temporally extended to include the evening transition of the convective boundary layer. On August 8th, although synoptically similar, a clearly deeper boundary layer developed during day compared to the 23rd.

During the transition to the new Hawk system, we changed to the more recent version 4.2.1 of WRF. The simulations were started at 06 UTC (01 a.m. local time) and run for 24 hours within the 24 hour wall time limit on the HAWK system. During the last report period, we tested the influence of the grid ratio during the nesting to the target resolution of 100 m. It was found that the results in the innermost domain are almost identical and the transition to the finer scales performs well in both setups. Therefore, we focus on the grid ratio of five in the new simulations described in this report.

WRF was set-up using three domains with 2500 m, 500 m and 100 m resolution. The outer domain was driven by the operational analysis of the European Centre of Medium Range Weather Forecasting (ECMWF). The size of the domains is 1000×1000 grid cells in the outer domains and 1201×1001 grid cells in the innermost 100 m domain. The vertical resolution is the same in all domains with 100 vertical levels up to a height of 50 hPa with 30 levels in the lowest 1500 m of the troposphere. Figure 1 shows the third domain with 100 m and the fourth domain with 20 m horizontal resolution.

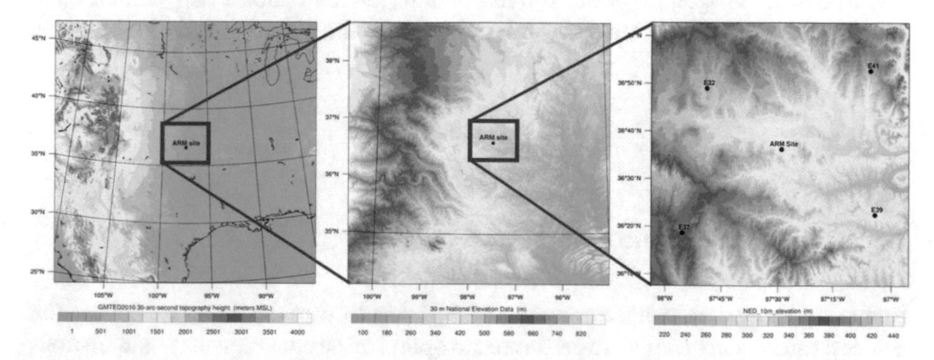

Fig. 1: Domain configuration of the first LAFE simulation. From left to right the domains with 2500 m, 500 m and 100 m resolution. The outer and middle domain have sizes of 1000×1000 grid cells and the inner domain consists of 1201×1001 points.

In addition to this standard configuration, a fourth nest with a horizontal resolution of 20 m was added. So far it is not nested "online" by performing all 4 domains in one simulation. Rather, the NDOWN capability of WRF is applied to calculate the lateral and initial forcing for this 20 m domain from the 100 m nest. Advantage of this procedure is that the vertical resolution can be increased for this nest during NDOWN. In our case we double the vertical resolution to 200 levels ending with a representation of the lowest 1500 m of the troposphere by a approx. 60 levels. Furthermore, the expensive simulation of the 20 m domain runs separately and can be reduced to a part of the full simulation period. In the presented first test, the 20 m domain is only simulated for the 3-hour period 18 to 21 UTC (13 to 16 LT in Oklahoma). Figure 2 shows the third domain with 100 m and the 20 m nest created with NDOWN.

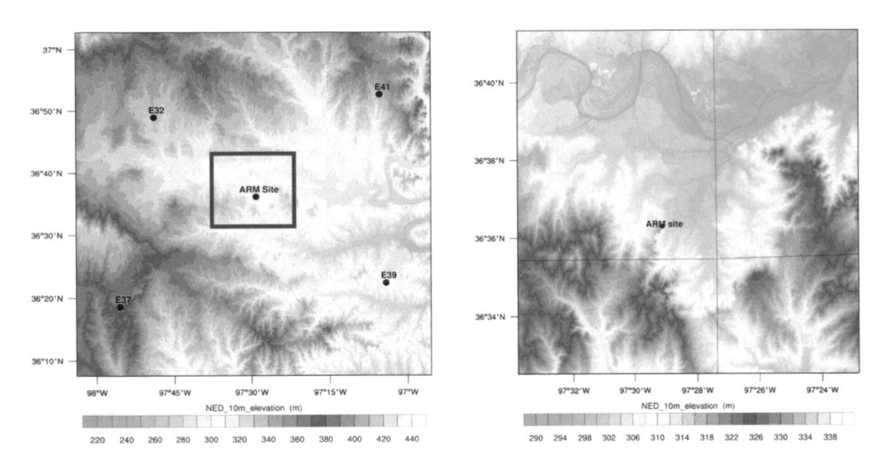

Fig. 2: Surface orography of the 100 m domain (left) and additionally added 20 m domain (right). Thc location of thc 20 m domain in the 100 m domain is shown by the red box.

2.2 Optimization of the model setup on Hawk

Before repeating the simulations for the 23rd of August with the newer model version and new case studies, tests were performed to optimize the operation of the simulations on the new Hawk system.

To estimate the performance of the WRF model on the new HAWK system, strong scaling tests for the configuration applied in LES_PROC subproject were performed. The scaling tests were done for the three-domain configuration 200 m – 500 m – 100 m described above. Table 1 summarizes the results. The times mentioned are averaged values, since the numbers vary slightly from timcstcp to timestep.

The smallest possible configuration was six nodes using 32 MPI tasks per node with four OpenMP threads each. According to our experience with the system, this is the best combination of MPI tasks and OpenMP threads to fully use the number of cores available per node with the applied WRF model. With less nodes, the simulation crashes because more memory than available is needed. More than 128 nodes with

32 MPI tasks per node is not possible with the selected configuration because then the sub-domain per MPI task is getting too small. Figure 3 shows that WRF scales nicely on the HAWK system.

Table 1: Tested configurations for the strong scaling test.

Nodes	MPI tasks per Node	OpenMP threads per MPI task	Number of cores	Factor cores	Needed time per timestep (s)	Speedup timestep	Needed time per radiation timestep (s)
6	32	4	768	1	1.58	1	9.8
8	32	4	1024	1.33	1.17	1.35	7.57
10	32	4	1280	1.67	0.88	1.8	5.45
12	32	4	1536	2	0.7	2.26	4.45
16	32	4	2048	2.67	0.5	3.16	3.3
24	32	4	3072	4	0.32	4.94	2.1
32	32	4	4096	5.33	0.26	6.08	1.7
64	32	4	8192	10.67	0.14	11.29	0.9
128	32	4	16384	21.33	0.08	19.75	0.56

To minimize the time needed for I/O, the WRF model is compiled with parallel NetCDF support and in addition I/O quilting is applied. The latter reserves some cores for I/O and reduces the time, the computation is interrupted for I/O to at most a few seconds even if several files larger than 15 GB are written at the same time.

2.3 Results of the simulations

First, the well-developed boundary layer for both cases is illustrated with the horizontal distribution of vertical velocity and water vapor mixing ratio interpolated to 1000 m above sea level. Figure 4 shows the two fields for 2 p.m. LT and the two case studies. For a better illustration of the evolution of turbulence, a subregion of the 100 m domain is presented.

Both cases show a predominant northeasterly to easterly circulation. On the 23rd of August, clearly drier air is advected into the region. The well-developed turbulence is clearly seen. The size of the updraft plumes (blue) and the compensating downdrafts (red) are nicely captured and correspond in size to observed eddy sizes in the region (e.g. [1,6,7]). With this updraft plumes moister air is transported from near surface

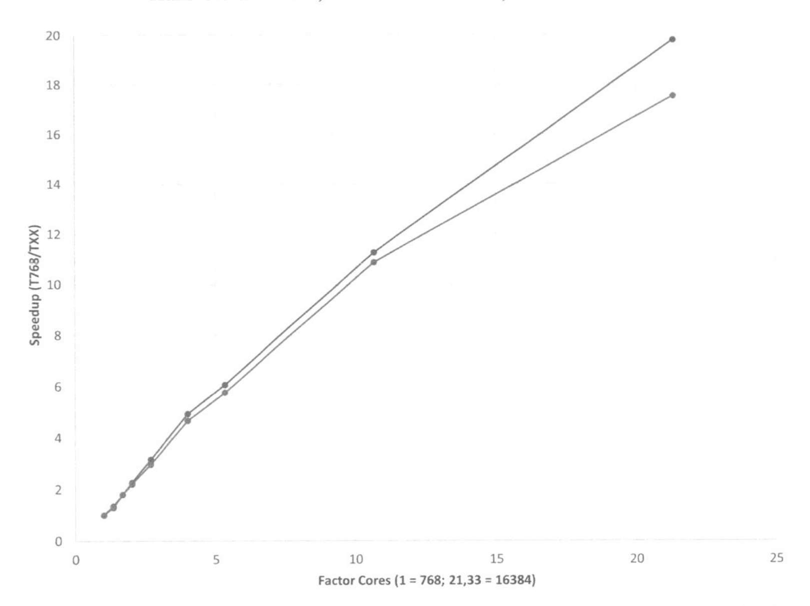

Fig. 3: Scaling of the different configurations presented in Table 1. T768 is the time needed with 768 cores, the smallest possible configuration, and TXX is the time needed by the larger configurations.

to higher levels in the boundary layers seen as green, orange and red plumes on the right panels of the plot. The compensating downdrafts, on the other hand, transport drier air downwards. The stronger turbulence developing on August 8 is clearly seen, transporting larger amounts of moisture upwards.

Figure 5 shows time-height cross sections of water vapor mixing ratio and horizontal wind velocity to illustrate the temporal evolution of the boundary layer. The forecast length of 24 hour allows the analysis of a full diurnal cycle of the temporal evolution of the daytime boundary layer. To increase the temporal resolution, time series output at selected model grid points were written in addition to provide data in 10 s resolution.

The different stages of the development of the boundary layer are clearly seen. During night, a shallow nighttime boundary layer is seen, overlaid by the residual layer, namely the convective boundary layer of the previous day. In the morning local time, with the onset of turbulence, the new convective boundary layer is growing. The turbulent fluctuations are typical for the development of an undisturbed convective boundary layer. In the evening after sunset, turbulence diminishes and finally a new nighttime stable boundary layer develops.

Another interesting feature, commonly found in this region is the nighttime low-level jet, prominently shown by the lower panel, transporting moist air from the Gulf of Mexico to the region. During daytime, the jet is replaced by the developing turbulent eddies transporting heat and moisture vertically.

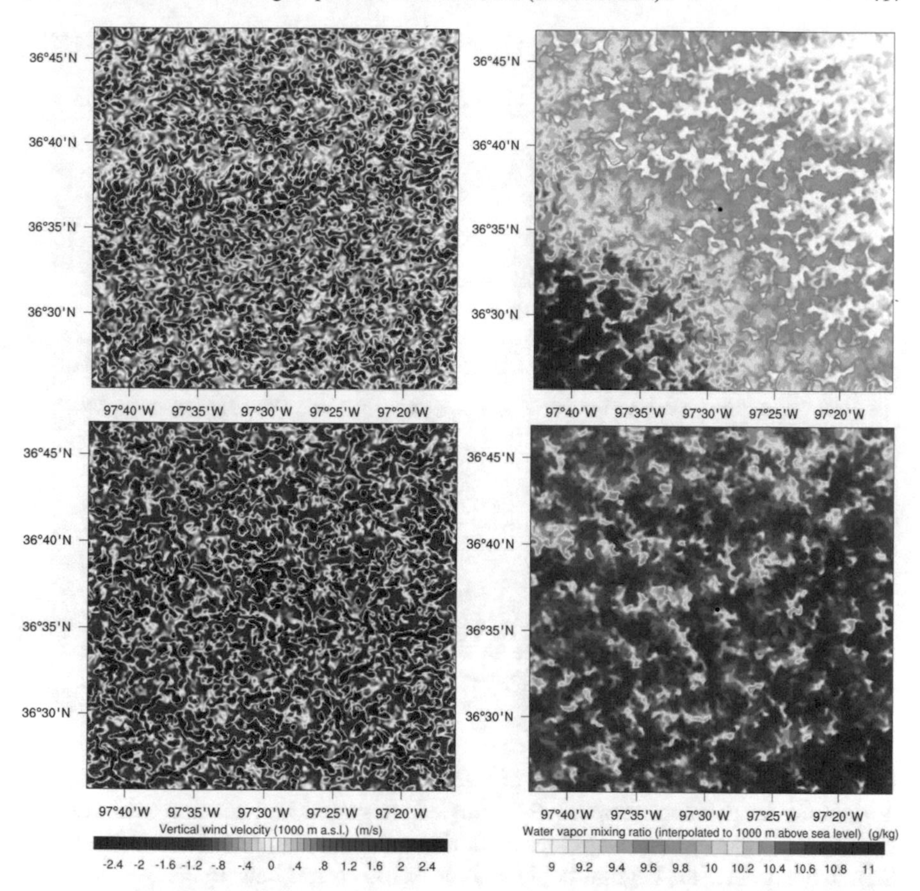

Fig. 4: Representation of the convective boundary layer zoomed into a small region around the ARM SGP site (black dot) at 2 p.m. LT on 23 August 2017 (top) and for 8 August 2017 (bottom). Vertical velocity m/s 1000 m above sea level is shown in the left column and water vapor mixing ratio g/kg 1000 m above sea level in the right column.

Another new result is the higher 20 m resolution inner domain. Figure 6 compares the same fields vertical velocity and water vapor mixing ratio interpolated to 1000 m above sea level for the 100 m and the 20 m resolutions over the domain of the 20 m simulation at 19 UTC (15 LT in Oklahoma) during the well-established turbulent boundary layer.

The comparison demonstrates that the finer resolution reveals more details of the evolution of turbulence. Especially much more movement is seen along the edges of the turbulent updrafts and outflow boundaries are seen in the downdraft regions.

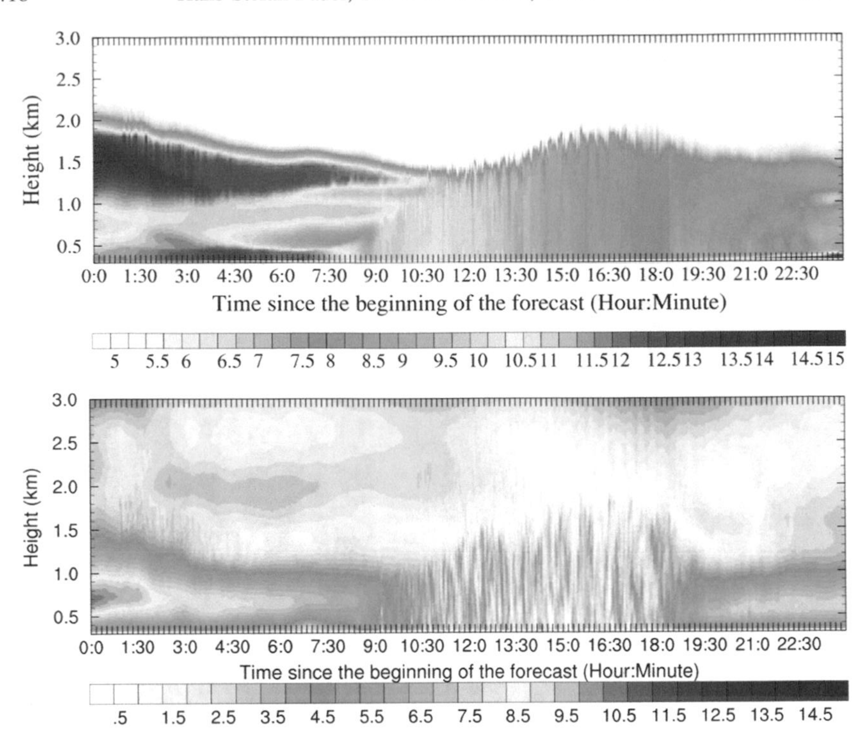

Fig. 5: Time-height cross sections of water vapor mixing ratio $[gkg^{-1}]$ (top) and horizontal wind velocity $[m/s]$ (bottom) on August 23 rd for the grid cell the Hohenheim lidar systems were located during the campaign. The X-Axis marks the time in hours since the beginning of the forecast (00 corresponds to 06 UTC or 01 LT in Oklahoma).

Finally, the vertical representation of turbulence in the two resolutions is compared. This is done in Figure 7 showing the time-height cross sections of vertical velocity for the three-hour time window 18 to 21 UTC (13 to 16 LT) for the grid point the ARM Doppler lidar was located. The two simulated representations are compared with the observation of the ARM lidar.

Comparing the two model simulations reveals the finer details simulated by the 20 m resolution. This is more in accordance with the lidar observation leading to stronger variability of vertical velocity in the boundary layer (see e.g. Bauer et al., 2020). The PBL height, on the other hand is represented by both resolutions. At the moment, it is investigated in detail of whether the higher horizontal or the doubled vertical resolution is the major cause for the more realistically represented boundary layer.

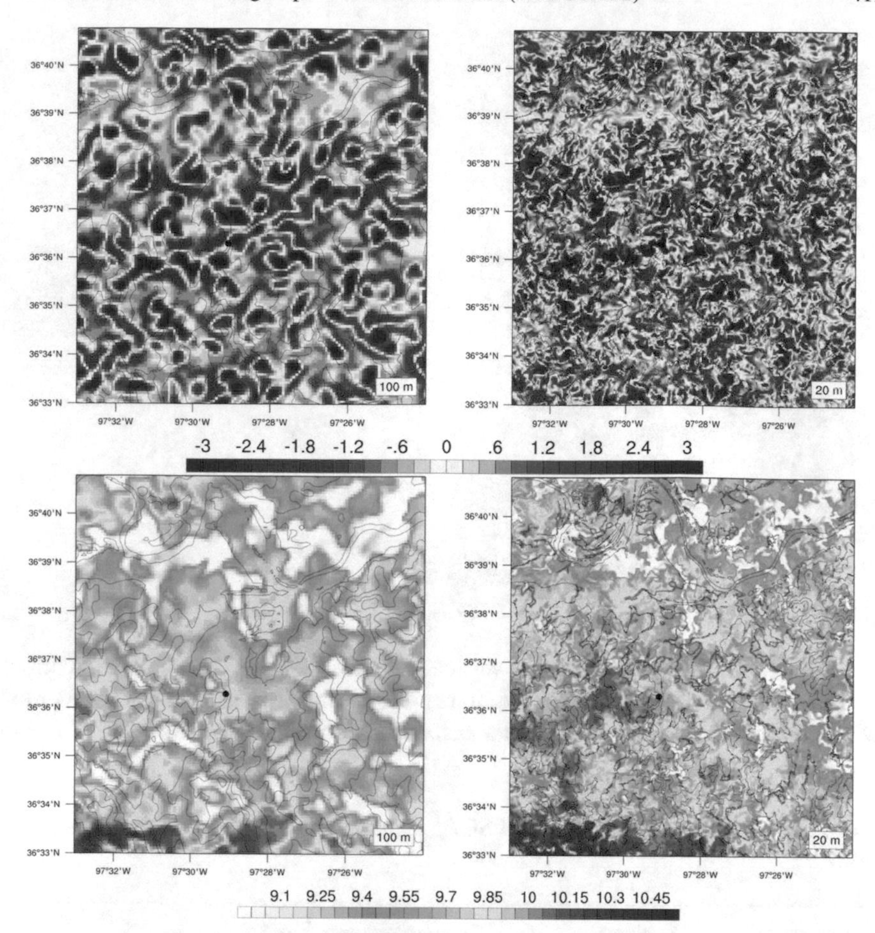

Fig. 6: Comparison of vertical wind velocity $[m/s]$ (top) and water vapor mixing ratio $[gkg^{-1}]$ 1000 m above sea level. Shown are results with 100 m resolution (left) and 20 m resolution (right) for the region covered by the 20 m domain.

3 Used resources

Table 2 lists the resources used during the report period March 2020 to June 2021. The term "clean up" also contains the move of all needed data from the old work space "ws9" to the new "ws10" and "testing" refers to the optimization of WRF on the new HAWK system.

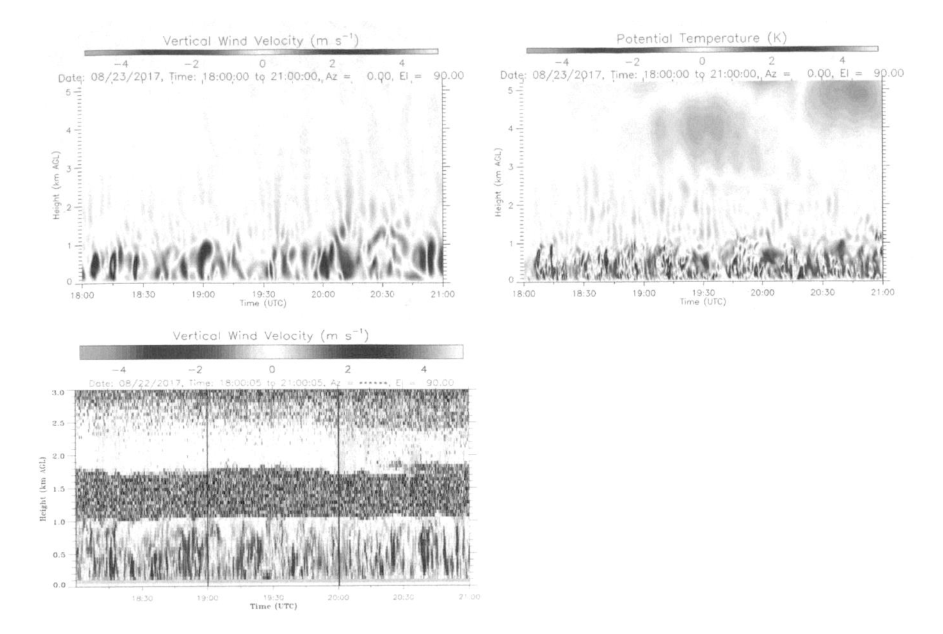

Fig. 7: Time-height cross section of vertical velocity $[m/s]$ of the model simulations with 100 m resolution (upper left), 20 m resolution (upper right) and the observation of the Doppler lidar (lower left) for the three-hour time window 18 to 21 UTC (13 to 16 LT).

Table 2: Resources used by the WRFSCALE project between March 2020 and June 2021.

Subproject	User	Computing time (Rth.)	Purpose
LES_PROC	Hans-Stefan Bauer	10899.2848	Clean up, testing and case study simulations mesoscale to LES
UAE-1	Oliver Branch	1373.6066	Clean up + testing
VAP-DA	Rohith Thundathil	46.7555	Clean up
Open Forecast	Thomas Schwitalla	574.9204	Clean up
Total		12894.5673	

References

1. H.-S. Bauer, S.K. Muppa, V. Wulfmeyer, A. Behrendt, K. Warrach-Sagi, and F. Späth. Multi-nested WRF simulations for studying planetary boundary layer processes on the turbulence-permitting scale in a realistic mesoscale environment. *Tellus A*, 72:1–28, 2020. doi:10.1080/16000870.2020.1761740.
2. O. Branch, T. Schwitalla, M. Temimi, R. Fonseca, N. Nelli, M. Weston, J. Milovac, and V. Wulfmeyer. Seasonal and diurnal performance of daily forecasts with WRF v3.8.1 over the united arab emirates. *Geosci. Model Dev.*, 14:1615–1637, 2021. doi:10.5194/gmd-14-1615-2021.
3. T. Schwitalla, H.-S. Bauer, K. Warrach-Sagi, T. Bönisch, and Wulfmeyer. V. Turbulence-permitting air pollution simulation for the stuttgart metropolitan area. *Atmos. Chem. Phys.*, 21:4575–4597, 2021. doi:10.5194/acp-21-4575-2021.
4. W.C. Skamarock, J.B. Klemp, J. Dudhia, O. Gill, Z. Liu, J. Berner, W. Wang, J.G. Powers, M.G. Duda, D.M. Barker, and X.-Y. Huang. A description of then advanced research WRF version 4. *NCAR Tech. Note*, NCAR/TN-556+STR:145pp, 2019. doi:10.5056/1dfh-6p97.
5. R. Thundathil, T. Schwitalla, A. Behrendt, and V. Wulfmeyer. Impact of assimilating lidar water vapour and temperature profiles with a hybrid ensemble transform kalman filter – three-dimensional variational analysis on the convection permitting scale. *Quart. J. Roy. Meteorol. Soc.*, 147:1–23, 2021. doi:10.1002/qj.4173.
6. D.D. Turner, V. Wulfmeyer, L.K. Berg, and J.H. Schween. Water vapor turbulence profiles in stationary continental convective mixed layers. *J. Geophys. Res. Atmos.*, 119:151–165, 2014. doi:10.1002/2014JD022202.
7. V. Wulfmeyer, D.D. Turner, B. Baker, A. Behrendt, T. Bonin, A. Brewer, M. Burban, A. Choukulkar, E. Dumas, R.M. Hardesty, T. Heus, J. Ingwersen, D. Lange, T.R. Lee, S. Metzendorf, S.M. Muppa, T. Meyers, R. Newsom, M. Osman, and S. Raasch. A new research approach for obswerving and characterizing land-atmosphere feedback. *Bull. Amer. Meteorol. Soc.*, 99:1639–1667, 2018.

Computer Science

The following part of this volume typically is a smaller one, and it deals with research labelled as "Computer Science", mostly due to the fact that the respective groups are affiliated with Computer Science departments. As in previous years, however, if we look at the topics addressed in the five annual reports submitted this year, the impression is a bit different: we find both classical informatics topics (such as load balancing, machine learning, or discrete algorithms) and classical application domains (such as molecular dynamics or multi-physics problems). Nevertheless, the common theme in the projects and the reports is their focus on state-of-the-art informatics topics related to HPC.

Out of those five submissions undergoing the usual reviewing process, two project reports were selected for publication in this volume: GCS-MDDC and SDA.

The contribution *Dynamic Molecular Dynamics Ensembles for Multiscale Simulation Coupling* by Neumann, Wittmer, Jafari, Seckler and Heinen reports on recent progress in the GCS-MDDC project. The project's basis are, first, the software *ls1 mardyn*, a molecular dynamics (MD) framework for multi-phase and multi-component studies at small scales with applications in process engineering, and, second, the coupling software *MaMiCo* that addresses the challenge of efficiently sampling hydrodynamic quantities from MD. The overall goal is to couple CFD solvers with *ls1 mardyn*, thus allowing for making larger scales accessible. The report presented in this volume deals with a first step in that direction, an extension of *MaMiCo* to handle dynamic ensembles. The system used for the computations of this project was HAWK at HLRS.

The second report, *Scalable Discrete Algorithms for Big Data Applications* by Hespe, Hübschle-Schneider, Sander, Schreiber and Hübner is definitely not on the usual suspects side. In contrast to classical HPC applications based on continuous (numerical) algorithms, the SDA project focuses on discrete algorithms, to be precise SAT solving, malleable job scheduling, load balancing, and fault tolerance. In the period to be reflected in this volume, emphasis was put on SAT solving and fault-tolerant algorithms. The system used for that was ForHLR II at KIT.

Together, both papers nicely reflect the breadth of HPC use in science: continuous and discrete, simulation as well as data engineering and analytics.

Technical University of Munich,
Department of Informatics,
Chair of Scientific Computing,
Boltzmannstrasse 3,
D-85748 Garching,
Germany

Hans-Joachim Bungartz

Dynamic molecular dynamics ensembles for multiscale simulation coupling

Philipp Neumann, Niklas Wittmer, Vahid Jafari, Steffen Seckler and Matthias Heinen

Abstract Molecular dynamics (MD) simulation has become a valuable tool in process engineering. Despite our efforts in software developments for large-scale molecular simulations over several years, which have amongst others enabled record-breaking trillion-atom runs, MD simulations are—as stand-alone simulations—limited to rather small time and length scales. To make bigger scales accessible, we propose to work towards a coupling of CFD solvers with our efficient MD software *ls1 mardyn*. As a first step, we discuss extensions to our coupling software *MaMiCo*, addressing the challenge of efficiently sampling hydrodynamic quantities from MD: due to the high level of thermal fluctuations, MD ensemble considerations are required for sampling. We propose an extension of *MaMiCo* that allows to handle dynamic ensembles, i.e. to launch and remove MD simulations on-the-fly over the course of a coupled simulation. We explain the underlying implementation and provide first scalability results.

1 Introduction

Molecular dynamics (MD) simulation has become a valuable tool for various application fields, in particular also for process engineering. Over several years, we have developed the software *ls1 mardyn* [13] for this purpose, i.e. *ls1 mardyn* specializes on the handling of large numbers of rather small molecules. Recently, *ls1 mardyn* was extended by (1) auto-tuning at node-level, always choosing the

Philipp Neumann, Niklas Wittmer and Vahid Jafari
Helmut Schmidt University, Chair for High Performance Computing, Holstenhofweg 85, 22043 Hamburg, e-mail: philipp.neumann@hsu-hh.de

Steffen Seckler
Technical University of Munich, Chair of Scientific Computing, Boltzmannstr. 3, 85748 Garching

Matthias Heinen
Technical University of Berlin, Chair of Thermodynamics, Ernst-Reuter-Platz 1, 10587 Berlin

W. E. Nagel et al. (eds.), *High Performance Computing in Science and Engineering '21*,
https://doi.org/10.1007/978-3-031-17937-2_26

optimal combination of particle data container, traversal scheme and corresponding shared-memory parallelization scheme on-the-fly during a simulation [2, 14] and (2) an improved dynamic load balancing approach to tackle highly heterogeneous loads such as liquid-gas systems at inter-node level [14].

This software development, in conjunction with the provision of computational resources in the scope of a GCS Large Scale Project, allowed us to simulate nanoscale droplet coalescence at unprecedented size, cf. Figure 1. Our biggest study with two 200nm-sized (in terms of diameter; twice as big as the example shown in Fig. 1) droplets using the dynamic load balancing of *ls1 mardyn* consumed approx. 11 million core hours. It also enabled direct comparison with coarse-grained modelling

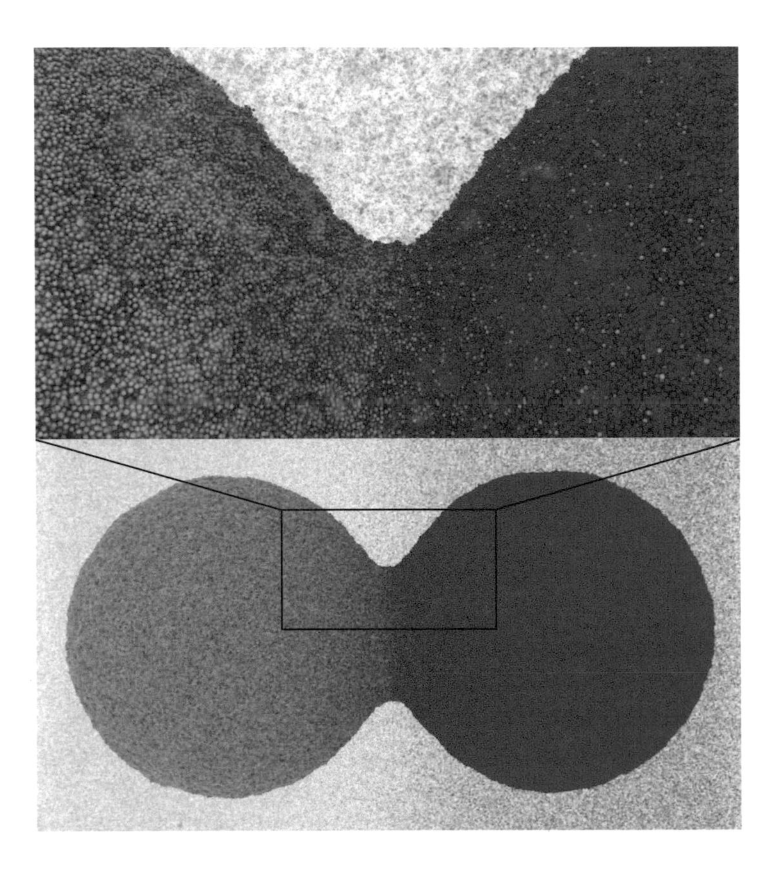

Fig. 1: Snapshot, rendered by *MegaMol* [3], of two coalescing argon droplets with an initial diameter of 100 nm (bottom). The molecular system consists of $\approx$ 25 million particles in total. Particles constituting the initial left and right droplet were colored green and red, respectively. The magnified view (top) shows the upper part of the growing bridge between the droplets.

approaches such as phase-field theory which showed very good agreement between both modelling approaches, see amongst others [1].

These findings on consistency between coarse- and fine-scale models suggest that a coupling of coarse-scale, continuum models and fine-scale, molecular models could be a promising option to resolve molecular behavior only when and where necessary. Molecular-continuum methods have been established for this purpose in the past. We believe that this will make even bigger scenarios accessible through available supercomputing resources, such as the current machine HAWK hosted at HLRS. One major challenge for these methods lies, however, in efficient sampling of the highly fluctuating thermodynamics at the molecular scale.

In this regard, we present preparatory results that we have worked on to push our software technology towards multiscale molecular-continuum simulations. Over the past years, much development effort also went into our macro-micro-coupling tool (*MaMiCo*) [10, 11] which allows the coupling of arbitrary continuum flow solvers and MD packages. In the mid term, we strive for coupling *ls1 mardyn* with *MaMiCo* and thus bring together both highly efficient molecular simulation software and the multiscale approach. To address the challenge of sampling the highly fluctuating quantities on the molecular scale, we present work that extends *MaMiCo* towards dynamic MD ensembles in the following.

We describe the underlying idea of molecular-continuum simulations in Sect. 2, including a short recap of related work and the involved software. The implementation of dynamic MD ensembles is discussed in Sect. 3. In Section 3, we validate this ensemble handling in parallel simulations, considering a Couette flow scenario. We also present scalability results in the same section. Findings and next steps are summarized in Sect. 5. Parts of this report have been accepted for publication [5].

2 Molecular-continuum coupling

2.1 Short-range Molecular Dynamics

In the field of molecular dynamics (MD) we investigate fluids, suspensions, and materials at the nanoscale. The system under consideration is modelled by particles which are characterized by their positions $\mathbf{x}_i$, velocities $\mathbf{v}_i$ and forces $\mathbf{F}_{ij}$.

Forces typically arise from pairwise inter-molecular potentials. In the following, we will restrict our considerations to the Lennard–Jones potential from which the following inter-molecular force arises between two particles i and j:

$$\mathbf{F}_{ij} = \frac{48}{\sigma^2}\left[\left(\frac{\sigma}{r_{ij}}\right)^{14} - \frac{1}{2}\left(\frac{\sigma}{r_{ij}}\right)^{8}\right](\mathbf{x_i} - \mathbf{x_j}), \tag{1}$$

where $r_{ij} := ||\mathbf{x}_i - \mathbf{x}_j||$. This distance is additionally limited to a cut-off radius r_c, such that $\mathbf{F}_{ij} = 0$ for $r_{ij} > r_c$ (short-range molecular dynamics). ϵ, σ are material-dependent parameters. Furthermore, we restrict our considerations to NVT ensembles, this means that beside a fixed density, the temperature of the investigated systems is maintained to be constant via thermostats.

2.2 Computational fluid dynamics

In the following, we make use of a simple Lattice Boltzmann solver with the standard BGK collision model, see e.g. [9], to compute the fluid flow on the continuum scale:

$$f_i(\mathbf{x} + \mathbf{c}_i \Delta t, t + \Delta t) = f_i(\mathbf{x}, t) - \frac{1}{\tau}(f_i(\mathbf{x}, t) - f_i^{eq}(\mathbf{x}, t)), \quad i = 1, ..., Q. \qquad (2)$$

Here $f_i(\mathbf{x}, t)$ denotes the probability to find particles in a cell with midpoint $\mathbf{x}$ and time t moving with *lattice velocity* $\mathbf{c}_i$. The set of lattice velocities is fixed, we rely on the well-known D3Q19 discretization with 19 lattice velocities in 3D. The fluid density $\rho(\mathbf{x}, t)$, pressure $p(\mathbf{x}, t)$ and velocity $\mathbf{u}(\mathbf{x}, t)$ evolve locally from:

$$\rho(\mathbf{x}, t) = \sum_{i=1}^{Q} f_i(\mathbf{x}, t),$$

$$p(\mathbf{x}, t) = \tfrac{1}{3}\rho(\mathbf{x}, t), \qquad (3)$$

$$\rho(\mathbf{x}, t)\mathbf{u}(\mathbf{x}, t) = \sum_{i=1}^{Q} f_i(\mathbf{x}, t)\mathbf{c}_i.$$

As boundary conditions, we rely on periodic conditions as well as half-way bounce back modeling of solid stationary and moving walls.

LB methods are very efficient and are known to scale very well up to full supercomputer machine size due to their local nature in computations and since they operate typically on standard Cartesian grids (or adaptive variants thereof). Note that the choice of the fluid flow solver is however more or less arbitrary from a technical perspective, as long as density and velocity values are available and can be provided to our coupling tool *MaMiCo*, cf. Sect. 2.4, on a mesh. From a physical perspective, the choice of the CFD solver may impact accuracy and stability of the arising coupled scheme.

2.3 Molecular-continuum algorithm in a nutshell

In the following, we will restrict views to concurrent transient coupling of MD and CFD. The method sketched in this section follows the algorithm described in [12] and slightly adopted in [10]; cf. the latter reference for further details.

Our coupling method makes use of an overlapping domain decomposition, i.e. both MD and CFD domain overlap to some extent in space and this overlapping region is used to couple both simulations. In our case, the MD domain is basically embedded in the CFD domain, and overlapping is assumed in a three Cartesian grid cell-wide section within the MD domain. Mass and momentum are transferred from the CFD solver to the MD solver inside this overlap region in separate sub-regions: from the MD perspective, the outermost layer of overlapping Cartesian grid cells is used to transfer mass using particle insertion and removal mechanisms. The following two cell layers are used to exchange momentum by imposing the average acceleration from the CFD solution onto the MD system. Boundaries of the MD region are either modelled with periodic boundaries where appropriate or with reflecting boundaries to hinder molecules from escaping, together with additional boundary forcing to impose the correct average hydrodynamic pressure on the MD system.

This approach is sufficient for *one-way coupling*, i.e. to couple the CFD solver to the MD solver, which is subject to the current work. Note that, to enable *two-way coupling*, MD data such as average flow velocities are sampled cellwise in the inner region of the MD simulation and are imposed on CFD side as boundary condition.

2.4 MaMiCo

MaMiCo stands for macro-micro coupling tool and shall facilitate the coupling of particle- and mesh-based solvers for multiscale fluid flow simulations, with focus on molecular-continuum coupling [6,7,10,11]. It provides both steady-state and transient coupling schemes, with the latter given in the scope of the algorithm presented in Sect. 2.3.

MaMiCo especially provides the ability to couple CFD solvers to ensembles of MD simulations: let a MD domain be embedded somewhere in a big CFD domain. Then, the MD simulation is computed multiple times starting from random initial configurations. This yields an ensemble of MD simulations whose average hydrodynamic data such as flow velocities can be computed locally (i.e. per CFD-related grid cell). An averaging procedure is essential, since MD exhibits strong thermal fluctuations in typical scenarios of interest. These fluctuations may be critical for the stability of the CFD solver, if this solver is not compressible and incorporating thermal fluctuations in a comparable way. Ensemble averaging (1) removes potential biases from time-dependent averaging, (2) allows to investigate scenarios on shorter time scales, yet relying on a coarse-scale CFD view and (3) is preferable in terms of leveraging supercomputing capacities due to the embarrassingly parallel nature of running the different ensemble members.

MaMiCo is parallelized with MPI, supporting arbitrary domain decompositions on CFD side, blockwise domain decomposition on MD side and ensemble-wise parallelism for the MD simulations. The tool further provides coupling interfaces to a multitude of MD codes such as *LAMMPS* and *ls1 mardyn* [11]. For the current work, we focus on the use of *SimpleMD*, a built-in MD simulation that is shipped with *MaMiCo*, and a built-in LB solver.

2.5 Related work

Molecular dynamics-continuum coupled simulations have been the subject of numerous studies; we will only review selected approaches which make use of ensembles or well-chosen sets of MD simulations.

In [15], a molecular-continuum method is introduced in which MD simulations are dynamically launched to investigate unsteady channel flow. MD simulations are executed at different cross-sections of the channel, their data is gathered in a database and interpreted via Gaussian process regression to steer accuracy of the flow over long time scales. This is different to our approach, in which we currently only resolve one particular sub-region of the considered computational domain with MD.

Besides, various sampling methods for MD systems have been developed in the past to assess long time scale behavior or large molecular systems. Examples are given by equation of state sampling [8] or the investigation of rare events [4]. In our work, we consider quasi-identical MD simulations to explore a single thermodynamic non-equilibrium state.

To the best of our knowledge, no HPC-aware, massively parallel molecular-continuum simulation approach with dynamic ensemble handling has been developed until now.

3 Implementation of dynamic MD ensemble handling

The MultiMDCellService of *MaMiCo* is the central class to incorporate the coupling of an MD ensemble and a single instance of a CFD solver [10]. This service manages an array of MacroscopicCellService instances, each being responsible for a single MD simulation in the coupling [11]. Each MPI rank of a running simulation holds one instance of the MultiMDCellService, which in turn manages a MacroscopicCellService for every available MD simulation. Two implementations of the MacroscopicCellService are used: MacroscopicCellServiceImpl manages the interactions between the CFD solver and MD solvers. The other implementation, MacroscopicCellServiceMacroOnly, is necessary if the MPI parallelization of the respective CFD solver requires communication with an MD simulation that does not have a sub-domain to be computed on the CFD solver ranks.

During the setup of *MaMiCo*, the MPI ranks are being grouped into equally sized blocks. The size of these groups corresponds to the number of processes per MD instance, which is the same across all instances. These groups of MPI ranks are also ordered according to their respective ranks; e.g., assuming a $2 \times 2 \times 2$ block-wise decomposition of the MD domain and a total of 24 processes being available, we obtain 3 groups with ranks 0-7, 8-15, 16-23.

The initialization of *MaMiCo* works such that the MD simulations are homogeneously distributed across the process groups. Oversubscription is possible, i.e. multiple MD simulations can execute within a process group. For example, in the constellation from above, we could run 6 MD simulations, with each process group handling 2 MD simulations.

Inserting or removing arbitrary numbers of MD simulations would break this homogeneity in MD simulation load. We therefore add a layer of abstraction by introducing slots in *MaMiCo*. The number of slots in a process group corresponds to the size of the MacroscopicCellService array of the MultiMDCellService. It can furthermore be active or inactive. An active slot's MacroscopicCellService takes part in the running simulation. If it is inactive, no work or communication will be performed.

For management of the slots we introduce two new classes in *MaMiCo*. First, InstanceHandling can be used to centralize all tasks regarding the use of MD instances. The InstanceHandling holds an STL vector of MD simulations and one for MicroscopicSolverInterfaces. Initialization, execution of MD time steps and shutdown are abstracted into this class. Second, the class MultiMDMediator manages the slots; adding new slots as well as activation/ deactivation is handled. The MultiMDMediator is closely coupled with the MultiMDCellService and the InstanceHandling.

The process of launching a new MD simulation works as follows. A slot is chosen either manually by declaring the exact slot or the MPI process group from which a slot should be chosen, or according to the MultiMDMediator, which tries to keep the number of MD simulations per rank balanced by applying a round-robin scheme.

After a slot has been selected, the MultiMDMediator delegates to the MultiMDCellService which initializes a new MacroscopicCellService which in turn corresponds to the selected slot. In the activated slot, a new MD simulation is launched, together with a corresponding communicator (related to the process group).

Each process group regularly saves checkpoints. Upon launch, the new MD instance is initiated from the last available checkpoint. As this checkpoint will basically always reflect a deprecated state of the flow behavior, the new MD instance first needs to be equilibrated and pushed towards the current CFD state over a defined number of coupling cycles. Currently, we use 10 coupling cycles, with each coupling cycle comprising 50 MD time steps, in the concurrently coupled simulation. During this equilibration phase, the rest of the coupling simulation continues. The new simulation receives coupling information from the CFD solver, but does not take part in the ensemble sampling or sending of information to the CFD solver; note that this sampling is also relevant in the one-way coupling to extract flow information from the MD system for post-processing. After the equilibration phase, the new simulation is activated and starts behaving like all other MD simulations, i.e. it is fully integrated into the MD ensemble.

On the one hand, to achieve high performance, the frequency at which the required checkpoints are written should be low to avoid frequent I/O. On the other hand, for the sake of physical consistency and rapid equilibration, checkpoints should be written as often as possible. This opposition will require further research and evaluations to find a good compromise in the future.

Furthermore, the checkpoint used for initialization of the simulation resembles the state of another running MD simulation from an unknown number of coupling cycles before; the checkpoint might even originate from the very last finished cycle—which is good in terms of physical consistency with regard to the CFD state, but which is very bad in terms of adding an *entirely independent* ensemble member to the ensemble of MD simulations. In addition to the equilibration phase, we therefore vary the particles' velocities of the checkpoint. For particles residing in one CFD grid cell, we set their new velocities to the mean flow velocity in this grid cell and add Gaussian noise, resembling the corresponding temperature. This will introduce enough chaos into the system so that the state of the new MD simulation will rapidly diverge from the original MD simulation which produced the checkpoint.

Removal of simulations is simpler. The MD simulation and its corresponding MDSolverInterface are shut down. Then, the respective instance of the Macroscopic-CellService is removed. Finally, the selected slot will be set to inactive. This slot is now available again for the launch of a new MD instance in the future.

4 Results

In the following subsections, the name MD30 refers to a specific single-site Lennard–Jones (LJ) MD domain and simulation configuration shown in Tab. 1. This MD configuration is coupled with the LB solver in order to simulate Couette flow: in this scenario, we consider the 3D flow between two plates, with the lower plate moving at fixed velocity. Although the evolving flow profile is typically one-dimensional, we run the full 3D simulation for the sake of preparing our entire simulation methodology for more complex, potentially fully 3D flow scenarios in the future. From the configuration MD30, we derive MD60 by doubling all MD and CFD domain sizes, while keeping the other parameters fixed, cf. also [10] for more details.

We first validate the correctness of our implementation. For this purpose we prepare an MD30 scenario using *MaMiCo*'s integrated MD simulation code *SimpleMD* in Sect. 4.1. We then investigate the scalability of the implementation running two distinctive scenarios on the HAWK system in Sect. 4.2.

Table 1: Configuration parameters for MD30 setup.

Parameter	Value	Parameter	Value
molecular mass (m)	1	density ρ	0.81
σ	1	kinematic viscosity	$2.64/\rho$
ϵ	1	channel size	(50,50,50)
Boltzmann constant	1	MD domain size	(30,30,30)
temperature	1.1	moving wall velocity	0.5
cut-off radius r_c	2.2	MD domain center	(25,25,17.5)
MD time step	0.005	CFD time step	0.25

4.1 Validation

We validate our implementation in a MD30 Couette flow scenario, cf. Tab. 1 for the parametrization. The coupling is initialized using 128 SimpleMD simulations that are coupled to the LB solver. The simulation is executed over 1000 coupling cycles, i.e. CFD time steps. After the first 100 cycles, dynamic launch and removal is activated. For this purpose, every 20 coupling cycles, we choose a random number $r \in (-50, 50)$ which corresponds to the number of MD simulations that shall be removed from ($r < 0$) or added to ($r > 0$) the ensemble.

This study was performed using one full node on the Hawk system. We applied a domain decomposition scheme with $2 \times 2 \times 2$ processes on LB side and $2 \times 2 \times 2$ processes for every MD simulation. We use the notation LB$2 \times 2 \times 2$-MD$2 \times 2 \times 2$ for these parallel configurations in the following. The entire simulation consumed approx. 105 core-h.

Figure 2 shows the Couette flow profile between the plates after different coupling cycles. The velocity is measured across a 1D cross-section between the plates. It can be observed, that the MD ensemble solution follows the analytical solution very well. It is further demonstrated that varying the size of the MD ensemble affects the degree of deviation from the analytical profile: before the start of the dynamic MD simulation handling, the measured state fits the expected state well. At cycle 130, we can see some slightly higher fluctuations around the expected state. This fluctuation is even larger at cycle 410, whereas at cycle 640 it has decreased again and is thus closer to the expected state.

Comparing these observations to Fig. 3, which shows the number of MD simulations M over the course of the simulation, we see that the number of active MD instances first decreases and reaches a number of 32 at cycle 410. After this cycle, the number of MD instances within the ensemble starts to increase again, reaching its

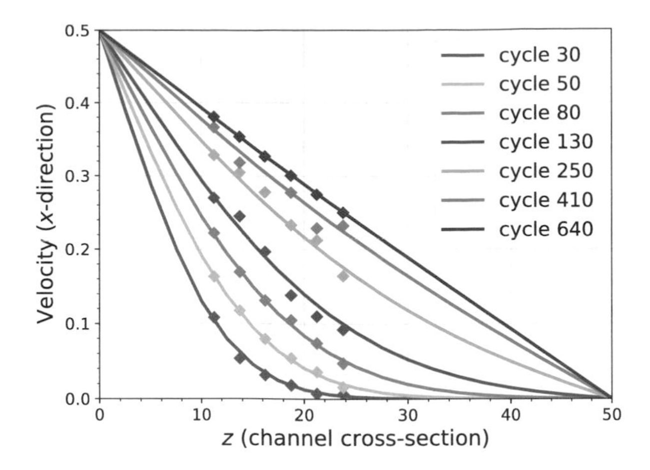

Fig. 2: Couette flow profile of the simulation after different coupling cycles. The profile is sampled across a 1D cross-section between the plates. Lines depict the analytical flow profile, diamonds depict the state sampled from the MD ensemble.

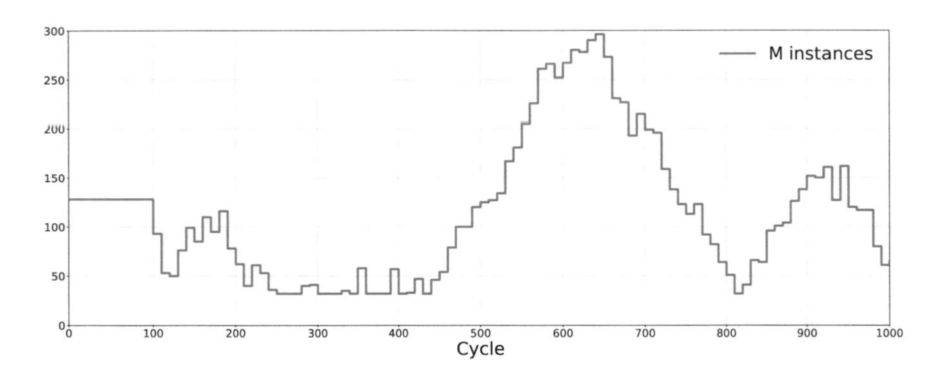

Fig. 3: Dynamic change of the number of MD instances over the course of the entire coupled simulation.

maximum of 298 after cycle 640. We can thus conclude that our dynamic coupling scheme qualitatively works as expected. A more rigorous error estimator is under current development and has already confirmed our findings.

4.2 Scalability

We investigate the strong scalability on Hawk in two Couette scenarios, considering MD30 and MD60. For MD30, we run each simulation over 100 coupling cycles and perform random launch/removal of MD instances every 10 cycles.

In the second scenario based on MD60, we run a coupling simulation over a total of 20 coupling cycles. Random launch or removal happens every 5 cycles.

To retain comparability among all runs, we impose a lower limit of 32 active MD instances. This is necessary due to the fact that the actual reasonable minimum number of MD simulations is defined by the number of MPI communicators: once a process group becomes entirely empty, we waste these idle resources and can—without significant loss in performance—execute at least one MD simulation in the corresponding slots, since there will be at least one other process group that executes at least one MD simulation. As we investigate strong scalability, the number of communicators varies across setups. The number of communicators effectively sets the minimal number of MD simulations running at any time. To circumvent this behavior and thus ensure comparability between the runs, we enforce a minimal number of 32 instances on setups that run with less communicators. This furthermore limits the range of possible setups to those consisting of no more than 32 communicators.

Table 2: Time-to-solution (in seconds) on the MD30 scenario. A configuration of (LB$M \times M \times M$-MD$N \times N \times N$) corresponds to a decomposition of the LB solver into $M \times M \times M$ equally sized sub-domains and of the MD domain into $N \times N \times N$ sub-domains.

Topology	# cores									
	1	2	4	8	16	32	64	128	256	512
LB $1 \times 1 \times 1$-MD$1 \times 1 \times 1$	19608	10188	5174	3098	2044	1037	461			
LB $1 \times 1 \times 1$-MD$2 \times 2 \times 2$				3894	2059	1034	680	573	286	120
LB $2 \times 2 \times 2$-MD$2 \times 2 \times 2$				3920	1982	1030	675	562	285	117

Table 2 shows the runtime results of the MD30 configuration. The 1-core topology was run in sequential mode, i.e. without MPI parallelization. This study consumed about 280 core hours.

The run time results of MD60 are shown in Tab. 3. This study consumed about 920 core hours.

The speedups for both MD30 and MD60 are displayed in Fig. 4. We observe that both scenarios scale reasonably well. In the MD30 scenario, there is a drop in efficiency visible for the larger setups of the LB$2 \times 2 \times 2$ decomposition. This is due to the relatively small-sized MD domain and the actual communication pattern: with

Table 3: Time-to-solution (in seconds) on different parallel topologies of MD60. Nomenclature follows that of MD30 (cf. Table 2).

Topology	# cores												
	1	2	4	8	16	32	64	128	256	512	1024	2048	4096
LB1 × 1 × 1-MD 1 × 1 × 1	48933	26897	14008	7589	4574	2320	1290						
LB1 × 1 × 1-MD2 × 2 × 2				7908	3972	2084	1202	889	453	225			
LB2 × 2 × 2-MD2 × 2 × 2				7877	3966	2085	1191	901	466	231			
LB2 × 2 × 2-MD4 × 4 × 4							1688	1053	567	301	177	91	50
LB4 × 4 × 4-MD2 × 2 × 2							1677	937	526	299	178	90	46

a domain decomposition of MD2 × 2 × 2, each sub-domain contains around 2740 particles. Considering the initial number of 2 MD simulations per communicator, the load per process is relatively small. Additionally, each process has to execute a relatively high number of small MPI send operations. In contrast, in the scenario LB1 × 1 × 1-MD2 × 2 × 2, the same amount of data needs to be sent on average via MPI, but each process needs to execute less send operations per MD instance.

For MD60, we do not observe this drop in performance. As discussed above, it was not possible with this setup to scale MD60 any further without increasing the minimum number of MD instances. Larger simulations are to be executed in the future.

We also investigated the time spent in the actual launch/removal phases of MD30. We achieve a mean of 1.7% of the overall runtime. None of the runs exceeded a total launch/removal time fraction of 4.5%.

5 Summary and outlook

We have introduced a mechanism for dynamic launch and removal of coupled MD ensembles. Our results show that our implementation works in reasonable time frames, even when applying it in drastic scenarios of dynamically launching/ removing large numbers of MD simulations at short time intervals. We have yet to perform further scalability studies at larger scale, especially with regard to weak scaling, and also in combination with larger setups. This is subject to a HLRS/GCS project extension that we recently applied for.

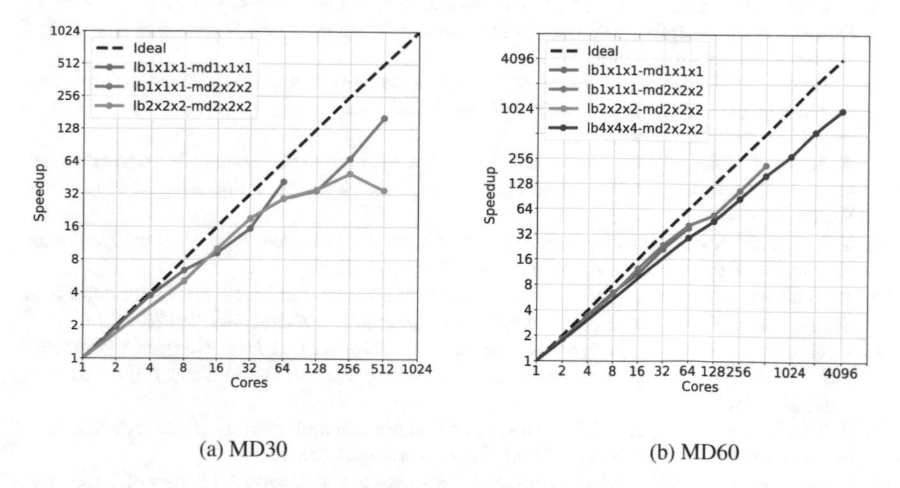

(a) MD30

(b) MD60

Fig. 4: Scalability of dynamic launch/removal for scenarios (a) MD30 and (b) MD60.

This work forms the ground for the ongoing research at Helmut-Schmidt-University Hamburg to enable error estimation and fault tolerance in molecular-continuum simulations, exploiting the MD ensemble approach.

Acknowledgements We thank HLRS and GCS for providing computational resources in the scope of the project GCS-MDDC, ID 44130. We further thank HSU for supporting our project through the HSU-internal research funding program (IFF), project "Resilience and Dynamic Noise Reduction at Exascale for Multiscale Simulation Coupling".

References

1. F. Diewald, M.P. Lautenschlaeger, S. Stephan, K. Langenbach, C. Kuhn, S. Seckler, H.-J. Bungartz, H. Hasse, and R. Müller. Molecular dynamics and phase field simulations of droplets on surfaces with wettability gradient. *Computer Methods in Applied Mechanics and Engineering*, 361:112773, 2020.
2. F.A. Gratl, S. Seckler, H.-J. Bungartz, and P. Neumann. N Ways to Simulate Short-Range Particle Systems: Automated Algorithm Selection with the Node-Level Library AutoPas. *Submitted*, 2020.
3. S. Grottel, M. Krone, C. Muller, G. Reina, and T. Ertl. Megamol – a prototyping framework for particle-based visualization. *Visualization and Computer Graphics, IEEE Transactions on*, 21(2):201–214, 2015.
4. S. Hussain and A. Haji-Akbaria. Studying rare events using forward-flux sampling: Recent breakthroughs and future outlook. *Journal of Chemical Physics*, 152:060901, 2020.
5. V. Jafari, N. Wittmer, and P. Neumann. Massively parallel molecular-continuum flow simulation with error control and dynamic ensemble handling. *International Conference on High Performance Computing in Asia-Pacific Region*, page 52–60, 2022.

6. P. Jarmatz, F. Maurer, and P. Neumann. MaMiCo: Non-Local Means Filtering with Flexible Data-Flow for Coupling MD and CFD. *Lecture Notes in Computer Science (ICCS 2021 proceedings)*, pages 576–589, 2021.
7. P. Jarmatz and P. Neumann. MaMiCo: Parallel Noise Reduction for Multi-Instance Molecular-Continuum Flow Simulation. *Lecture Notes in Computer Science (ICCS 2019 proceedings)*, pages 451–464, 2019.
8. A. Köster, T. Jian, G. Rutkai, C. Glass, and J. Vrabec. Automatized determination of fundamental equations of state based on molecular simulations in the cloud. *Fluid Phase Equilibria*, 425:84–92, 2016.
9. T. Krüger, H. Kusumaatmaja, A. Kuzmin, O. Shardt, G. Silva, and E.M. Viggen. *The Lattice Boltzmann Method. Principles and Practice*. Springer, 2016.
10. P. Neumann and X. Bian. MaMiCo: Transient Multi-Instance Molecular-Continuum Flow Simulation on Supercomputers. *Comput. Phys. Commun.*, 220:390–402, 2017.
11. P. Neumann, H. Flohr, R. Arora, P. Jarmatz, N. Tchipev, and H.-J. Bungartz. MaMiCo: Software design for parallel molecular-continuum flow simulations. *Comput. Phys. Commun.*, 200:324–335, 2016.
12. X. Nie, S. Chen, W. E, and M. Robbins. A continuum and molecular dynamics hybrid method for micro-and nano-fluid flow. *J. Fluid Mech.*, 500:55–64, 2004.
13. C. Niethammer, S. Becker, M. Bernreuther, M. Buchholz, W. Eckhardt, A. Heinecke, S. Werth, H.-J. Bungartz, C.W. Glass, H. Hasse, J. Vrabec, and M. Horsch. ls1 mardyn: The massively parallel molecular dynamics code for large systems. *Journal of Chemical Theory and Computation*, 10(10):4455–4464, 2014.
14. S. Seckler, F. Gratl, M. Heinen, J. Vrabec, H.-J. Bungartz, and P. Neumann. AutoPas in ls1 mardyn: Massively Parallel Particle Simulations with Node-Level Auto-Tuning. *Journal of Computational Science*, 50:101296, 2021.
15. D. Stephenson, J.R. Kermode, and D.A Lockerby. Accelerating multiscale modelling of fluids with on-the-fly Gaussian process regression. *Microfluidics and Nanofluidics*, 22:139, 2018.

Scalable discrete algorithms for big data applications

Demian Hespe, Lukas Hübner, Lorenz Hübschle-Schneider, Peter Sanders and Dominik Schreiber[†]

Abstract In the past year, the project "Scalable Discrete Algorithms for Big Data Applications" dealt with *High-Performance SAT Solving*, *Malleable Job Scheduling and Load Balancing*, and *Fault-Tolerant Algorithms*. We used the massively parallel nature of ForHLR II to obtain novel results in the areas of SAT solving and fault-tolerant algorithms.

1 Introduction

Developing scalable algorithms is a challenging task that requires careful analysis and extensive experimental evaluation. CPU technology shifts to deliver increasing amounts of cores with relatively low clock rates, since these are cheaper to produce and operate. Parallelizing computationally intensive algorithms at the core of all applications is therefore an important research topic. Developing distributed algorithms on cluster computers such as the ForHLR II is an integral part of this scalability challenge. Our focus is especially on discrete algorithms, such as graph partitioning, text search, and propositional satisfiability (SAT) solving.

In the past year, we used the ForHLR II cluster in the scope of three subprojects. First, we designed and implemented a massively-parallel and distributed SAT solving system [22, 24] (Section 2.1). The central novelty of our solver is a succinct and communication-efficient exchange of information among the core solvers which helps

[†] All authors contributed equally and are sorted alphabetically.

Demian Hespe, Lukas Hübner, Lorenz Hübschle-Schneider, Peter Sanders and Dominik Schreiber
Institute for Theoretical Informatics: Algorithms II, Karlsruhe Institute of Technology (KIT), Am Fasanengarten 5, Karlsruhe, 76131, Germany, e-mail: demian.hespe@kit.edu, huebner@kit.edu, huebschle@kit.edu, sanders@kit.edu, schreiber@kit.edu

Lukas Hübner
Computational Molecular Evolution Group, Heidelberg Institute for Theoretical Studies, Schloss-Wolfsbrunnenweg 35, Heidelberg, 69118, Germany, e-mail: huebner@kit.edu

W. E. Nagel et al. (eds.), *High Performance Computing in Science and Engineering '21*,
https://doi.org/10.1007/978-3-031-17937-2_27

to speed up the solving process. Secondly, in order to improve resource efficiency and to reduce scheduling times for the resolution of difficult SAT problems in cloud-like HPC environments, we explored novel dynamic scheduling and load balancing approaches (Section 2.2). For this means, we exploit *malleability* of tasks; that is, the task's capability of handling a fluctuating number of processing elements during its execution. Thirdly, we explored parallel fault-tolerance mechanisms and algorithms, the software modifications required, and the performance penalties induced via enabling parallel fault-tolerance by example of RAxML-NG, the successor of the widely used RAxML tool for maximum likelihood based phylogenetic tree inference (Section 2.3).

In previous years, we studied distributed online sorting and string sorting in our Big Data toolkit Thrill, developed a scalable approach to edge partitioning, developed and evaluated algorithms for maintaining uniform and weighted samples over distributed data streams (reservoir sampling), and designed new approaches to massively parallel malleable job scheduling applied to propositional satisfiability (SAT) solving. To conclude our project on scalable discrete algorithms in the scope of ForHLR II, we provide a compact project retrospective in Section 6.

2 Scientific work accomplished and results obtained

2.1 High-Performance SAT Solving

Satisfiability (SAT) solving deals with one of the most famous $\mathcal{NP}$-complete problems and has many interesting applications in software verification, theorem proving and automated planning. We designed and implemented a massively-parallel and distributed SAT solving system [22, 24] which is also able to gracefully handle fluctuating computational resources (see Section 2.2). The central novelty of our solver is a succinct and communication-efficient exchange of information among the core solvers which helps to speed up the solving process.

Experiments on up to 128 compute nodes of the ForHLR II showed that our solving system named Mallob significantly outperforms its precursor HordeSat [2] and shows much better scaling properties [24]. Fig. 1 shows a direct comparison between HordeSat and our new solver Mallob whereas both make use of exactly the same backend solvers. As HordeSat fails to scale beyond 32 nodes, Mallob on 32 nodes outperforms HordeSat on 128 nodes. Furthermore, Mallob can make effective use of up to 128 nodes. We provide more detailed scaling results in Table 1. To the best of our knowledge, the speedups achieved by Mallob are the best speedups reported by any SAT solver in an HPC environment so far.

Our solver scored the first place in the first *Cloud Track* of the International SAT Competition 2020 [1] where Mallob-mono was executed on 100 8-core nodes of an AWS infrastructure and solved more formulae than any other solver in the competition.

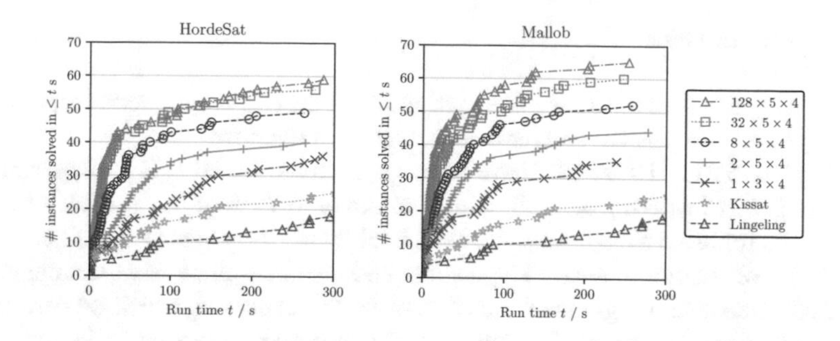

Fig. 1: Scaling behaviour of HordeSat (with updated solvers) and untuned Mallob compared to two sequential solvers [24].

Table 1: Parallel speedups for HordeSat (H) and Mallob (M). In the left half, "#" denotes the number of instances solved by the parallel approach and S_{med} (S_{tot}) denotes the median (total) speedup for these instances compared to Lingeling / Kissat. In the right half, only instances are considered for which the sequential solver took at least (num. cores of parallel solver) seconds to solve. Here, "#" denotes the number of considered instances for each combination [24].

	All instances					Hard instances					
		Lingeling		Kissat			Lingeling			Kissat	
Config.	#	S_{med}	S_{tot}	S_{med}	S_{tot}	#	S_{med}	S_{tot}	#	S_{med}	S_{tot}
H1×3×4	36	3.84	51.90	2.22	29.55	32	4.39	52.01	31	4.03	32.49
H2×5×4	40	12.00	95.80	5.06	64.44	35	12.27	96.83	33	9.11	69.63
H8×5×4	49	22.83	135.55	9.76	90.08	38	32.00	142.76	32	24.88	105.94
H32×5×4	56	42.12	203.66	15.25	112.14	34	97.61	231.77	19	114.86	208.68
H128×5×4	59	50.35	204.10	17.38	111.46	21	356.33	444.12	10	243.42	375.04
M1×3×4	35	4.83	58.15	3.62	64.66	31	5.37	58.24	30	5.29	66.08
M2×5×4	44	12.98	94.44	10.52	67.71	39	14.37	95.28	37	11.54	69.25
M8×5×4	52	28.38	154.62	12.06	89.61	41	34.29	162.23	34	23.43	106.85
M32×5×4	60	53.75	220.92	23.41	148.57	37	152.19	245.54	23	134.07	262.04
M128×5×4	65	81.60	308.48	25.97	175.58	25	363.32	447.97	12	363.32	483.11

We submitted a new version of Mallob to the upcoming International SAT Competition 2021 as well [23] and are excited to see how our system performs this year. Mallob is open-source and can be obtained from https://github.com/domschrei/mallob.

2.1.1 Usage statistics

The following measures also subsume Section 2.2, as the two subprojects are integrated in a single software system and have been evaluated together.

In total, 675 219 TRES hours have been spent on the ForHLR II for experiments and evaluations involving Mallob, which amounts to 71% of our project's overall usage in the reported time frame. We used twelve to 2560 cores (1 to 128 compute nodes). As we are committed to responsible and resource-efficient experiments, we identified a statistically significant selection of SAT instances on which we performed most experiments as opposed to running all experiments on a much larger set of benchmarks [24]. Furthermore, we limited the run time of parallel solvers to 300 s per formula in most cases. For the experiments with sequential solvers, we scheduled multiple solvers to run in parallel on a single compute node in order to make efficient use of resources.

2.2 Malleable job scheduling and load balancing

In order to improve resource efficiency and to reduce scheduling times for the resolution of difficult problems in cloud-like HPC environments, we explored novel dynamic scheduling and load balancing approaches. For this means, we exploit *malleability* of tasks: A malleable task is capable of handling a fluctuating number of processing elements during its execution. We developed a novel system named Mallob for the scheduling and load balancing of malleable tasks [24]. New jobs entering the system are scheduled virtually immediately (mostly within tens of milliseconds) and balanced through a fully decentralized load balancing protocol: Lightweight asynchronous and event-driven message passing ensures a fair distribution of resources according to the priorities and demands of active jobs. As mentioned in Section 2.1, we developed a scalable distributed SAT solver which can handle malleability and integrated it into our system as an exemplary application.

Experiments on up to 128 nodes on the ForHLR II [24] showed that our job scheduling and load balancing framework imposes minimal computational and communication overhead and dynamically assigns active processing elements to jobs in a fair manner. Most jobs arriving in the system are initiated within tens of milliseconds. In the context of our application, we showed that our system is able to find an appealing trade-off between the "trivial" resource efficiency of solving many formulae at once and the speedups of flexible parallel SAT solving. For instance, we experimentally compared Mallob's malleable processing of 400 SAT jobs with an "embarrassingly parallel" processing using 400 sequential SAT solvers in parallel and with a hypothetical optimal sequential scheduling (HOSS) where 400 corresponding runs of 128-node Mallob-mono are sorted ascendingly by their run time. As Fig. 2 shows, Mallob with malleable scheduling outperforms any of these extremes and achieves low response times and a high number of solved instances.

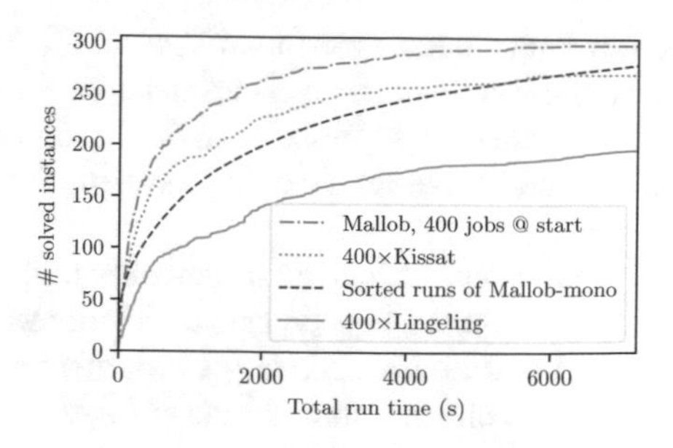

Configuration	R_{all} avg.	R_{all} med.	R_{slv} avg.	R_{slv} med.
Mallob $J = \infty$	2422.4	679.8	808.6	260.6
400×Kissat	2998.4	1362.5	975.5	355.5
HOSS	2774.7	2024.5	1396.4	937.3
400×Lingeling	4436.0	7200.0	1559.2	819.9

Fig. 2: Cumulative solved instances by different scheduling approaches on 128 compute nodes within two hours. The table shows average and median response times, calculated for all 400 instances (R_{all}) and for the solved instances per approach (R_{slv}). Each unsolved instance leads to a response time of 7200 s [24].

2.3 Fault-tolerance for massively parallel phylogenetic inference

Failing hardware is projected to be one of the main challenges in future exascale systems [25]. In fact, one may expect that a hardware failure will occur in exascale-systems every 30 to 60 min [5, 8, 26]. The reasons for failures are manifold: HPC systems can fail due to core hangs, kernel panics, file system errors, file server failures, corrupted memories or interconnects, network outages, air conditioning failures, or power halts [12, 18].

We apply fault-tolerance to RAxML-NG [17], a phylogenetic tree inference tool. Phylogeny is one of the central applications in bioinformatics. Given a set of aligned input sequences, the task is to reconstruct the phylogenetic tree which maximizes the (negative) log-likelihood of this tree showing the evolutionary relationships among the input entities. Phylogenetic trees are now routinely inferred on large scale HPC systems with thousands of cores as the parallel scalability of phylogenetic inference tools has improved over the past years to cope with the molecular data avalanche [17]. Thus, the parallel fault-tolerance of phylogenetic inference tools has become a relevant challenge. To this end, we explore parallel fault-tolerance mechanisms and algorithms, the software modifications required, and the performance penalties induced via enabling parallel fault-tolerance by example of the RAxML-NG software [17], the successor of the widely used RAxML tool for phylogenetic inference [27]. We find that the slowdown induced by the necessary additional recovery mechanisms in RAxML-NG is on average 1.00 ± 0.04. The overall slowdown by using these recovery mechanisms in conjunction with a fault-tolerant MPI implementation amounts to on average 1.7 ± 0.6 for large empirical datasets [14]. Via failure simulations, we show that RAxML-NG can successfully recover from multiple simultaneous failures, subsequent failures, failures during recovery, and failures during checkpointing. Recoveries are automatic and transparent to the user [14].

We developed, tested, and benchmarked this fault-tolerant version of RAxML-NG (FT-RAxML-NG) on the ForHLR II. We used the results of these experiments in our first submission of this work to the Oxford University Press Bioinformatics journal. As the ForHLR II was then discontinued we had to use another HPC system for further experiments as part of the revision.

In the scope of this project, we deployed and tested ULFM, a failure-mitigating MPI implementation on the ForHLR II. This also included experiments to determine the most reliable failure detection method and optimizing recovery times after a failure. Various methods to simulate failures on the cluster without interfering with other jobs or needing administrative intervention were evaluated [14].

2.3.1 Usage statistics

For the experiments described in Section 2.3 we used 20 to 400 cores. We did some preliminary tests with large jobs using 512 nodes. As we saw runtime fluctuations of over 300 % in these tests and each job took around two weeks to be scheduled, we decided against using the ForHLR II for such large jobs.

RAxML-NG supports parallelization at three levels. At the single thread level, it uses parallelism as provided by the x86 vector intrinsics (SSE3, AVX, AVX2). At the single node level, RAxML-NG leverages the available cores by parallelization using PThreads. If we run RAxML-NG on a distributed memory HPC system, it uses parallelization via message passing (using MPI) [19, 28]. We can enable all three levels of parallelism at the same time.

As this was not the focus of this work, we did not perform any scaling experiments using FT-RAxML-NG.

3 Publications with project results

The technical report describing our winning submission of our SAT solving system (Section 2.1) to the Cloud Track of the SAT Competition 2020 was published in the Proceedings of SAT Competition 2020 [22]. A full paper introducing the Mallob system for job scheduling and load balancing (Section 2.2) and analyzing the scalability of our SAT solving system has been accepted at the 24th International Conference on Theory and Applications of Satisfiability Testing [24]. We are currently in the process of editing another publication which will describe our scheduling and load balancing approaches in greater detail. Our new version of Mallob (Section 2.2) competes in the upcoming International SAT Competition 2021 [23]. Our work on fault-tolerant phylogenetic inference (Section 2.3) has been accepted for publication at the Bioinformatics Journal [14].

4 Ongoing theses within the project

We are supervising a promising ongoing master thesis project that used the ForHLR II cluster:

4.1 Low overhead fault-tolerant MapReduce for HPC clusters

The MapReduce framework [7] is a popular solution for executing tasks in parallel among multiple machines by formulating the task using *map* and *reduce* operations. The framework then takes care of parallelization, load balancing and fault-tolerance, i.e. continuing work when one of the machines used stops working. In the past years there have been several implementations of MapReduce both for cloud computing [7, 9, 13] and for HPC clusters using MPI [10, 11, 20]. However, fault-tolerance is usually implemented by storing checkpoints to a distributed file system [7, 9, 11, 13] or is omitted entirely [10, 20]. The master thesis of Charel Mercatoris aims at developing a MapReduce implementation in MPI that achieves fault-tolerance by storing all relevant information redundantly in memory: During normal computation, all messages sent remain stored on the sending machine. Additionally, all information that would be lost in case of a failure is sent to a different machine as backup. After a failure, only the most recent map and reduce functions have to be re-executed *only* on the data that resided on the machine that stopped working. Due to the type of data flow defined through the map and reduce functions, we only have to store redundant data for one round of map and reduce functions. As soon as the next backup-cycle is complete, the previous backup data can be discarded. Preliminary experiments have been performed on the ForHLR-II cluster and are now continued on a different system since ForHLR-II is no longer available. Figure 3 shows preliminary scaling experiments executed on ForHLR-II. We can see that the performance and scaling behavior virtually does not change when activating fault-tolerance mechanisms (like storing backup data redundantly). Even when we simulate a failure of 10% of all MPI ranks, we only observe a small slowdown.

5 Theses completed within the project

The master's thesis of Lukas Hübner looked into load-balance and fault-tolerance for massively parallel phylogenetic inference. After completion of this Master's Thesis we continued to look into this topic; our work is described in Section 2.3.

The dissertation of Lorenz Hübschle-Schneider [15] considers communication-efficient probabilistic algorithms for three fundamental Big Data problems: selection, sampling, and checking. ForHLR II was used to evaluate the weighted reservoir sampling algorithm, a batched distributed streaming algorithm for maintaining a

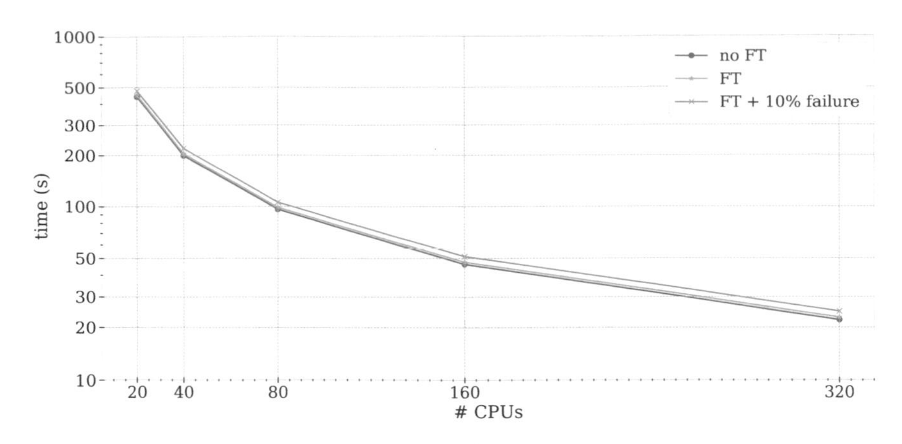

Fig. 3: Preliminary scaling experiments of our MapReduce implementation calculating PageRanks on a random grid graph with 2^{24} vertices. *no FT*: No fault-tolerant mechanisms enabled *FT*: Fault-tolerance mechanisms enabled but no faults occurred. *FT + 10% failure*: Fault-tolerance mechanisms enabled and the failure of 10% of the MPI ranks is simulated.

weighted random sample over an input that arrives over time. The results and which we previously described in last year's report, show good scalability on up to 256 nodes [16].

6 Project retrospective

In the following, we conclude our report by briefly reiterating the most notable project results which we achieved with the help of ForHLR II over the past years.

- **Thrill – A High-Performance Big Data Framework in C++.** We improved the sorting algorithm of Thrill [6], our next-generation C++ framework for distributed Big Data batch processing on a cluster of homogeneous machines which enables writing distributed applications conveniently using "dataflow" graph-like computations [3].
- **Graph Partitioning.** We developed a fast distributed graph construction algorithm and showed that combining it with advanced parallel node partitioning algorithms yields high-quality edge partitions scaling to networks with billions of edges on thousands of processors on large real-world networks [21].

- **Communication-Efficient Distributed String Sorting.** We proposed two new distributed algorithms for string sorting, Distributed String Merge Sort (MS) and Distributed Prefix-Doubling String Merge Sort (PDMS); experiments on up to 1280 cores revealed that the new algorithms are often more than five times faster than previous algorithms [4].
- **Reservoir Sampling.** We developed algorithms for uniform and weighted reservoir sampling in a distributed mini-batch model which scaled up to 256 nodes (5120 cores) and handily outperformed a centralized comparison algorithm [16].
- **Malleable Job Scheduling and Load Balancing.** We developed a novel system named Mallob for the scheduling and load balancing of malleable tasks [24], imposing minimal computational and communication overhead, dynamically assigning active processing elements to jobs fairly and scheduling most jobs within tens of milliseconds on 128 nodes.
- **High-Performance SAT Solving.** We designed and implemented a massively parallel and distributed SAT solving system [22, 24] which won the Cloud Track of the International SAT Competition 2020 [1] and which enables our Mallob system to resolve SAT jobs in a scalable manner.
- **Fault-Tolerance for Massively Parallel Phylogenetic Inference.** We applied fault-tolerance to RAxML-NG[17], a phylogenetic tree inference tool, and showed that the resulting software can successfully recover from multiple simultaneous failures, subsequent failures, failures during recovery, and failures during checkpointing [14].

For the mentioned subprojects, ForHLR II proved to be an invaluable resource for testing and scientifically evaluating our software. As such, we close with thanking the maintainers and the steering committee of the ForHLR II cluster for their kind support.

References

1. Tomáš Balyo, Nils Froleyks, Marijn JH Heule, Markus Iser, Matti Järvisalo, and Martin Suda. SAT competition, 2020. Accessed: 2021-03-19.
2. Tomáš Balyo, Peter Sanders, and Carsten Sinz. Hordesat: A massively parallel portfolio SAT solver. In *International Conference on Theory and Applications of Satisfiability Testing*, pages 156–172. Springer, 2015. Preprint arXiv:1505.03340 [cs.LO].
3. Timo Bingmann, Michael Axtmann, Emanuel Jöbstl, Sebastian Lamm, Huyen Chau Nguyen, Alexander Noe, Sebastian Schlag, Matthias Stumpp, Tobias Sturm, and Peter Sanders. Thrill: High-performance algorithmic distributed batch data processing with C++. In *2016 IEEE International Conference on Big Data*, pages 172–183. IEEE, 2016. Preprint arXiv:1608.05634 [cs.DC].
4. Timo Bingmann, Peter Sanders, and Matthias Schimek. Communication-Efficient String Sorting. In *34th IEEE International Parallel and Distributed Processing Symposium (IPDPS)*. IEEE, May 2020. to appear, preprint arXiv:2001.08516 [cs.DC].
5. Franck Cappello, Al Geist, William Gropp, Sanjay Kale, Bill Kramer, and Marc Snir. toward exascale resilience: 2014 update. *Supercomputing Frontiers and Innovations*, 1(1), September 2014.

6. Jonas Dann. improving distributed external sorting for big data in Thrill. Master Thesis. Karlsruhe Institute of Technology, Germany, September 2019.
7. Jeffrey Dean and Sanjay Ghemawat. MapReduce: simplified data processing on large clusters. In Eric A. Brewer and Peter Chen, editors, *6th Symposium on Operating System Design and Implementation*, pages 137–150. OSDI, 2004.
8. Jack Dongarra, Thomas Herault, and Yves Robert. Fault tolerance techniques for high-performance computing. `https://www.netlib.org/lapack/lawnspdf/lawn289.pdf`, 2015.
9. Jaliya Ekanayake, Hui Li, Bingjing Zhang, Thilina Gunarathne, Seung-Hee Bae, Judy Qiu, and Geoffrey Fox. Twister: a runtime for iterative mapreduce. In *Proceedings of the 19th ACM international symposium on high performance distributed computing*, pages 810–818, 2010.
10. Tao Gao, Yanfei Guo, Boyu Zhang, Pietro Cicotti, Yutong Lu, Pavan Balaji, and Michela Taufer. Mimir: memory-efficient and scalable MapReduce for large supercomputing systems. In *2017 IEEE international parallel and distributed processing symposium (IPDPS)*, pages 1098–1108. IEEE, 2017.
11. Yanfei Guo, Wesley Bland, Pavan Balaji, and Xiaobo Zhou. Fault-tolerant MapReduce-MPI for HPC clusters. In *Proceedings of the International Conference for High Performance Computing, Networking, Storage and Analysis*, pages 1–12, 2015.
12. Saurabh Gupta, Tirthak Patel, Christian Engelmann, and Devesh Tiwari. Failures in large scale systems. In *Proceedings of the International Conference for High Performance Computing, Networking, Storage and Analysis*, pages 1–12, November 2017.
13. Apache Hadoop website. `https://hadoop.apache.org/`. Accessed: 2021-06-24.
14. Lukas Hübner, Alexey M. Kozlov, Demian Hespe, Peter Sanders, and Alexandros Stamatakis. Exploring Parallel MPI Fault-Tolerance Mechanisms for Phylogenetic Inference with RAxML-NG.
15. Lorenz Hübschle-Schneider. *communication-efficient probabilistic algorithms: selection, sampling, and shecking*. PhD thesis, Karlsruher Institut für Technologie (KIT), 2020.
16. Lorenz Hübschle-Schneider and Peter Sanders. brief announcement: communication-efficient weighted reservoir sampling from fully distributed data streams. In *32nd ACM Symposium on Parallelism in Algorithms and Architectures (SPAA)*, pages 543–545, 2020. extended preprint: arXiv:1910.11069 [cs.DS].
17. Alexey M. Kozlov, Diego Darriba, Tomáš Flouri, Benoit Morel, and Alexandros Stamatakis. RAxML-NG: a fast, scalable and user-friendly tool for maximum likelihood phylogenetic inference. *Bioinformatics*, 35(21):4453–4455, May 2019.
18. Charng-Da Lu. failure data analysis of HPC systems. *Computer Science*, February 2013.
19. Wayne Pfeiffer and Alexandros Stamatakis. hybrid MPI/Pthreads parallelization of the RAxML phylogenetics code. In *2010 IEEE International Symposium on Parallel & Distributed Processing, Workshops and PhD Forum (IPDPSW)*, pages 1312–1313, 2010.
20. Steven J. Plimpton and Karen D. Devine. MapReduce in MPI for large-scale graph algorithms. *Parallel Computing*, 37(9):610–632, 2011.
21. Sebastian Schlag, Christian Schulz, Daniel Seemaier, and Darren Strash. Scalable edge partitioning. In *2019 Proceedings of the Twenty-First Workshop on Algorithm Engineering and Experiments (ALENEX)*, pages 211–225. SIAM, 2019.
22. Dominik Schreiber. Engineering HordeSat towards malleability: mallob-mono in the SAT 2020 cloud track. *SAT COMPETITION 2020*, page 45, 2020.
23. Dominik Schreiber. Mallob in the SAT competition 2021. *SAT COMPETITION 2021*, 2021. 38–39.
24. Dominik Schreiber and Peter Sanders. Scalable SAT solving in the cloud. In *International Conference on Theory and Applications of Satisfiability Testing*, pages 518–534. Springer, 2021.
25. John Shalf, Sudip Dosanjh, and John Morrison. Exascale computing technology challenges. In *Lecture Notes in Computer Science*, pages 1–25. 2011.
26. Marc Snir, Robert Wisniewski, Jacob Abraham, Sarita Adve, Saurabh Bagchi, Pavan Balaji, Jim Belak, Pradip Bose, Franck Cappello, Bill Carlson, Andrew Chien, Paul Coteus, Nathan DeBardeleben, Pedro Diniz, Christian Engelmann, Mattan Erez, Saverio Fazzari, Al Geist,

Rinku Gupta, Fred Johnson, Sriram Krishnamoorthy, Sven Leyffer, Dean Liberty, Subhasish Mitra, Todd Munson, Rob Schreiber, Jon Stearley, and Eric Van Hensbergen. Addressing failures in exascale computing. *The International Journal of High Performance Computing Applications*, 28(2):129–173, March 2014.
27. Alexandros Stamatakis. RAxML version 8: a tool for phylogenetic analysis and post-analysis of large phylogenies. *Bioinformatics*, 30(9):1312–1313, January 2014.
28. Alexandros Stamatakis, T. Ludwig, and H. Meier. Computing large phylogenies with statistical methods: Problems & solutions. In *Proceedings of 4th International Conference on Bioinformatics and Genome Regulation and Structure (BGRS2004)*, 2014.

Miscellaneous Topics

The Miscellaneous Topics section documents, besides the beauty, the breadth of numerical simulations. It supports research in many fields other than just the topics like fluid dynamics, aerodynamics, structure mechanics, and so forth. The following five articles show that today's computational approaches are by far not complete from a numerical or from a modeling point of view. However, the physically correct focus will lead to reliable prediction methods derived from data-driven or physics-driven approaches to substantiate new theories.

The report of the Goethe Universität Frankfurt is a first performance analysis of the software framework UG4 on the Hawk Apollo supercomputer. In Computational Science and Engineering (CSE) the access to high-end computational systems naturally sparks the desire to match the computational resources and to solve larger problems. Overall increasing complexity requires scalable and efficient computational methods for very large problems. Solving intricate applications requires a flexible and robust software infrastructure. In a classic setting the numerical solution of transient PDEs is obtained in a pipeline including the following steps. The governing equations are discretized in time and space. In a next step, the problem is linearized, before a resulting linear system can be solved. This strategy is incorporated in many open source projects for general purpose simulations. In the article, the focus is on the software UG4. A scaling study is provided for benchmark problems on Hawk. Moreover, results for a thermohaline flow problem are discussed.

The contribution from the Technische Universität Berlin addresses scaling issues related to the field of molecular dynamics simulations focusing on computational details. In the beginning, finite size effects in the context of multicomponent diffusion are investigated. Different methods to correct the influence of the finite simulation domain are compared. Next, the structure of a fluid near its critical point is discussed in the context of the strong scaling behavior. Subsequently, droplet coalescence dynamics determined by large molecular dynamics simulations and a macroscopic phase field model is analyzed. The performance of the new supercomputer Hawk is compared to that of Hazel Hen. Finally, the influence of the direct sampling of the energy flux on the performance of the code ls1 mardyn is described. Various speed tests are carried out and analyzed in detail.

The third article is a joint contribution of the Ruhr Universität Bochum and the Universität Hamburg. The topic is scalable multigrid for fluid dynamics shape optimization. To be more precise, a parallel approach for the shape optimization of an obstacle in an incompressible Navier–Stokes flow is investigated. A self-adapting nonlinear extension equation within the method of mappings is used, which links a boundary control to a mesh deformation. It is demonstrated how the approach preserves mesh quality and allows for large deformations by controlling nonlinearity in critical regions of the domain. Special focus is on reference configurations, where the transformation has to remove and create obstacle boundary singularities. This is particularly relevant for the employed multigrid discretizations. Aerodynamic drag optimizations for 2d and 3d configurations define benchmark problems. The efficiency of the algorithm is demonstrated in weak scalability studies performed on the supercomputer HPE Hawk for up to 5,120 cores.

The fourth article "Numerical calculation of Lean-Blow-Out (LBO) of a premixed and turbulent burner consisting of an array of jet flames" comes from the Engler-Bunte-Institute of Combustion Technology, Karlsruhe Institute of Technology. Swirl-stabilized flames have been the preferred choice in gas turbine combustors due to their aerodynamic means of stabilization. However, swirl flames can be susceptible to the issue of combustion instability in the lean combustion regime leading to Lean-Blow-Out and therefore damaging the hardware. The main task of this project is improving the low load capability of stationary gas turbines.

The focus is on new numerical calculations of the Lean-Blow-Out limit of a premixed and turbulent burner consisting of an ensemble of jet flames using large eddy simulations and two different approaches of modeling the turbulent flame interaction. The solver used for calculating the reactive flow is based on an OpenFOAM solver. The turbulent-chemistry interaction are treated by two different combustion models. In the first model the calculation of the source term is done via a density function of the reaction progress variable. The second model calculates the source term on the basis of the turbulent flame speed.

Simulations are validated through corresponding experiments. The numerically calculated Lean-Blow-Out limit is in good agreement with the experimental values

In the fifth contribution "Data-driven multiscale modeling of self-assembly and hierarchical structural formation in biological macro-molecular systems" the authors present and evaluate a multiscale modelling framework called the "Molecular Discrete Element Method" where interaction potentials are calculated via "Universal Kriging" from fine scale molecular dynamics simulations. The approach is tested for the hepatitis B viral protein HBcAg.

Wolfgang Schröder

Institute of Aerodynamics,
RWTH Aachen University,
Wüllnerstr. 5a,
52062 Aachen,
Germany,
e-mail: office@aia.rwth-aachen.de

Large scale simulations of partial differential equation models on the Hawk supercomputer

Ruben Buijse, Martin Parnet and Arne Nägel

Abstract This work presents a first performance analysis of the software framework UG4 on the Hawk Apollo supercomputer. The software demonstrated excellent scaling properties before. It has now been demonstrated that this also holds true for HLRS's new flagship and its architecture. Three aspects are emphasized: (i) classic weak scaling of a multi-grid solver for the Poisson equation, (ii) strong scaling for the heat equation using multi-grid-in-time, and (iii) application to a thermo-haline-flow problem in a fully-coupled fashion.

1 Introduction

Today, access to high-end computational systems is widely spread and available in ever increasing amounts. In Computational Science and Engineering (CSE) this naturally sparks the desire to match the computational resources and to solve larger and larger problems in the applications. Overall complexity increases not only with the mesh size, but also with the level of interactions and couplings between the physical unknowns. This requires scalable and efficient computational methods for very large problems.

Solving complex applications requires a flexible and robust software infrastructure. In a classic setting the numerical solution of transient PDE is obtained in a pipeline including the following steps: starting from the governing equations, (i) the problem is discretized in time and space first. In a next step (ii) the problem is linearized, before (iii) a resulting linear system can be solved. This strategy is incorporated in many open source projects for general purpose simulations such as deal.II [1, 18], Dune [3, 4], DuMu^x [9], FEFLOW [6], FEniCS[8, 20], IPARS [16], the Matlab Reservoir Toolbox (MRST) [22], or OpenGeoSys [19]. In this work, we focus on the software UG4

Arne Nägel
Goethe-Universität Frankfurt, G-CSC, Kettenhofweg 139, 60325 Frankfurt,
e-mail: naegel@gcsc.uni-frankfurt.de

W. E. Nagel et al. (eds.), *High Performance Computing in Science and Engineering '21*,
https://doi.org/10.1007/978-3-031-17937-2_28

[25, 30]. We provide a scaling study for benchmark problems on Hawk. Moreover, we provide results for a thermohaline flow problem. This PDE couples fluid flow, substance transport and heat transport and is an example for a complex PDE problem solved using parallel computing.

2 Methods

2.1 Multilevel solvers

For a large-scale simulations, the solution of the linear systems, turns out to be the limiting factor. In order to benefit from high performance computing, this step is performed using multigrid methods, as introduced in greater detail, e.g. in the monographs [14, 29]. These methods have proven to be highly scalable to 100,000 processes and beyond, e.g. [2, 11, 12, 17, 25, 25, 27]. Details of the MPI based parallelization of UG4's multigrid solver, highly scalable on hundreds of thousands of processes, have been also been described in previous reports, e.g. [15, 26]. Results on Hawk are provided in Section 3.1.

2.2 Time integration

Classic time stepping by one-step or multi-step-methods is a serial procedure. On modern and future architectures, clock speeds are stagnant and speedups will be likely only be available through greater concurrency. Hence researchers face a serial time-integration bottleneck. To that end, parallel in time methods, which have been developed for more than 50 years, recently have regained popularity [10] provides an overview of existing methods. An attractive approach is the multigrid in time method (MGRIT) [7]. This has been implemented in form of the XBraid library [21], which has been coupled with UG4 successfully [24]. This approach is investigated in Section 3.2.

2.3 Hawk Apollo supercomputer

The Hawk Apollo is a 26 Petaflop supercomputer at HLRS Stuttgart [1]. It consists of 5,632 compute nodes which are connected by an InfiniBand HDR200 interconnect network in a 9D-Hypercube. Each node consists of two AMD EPYC™ 7742 processors. This provides $2 \times 64 = 128$ cores per node, and a total of $720,896$ cores. The structure of the nodes is hierarchical: 4 cores form a core complex (CCX) sharing a 16 MB L3

[1] `https://kb.hlrs.de/platforms/index.php/HPE_Hawk`

cache [2]. 2 CCXs are combined to a cluster complex die (CCD), which forms a group of cores with a common interface to the I/O die. Each node is equipped with 256 GB DRAM memory. This structure will be investigated in the performance results later. All results are obtained using the `aocc/clang` compiler (version 2.1.0/9.0.0) in combination with the `mpt`-toolkit (version 2.23) and `mkl`-library (version 19.1.0).

3 Results

In this section we present results for three different simulation scenarios in 3D space:

- Weak scaling study for Poisson's equation,
- Strong scaling study for the heat equation for MGRIT,
- Results for a thermo-haline transport problem.

3.1 Weak scaling for Poisson's equation

We use a classic setup that has been used before[25]: Poisson's equation has been is solved on the unit cube with Dirichlet boundary conditions. The equation is discretized using finite elements on a structured hexagonal grid. Afterwards, the system is solved with a geometric multigrid solver. Damped Jacobi is used as a smoother, with 1 pre-/post smoothing step in a V-cycle respectively. Times for three different phases have been measured:

- The `assemble`-phase computes the finite element stiffness matrix. This is primarily dominated by computation, but memory access is required for reading element information and writing element stiffness matrix into a sparse-matrix format. No communication via MPI is taking place in this phase.
- The `init`-phase refers to the setup of the multilevel solver. Here the communication interfaces are constructed. Moreover, data for a proper application of matrix-vector operations is exchanged.
- The `apply`-phase involves primarily computations on vectors and sparse-matrix data structures. MPI communications occurs along vertical and horizontal process interfaces. For this phase, timings for a fixed number of five BiCGStab-sweeps are reported. This roughly corresponds to a reduction of the residual by eight orders of magnitude.

In a classic weak scaling, the number of cores and the number of degrees of freedom is increased simultaneously. Starting from a 3D grid with $2\times2\times2$ cubes, the grid has been refined uniformly. This results in an eightfold increase of the number

[2] `https://kb.hlrs.de/platforms/upload/Processor.pdf`

of degrees of freedom, which is matched by an increase of number of cores by a factor of eight. This strategy allows for an optimal distribution among processes. (cf. Tab. 1).

The scaling study starts with 8 cores on refinement level 6, which corresponds to roughly $250,000$ degrees of freedom per core[3]. On the levels ≤ 5, the number of processes is limited. The first two levels are stored on a single process. Afterwards, the number of increased processes is increased by a factor of 64 after every second refinement. This strategy limits the number of processes a single process needs to communicate with to a maximum of 64.

Table 1: Number of degrees of freedom.

Level	Processes used	Degrees of freedom	Process limit
0	n.a.	27	1
1	n.a.	125	1
2	n.a.	729	≤ 64
3	n.a.	4,913	≤ 64
4	n.a.	35,937	$\leq 4,096$
5	n.a.	274,625	$\leq 4,096$
6	8	2,146,689	none
7	64	16,974,593	
8	512	135,005,697	
9	4,096	1,076,890,625	
10	32,768	8,602,523,649	
11	262,144	68,769,820,673	

3.1.1 Standard process assignment

In the first test, we consider a fixed number m of cores per node. If n denotes the number of nodes, the number of processes is given by $n * m$. This corresponds to the setting `select=n:node_type=rome:mpiprocs=m` .

As shown in Tab. 2, the wallclock times in the range 64–32,768 processes are almost constant. This is valid for all three code regions (`assemble`, `init`, `apply`). When only eight cores are used, the times reduce slightly. and almost identical in all

[3] One core corresponds to one MPI process.

Table 2: Standard process assignment (w/o stride): Weak scaling with 16, 32, 64, and 128 MPI processes per node (ppn).

16-ppn	8	64	512	4,096	32,768	262,144
assemble	2.403	2.463	2.548	2.572	3.514	[a]
init	0.699	0.975	0.986	1.091	1.606	
apply	1.783	2.932	2.987	2.982	3.072	
32-ppn	**8**	**64**	**512**	**4,096**	**32,768**	**262,144**
assemble	2.417	2,757	2,705	2,675	3,397	[a]
init	0.702	0,931	1,021	1,081	1,251	
apply	1.789	2.931	2.989	3.013	3.085	
64-ppn	**8**	**64**	**512**	**4,096**	**32,768**	**262,144**
assemble	2.418	3.311	2.963	3.293	3.307	[b]
init	0.686	0.956	1.063	1.044	1.100	
apply	1.780	2.945	3.010	3.003	3.093	
128-ppn	**8**	**64**	**512**	**4,096**	**32,768**	**262,144**
assemble	2.394	3.388	3.402	3.436	3.482	4.803
init	0.715	0.979	1.056	1.004	2.186	2.631
apply	1.796	2.944	3.006	2.997	3.224	3.281

[a] A job with this configuration exceeds the maximum number of 5,632 compute nodes.
[b] Calls of `mpirun` terminated In this configuration before the custom executable could be launched.

four cases, which can be expected. One exception is the case with fully populated nodes (128 cores per node): Here the run time for 32,768 processes is slightly increased. When 262,144 processes are used, time increases significantly again. These results must be taken with a grain of salt and require further analysis and detailed profiling.

3.1.2 Topology aware process assignment

As mentioned earlier, the *Rome* nodes feature an internal hierarchical structure. As four cores share a common L3 cache, collisions in the previous test are very likely. In an additional test, we tried to address the topology of the nodes by enforcing a limited number of MPI processes per node. Thus, we repeat the previous experiment using only every eighth, fourth and second core, respectively. The distribution is achieved by the commands `omplace -c 0-127:st=s` with a stride $s = 8, 4, 2$. This corresponds to using 2 ($s = 2$) or 1 ($s = 4$) cores per CCX, and 1 core per CCD ($s = 8$), respectively.

Table 3: Topology aware process assignment: Weak scaling with 16, 32, and 64 MPI processes per node (ppn). The stride was $s = 8, 4$, and $s = 2$ respectively.

16-ppn ($s = 8$)	8	64	512	4096	32,768
assemble	2.254	2.276	2.292	2.480	4.000
init	0.521	0.527	0.533	0.533	0.621
apply	0.605	0.618	0.627	0.642	0.705
32-ppn ($s = 4$)	**8**	**64**	**512**	**4096**	**32,768**
assemble	2.330	2.287	2.290	2.375	2.946
init	0.559	0.555	0.622	0.677	0.887
apply	0.796	0.796	0.808	0.881	0.903
64-ppn ($s = 2$)	**8**	**64**	**512**	**4096**	**32,768**
assemble	2.387	2.585	2.601	2.755	2.882
init	0.660	0.717	0.718	0.769	0.824
apply	1.418	1.437	1.463	1.478	1.558

As shown in Tab. 3, the impact of the topology aware process assignment w.r.t the code region:

- For the `assemble`-phase, a small acceleration of about 10% is observed. This phase includes a few read/write memory accesses, but is primarily computation dominated. In addition, no communication is involved. Hence, this result can be expected.
- In the `init`-phase communication becomes more important. In this case a slightly more pronounced acceleration is observed, and run times reduce by 20-30%.

- The biggest gains are achieved in the apply-phase. This phase features many memory accesses to vector and sparse-matrix data structures. Moreover, the network is used heavily for all-reduce operations (e.g., for stopping criteria) as well as nearest neighbor communication (e.g., master-slave-exchange along the process boundaries). The acceleration approaches the optimal factors, i.e., 2 when using 2 out of 4 cores per CCX ($s = 2$), and 4 when using only a single core per CCX ($s = 4$). Restricting the resources to a single core per CCD ($s = 8$) only, yields additional acceleration.

3.1.3 Summary

For the benchmark problem, weak scaling capabilities have been demonstrated for a wide range of MPI processes. The results show, that using all available 128 cores per node is reasonable and economic. In our experiments, using a reduced number of processes per node, only provides extra benefit in the apply-phase. At the same time, the number of allocated nodes grows and resources by a factor of four. Hence, this is not a very economic alternative. From a practical perspective, this should be considered only when small wall-clock timings are crucial.

3.2 Strong scaling for the heat equation

The previous section demonstrated the excellent weak scaling properties of the multilevel method for a steady state problem. In this case, the grid resolution in space tends to zero. However, in particular for some transient problems, it is sometimes desirable to work with a spatial grid with a given resolution, which corresponds to a fixed number of degrees of freedom in space.

The following test, which was first suggested by [7], is an extension of the tests conducted in [24] on a small cluster in 2D. We consider the heat equation

$$\partial_t u(\mathbf{x}, t) - \alpha \Delta u(\mathbf{x}, t) = f(\mathbf{x}, t)$$

for $(\mathbf{x}, t) \in (0, 1)^3 \times (0, 4\pi)$ and $\alpha = 0.1$. Initial value, Dirichlet boundary conditions and right hand side f are chosen such that the solution

$$u(\mathbf{x}, t) = \sin(\pi x_1) \cdot \sin(\pi x_2) \cdot \sin(\pi x_3) \cdot \cos(t)$$

is obtained. The equation is discretized using an implicit Euler method in time and Q_1 finite elements in space. Assuming that p MPI processes are available, two setups are compared:

- In *serial* time integration, all processes are dedicated to the spatial domain. We consider the sequence $p \in \{8, 64, 512\}$. The time domain is split into 16384 equidistant intervals that are treated sequentially.

- The parallel-in-time integration uses $p = p_s \times p_t \in \{512, \ldots, 16384\}$ processes. A fixed number of $p_s = 64$ processes is dedicated to spatial domain. For the time domain a variable number $p_t \in \{8, \ldots, 256\}$ is used. Each process then owns an interval with $16,384 : p_t$ equidistant time points.

On the finest level, the unit cube $\Omega = (0, 1)^3$ is split into $128 \times 128 \times 128$ elements, with 2, 146, 689 degrees of freedom. Splitting this onto eight processes, this means that workload is in a similar order of magnitude as in the previous example for the steady state. Note, however, that due to the fixed time step size, assembling the operator is only required once. Instead of a BiCGStab, a linear iteration is used. On each node, every second core has been used (cf. previous section with a stride $s = 2$).

Results are provided in Fig. 1: We observe that for serial time stepping the strong scaling w.r.t the spatial variables is limited: From eight to 64 processes, a speedup of ≈ 5.5 is observed, which is already smaller than the optimal value of eight. For 512 processes the run time even deteriorates. In this case, there is only roughly 4, 000 degrees of freedom per process left, which means that computation is only of little importance, but communication becomes the dominant factor. Since the number of process increases, more communication is required.

The MGRIT method initially suffers from the computational overhead of being an iterative method. Several iterations are required to achieve the discretization error of the serial method. To that end, observe that using 512 processes for $p_t = 8 = 512/64$ equidistant time intervals roughly takes the same time as the serial computation using $p = 8$ processes. However, from that point on, adding more processes in the time dimension always lead to a speedup. Doubling the number of processes in each step, the observed speedup is $1.88 \approx 2$ initially, but still 1.52 in the last step. With respect to wall-clock time, the break even point is roughly at 4, 096 cores. From that point on, MGRIT is faster than the fastest run with serial time integration.

The results are in agreement with results [7, 24]. However, problems of the size investigated here were not computed, but only predicted. It is important to emphasize that the MGRIT is not attractive as long as a scaling from Section 3.1 is to be sought. However, when all solutions w.r.t. the temporal dimension, are to be computed simultaneously, as is, e.g. important for optimization, multigrid-in-time can provide an attractive alternative which is not intrusive w.r.t. software design.

3.3 Thermohaline transport

The previous two sections focused primarily on scaling properties of the algorithms for select models problems. In addition to the parallel capabilities, one major benefit of UG4 is, however, the versatile applicability for a wide range of problems. in particular for coupled PDE problems. This includes, e.g., density-driven flow, computational neuro-science, poroelasticity.

To that end, this report is concluded by results for a thermohaline flow problem. In this special instance of density-driven flow, where the fluid density is modeled as a variable depending on both temperature and fluid composition. Hot fluids, e.g.,

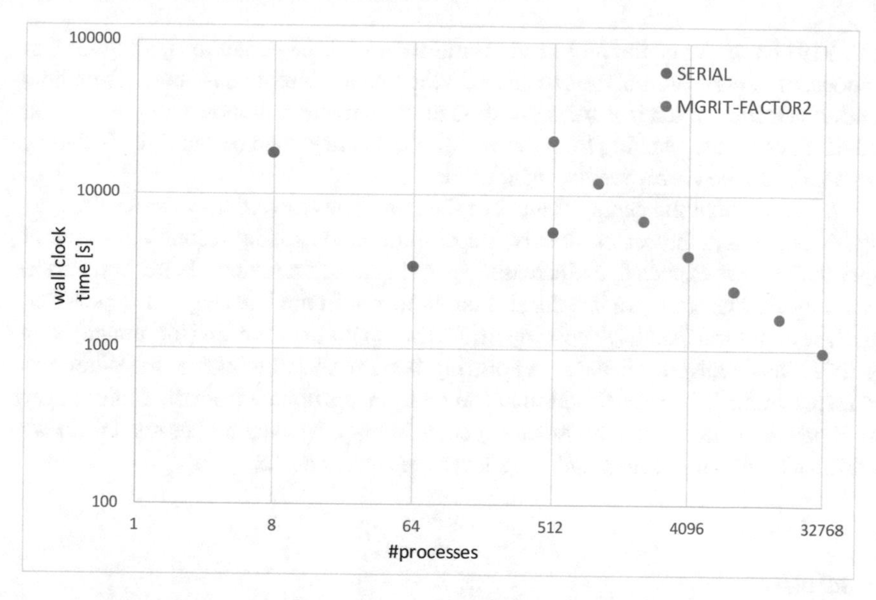

Fig. 1: Strong scaling for heat equation. Comparison of a classic serial time integration (SERIAL) and multigrid in time (MGRIT-FACTOR2) with factor of 2 coarsening.

typically have a lower density than cold fluids. Similarly in solutions, an increase of solute concentration yields a higher density. Systems of this type are important, e.g., for modeling transport of CO2 or NaCl in repositories and deep geological layers.

In this work, we concentrate on a *parcel benchmark* problem suggested in [13]: In this test, a hot parcel with saline solution is placed in a rock matrix of lower temperature. The behavior of the system then depends on the configuration of the parcel, e.g., its temperature T_p and its salt mass fraction ω_p. As outlined above, for very hot parcels, the density of the parcel is lower than in the surrounding rock. This yields a configuration with *positive initial buoyancy*, i.e, the parcel initially starts moving upward, before the fluid is cooled down by the surrounding rock. If, on the contrary, the salt mass fraction is very high, this increases density and yields a case with *negative initial buoyancy*, that is the parcel immediately starts sinking down.

As observed in [13], one important feature of the negative buoyancy case is, that this configuration leads to a fingering effect. However the structure of the convection cells, or, question whether the number of fingers is finite, have not been resolved yet. Computers like HLRS's Hawk now allow to address these research topics. In addition to the massive computing resources, algorithmically a proper control of the discretisation error is crucial. To that end, novel time integration schemes have been introduced [5, 23].

The following images provides preliminary results for the parcel benchmark with negative buoyancy. We employ the implementation provided by the d^3f-framework [28]. Figs. 2 and 3 show the evolution of the parcel at an early and late stage for different grid resolutions respectively. The simulation were performed using 128,

1024, 8192 cores. Since the problem is symmetric w.r.t the center axis, all fields can be shown in a single plot. In the background, the temperature field is shown from blue to red. A selection of ten isosurfaces of the salt mass fraction is shown in greyscale on the left. Streamlines resulting from the velocity field are shown on the right. Rainbow colors indicate the corresponding magnitude.

Fig. 2 shows that the center of mass of the parcel has moved downwards from its original position at the center already. Depending on the spatial resolution, different fingers evolve. As expected, an increased spatial resolution yields a better separation of the fingers. The structure is quite similar, however. In the late stage, as depicted in Fig. 3, several branches have developed. At low spatial resolution (top image), one big finger has evolved. However, a splitting becomes visible at the tip. When the spatial resolution is increased, additional smaller fingers branch from the center finger, which persists with a smaller diameter (center image). At even higher resolution, an additional branching occurs, and three layers of fingers occur.

Conclusion

This report provided first results for the performance of the software UG4 on the Hawk Supercomputer. The results indicate that the highly integrated CPUs yield an excellent weak scaling. It is demonstrated that a topology aware scheduling of the MPI processes leads to an increased performance. This seem to be related to the hierarchical organization of the nodes an the internal NUMA architecture. Results at this stage use about 1/3 of the full machine. One goal for the next reporting period is to extend this to the full machine. For transient problems, strong scaling can be improved using the multigrid in time approach. The presented scaling analysis focuses on scalar linear problems. With respect to real-world applications with complex physics, it is equally important to address transient non-linear coupled problems requiring high spatio-temporal resolution. To that end, preliminary results for a thermohaline flow have been presented. This should be extended to adaptive grid refinement and time-stepping. Earlier studies, e.g. for poroelasticity or density-driven flow without transfer of heat [23], showed that the linearly-implicit extrapolation method can provide a useful tool. This will be investigated in greater detail in the next steps of this project.

Acknowledgements The authors would like to thank Björn Dick, Bernd Krischok and all members the HLRS staff for all technical support, guidance, and advice regarding the Hawk system.

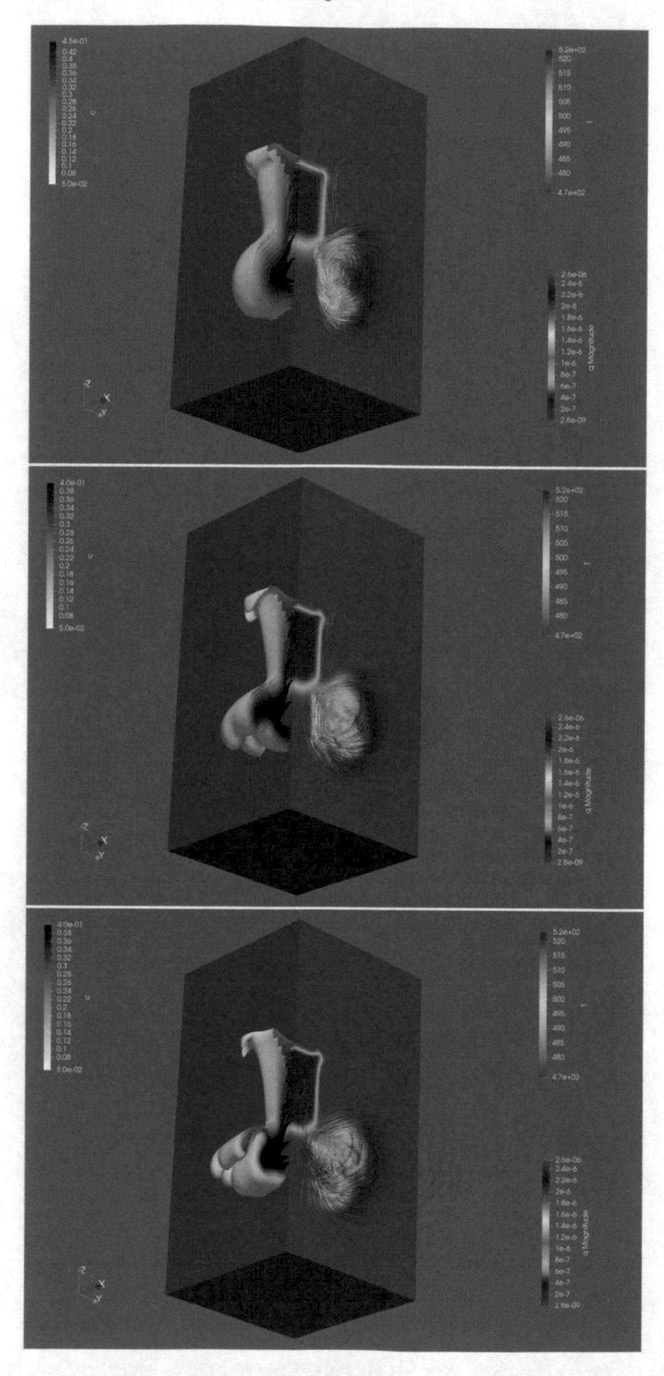

Fig. 2: Early stage (three different levels of spatial refinement from top to bottom): The parcel sinks down. Depending on the spatial resolution, a branching of the central finger becomes visible.

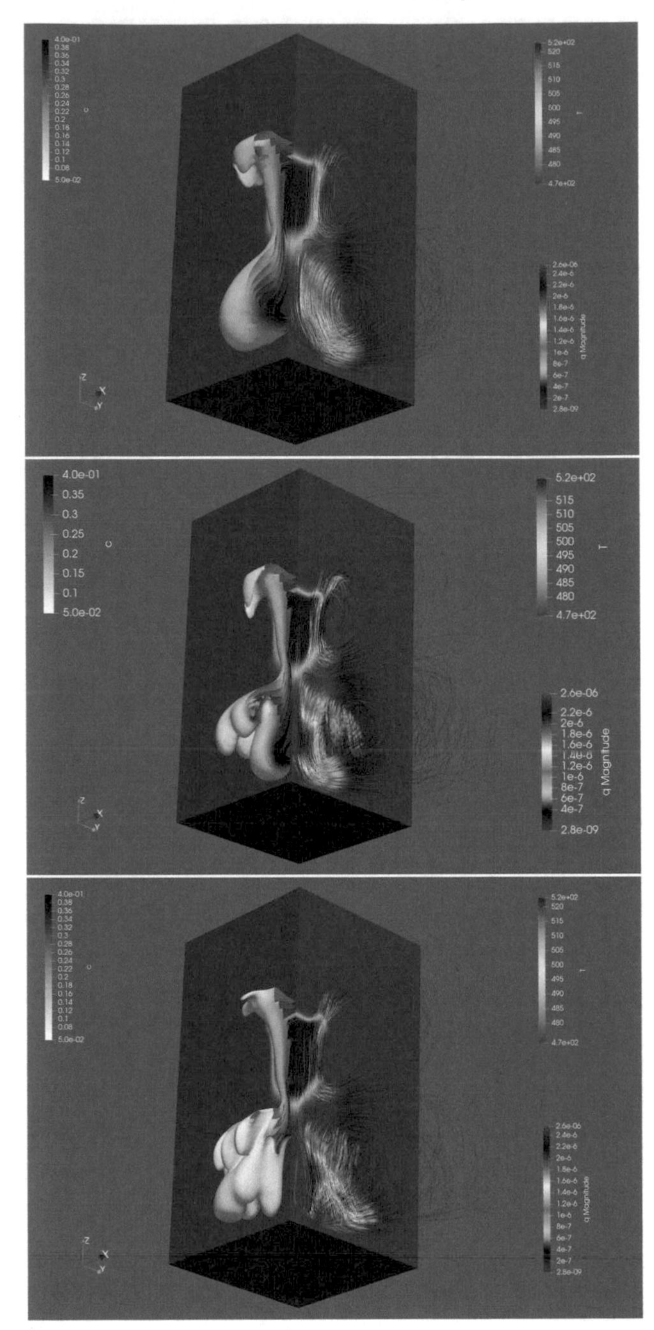

Fig. 3: Late stage (three different levels of refinement): A layered fingering evolves. In the highest resolution, three layers of fingers are visible.

References

1. G. Alzetta et al. The deal.II Library, Version 9.0. *Journal of Numerical Mathematics* **26**(4), pp. 173–183 (2018). https://doi.org/10.1515/jnma-2018-0054.
2. Allison H. Baker et al. Scaling Hypre's Multigrid Solvers to 100,000 Cores. In: *High-Performance Scientific Computing: Algorithms and Applications*. Ed. by Michael W. Berry et al., pp. 261–279, Springer, London, (2012). isbn: 978-1-4471-2437-5. https://doi.org/10.1007/978-1-4471-2437-5_13.
3. Peter Bastian. *DUNE – the Distributed and Unified Numerics Environment*. https://www.dune-project.org/.
4. Markus Blatt et al. The Distributed and Unified Numerics Environment, Version 2.4. eng. In: *Archive of Numerical Software* Vol 4 (2016), Starting Point and Frequency: Year: 2013. https://doi.org/10.11588/ans.2016.100.26526.
5. Ruben Buijse. *Numerische Verfahren zur Simulation thermohaliner Strömungen in der Software d3f++*. B.Sc. thesis; Goethe Universität Frankfurt (2018).
6. Hans-Jörg G. Diersch. *FEFLOW: Finite Element Modeling of Flow, Mass and Heat Transport in Porous and Fractured Media*. Springer-Verlag Berlin Heidelberg (2014).
7. R.D. Falgout et al. Parallel time integration with multigrid. *SIAM Journal on Scientific Computing* **36**(6), pp. C635–C661 (2014). issn: 1064-8275. https://doi.org/10.1137/130944230.
8. FEniCS Project. *The FEniCS computing platform*. https://fenicsproject.org/.
9. B. Flemisch et al. DuMux: DUNE for multi-phase, component, scale, physics ,... flow and transport in porous media. *Advances in Water Resources* **34**(9), 1102–1112 (2011). issn: 0309-1708. https://doi.org/10.1016/j.advwatres.2011.03.007.
10. Martin J. Gander. 50 years of Time Parallel Time Integration. In: *Multiple Shooting and Time Domain Decomposition*. Springer (2015).
11. Björn Gmeiner et al. Towards textbook efficiency for parallel multigrid. *Numerical Mathematics: Theory, Methods and Applications* **8**(1), 22–46 (2015).
12. Björn Gmeiner et al. A quantitative performance study for Stokes solvers at the extreme scale. *Journal of Computational Science* **17**, pp. 509–521 (2016).
13. Alfio Grillo, Michael Lampe, and Gabriel Wittum. Three-dimensional simulation of the thermohaline-driven buoyancy of a brine parcel. *Comput Visual Sci* **13**, pp. 287–297 (2010).
14. W. Hackbusch. *Multi-Grid Methods and Applications*. Springer, Berlin (1985). isbn: 3-540-12761-5.
15. Myra Huymayer. First steps towards a scaling analysis of a fully resolved electrical neuron model. In: *High Performance Computing in Science and Engineering '19* (2021).
16. Mary F. Wheeler. IPARS: A New Generation Framework for Petroleum Reservoir Simulation. http://csm.ices.utexas.edu/ipars/.
17. O. Ippisch and M. Blatt. Scalability Test of $\mu\varphi$ and the Parallel Algebraic Multigrid solver of DUNE-ISTL. In: *Jülich Blue Gene/P Extreme Scaling Workshop 2011*, Technical Report FZJ-JSC-IB-2011-02, April 2011. Ed. by B. Mohr and Wolfgang Frings (2011).
18. Guido Kanschat. *Web site: deal.II – an open source finite element library*. http://www.dealii.org/.
19. O. Kolditz, S. Bauer, L. Bilke et al. OpenGeoSys: an open-source initiative for numerical simulation of thermo/hydro/mechanical/chemical (THM/C) processes in porous media. *Environ Earth Sci* **67**(2), 589–599 (2012).
20. Hans Petter Langtangen and Anders Logg. *Solving PDEs in Python – The FEniCS Tutorial I*. Springer International Publishing (2016). https://doi.org/10.1007/978-3-319-52462-7.
21. Lawrence Livermore National Laboratory. *Web site: XBraid: Parallel multigrid in time*. http://llnl.gov/casc/xbraid.
22. Knut-Andreas Lie et al. Open-source MATLAB implementation of consistent discretisations on complex grids. *Computational Geosciences* **16**(2), 297–322 (Mar. 2012). issn: 1573-1499. https://doi.org/10.1007/s10596-011-9244-4.

23. Arne Nägel and Gabriel Wittum. Scalability of a Parallel Monolithic Multilevel Solver for Poroelasticity. In: *High Performance Computing in Science and Engineering '18*. Springer Nature Switzerland AG (2019).
24. Martin Parnet. *Zeitparallele Lösungsverfahren für die Wärmeleitungsgleichung* (2020).
25. Sebastian Reiter et al. A massively parallel geometric multigrid solver on hierarchically distributed grids. *Computing and Visualization in Science* **16**(4), 151–164 (Aug. 2013). https://doi.org/10.1007/s00791-014-0231-x.
26. S. Reiter et al. A massively parallel multigrid method with level dependent smoothers for problems with high anisotropies. *High Performance Computing in Science and Engineering '16*, pp. 667–675, Springer (2017).
27. Johann Rudi et al. An extreme-scale implicit solver for complex PDEs. In: *Proceedings of the International Conference for High Performance Computing, Networking, Storage and Analysis – SC '15*. ACM Press (2015). https://doi.org/10.1145/2807591.2807675.
28. Anke Schneider, ed. *Modeling of Data Uncertainties on Hybrid Computers*. GRS Bericht 392 (2016). isbn: 978-3-944161-73-0.
29. U. Trottenberg, C.W. Oosterlee, and A. Schüller. *Multigrid*. Contributions by A. Brandt, P. Oswald and K. Stüben. Academic Press, San Diego, CA (2001).
30. Andreas Vogel et al. UG 4: A novel flexible software system for simulating PDE based models on high performance computers. *Computing and Visualization in Science* **16**(4), 165–179 (Aug. 2013). https://doi.org/10.1007/s00791-014-0232-9.

Scaling in the context of molecular dynamics simulations with *ms2* and *ls1 mardyn*

Simon Homes, Robin Fingerhut, Gabriela Guevara-Carrion, Matthias Heinen and Jadran Vrabec

Abstract This chapter covers scaling issues related to our recent work in the field of molecular dynamics simulations with a focus on computational details. The first section deals with finite size effects in the context of multicomponent diffusion. Different methods to correct the influence of the finite simulation domain are compared. In the second section, the structure of a fluid near its critical point is discussed in the context of the strong scaling behavior of the code *ms2*. The third section discusses droplet coalescence dynamics investigated by large molecular dynamics simulations and a macroscopic phase field model, respectively. The performance of the new supercomputer *Hawk* is compared to that of *Hazel Hen*. The last section describes the influence of the direct sampling of the energy flux on the performance of the code *ls1 mardyn*. Various speed tests are carried out and analyzed in detail.

1 Finite size effects of multicomponent diffusion coefficients

Diffusion processes in multicomponent mixtures are highly complex and still not well understood due to the presence of coupling effects. Moreover, experiments to measure transport diffusion coefficients are difficult and time consuming. Molecular dynamics simulation has become a powerful alternative to accurately predict diffusion coefficients and thus to improve data availability.

Molecular dynamics simulations under periodic boundary conditions are typically performed employing a small number of molecules which is far away from the thermodynamic limit. In this context, systematic errors associated with the system size are present when diffusion coefficients are calculated. It is thus necessary to correct the simulation data to account for such effects. The most widely employed correction method is based on the shear viscosity η and the edge length of the simulation volume

Simon Homes, Robin Fingerhut, Gabriela Guevara-Carrion, Matthias Heinen and Jadran Vrabec
Lehrstuhl für Thermodynamik und Thermische Verfahrenstechnik, Technische Universität Berlin, Ernst-Reuter-Platz 1, 10587 Berlin, Germany, e-mail: vrabec@tu-berlin.de

W. E. Nagel et al. (eds.), *High Performance Computing in Science and Engineering '21*,
https://doi.org/10.1007/978-3-031-17937-2_29

L, i.e. $2.837297 \cdot k_B T/(6\pi\eta L)$, and does not require additional simulation runs. This method was proposed originally by Yeh and Hummer [21] to correct self-diffusion coefficients has been demonstrated to not always be adequate [12].

In case of multicomponent mixtures, finite size effects are observed not only for self-diffusion coefficients but also for mutual diffusion coefficients according to Maxwell-Stefan or Fick. Typically, the Fick diffusion coefficient matrix of a multicomponent mixture with n components is not obtained directly by equilibrium molecular dynamics simulation, but it is calculated from the sampled phenomenological diffusion coefficients L_{ij} and the thermodynamic factor matrix $\mathbf{\Gamma}$, employing the relation $[\mathbf{D}] = [\mathbf{B}]^{-1}[\mathbf{\Gamma}]$, in which all three symbols represent $(n-1) \times (n-1)$ matrices and $[\boldsymbol{B}] = [\mathbf{\Delta}]^{-1}$, where

$$\Delta_{ij} = (1 - x_i)\left(\frac{L_{ij}}{x_j} - \frac{L_{in}}{x_n}\right) - x_i \sum_{k=1 \neq i}^{n} \left(\frac{L_{kj}}{x_j} - \frac{L_{kn}}{x_n}\right). \tag{1}$$

In fact, the observed system size dependence of the Fick diffusion coefficient matrix elements is directly associated with the corresponding effects of the underlying phenomenological coefficients [8]. Therefore, a finite size correction approach based on the correction of the phenomenological coefficients is more adequate than a direct correction of Fick diffusion coefficients [14].

Here, this system size dependence was studied for the ternary mixture methanol + ethanol + isopropanol, performing series of simulations with varying system size containing 512 to 8000 molecules. The infinite size values were obtained from the extrapolation of the respective diffusion coefficients to the thermodynamic limit $L^{-1} \to 0$. Figure 1 shows the predicted intra-diffusion coefficients and their values corrected with the Yeh and Hummer term [21] as a function of the system size. It can be seen that this correction yields an overestimation between 4% and 10% from the extrapolated values for systems containing 1000 and 8000 molecules, respectively.

To study the finite size effects of the mutual diffusion coefficients, the values for an infinite system size were calculated for all main L_{ii} and cross phenomenological coefficients L_{ij} of the studied ternary mixture as depicted in Figure 2. The proposed fast correction procedure based on normalized coefficient values [8, 9] leads to corrections of the main and cross phenomenological coefficients for simulations with 6000 molecules of approximately 5% and 4%, respectively. Infinite size extrapolated and corrected diffusion values exhibit a good agreement, with relative deviations below 4%.

Because the phenomenological coefficients are underlying to the Fick diffusion coefficient coefficients, a propagation of finite size effects is expected. Figure 3 shows the observed system size dependence for all elements of the Fick diffusion coefficient matrix of the studied mixture. The extrapolated values in the thermodynamic limit are compared with values that were obtained with the proposed correction procedure based on the phenomenological coefficients [8,9] and those corrected with the method by Jamali et al. [14]. Note that this method only considers corrections for the main elements of the Fick diffusion coefficient matrix. The proposed correction method based on the phenomenological coefficients agrees on average within 2% with the

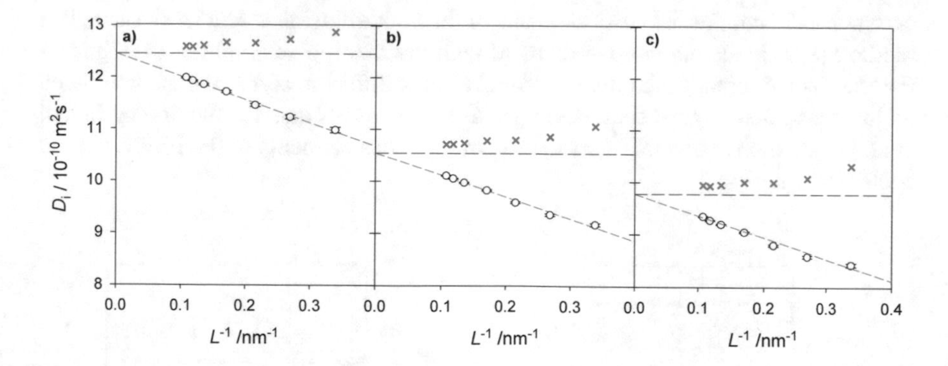

Fig. 1: Intra-diffusion coefficients of methanol (a), ethanol (b) and isopropanol (c) in their ternary mixture (x_{CH4O} = 0.125, x_{C2H6O} = 0.625 and x_{C3H8O} = 0.25 mol·mol^{-1}) as a function of the inverse edge length of the simulation volume L^{-1} at 298.15 K and 0.1 MPa. The uncorrected simulation results (empty circles) are shown together with the corrected values using the Yeh and Hummer approach [21] (green crosses). The gray dashed line is a linear fit to the uncorrected simulation results (empty circles) and the blue line represents the extrapolated value in the thermodynamic limit.

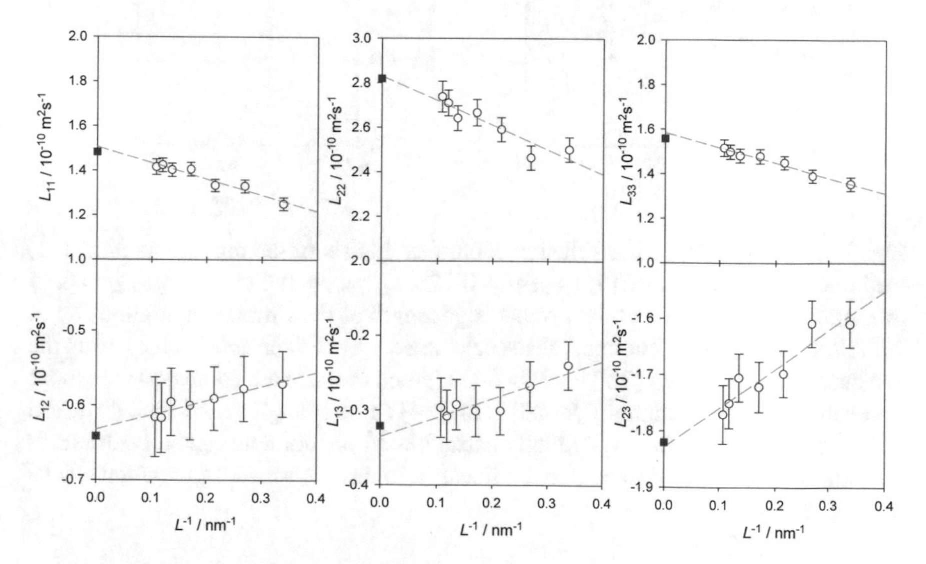

Fig. 2: Phenomenological coefficients of the mixture methanol (1) + ethanol (2) + isopropanol (3) (x_{H2O} = 0.125, x_{C2H6O} = 0.625 and x_{C3H8O} = 0.25 mol·mol^{-1}) as a function of the inverse edge length of the simulation volume L^{-1} at 298.15 K and 0.1 MPa. The uncorrected simulation results (empty circles) are shown together with the corrected values using the fast correction procedure [8,9] for N = 8000 (red squares). The gray dashed line is a linear fit to the uncorrected simulation results (empty circles).

extrapolated data. For the cross elements of the Fick diffusion coefficient matrix, this method [8, 9] is also in good agreement with the extrapolated values, cf. Figure 3. For the studied ternary mixture, the proposed Fick diffusion correction method based on the phenomenological coefficients yields better results than the method by Jamali et al. [14], which neglects the correction of the cross elements of the Fick diffusion coefficient matrix.

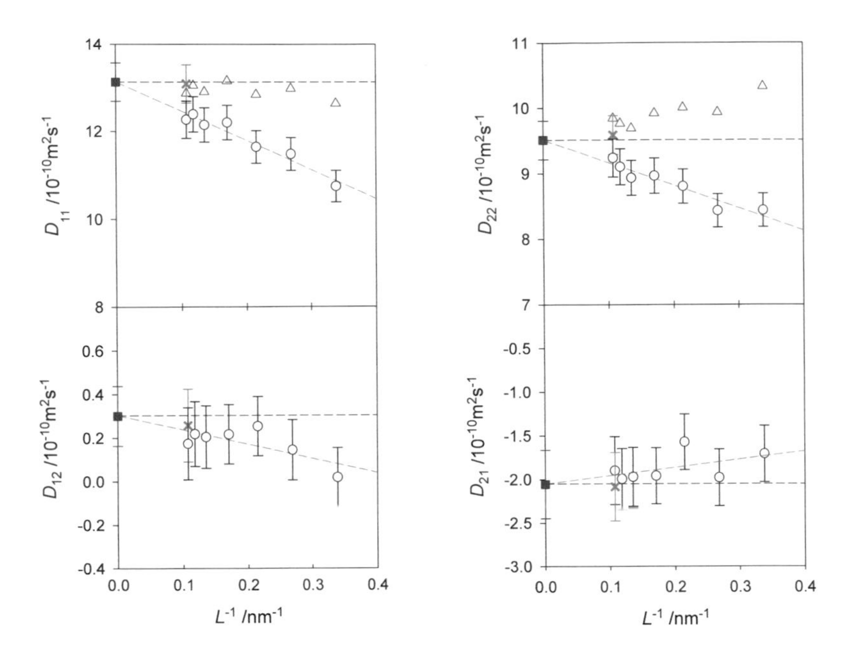

Fig. 3: Elements of the Fick diffusion coefficient matrix of the mixture methanol (1) + ethanol (2) + 2-propanol (3) ($x_{CH4O} = 0.125$, $x_{C2H6O} = 0.625$ and $x_{C3H8O} = 0.25$ mol·mol^{-1}) as a function of the inverse edge length of the simulation volume L^{-1} at 298.15 K and 0.1 MPa. The coefficients calculated with the corrected values using the fast correction procedure [8,9] for $N = 8000$ (green crosses) are compared with those according to the procedure by Jamali et al. [14] (blue triangles) and Fick diffusion coefficients based on the individually extrapolated phenomenological coefficients (red squares). The gray dashed line is a linear fit to the uncorrected simulation results (empty circles).

2 Strong scaling of *ms2* when sampling the structure

It is well known that sampling thermodynamic equilibrium properties with standard system sizes of a few thousand molecules may lead to challenges in regions close to the critical point. Because structure and dynamics of such states exhibit pronounced fluctuations, small systems may not cover the relevant physics.

A recent study [16] has shown that molecular dynamics (MD) simulations combined with sampling the radial distribution function (RDF) $g(r)$ are well suited to analyze structural phenomena. Here, the Lennard-Jones (LJ) potential was applied in MD simulations utilizing the program *ms2* [1,5,6,18]. Three isotherms at $T = 1.4, 1.5$ and 1.6 were studied somewhat above the critical temperature $T_c = 1.32$. Each isotherm was sampled by varying the density from ±5 % to ±20 % around the critical density $\rho_c = 0.31$ in steps of 5 %. MD simulations were carried out with $N = 120{,}000$ particles and $4 \cdot 10^5$ equilibration as well as 10^6 production steps in the canonical ensemble with a time step of $\Delta t = 0.004$. RDF were sampled over the entire production period up to the maximum radius r given by half of the simulation volume edge length $L/2$, which corresponds to $r/\sigma = 39.25$ for the smallest and $r/\sigma = 34.29$ for the largest density. A discretization of 8,000 shells over the radius was specified. Fig. 4 shows a magnified view on the RDF.

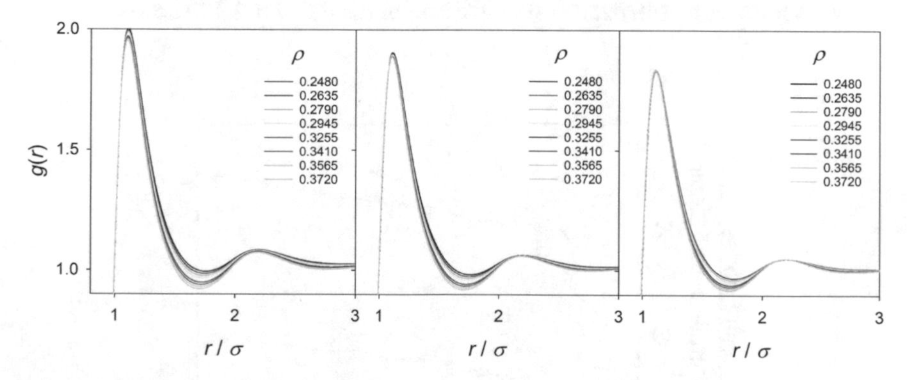

Fig. 4: Magnified view on the RDF sampled by MD simulations with *ms2*; left: $T = 1.4$, center: $T = 1.5$, right: $T = 1.6$.

It can be clearly seen that the first peak of the RDF decreases with rising temperature, whereas its position shifts only slightly towards smaller values of r. Moreover, the area below $g(r)$ decreases with rising density or temperature.

The parallel performance of *ms2* was assessed with respect to strong scaling, where the problem size is fixed while the number of processing elements is increased. The strong scaling efficiency of *ms2* was analyzed for its hybrid MPI + OpenMP parallelization. This scheme allows simulations with a larger number of particles N because of more efficient data handling. The vertical axis of Fig. 5 shows the computing power (nodes) times computing time per computing intensity (problem size). A horizontal line would indicate a perfect strong scaling efficiency. Here, concurrent sampling of several state points in one program execution was allocated with *ms2*. In each ensemble, MD simulations with $N = 120{,}000$ LJ particles with a cutoff radius of $r_c/\sigma = 6$ and $r_c/\sigma = 35.3$ were chosen. From Fig. 5, it becomes clear that *ms2* is close to optimal strong scaling, however, 100 % efficiency is not reached. In *ms2*, the computing intensity for traversing the particle matrix is proportional to N^2, but the intermolecular interactions are calculated for particles that are in the cutoff sphere only. Thus, the computational cost of *ms2* should be in-between N and N^2. The effective proportionality depends on the computing intensity of the intermolecular interactions. Because single LJ particles were considered here, the cost of calculating these interactions is not much higher than that of traversing the particle matrix. Fig. 5 compares simulations with a cutoff radius of $r_c/\sigma = 6$ and simulations with the maximum cutoff radius of $r_c/\sigma = 35.3$, which have a ratio of around 5.883. The number of particles in the cutoff sphere is proportional to its volume. Thus, a radius ratio of 5.883 indicates a volume ratio of $5.883^3 = 203.64$, which is proportional to the computational load for the intermolecular interactions. However, increasing r_c/σ from 6 to 35.3 leads to an increase of execution time of only 4.38 (comparison for 300 nodes). Thus, traversing the particle matrix is dominating in the present LJ particle scenarios. Because one of the parallelization schemes in *ms2* is implemented on the particle matrix traversing level, an execution time ratio of around 2.52 was achieved for larger parallelization (comparison of the data for 2,400 nodes).

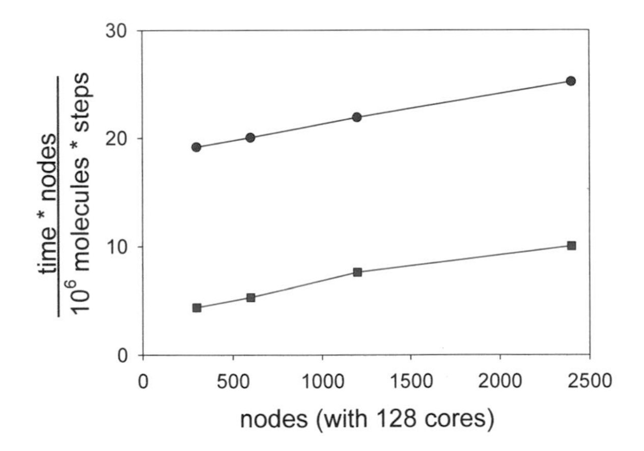

Fig. 5: Strong scaling efficiency of MD simulations with *ms2* for a LJ fluid with $N = 120{,}000$ particles measured on *HPE Apollo (Hawk)* for the hybrid MPI + OpenMP parallelization (each MPI process with four OpenMP threads); red circles: $r_c/\sigma = 35.3$; blue squares: $r_c/\sigma = 6$.

3 Droplet coalescence on *Hawk* and *Hazel Hen*

Large scale MD simulations were conducted to investigate the coalescence process of two initially resting argon droplets, considering three sizes, i.e. with diameter $d = 50$, 100 and 200 nm. The droplets were arranged on the central axis of a cuboid system with a distance of only a single particle diameter, cf. Fig. 6. They were surrounded by saturated vapor at a temperature of 110 K ($\approx$ 74 % of the critical temperature). The systems consisted of approximately 3, 25 and 200 million particles, respectively, denoted with 3M 25M and 200M in the following. To cope with such a large number of particles, the MD code *ls1 mardyn* [17] was employed for all simulations, which is continuously developed and optimized for massively parallel execution [20].

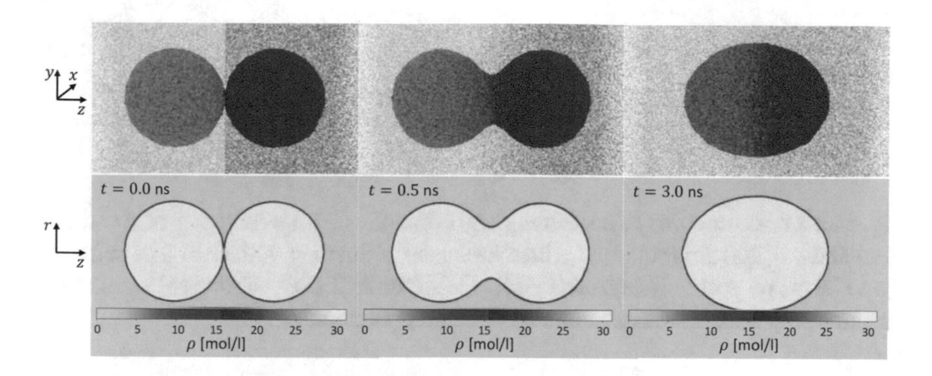

Fig. 6: Snapshots, rendered by MegaMol [7], from the MD simulation with two argon droplets of diameter $d = 50$ nm (top) and two-dimensional density fields sampled during simulation, depicted with a color map identifying the vapor phase (light blue), the liquid phase (yellow) and the interface in-between (bottom color map).

During simulation, two-dimensional density fields where sampled by employing a rotationally symmetric binning scheme. For this purpose, the system was divided by equidistant cylinder shells around and equidistant slices along the z axis. The sampled density field $\rho(z,r)$, with r being the distance to the z axis, represents a centered cut trough the droplets and can be visualized by a color map to identify vapor, liquid and the interface in-between, cf. Fig. 6.

These data sets were used as a benchmark for a macroscopic solution based on a phase-field model, i.e. results obtained from computational fluid dynamics (CFD) simulations, considering exactly the same scenarios. This model is described in detail in Ref. [2] and was already successfully used to simulate the motion of a droplet on a surface controlled by a wettability gradient [3].

For a comparison of the results of MD and CFD simulations, the contour of the interface of the coalescing droplets was extracted from the sampled (MD) or calculated (CFD) density fields and superimposed in the same plot, cf. Fig. 7. This representation shows a very good agreement of the results. A more elaborate comparison, e.g. based on the growth rate of the bridge that forms between the coalescing droplets, was recently published in [11].

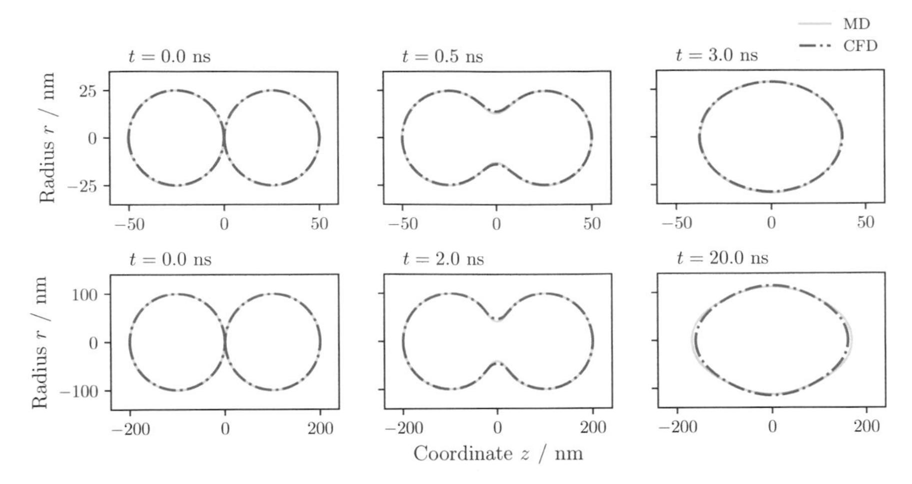

Fig. 7: Interface contour of coalescing argon droplets with a diameter of d = 50 nm (top) and d = 200 nm (bottom) at three instances of time t = 0, 0.5 and 3 ns and t = 0, 2 and 20 ns, respectively. Results of MD (solid) and CFD (dash-dotted) simulations are superimposed, showing an almost perfect agreement.

For the present MD simulation series, comprehensive scaling experiments were already conducted in the framework of the project *TaLPas*, funded by the Federal Ministry of Education and Research (BMBF), at that time on the machine *Hazel Hen*, the previous supercomputer of the High Performance Computing Center Stuttgart (HLRS). In this work, after a revision of the initial configurations, composed from carefully equilibrated droplets, the simulations were conducted on the new machine *Hawk*. For these production runs, the results of the scaling experiments on *Hazel Hen* were used to estimate a good choice for the allocation of computational resources. Since the entire coalescence process up to the complete fusion of the droplets had to be considered, the *time-to-solution* should be as short as possible. However, in order to use the resources efficiently, the number of cores to be allocated was set in a range before the course of the scaling curves starts to diverge substantially from perfect strong scaling.

Results of the present *time-to-solution* measurements (*Hawk*) compared to the previous ones (*Hazel Hen*) are depicted in Fig. 8. They show a speed-up of up to a factor of 2.7 for the 200M scenario (factors 2.2 and 2.1 for the 3M and 25M scenarios). These results underline the excellent performance of *Hawk*, pushing forward the MD simulations' accessible scales towards micrometers and microseconds.

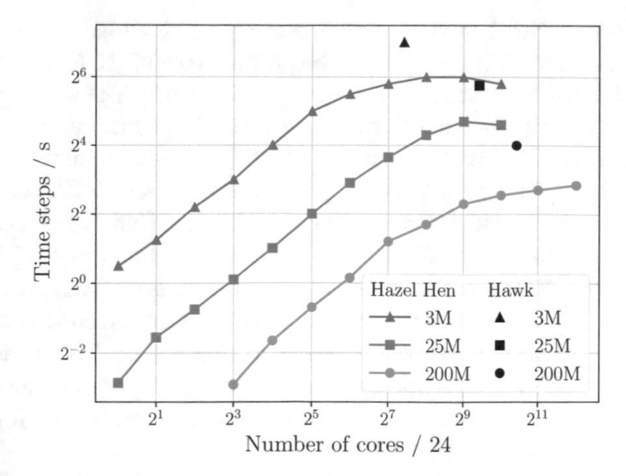

Fig. 8: Performance of *ls1 mardyn* for the droplet coalescence scenario measured on *Hawk* compared to previous results of scaling experiments conducted on *Hazel Hen*.

4 Scaling of energy flux implementation in *ls1 mardyn*

The energy flux plays a crucial role in many scientific and engineering problems. Until now, our open-source MD code *ls1 mardyn* [17] was not capable to sample the energy flux directly. Instead, it had to be calculated by post-processing as it was done in Ref. [13]. The utilized equation results from the first law of thermodynamics

$$j_\mathrm{e} = (h + e_\mathrm{kin}) j_\mathrm{p} + \dot{q}, \tag{2}$$

with j_e being the total energy flux, h the enthalpy, e_kin the kinetic energy, j_p the particle flux and $\dot{q}$ the heat flux. Until now, only the kinetic energy and the particle flux could be sampled directly. For the calculation of the remaining quantities, external data like correlations for the thermal conductivity [15] or equations of state [10] had to be used. In order to overcome this, the direct sampling of the energy flux was implemented into our code. The governing equation reads [4]

$$\boldsymbol{J}_e = \frac{1}{2} \sum_{i=1}^{N} m_i v_i^2 \boldsymbol{v}_i - \sum_{i=1}^{N} \sum_{j>i}^{N} \left[\boldsymbol{r}_{ij} \otimes \frac{\partial u_{ij}}{\partial \boldsymbol{r}_{ij}} - \boldsymbol{I} \cdot u_{ij} \right] \cdot \boldsymbol{v}_i . \tag{3}$$

Several quantities, like the particle mass m and the particle velocity v, are summed up over N particles, while $\boldsymbol{r}_{ij}$, u_{ij} and $\partial u_{ij}/\partial \boldsymbol{r}_{ij}$ are the distance vector, potential energy and virial between two particles i and j, respectively, and $\boldsymbol{I}$ stands for the identity matrix.

Since the off-diagonal elements of the virial tensor are needed, additional code had to be added to the core of *ls1 mardyn*, which adds significant additional computational load. In order to get an idea about the impact, multiple tests were conducted. First, the fastest compiler/MPI module combination was identified and the impact of vectorization analyzed. In a second step, the performance of the new implementation was compared to the legacy code to determine its influence. The last step was a short evaluation of the strong scaling behavior of the new implementation.

There are several different compiler and MPI modules available on *Hawk*. In this work, the compilers *aocc* (AMD), *gcc* (GNU) and *icc* (Intel) were combined with the MPI implementations *mpt*, *openmpi* and *impi* (Intel). Some of the possible combinations are not operable as they lead to building errors. The following five combinations were tested: *aocc+aocc*, *aocc+aocc*, *aocc+aocc*, *aocc+aocc* and *aocc+aocc*. All of these were used to run the same test case which stands exemplarily for a study in which stationary evaporation was investigated. This typical test case consisted of about $2 \cdot 10^6$ LJ particles, constituting one liquid and one vapor phase. The simulations were conducted on $8 \cdot 128 = 1024$ cores for 25,000 time steps and the execution time was measured.

The results of the compiler/MPI study are shown in Fig. 9. It can be seen that vectorization leads to a speed-up of about 30%, regardless of the chosen compiler/MPI combination. Furthermore, the tests show that the *gcc* compiler produces the fastest code. The choice of the MPI implementation has little effect on the performance, since the two fastest simulations were both performed with *gcc*-compiled code. Nevertheless, choosing *openmpi* for MPI communication can speed-up the test simulation for another 2%. The poorest performance was achieved when utilizing the *icc* (Intel) compiler. One explanation for this finding may be that *Hawk* consists of AMD processors for which the *icc* compiler may be not well optimized.

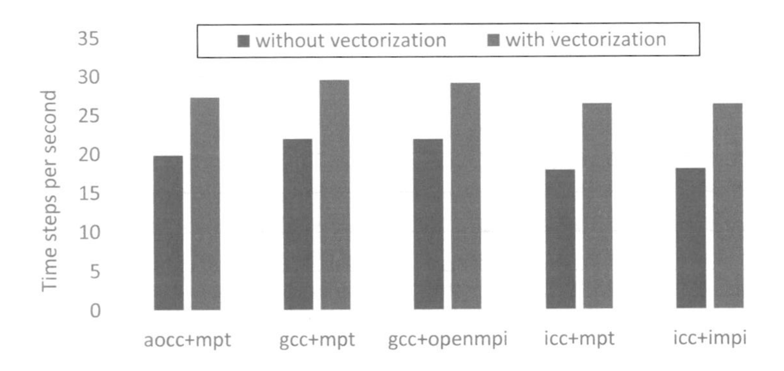

Fig. 9: Performance of different compiler/MPI combinations with and without vectorization. Performance was measured in time steps per second as an average over 25,000 time steps.

In a second study, the impact of the implementation of the direct sampling of the energy flux was investigated. The most performant compiler/MPI combination was used to conduct the simulation runs. Again, the simulations were run for 25,000 time steps, respectively, and the average number of time steps per second was calculated. Even though significantly more calculations have to be conducted in order to sample the energy flux directly, the new code takes just about 25% longer to execute the simulation of the same test scenario. Furthermore, with enabled vectorization, the new code is still faster compared to the old one without vectorization.

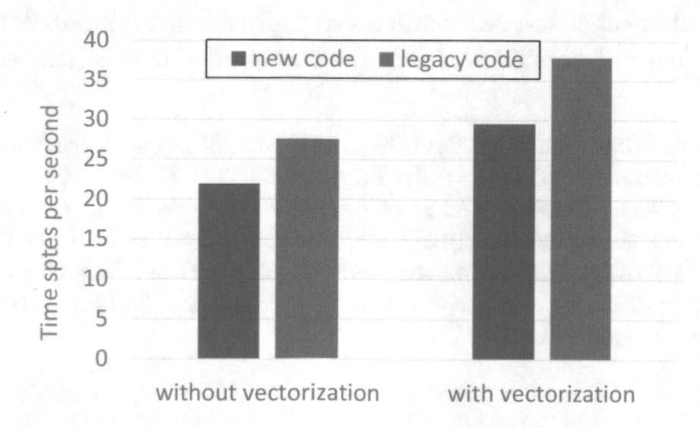

Fig. 10: Performance comparison of *ls1 mardyn* with and without the implementation of direct energy flux sampling. Performance is measured in time steps per second as an average over 25,000 time steps.

In a last test, the strong scaling behavior of *ls1 mardyn* was investigated for the aforementioned test case. Three scaled-up simulations were run in total. The smallest run utilized $2 \cdot 128 = 256$ cores, the mid-sized one $8 \cdot 128 = 1024$ cores and the biggest one $16 \cdot 128 = 2048$ cores. The speed-up of all three simulations was compared to a run on a single node, i.e. 128 cores. Results are shown in Fig. 11. Ideal scaling as well as the speed-up of the test case simulations are plotted over the number of cores.

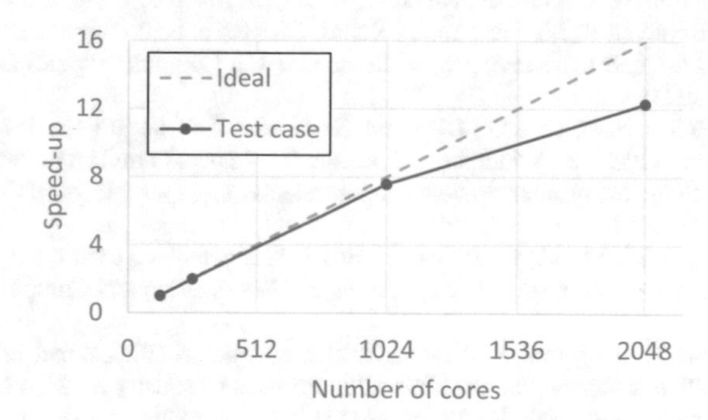

Fig. 11: Strong scaling of a test case with $N = 2 \cdot 10^6$ particles executed with *ls1 mardyn* on a varying number of cores.

It became apparent that the test case scales almost perfectly up to a total number of 1024 cores. Utilizing even more cores speeds up the simulation, although the scaling is not close to ideal anymore. This is a consequence of the investigated special test case which consists of a very elongated domain in combination with simple domain decomposition. For other test cases and more sophisticated domain decompositions, the scaling behavior of our code *ls1 mardyn* was studied in detail as well [19].

Acknowledgements The co-authors R.F., G.G.-C., M.H. and J.V. acknowledge funding by Deutsche Forschungsgemeinschaft (DFG) through the Project SFB-TRR 75, Project number 84292822 - "Droplet Dynamics under Extreme Ambient Conditions". This work was carried out under the auspices of the Boltzmann-Zuse Society of Computational Molecular Engineering (BZS), and it was facilitated by activities of the Innovation Centre for Process Data Technology (Inprodat e.V.), Kaiserslautern. The simulations were performed on the *HPE Apollo (Hawk)* at the High Performance Computing Center Stuttgart (HLRS).

References

1. Deublein, S., Eckl, B., Stoll, J., Lishchuk, S.V., Guevara-Carrion, G., Glass, C.W., Merker, T., Bernreuther, M., Hasse, H., Vrabec, J.: ms2: A molecular simulation tool for thermodynamic properties. Computer Physics Communications **182**, 2350–2367 (2011)
2. Diewald, F.: Phase field modeling of static and dynamic wetting, *Forschungsbericht / Technische Universität Kaiserslautern, Lehrstuhl für Technische Mechanik*, vol. 19 (2020)
3. Diewald, F., Lautenschlaeger, M.P., Stephan, S., Langenbach, K., Kuhn, C., Seckler, S., Bungartz, H.J., Hasse, H., Müller, R.: Molecular dynamics and phase field simulations of droplets on surfaces with wettability gradient. Computer Methods in Applied Mechanics and Engineering **361**, 112773 (2020)
4. Fernández, G., Vrabec, J., Hasse, H.: A molecular simulation study of shear and bulk viscosity and thermal conductivity of simple real fluids. Fluid Phase Equilibria **221**, 157–163 (2004)
5. Fingerhut, R., Guevara-Carrion, G., Nitzke, I., Saric, D., Marx, J., Langenbach, K., Prokopev, S., Celný, D., Bernreuther, M., Stephan, S., Kohns, M., Hasse, H., Vrabec, J.: ms2: A molecular simulation tool for thermodynamic properties, release 4.0. Computer Physics Communications **262**, 107860 (2021)
6. Glass, C.W., Reiser, S., Rutkai, G., Deublein, S., Köster, A., Guevara-Carrion, G., Wafai, A., Horsch, M., Bernreuther, M., Windmann, T., Hasse, H., Vrabec, J.: ms2: A molecular simulation tool for thermodynamic properties, new version release. Computer Physics Communications **185**, 3302–3306 (2014)
7. Grottel, S., Krone, M., Müller, C., Reina, G., Ertl, T.: Megamol – a prototyping framework for particle-based visualization. IEEE Transactions on Visualization and Computer Graphics **21**, 201–214 (2015)
8. Guevara-Carrion, G., Fingerhut, R., Vrabec, J.: Fick diffusion coefficient matrix of a quaternary liquid mixture by molecular dynamics. Journal of Physical Chemistry B **124**, 4527–4535 (2020)
9. Guevara-Carrion, G., Fingerhut, R., Vrabec, J.: Diffusion in multicomponent aqueous alcoholic mixtures. Scientific Reports **11**, 12319 (2021)
10. Heier, M., Stephan, S., Liu, J., Chapman, W.G., Hasse, H., Langenbach, K.: Equation of state for the Lennard-Jones truncated and shifted fluid with a cut-off radius of 2.5 sigma based on perturbation theory and its applications to interfacial thermodynamics. Molecular Physics **116**, 2083–2094 (2018)
11. Heinen, M., Hoffman, M., Diewald, F., Seckler, S., Langenbach, K., Vrabec, J.: Droplet coalescence by molecular dynamics and phase-field modeling. Physics of Fluids **34**, 042006 (2022)

12. Heyes, D.M., Cass, M.J., Powles, J., Evans, W.A.B.: Self-diffusion coefficient of the hard-sphere fluid: System size dependence and empirical correlations. Journal of Physical Chemistry B **111**, 1455–1464 (2007)
13. Homes, S., Heinen, M., Vrabec, J., Fischer, J.: Evaporation driven by conductive heat transport. Molecular Physics, in press (2021). DOI 10.1080/00268976.2020.1836410
14. Jamali, S.H., Bardow, A., Vlugt, T.J.H., Moultos, O.A.: Generalized form for finite-size corrections in mutual diffusion coefficients of multicomponent mixtures obtained from equilibrium molecular dynamics simulation. Journal of Chemical Theory and Computation **16**, 3799–3806 (2020)
15. Lautenschlaeger, M.P., Hasse, H.: Transport properties of the Lennard-Jones truncated and shifted fluid from non-equilibrium molecular dynamics simulations. Fluid Phase Equilib. **482**, 38–47 (2019)
16. Mausbach, P., Fingerhut, R., Vrabec, J.: Structure and dynamics of the Lennard-Jones fcc-solid focusing on melting precursors. Journal of Chemical Physics **153**, 104506 (2020)
17. Niethammer, C., Becker, S., Bernreuther, M., Buchholz, M., Eckhardt, W., Heinecke, A., Werth, S., Bungartz, H.J., Glass, C.W., Hasse, H., Vrabec, J., Horsch, M.: ls1 mardyn: The massively parallel molecular dynamics code for large systems. Journal of Chemical Theory and Computation **10**, 4455–4464 (2014)
18. Rutkai, G., Köster, A., Guevara-Carrion, G., Janzen, T., Schappals, M., Glass, C.W., Bernreuther, M., Wafai, A., Stephan, S., Kohns, M., Reiser, S., Deublein, S., Horsch, M., Hasse, H., Vrabec, J.: ms2: A molecular simulation tool for thermodynamic properties, release 3.0. Computer Physics Communications **221**, 343–351 (2017)
19. Seckler, S., Gratl, F., Heinen, M., Vrabec, J., Bungartz, H.J., Neumann, P.: AutoPas in ls1 mardyn: Massively parallel particle simulations with node-level auto-tuning. Journal of Computational Science **50**, 101296 (2021)
20. Tchipev, N., Seckler, S., Heinen, M., Vrabec, J., Gratl, F., Horsch, M., Bernreuther, M., Glass, C.W., Niethammer, C., Hammer, N., Krischok, B., Resch, M., Kranzlmüller, D., Hasse, H., Bungartz, H.J., Neumann, P.: TweTriS: Twenty trillion-atom simulation. International Journal of High Performance Computing Applications **33**, 838–854 (2019)
21. Yeh, I.C., Hummer, G.: System-size dependence of diffusion coefficients and viscosities from molecular dynamics simulations with periodic boundary conditions. Journal of Physical Chemistry B **108**, 15873–15879 (2004)

Scalable multigrid algorithm for fluid dynamic shape optimization

Jose Pinzon, Martin Siebenborn and Andreas Vogel

Abstract We investigate a parallel approach for the shape optimization of an obstacle in an incompressible Navier–Stokes flow. For this purpose, we used a self-adapting nonlinear extension equation within the method of mappings, which links a boundary control to a mesh deformation. It is demonstrated how the approach preserves mesh quality and allows for large deformations by controlling nonlinearity in critical regions of the domain. Special focus is given to reference configurations, where the transformation has to remove and create obstacle boundary singularities, which is particularly relevant for the employed multigrid discretizations. As benchmark problems, we demonstrate the aerodynamic drag optimization in 2d and 3d configurations. The efficiency of the algorithm is demonstrated in weak scalability studies performed on the supercomputer *HPE Hawk* for up to 5,120 cores.

1 Introduction

Our project aims at furthering our understanding of optimization schemes for domains that experience large deformations. With this purpose, we employ shape optimization to obtain the optimal shape of an obstacle in terms of a physical quantity, e.g. the aerodynamic drag or lift.

Our main focus is on extension operators within the method of mappings [2, 8, 13], with which a scalar variable defined on the surface of an obstacle is related to a deformation field defined on the surrounding domain, c.f. fig. 1. This approach has been implemented with linear elastic extension equations in [7], however it

A. Vogel
High Performance Computing, Ruhr University Bochum, Universitätsstraße 150, 44801 Bochum, Germany, e-mail: a.vogel@rub.de

J. Pinzon and M. Siebenborn
Department of Mathematics, University Hamburg, Bundesstraße 55, 20146 Hamburg, Germany, e-mail: {jose.pinzon,martin.siebenborn}@uni-hamburg.de

W. E. Nagel et al. (eds.), *High Performance Computing in Science and Engineering '21*,
https://doi.org/10.1007/978-3-031-17937-2_30

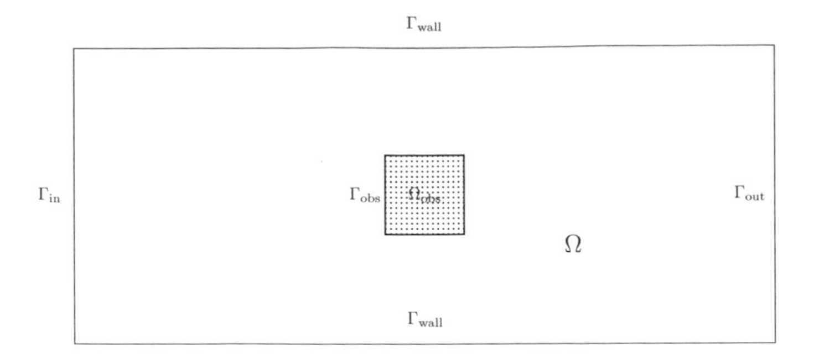

Fig. 1: Sketch of the domain used in the 2d simulations. An obstacle Γ_{obs} is located inside a flow tunnel.

suffers under large deformations. To allow for large deformations, it turns out to be advantageous to add nonlinear terms to the extension model. For instance, in [14] a nonlinear advective term leads the obtained displacements on the major directions of deformations, thus preserving mesh quality even under large displacements with respect to the reference configuration.

The domain of the underlying optimization experiment is sketched in fig. 1. It represents an obstacle Ω_{obs} and its surface Γ_{obs} inside a holdall domain Ω, which represents a flow tunnel. The obstacle is subject to an incompressible flow with inflow at Γ_{in}. The objective functional is the drag, as expressed in [12]. We follow the continuous adjoint approach (see for instance [6, 9, 12]) where the determination of the fixed dimensional multipliers is realized via an augmented Lagrangian approach. For an overview of the ongoing research on various approaches in the area of shape optimization in fluid dynamics, the reader is referred to [2, 4, 11, 17]. For a complete monograph on discretization schemes, solvers, and numerical stabilization techniques for the Navier–Stokes equations, please refer to [3] and the citations therein.

The core of the method of mappings is to rephrase the shape optimization problem into an optimal control problem over the set of admissible transformations acting on a reference domain. This is in contrast to defining a set of admissible shapes itself and to optimize within this. The advantage is that the appropriate choice of the extension equation allows to require properties of the optimal configuration such as a certain degree of mesh quality in the deformed domain or preventing self intersections in the discretization. This means that there are no mesh deformations performed throughout the algorithm, but all operators are traced back to the reference domain. In order to maintain mesh quality and prevent element overlappings, a threshold is enforced on the deformation gradient. We express this as an inequality constraint in the problem formulation, as in [7].

2 Mathematical background

The following strategy and optimization problem are outlined in detail in [15]. We start by introducing the shape optimization problem in a general form. Via the method of mappings it is formulated by means of an optimal control problem for a generic objective j and a PDE constraint e. Let further $X = L^2(\Gamma_{\text{obs}}) \times L^2(\Omega)$, $0 < u_{\text{lb}} < u_{\text{ub}}$, $\eta_{\text{det}} > 0$, $\alpha > 0$ and consider the problem

$$\min_{(u,\eta)\in X} \quad j(y, F(\Omega)) + \tfrac{\alpha}{2}\|u\|^2_{L^2(\Gamma_{\text{obs}})} + \tfrac{\chi}{2}\|\eta - \tfrac{1}{2}(\eta_{\text{ub}} + \eta_{\text{lb}})\|^2_{L^2(\Omega)} \tag{1}$$

$$\text{s.t.} \quad e(y, F(\Omega)) = 0 \tag{2}$$

$$F = \text{id} + \mathbf{w} \tag{3}$$

$$\mathbf{w} = S(\eta, u, \Omega) \tag{4}$$

$$\det(DF) \geq \eta_{\text{det}} \quad \text{in } \Omega \tag{5}$$

$$\eta_{\text{lb}} \leq \eta \leq \eta_{\text{ub}} \quad \text{in } \Omega \tag{6}$$

$$g(\mathbf{w}) = 0. \tag{7}$$

In the objective function (1), j refers to the quantity to be optimized, for instance the drag of the obstacle. y denotes the state variable of the PDE. The boundary control variable is denoted with u, $\mathbf{w}$ is its extension to a displacement field via the extension equation S (cf. (4)), which then defines the domain mapping function F. In (7), $g(\mathbf{w})$ represent geometric constraints, see [7, 14] for a complete explanation of how to treat these. The domain transformation $F(\Omega)$ in (3) is based on the perturbation of identity. Furthermore, (5) models a threshold on the deformation gradient towards local injectivity of F as investigated in [7]. Finally, (6) defines box constraints on the factor η, which limits the nonlinearity in S, taking into account that this affects the convergence of the iterative solver.
In the optimization problem (1) to (7), we consider the PDE constraint $e(y, F)$ to be the stationary, incompressible Navier–Stokes equations in terms of a velocity $\mathbf{v}$ and a pressure p given by

$$-\nu\Delta\mathbf{v} + (\mathbf{v}\cdot\nabla)\mathbf{v} + \nabla p = 0 \quad \text{in } F(\Omega) \tag{8}$$

$$\operatorname{div}\mathbf{v} = 0 \quad \text{in } F(\Omega) \tag{9}$$

$$\mathbf{v} = v_\infty \quad \text{on } \Gamma_{\text{in}} \tag{10}$$

$$\mathbf{v} = 0 \quad \text{on } \Gamma_{\text{obs}} \cup \Gamma_{\text{wall}} \tag{11}$$

$$pn - \nu\frac{\partial\mathbf{v}}{\partial n} = 0 \quad \text{on } \Gamma_{\text{out}}. \tag{12}$$

For compatibility, it is assumed that $\int_{\Gamma_{\text{in}}} \mathbf{v}_\infty \cdot n \, \mathrm{d}s = 0$ holds at the inflow. The extension $S(\eta, u, \Omega)$ is given as the solution operator of the PDE

$$\operatorname{div}(D\mathbf{w} + D\mathbf{w}^\top) + \eta_{\text{ext}}(\mathbf{w} \cdot \nabla)\mathbf{w} = 0 \quad \text{in } \Omega \tag{13}$$

$$(D\mathbf{w} + D\mathbf{w}^\top)n = u\mathbf{n} \quad \text{on } \Omega_{\text{obs}} \tag{14}$$

$$\mathbf{w} = 0 \quad \text{on } \Gamma_{\text{in}} \cup \Gamma_{\text{out}} \cup \Gamma_{\text{wall}}. \tag{15}$$

The condition (5) is approximated via a non-smooth penalty term. This results in the regularized objective function

$$J(u, \eta) = j(y, F(\Omega)) + \tfrac{\alpha}{2}\|u\|^2_{L^2(\Gamma_{\text{obs}})} + \tfrac{\chi}{2}\|\eta - \tfrac{1}{2}(\eta_{\text{ub}} + \eta_{\text{lb}})\|^2_{L^2(\Omega)} + \tfrac{\beta}{2}\|(\eta_{\det} - \det(DF))^+\|^2_{L^2(\Omega)}. \tag{16}$$

In contrast to the PDE constraints (3) to (5), the geometric constraints (7) are fixed dimensional (here it is $d + 1$ where $d \in \{2, 3\}$). Thus, the multiplier associated to these conditions is not a variable in the finite element space but an $d + 1$-dimensional vector. We incorporate this into our optimization algorithm in form of an augmented Lagrange approach. This leads to the augmented objective function

$$J_{\text{aug}}(u, \eta) = J(u, \eta) + \|g(\mathbf{w})\|_2^2 \tag{17}$$

and the corresponding Lagrangian

$$\begin{aligned}
&\mathcal{L}(\mathbf{w}, \mathbf{v}, p, u, \eta_{\text{ext}}, \psi_{\mathbf{w}}, \psi_{\mathbf{v}}, \psi_p, \psi_{\text{bc}}\psi_{\text{vol}}) = \\
&\quad \frac{\nu}{2}\int_\Omega (D\mathbf{v}(DF)^{-1}) : (D\mathbf{v}(DF)^{-1}) \det(DF)\mathrm{d}x + \frac{\alpha}{2}\int_{\Gamma_{\text{obs}}} u^2 \mathrm{d}s \\
&\quad + \frac{\beta}{2}\int_\Omega \left\|(\eta_{\det} - \det(DF))^+\right\|^2 \mathrm{d}x - \int_\Omega [\nu(D\mathbf{v}(DF)^{-1}) : (D\psi_{\mathbf{v}}(DF)^{-1}) \\
&\qquad + (D\mathbf{v}(DF)^{-1}) \cdot \psi_{\mathbf{v}} - p\operatorname{Tr}(D\psi_{\mathbf{v}}(DF)^{-1})] \det(DF)\mathrm{d}x \\
&+ \int_\Omega \psi_p \operatorname{Tr}(D\mathbf{v}(DF)^{-1}) \det(DF)\mathrm{d}x - \int_\Omega [(D\mathbf{w} + D\mathbf{w}^\top) : D\psi_{\mathbf{w}} + \eta_{\text{ext}}(D\mathbf{w} \cdot \mathbf{w})]\mathrm{d}x \\
&\quad + \int_{\Gamma_{\text{obs}}} u\mathbf{n} \cdot \psi_{\mathbf{w}}\mathrm{d}s - \psi_{\text{bc}} \cdot \int_\Omega F(x) \det(DF)\mathrm{d}x - \psi_{\text{vol}} \int_\Omega (\det(DF) - 1)\mathrm{d}x
\end{aligned} \tag{18}$$

under the assumption that the barycenter of Ω is $0 \in \mathbb{R}^d$.

The underlying numerical scheme is shown in algorithm 1. Conceptually, the algorithm consists of two nested loops. The outer one is the augmented Lagrange algorithm, which iteratively determines the fixed dimensional multipliers for the geometric constraints $g(\mathbf{w})$ (cf. (7)). For each approximation of λ_g and penalty factor ν_g the actual shape optimization problem is then solved in the inner loop where the descent direction is chosen according to a quasi-Newton update. Line 7 shows the termination condition for the iterated shape optimization problems, which is based on the $[\eta_{\text{lb}}, \eta_{\text{ub}}]$-box-projection of the gradient $(\nabla u, \nabla \eta)$. The termination of the entire algorithm is based on the change of u between two successive augmented Lagrange iterations, as shown in line 14.

Algorithm 1 Augmented Lagrange Method

Require: $\lambda_g = 0, \dots$
1: **repeat**
2: **repeat**
3: $(\mathbf{w}, \nabla u^{k,\ell}, \nabla \eta^{k,\ell}) \leftarrow$ SOLVE FOR$(g_{\text{def}}, u^{k,\ell}, \eta^{k,\ell})$
4: $(u^{k,\ell+1}, \eta^{k,\ell+1}) \leftarrow$ APPLY QUASI-NEWTON$(m, g_{\text{def}}, \nu_g, \nabla u^{k,\ell}, \nabla \eta^{k,\ell})$
5: $g_{\text{def}} \leftarrow g(\mathbf{w}) - g(0)$
6: $\ell \leftarrow \ell + 1$
7: **until** $\left\| \left(\nabla u^{k,\ell-1}, P_{(\eta^l,\eta^u)} \left(\eta^{k,\ell-1} - \nabla \eta^{k,\ell-1} \right) - \eta^{k,\ell-1} \right) \right\|_X < \epsilon_{\text{inner}}$
8: **if** $\|g_{\text{def}}\|_2 < \epsilon_g$ **then**
9: $\nu_g \leftarrow \nu_{\text{inc}} \nu_g$
10: **else**
11: $\lambda_g \leftarrow \lambda_g + \lambda_{\text{inc}} g_{\text{def}}$
12: **end if**
13: $k \leftarrow k + 1$
14: **until** $k \geq 1$ **and** $\|u^{k,\ell} - u^{k-1,\ell}\|_{L^2(\Gamma_{\text{obs}})} < \epsilon_{\text{outer}}$

The Lagrange multipliers λ_g are updated based on the norm of the geometrical defects, i.e. how close the current deformation field and its corresponding transformation fulfill the barycenter and volume constraints.

3 Numerical results

The simulations are performed with UG4 [19], a framework to solve PDE systems on massive parallel systems. The general parallel scalability of the software is reported in [16, 18]. The uncoupled equation systems are solved in a block-like fashion, making it necessary to pass the solution of a PDE system to another as integration point data. UG4 provides this functionality in the form of data imports. The computational meshes are created using GMSH [5] and the visualization of vector and scalar quantities, as well as of the deformed grids, employs ParaView [1].

Case studies for 2d and 3d domains are presented. The former uses a square obstacle as reference domain to illustrate the effects of algorithm 1 on grids with pronounced singularities. The latter demonstrates the successful generation of singularities using surface elements. In all cases, the simulations utilize 1 MPI process per core.

3.1 2d results

The 2d simulations are carried out using a discretization of the domain shown in fig. 1. A P_1 discretization is used for all PDEs, except for the solution of the Navier–Stokes equations, where $P_2 - P_1$ stable triangular elements are used. The grid has a total of 421,888 triangular elements, with 5 refinement levels. The simulations runs each

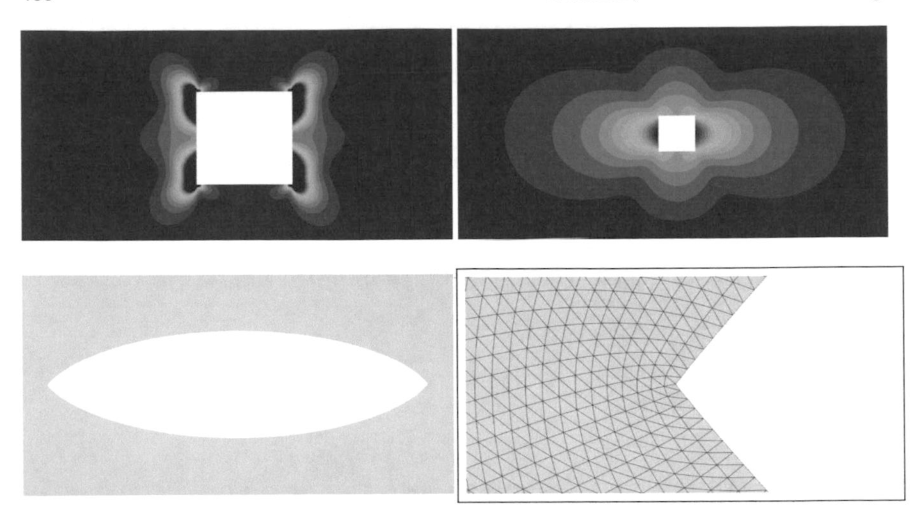

Fig. 2: At the top, the reference configuration is shown with the optimal η (left) and $\mathbf{w} = S(u)$. At the bottom, the transformed grid $F(\Omega)$ with resulting singularities (left) is shown, altogether with a zoom on the singularity where mesh quality is preserved due to the choice of S

on 320 cores, divided among 4 nodes. As a rule of thumb, we choose to use the number of refinements that better represent the used mathematical spaces. This can be empirically measured by the smoothing of grid singularities, e.g. edges, and by their creation. Nevertheless, it is shown in Sect. 3.2 that the obtained results are grid independent, with only minor variations proper of the iterative solution of PDE systems.

In fig. 2, the final results for a 2d simulation are shown. The deformation field obtained after 1000 steps is applied to transform the reference domain. In order to prove the efficiency of the optimization scheme shown in algorithm 1, a reference domain with pronounced edges is chosen. The algorithm is able to detect the edges, which are not part of the optimal configuration, to promote a concentration of both the extension factor η_{ext} and boundary control variable u. Likewise, it can promote domain transformations, which result in the creation of new boundary edges. As can be seen on the bottom of fig. 2, the front and back tips appear as a result of the optimization process, whereas the previously existing corners are smoothed down by the transformation.

The effects of the transformation on the reference domain, throughout several optimization steps, is shown in fig. 4. The transformation, i.e. the deformation field, is applied to the reference domain. In the initial steps, the deformation field incurs in notorious violations of the geometric constraints, i.e. there is shrinkage of the obstacle's area. This is related to the poor initial values for the deformation field, u

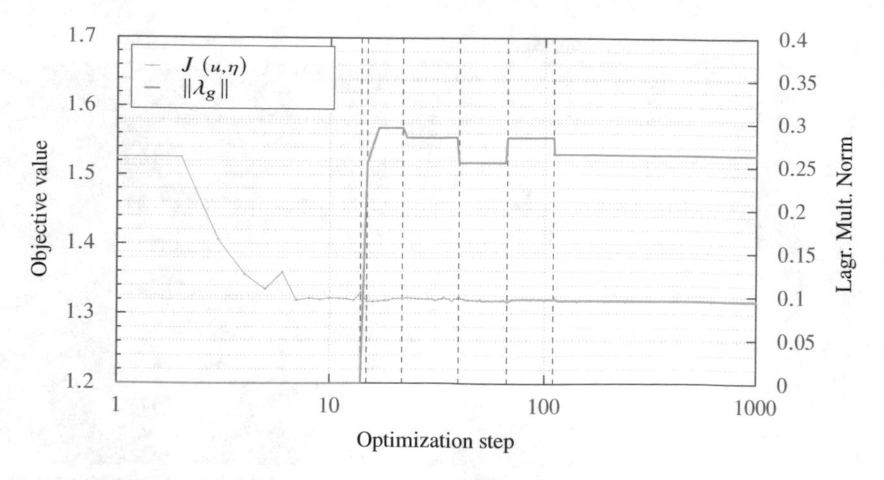

Fig. 3: The objective function per optimization step is shown in relation to the norm of the Lagrange multipliers. The vertical blue lines indicate major changes of the multipliers

scalar field, and η extension factor. As the simulation progresses, the front and back tips are formed, and the optimization scheme starts to smooth out the corners of the reference domain.

The extension factor η_{ext} is shown with respect to the reference domain. In (13), η_{ext} enriches the extension equation in order to allow for nonlinear deformations. While the extension factor could be chosen as a constant, refer to [14], the advantages of implementing it as a scalar grid function become evident from fig. 4. In the first steps of the simulation, the extension factor is already promoting the advective movement of the nodes which require large deformations. This can be appreciated both on the corners as well as the faces of the square obstacle.

The value of the objective function is shown for a 1000 steps in fig. 3. The dashed blue lines show the optimization steps where major Lagrange multiplier updated occurred. This is directly related to line 11 in algorithm 1, where the update is subject to the value of the small number ν_g. The first steps usually imply large and expected violations of the geometric constraints, due to the poor selection of an initial guess for λ_g. Upon each update, we see a peak in the value of the objective function. This is expected, since the optimization process can be considered global within the given values of the multipliers. Once the latter have converged, the value of the objective function reaches a minimum.

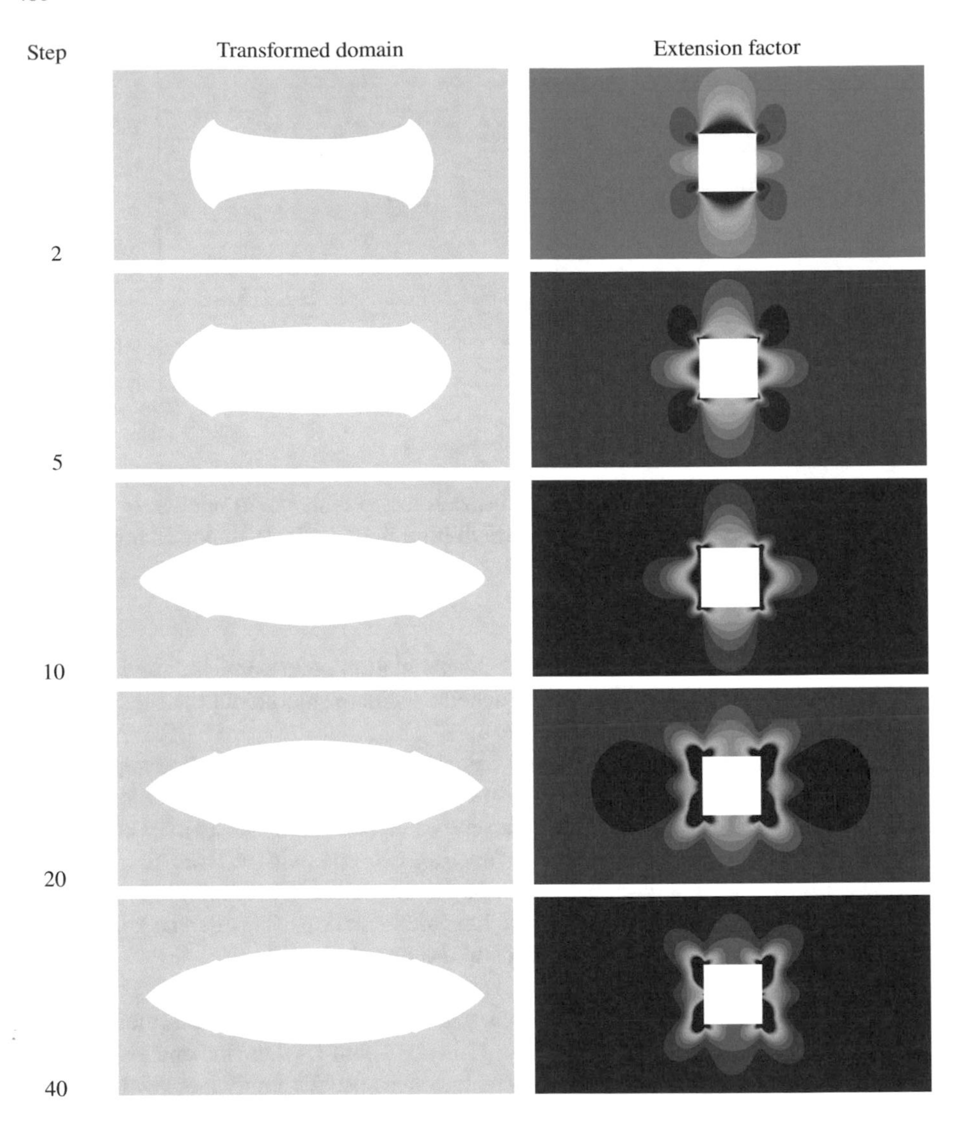

Fig. 4: Transformation of the reference domain by the application of the deformation field compared to the accumulation of the extension factor, given for steps 2,5,10,20,40

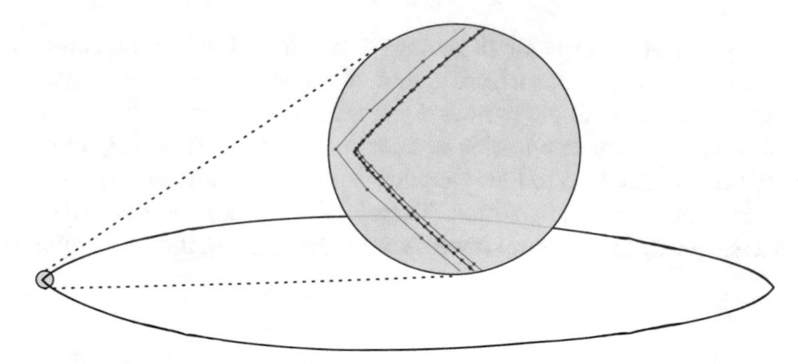

Fig. 5: The convergence of several refinement levels (indicated by the colored nodes) is shown with the superimposed grids

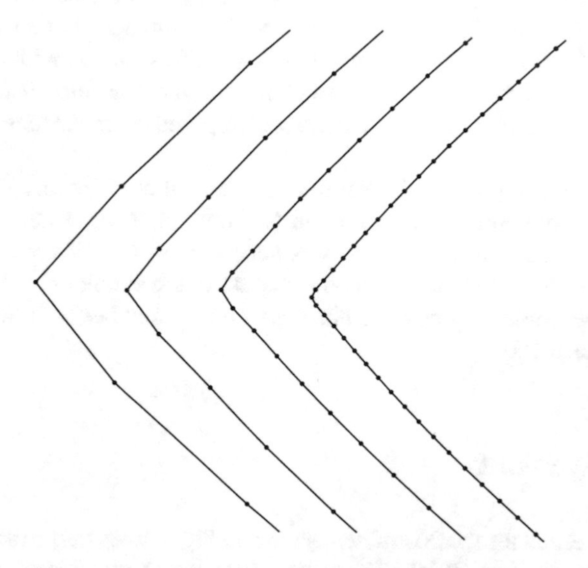

Fig. 6: Side-by-side comparison of the generated tips for several refinement levels. From left to right 2,3,4,5 refinements of the discretized domain shown in fig. 1

3.2 2d grid study

In this section, we provide results which demonstrate that the obstacle shape obtained by applying the transformation after a given number of steps is independent of the number of refinements used, and that the detection and generation of domain singularities is grid-independent. The coarse grid is refined from 6,592 to 421,888 elements. All simulations had the same settings, including viscosity $\nu = 0.02$. Figure 5 shows the final grid after 400 optimization steps. The number of MPI processes used went from 80 for the lowest refinement level to 320 for the highest, running an equal number of optimization steps for all settings to provide a meaningful comparison. It can be seen that the different surfaces across all refinement levels are superimposed to each other or have small differences at the magnified tip in fig. 5.

In addition to this, the side-to-side comparison in fig. 6 shows that, after applying the optimal transformation to the reference domain, the obtained geometry is the same in terms of shape and position. Therefore, the results are grid-independent which is a necessary fact to move towards more demanding, industrial applications.

3.3 3d results

In fig. 7, we present results for a 3d grid that consists of 12,296,192 tetrahedral elements with 4 levels of refinement. The obstacle is composed of 54,784 triangular faces. A $P_1 - P_1$ discretization is used across all PDE systems, with a stabilization term for the Navier–Stokes equations, which is necessary for numerical stability. The viscosity is set to 1, while the box constraints imposed on the extension factor η_{ext} are $0 \leq \eta_{\text{ext}} \leq 60$.

The reference configuration is chosen as a spherical obstacle inside a flow tunnel and the creation of singularities is demonstrated in fig. 7. Given the high levels of computation for these simulations, it was necessary to use an average of 4 nodes with at least 80 cores each. Moreover, the success of the simulations, in terms of the generated shape, could only be determined after a high number of optimization steps, usually more than 100.

4 Scalability results

Weak scaling results for the 2d problem shown in Fig. 2 were performed on the *HPE Hawk* supercomputer at HLRS. The machine features 5,632 nodes, each with two AMD EPYC 7742 CPUs consisting of 64 cores per CPU. In fig. 8, we present the accumulated timings and iteration counts for the first 3 optimization steps for up to 5,120 cores and more than 6 million elements. ParMETIS [10] is employed for load balancing. Following the recommendation from the *Hawk* online documentation, we chose to base our results on an 80 core count per node and increase the cores fourfold upon each mesh refinement.

We present results for the solution of the extension equation (13) to (15), for the shape derivative of (16), and the gradient of the extension factor system of equations, shown below:

$$M + \xi(\eta - \frac{1}{2}(\eta_{\text{lb}} + \eta_{\text{ub}})) - (D\mathbf{w} \cdot \mathbf{w}) \cdot \psi_{\mathbf{w}} = 0.$$

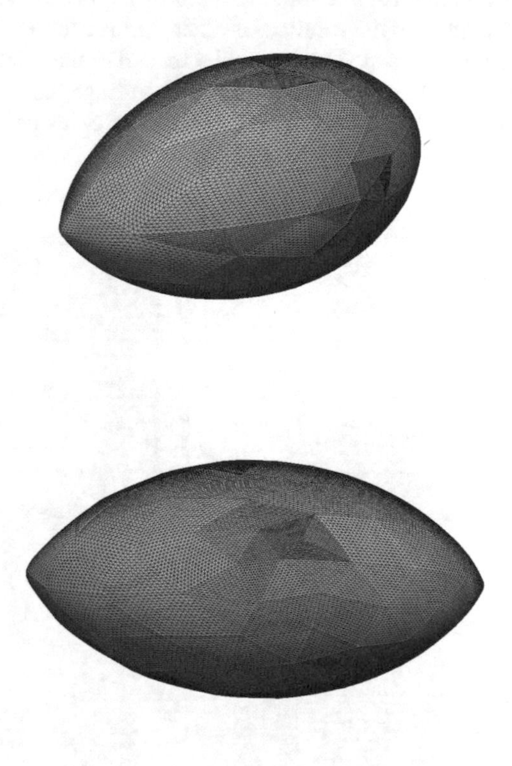

Fig. 7: High viscosity 3d simulation

The nonlinear extension equation is solved using a Newton method, while the linear problems are solved with a BiCGStab method preconditioned with a geometric multigrid method. A Jacobi smoother is used within the multigrid, with 3 pre- and post-smoothing steps in a V-cycle. The base level is solved on a single process by a serial LU solver in all cases. The convergence criteria for the linearization within the Newton solver are a maximum of 2000 iterations or an error reduction of 10^{-14}. Whereas, for the Newton method they are a maximum of 20 steps or an error reduction of 10^{-12}. The linear solvers must fulfill a maximum of 2000 steps or an error reduction of 10^{-16}.

Figure 8 presents timings and speedup for the solve phases of the Newton (newton) and linear solvers (solve), the multigrid initialization (init), and the fine matrix assembly (ass). A good weak scaling is found in general, but sub-optimal degradations are observed for the Newton solver of the extension equation, the assembly phases of the shape derivative and the extension gradient. The iteration counts remain constant for the Newton method steps and the linearization problem even for more than 6 million elements. The slight variation in the shape derivative could be related to the numerical differences caused by the PDE syste ; dependency on integration point data from the solution of other grid functions. A e same time, the extension gradient performance loss could be related to the selec of the cores per node within the same hypercube topology. The timings increa or 16 and 64 nodes, this might suggest that the sweet spot in terms of cores per e is yet to be achieved and must be further investigated.

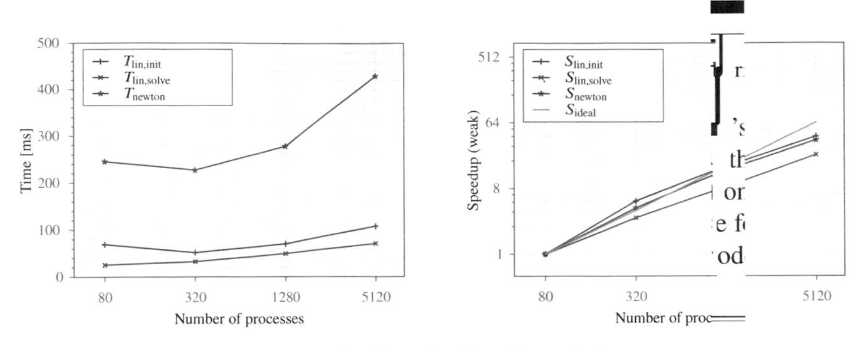

(a) Extension Equation Non-Linear Solver

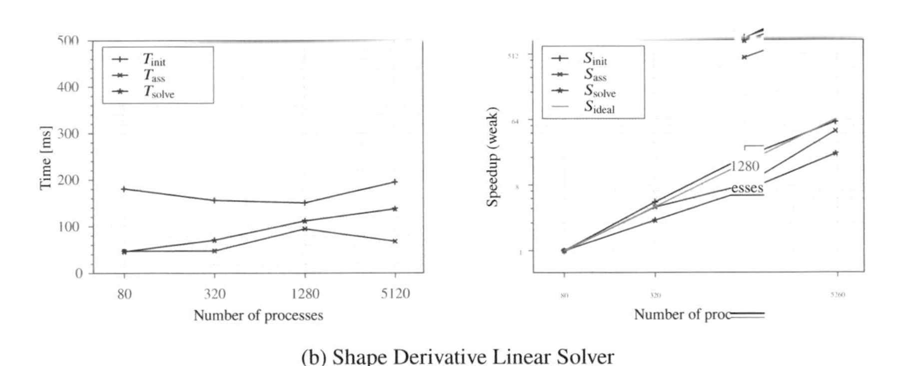

(b) Shape Derivative Linear Solver

Fig. 8: (continued on next page).

5 Conclusion and outlook

We presented an optimization result for a 2d square obstacle in a Na –Stokes flow channel. It has been demonstrated that the employed numerical sc ne is able to remove the edges present in the reference configuration, as well as i oducing new ones through the accumulation of the boundary control variable and ension factor.

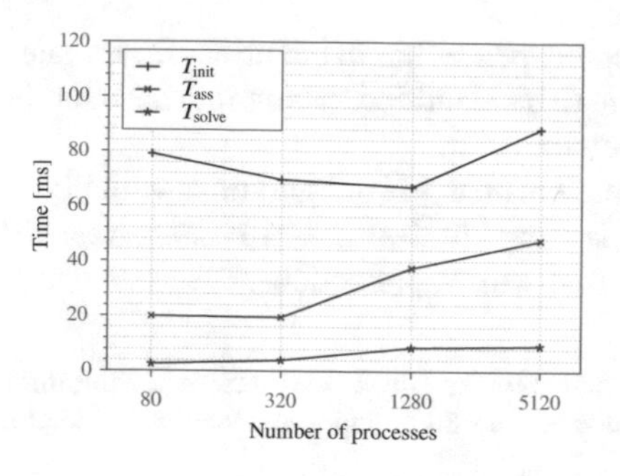

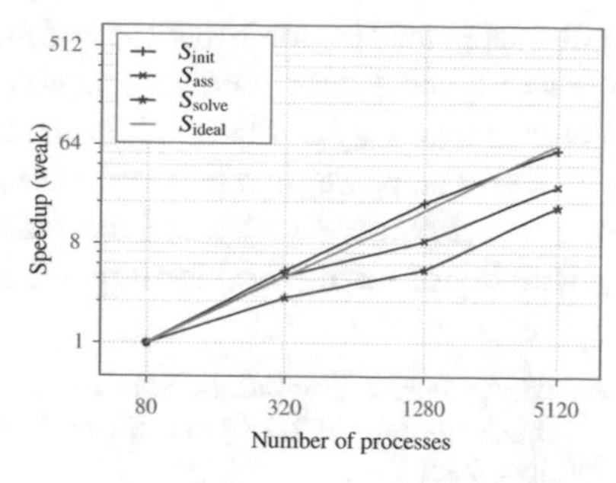

(c) Extension Gradient Linear Solver

Procs	Refs	NumElems	Linear solver (shape derivative)	Newton solver (deformation field)	Linear solver (extension gradient)
80	4	105,472	56	12	9
320	5	421,888	70	12	9
1,280	6	1,687,552	77	12	9
5,120	7	6,750,208	77	12	9

(d) Accumulated iteration counts for the weak scaling study

Fig. 8: Weak Scaling: For the first three optimization steps, the accumulated wallclock time is shown for: (a) the nonlinear extension equation, (b) the derivative of the objective function with respect to the deformation field, and (c) the extension factor gradient, equation given in Sec. 4. In (d), the accumulated iteration counts are presented for the geometric multigrid preconditioned linear solver of the shape derivative, the number of Newton steps and linear iterations necessary to solve the extension equation and its linearization.

The results for the objective function reflect the fact that, after the Lagrange multipliers have iteratively calculated within a certain tolerance, the functional converges to a minimum. Corresponding 3d results demonstrate that the algorithm is not restricted to creating singularities on obstacle contours with edges, but can also achieve similar patterns with surface elements.

A grid study emphasizes that the computed optimal shape is independent of the number of refinements. This allows the usage of the geometric multigrid method with grid-independent convergence. Although good weak scaling speedup results have been achieved for up to 5,120 cores and more than 6 million elements, an optimal parallel setup for Newton's method and assembly phases has to be further investigated. To this end, better core counts per node have to be identified, taking into account the hypercube topology of the *HPE Hawk* supercomputer.

As a next step, we plan to investigate more detailed 3d configurations with higher levels of refinement to extend the weak scaling studies and to demonstrate the applicability of the method for large-scale, real-world applications.

Acknowledgements Computing time on the supercomputer *Hawk* at HLRS under the grant ShapeOptCompMat (ACID 44171, Shape Optimization for 3d Composite Material Models) is gratefully acknowledged.
The current work is part of the research training group 'Simulation-Based Design Optimization of Dynamic Systems Under Uncertainties' (SENSUS) funded by the state of Hamburg under the aegis of the Landesforschungsförderungs-Project LFF-GK11.

References

1. URL `www.paraview.org`
2. Brandenburg, C., Lindemann, F., Ulbrich, M., Ulbrich, S.: A Continuous Adjoint Approach to Shape Optimization for Navier Stokes Flow. In: K. Kunisch, G. Leugering, J. Sprekels, F. Tröltzsch (eds.) Optimal Control of Coupled Systems of Partial Differential Equations, *Internat. Ser. Numer. Math.*, vol. 160, pp. 35–56. Birkhäuser, Basel (2009)
3. Elman, H., Silvester, D., Wathen, A.: Finite Elements and Fast Itertative Solvers With Applications in Incompressible Fluid Dynamics, vol. 1. Oxford Science Publications (2014)
4. Garcke, H., Hinze, M., Kahle, C.: A stable and linear time discretization for a thermodynamically consistent model for two-phase incompressible flow. Applied Numerical Mathematics **99**, 151–171 (2016)
5. Geuzaine, C., Remacle, J.F.: Gmsh: A 3-D finite element mesh generator with built-in pre- and post-processing facilities. International Journal for Numerical Methods in Engineering **79**(11), 1309–1331 (2009). DOI 10.1002/nme.2579
6. Giles, M., Pierce, N.: An introduction to the adjoint approach to design. Flow, turbulence and combustion **65**(3-4), 393–415 (2000)
7. Haubner, J., Siebenborn, M., Ulbrich, M.: A continuous perspective on shape optimization via domain transformations. SIAM Journal on Scientific Computing **43**(3), A1997–A2018 (2021). DOI 10.1137/20m1332050
8. Iglesias, J.A., Sturm, K., Wechsung, F.: Two-dimensional shape optimization with nearly conformal transformations. SIAM Journal on Scientific Computing **40**(6), A3807–A3830 (2018)
9. Jameson, A.: Aerodynamic shape optimization using the adjoint method. Lectures at the Von Karman Institute, Brussels (2003)
10. Karypis, G., Schloegel, K., Kumar, V.: Parmetis, parallel graph partitioning and sparse matrix ordering library (2013). URL `http://glaros.dtc.umn.edu/gkhome/metis/parmetis/overview`
11. Müller, P.M., Kühl, N., Siebenborn, M., Deckelnick, K., Hinze, M., Rung, T.: A novel p-harmonic descent approach applied to fluid dynamic shape optimization (2021)
12. Mohammadi, B., Pironneau, O.: Applied shape optimization for fluids. Oxford university press (2010)
13. Murat, F., Simon, J.: Etude de problèmes d'optimal design. In: J. Cea (ed.) Optimization Techniques Modeling and Optimization in the Service of Man Part 2: Proceedings, 7th IFIP Conference Nice, September 8–12, 1975, pp. 54–62. Springer-Verlag, Berlin, Heidelberg (1976)

14. Onyshkevych, S., Siebenborn, M.: Mesh quality preserving shape optimization using nonlinear extension operators. Journal of Optimization Theory and Applications **16**(5), 291—-316 (2021). DOI 10.1007/s10957-021-01837-8
15. Pinzon, J., Siebenborn, M.: Fluid dynamic shape optimization using self-adapting nonlinear extension operators with multigrid preconditioners (in preparation)
16. Reiter, S., Vogel, A., Heppner, I., Rupp, M., Wittum, G.: A massively parallel geometric multigrid solver on hierarchically distributed grids. Comp. Vis. Sci. **16**(4), 151–164 (2013)
17. Schmidt, S., Ilic, C., Schulz, V., Gauger, N.R.: Three-dimensional large-scale aerodynamic shape optimization based on shape calculus. AIAA journal **51**(11), 2615–2627 (2013)
18. Vogel, A., Calotoiu, A., Strube, A., Reiter, S., Nägel, A., Wolf, F., Wittum, G.: 10,000 performance models per minute – scalability of the UG4 simulation framework. in: Euro-Par 2015: Parallel Processing, J. L. Träff et al. (eds), Springer pp. 519–531 (2015)
19. Vogel, A., Reiter, S., Rupp, M., Nägel, A., Wittum, G.: UG 4: A novel flexible software system for simulating PDE based models on high performance computers. Comp. Vis. Sci. **16**(4), 165–179 (2013)

Numerical calculation of the lean-blow-out in a multi-jet burner

Alexander Schwagerus, Peter Habisreuther and Nikolaos Zarzalis

Abstract The report presents the results of the project TurboRe, which focuses on the numerical investigation of lean-blow-out phenomenon in a multi jet flame array. Successful calculations have been conducted for a model combustor and the developed numerical setup has been applied for the calculation of blow-out in a complex burner.

1 Introduction

The project TurboRe focuses on the numerical calculation of Lean-Blow-Out (LBO) of a premixed and turbulent burner consisting of an array of jet flames (matrix burner) using large eddy simulation and two different approaches of modelling the turbulence-flame-interaction. The main objective is to simulate turbulent lean premixed jet-flames and determine their LBO-limits as a function of basic geometrical and thermo-dynamical parameters. Simulations are validated through corresponding experiments and correlations are to be determined. The results of these investigations show the ability to numerically predict LBO and lead to a better understanding of the flame stability in matrix burner arrangements. The investigation provides crucial information for the design and scaling of the matrix burner. In the last several decades, a swirl-stabilized flame has been the preferred choice in gas turbine combustors due to their aerodynamic means of stabilization. However, swirl flames can be susceptible

---

Alexander Schwagerus
DVGW Research Center, Engler-Bunte-Institute of the Karlsruhe Institute of Technology, Karlsruhe, 76131, Germany, e-mail: alexander.schwagerus@kit.edu

Peter Habisreuther
Karlsruhe Institute of Technology, Institute for Technical Chemistry, Eggenstein-Lepoldshafen, 76344, Germany, e-mail: peter.habisreuther@kit.edu

Nikolaos Zarzalis
Karlsruhe Institute of Technology, Engler-Bunte-Institute, Combustion Technology, Karlsruhe, 76131, Germany, e-mail: nikolaos.zarzalis@kit.edu

W. E. Nagel et al. (eds.), *High Performance Computing in Science and Engineering '21*,
https://doi.org/10.1007/978-3-031-17937-2_31

to the issue of combustion instability in the lean combustion regime leading to lean-blow-out (LBO) and therefore damaging the hardware. The corresponding task of the EU-project Turbo-Reflex focuses on improving the low load capability of stationary gas turbines to reduce the number of hot starts. By having a higher load capability, plants can be run more flexible without being forced to shut-down/start-up completely, resulting in less cycling, and therefore leading to a longer lifespan of the turbine in the increasingly volatile energy demand environment. A matrix burner (arrangement of jet flames) is regarded as an alternative to the swirl burners mentioned earlier. However, since such matrix burners have only scarcely been investigated, further understanding of such systems is needed to correctly describe and scale them. This was the motivation to carry out a test campaign involving velocity, exhaust gas and LBO measurements at the combustion technology department of the Engler-Bunte-Institute of the KIT. A sketch of the used nozzles can be seen in Figure 1. For detailed information regarding the experiments and the investigated combustor, the reader is referred to previous publications [1, 6]. To complement the experimental investigation, detailed numerical simulations have been performed to further deepen the understanding of the dominating flame stability effects in this highly complex combustion system.

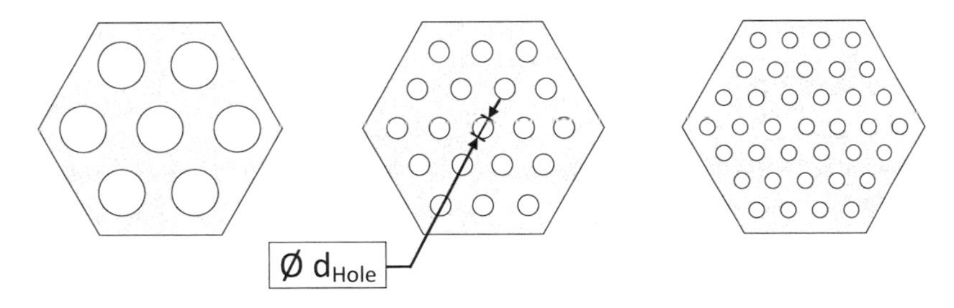

Fig. 1: Sketch of the investigated nozzles (from left: D_1, D_2 and D_3) [1]

For the simulation of the flame, a model for the calculation of the reactive source terms is required to calculate the distribution of temperature and species concentrations. Since detailed kinetic mechanisms usually contain hundreds of elementary reactions, this approach nowadays still surpasses the computation capacities generally available for complex, realistic geometries and therefore is limited to one-dimensional flames or simple laminar systems. Two methods to reduce the computational effort have been used and compared, that enable to calculate the reactive system by the introduction a quantity for the reaction progress.

2 Numerical setup

The main task is the numerical calculation of the lean-blow-off (LBO) limits. For that purpose, the open-source C++ libraries of *OpenFOAM* have been used. The solver used for calculating the reactive flow is based on the already existing *OpenFOAM* solver rhoPimpleFoam. It solves the compressible formulation of the Navier-Stokes equations. The pressure-velocity coupling is implemented through the PIMPLE method. Among the different approaches for the description of turbulence by numerical models, the LES (Large-Eddy-Simulation) technique was chosen, which enables the investigation of transient flows. This is considered useful, since LBO is also a transient phenomenon. Computational grids for four different nozzle have been created. The numerical grids consists of each around 1.3 mio. cells. The turbulence of the inlet has, as already shown in literature [5] and in preliminary simulations, a huge influence on the flow in the wake, but the specification of a suitable velocity field at the inlet is a non-trivial problem in LES-simulations. For this work, the turbulence generator proposed by Klein et al. [4] has been applied and implemented in OpenFoam. This method is based on digital filtering of a series of uncorrelated random data to generate correlated velocity fields according to user-defined turbulence properties. In order to reduce the computational effort and enable to the computation within a reasonable timeframe, a FGM (flame-generated manifold) approach is used, which reduces the reaction mechanism to a pre-tabulated reaction progress variable Θ. The turbulent-chemistry interaction (TCI) were captured by two different combustion models. In both models, a transport equation of a reaction progress variable is solved. The difference between the combustion models is in the source term modelling. The first model (JPDF model) involves the calculation of the source term via a presumed density function (PDF) of the reaction progress variable. In order to link the reaction state to the reaction progress variable a model reactor, like the one-dimensional premixed flame, is needed. This allows a tabulation of the source term as a function of the progress variable itself. The calculation of the necessary mean source term can then be conducted by integrating it with the PDF. The PDF is exactly defined by its mean and variance. Therefore two transport equation for the mean and the variance of the reaction progress variable need to be solved. This is already implemented in a solver developed at the Engler-Bunte-Institute and have been already successful used in different numerical investigations [2, 3]. For a more detailed description, the reader is referred to literature, e.g. the dissertation of Kern [3]. The second model (TFC model) calculates the source term through the KPP-theorem, which states that the source term can be calculated on basis of the turbulent flame speed, which is a measure of the turbulent volumetric conversion rate. The turbulent flame speed is calculated using the H.P. Schmid model [3] as a function of the Damkoehler number and the laminar flame speed.

3 General description of the flame

For comparison, the calculated distribution of the reaction progression variable of the flames under stable near stoichiometric conditions is shown in Fig. 2 for a median cut. Here, the unburned mixture (marked in blue) flows from left into the combustion chamber and is reacting then to fully burnt exhaust gases (light red color). Simulations of the JPDF model (top) and the TFC model (bottom) are compared at the same operating conditions. Snapshots are shown on the left and time averaged fields on the right side. In these simulations, only about half of the combustor length was discretized. The results of both combustion models show that the jets are very straight, similar to what was observed in the experiments. The simulations of the JPDF model shows separated reaction zones for each jet, while the TFC model predicts more overlaps of the individual jets. In addition, there are considerable differences in the width of the reaction zone: While the JPDF model shows very thin flame fronts, the flame zone of the TFC model extends up to the end of the combustion chamber. This is mainly because the second half of the reaction progress (the change from $\Theta = 0.5$ to 1) is predicted to be slower. The TFC model does not achieve complete burnout in the shown part of the combustion chamber, whereas the burnout is complete using the JPDF. Although the flames differ strongly in the main and post-flame zone, the flame root is predicted to be similar. This can be seen in the averaged images, in which both combustion models show the turquoise and green marked flame zones in almost identical axial positions. This fact is even more visible in Fig. 3 showing the iso-surface $\Theta = 0.25$. It can be seen that the predicted flame shape for the jets at their root is almost identical in both models. The flame root is in most cases an anchor point of the flame and is therefore decisive for flame stability. The beginning of the flame is predicted to look very similar in both models, while the flames differ particularly in the main and afterburning zone. The impact of these effects on the determination of LBO are investigated in the following section.

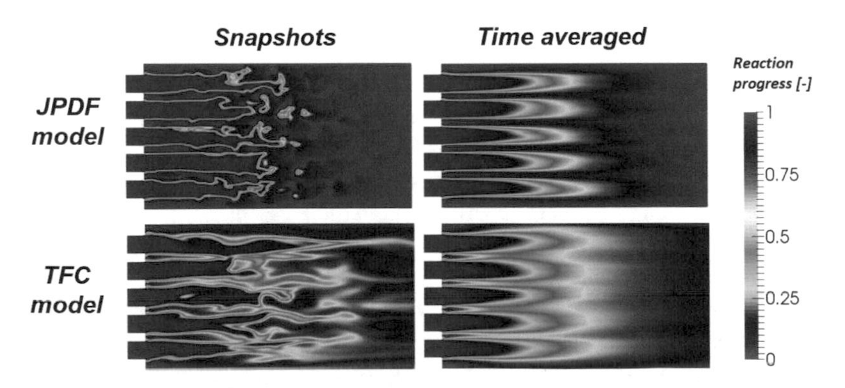

Fig. 2: Comparison of the simulated reaction progress distribution between the JPDF- and TFC-model for the nozzle D_2 at $T_0 = 100$ °C, $u_{Hole} = 35$ m/s and $\lambda = 1.15$

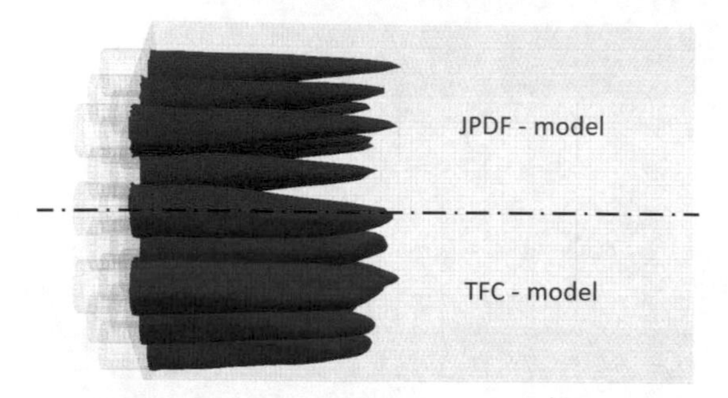

Fig. 3: Flame root indicated by the iso-surface $\Theta = 0.25$ simulated by the two combustion models for the nozzle D_2 at $T_0 = 100$ °C, $u_{Hole} = 35$ m/s and $\lambda = 1.15$

4 LBO results

Above, only flames at constant operating conditions have been investigated. In order to determine the explicit blow-off points, transient inlet conditions are set, which enables the determination of a blow-off point using only one simulation. There are two basic methods to induce LBO: On the one hand, the mass flow of fuel can be reduced, which leads to an increase of the air equivalence ratio λ and, consequently, a reduction of the flame speed until the flame blows off. LBO has been induced by increasing the total mass flow, characterized by the average inlet velocity u_{Hole}. A stepwise gradual increase in velocity was implemented. A schematic course of the input velocity of a simulation is shown in Fig. 4. In this example, the velocity is kept constant for $\Delta t = 40$ ms and is then increased by $\Delta u_{Hole} = 4$ m/s. In order to observe whether the time of constant velocity is long enough, global quantities are observed and checked whether a convergence is found within this time interval. The schematic plot also shows that the velocity increase does not occur instantaneously but in a steep rise. This prevents the creation of numerical pressure waves due to abrupt velocity changes, which could erroneously cause LBO too early. The overall LBO simulation always start at stable conditions (low velocity) and then are increased in the aforementioned stepwise way until LBO is observed.

As has already been shown, there are differences in the representation of the reaction zone and the jet flame interactions between the two combustion models. It is expected, that the flame shape is decisive for the behavior approaching unstable conditions and therefore also for the choice of the LBO criterion. In this section, it is examined how these two models exhibit LBO and how accurately they are able to reproduce the experimental data. For this purpose LBO simulations were conducted for the nozzle D_2 at a preheating temperatures of $T_0 = 100$°C. The corresponding experimental LBO data show a blow-off at an air equivalence ratio of $\lambda = 1{,}661$ at an inlet speed of $u_{Hole} = 53$ m/s. First the LBO simulations with the JPDF model are discussed. The simulations were initialized with an inlet velocity of $u_{Hole} = 35$ m/s and were increased by $\Delta u_{Hole} = 4$ m/s every $\Delta t = 30$ ms. Using these settings, the experimentally measured LBO condition is reached after 0.15 s simulated time.

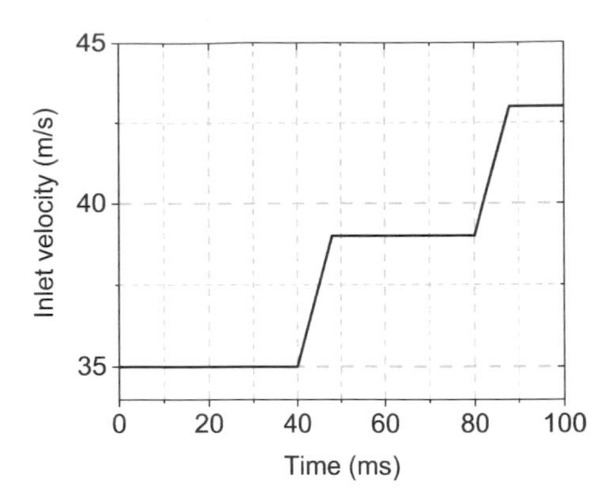

Fig. 4: Schematic change of the inlet velocity for the LBO procedure

Figure 5 shows snapshots of the reaction progress variable from the resulting LBO simulation at different inlet velocities. At these conditions, the flame forms a conical flame as shown in the upper picture. In the middle picture, where the inlet velocity is increased to u_{Hole} = 43 m/s, a qualitatively similar conical flame can be seen, which still burns stable but already extends considerably further into the combustion chamber. The bottom picture shows the effect of a further inlet velocity increase to u_{Hole} = 47 m/s: LBO occurs and the flame almost completely disappears. There is still ongoing reaction in the recirculation zones and at the edges of the outlet. The fact that reaction is still occurring at the edge of the combustion chamber outlet can be explained by the use of the outer domain (not shown in the pictures): This outflow zone causes a slow-down of the gas mixture due to the cross-sectional expansion. In reality, a dilution by sucked in air would take place, which would lead to a massive reduction of the flame speed and would prevent flame stabilization. Since the solver assumes a perfectly premixed composition, the flame always stabilizes in these low velocity regions and therefore is able to occasionally retaliate back into the combustion chamber. In spite of this fact, LBO can be still clearly detected, indicating that the flame stabilization in the outer domain is no problem for the simulation of LBO.

In summary, the JPDF model is able to reproduce a blow-off behavior, marked by a sudden extinction of the flame after reaching a critical velocity. The blow-off occurs between u_{Hole} = 43 m/s and 47 m/s, which is about 15 % lower than the measured value u_{Hole} = 53 m/s. This good agreement with the measured value is particularly surprising as no heat loss model is included in this simulation. It is to be expected that the calculated values will decrease and thus deviate even further from the measured values if heat losses are included. This is the reason no more LBO simulation were conducted with the inclusion of heat losses and is a strong indication that heat losses are not dominant for flame stability in the current multi jet burner system. The following section discusses the LBO simulation using the TFC model. The TFC-model predicts long reaction zones where complete burnout is not

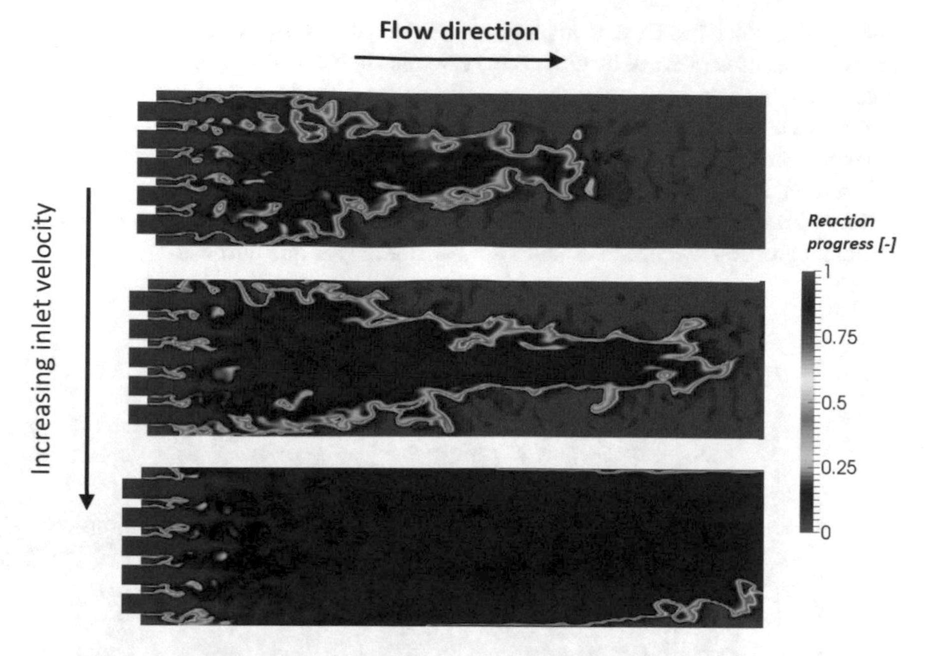

Fig. 5: Snapshots of the reaction progress variable inside the combustion chamber from the LBO simulation with the JPDF-model at different inlet velocities (from top u_{Hole} = 35 m/s, 43 m/s and 47 m/s)

reached until the end of the combustion chamber, even for velocity conditions far from experimentally measured LBO conditions. The effect of this flame pattern predicted by the TFC model, that is being different from the one observed with the JPDF model, on the approach of unstable operating conditions will be discussed in this section. For the LBO simulations with the TFC-model, the same conditions were chosen as for the JPDF-LBO simulation (nozzle D_2 at T_0 = 100 °C and λ = 1.661). The inlet speed u_{Hole} begins at 30 m/s and is increased incrementally by Δu_{Hole} = 5 m/s every Δt = 40 ms. The experimentally determined blow-off velocity is again u_{Hole} = 53 m/s, an inlet velocity which is reached after 0.2 s of simulated time. Snapshots of the reaction progress of the flame at different inlet velocities are shown in Figure 6. The top picture shows the starting flame at u_{Hole} = 30 m/s, characterized by long single jets, where the outer jets are significantly shorter than the inner ones. In the course of the velocity increase, two different stages are observed: when comparing the snapshots up to u_{Hole} = 50 m/s, it is noticeable that the flame root (illustrated by the end of the blue zone) remains almost at the same location. While the root of the flame is stationary at this stage, the reaction zone expands downstream, leading to a reduction of burnout at the exit. Afterwards, especially between the third and fourth image, which shows the increase in velocity at which LBO was measured experimentally, it can be observed that the flame root of the outer jets shifts significantly downstream (marked by black ellipses). From this point on, a displacement of the entire flame takes place, which can be seen in particular in comparison to the last image, in which the unburned area (marked by blue color) extends almost to the end of the combustion chamber. However, even under these conditions a stable flame root is still predicted at the end of the combustion chamber. A critical velocity that suddenly leads to a disappearance of the flame is, in contrast to the JPDF-model, not observed here. For this reason, it is difficult to find a suitable blow-off criterion in order to determine a clear blow-off velocity. If possible, a simple and global criterion should be used. Different globally averaged variables are considered for this purpose.

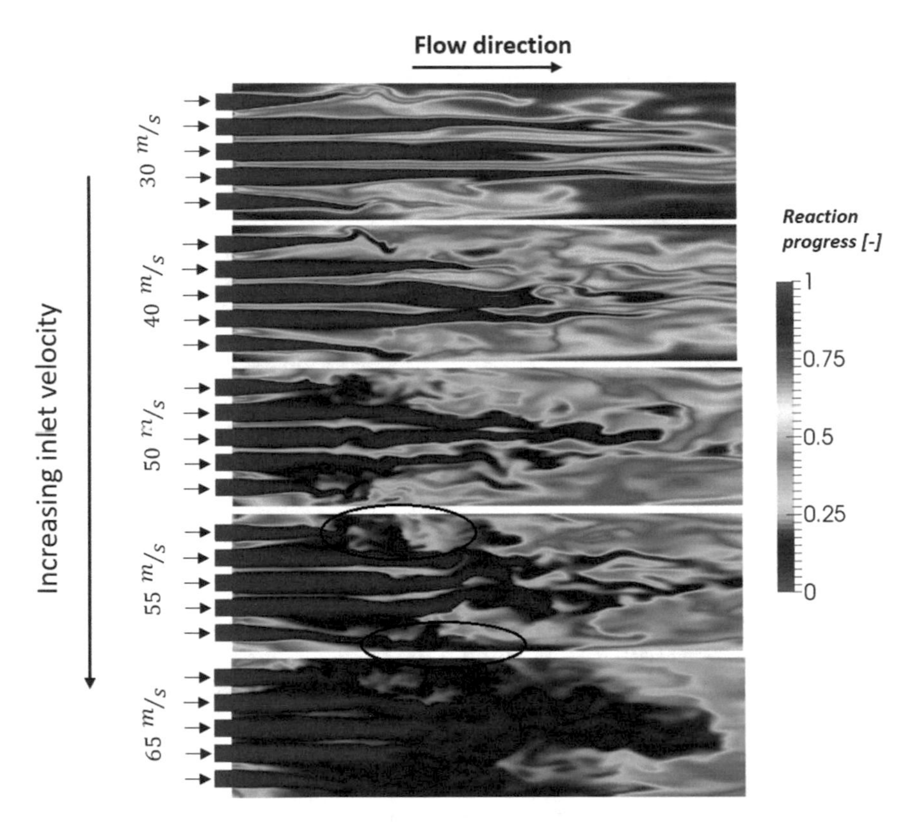

Fig. 6: Snapshots of the reaction progress variable inside the combustion chamber from the LBO simulation with the TFC-model of the nozzle D_2 at increasing inlet velocities

To give an overview, various possible globally averaged quantities are shown in Figure 7. In the upper two diagrams, the temperature and the reaction progress variable are shown followed by the mass fractions of CO2 and OH. As can be seen, all these quantities decrease continuously with an increase in velocity. The presence of OH is often used as an indicator for reaction and, as the graph shows, the mean OH concentration converges with increasing velocity towards zero. It is an obvious idea to define a limit value, where the flame is considered to be "blown off". However, it is difficult to define a general limit here and so a different LBO-criterion is required.

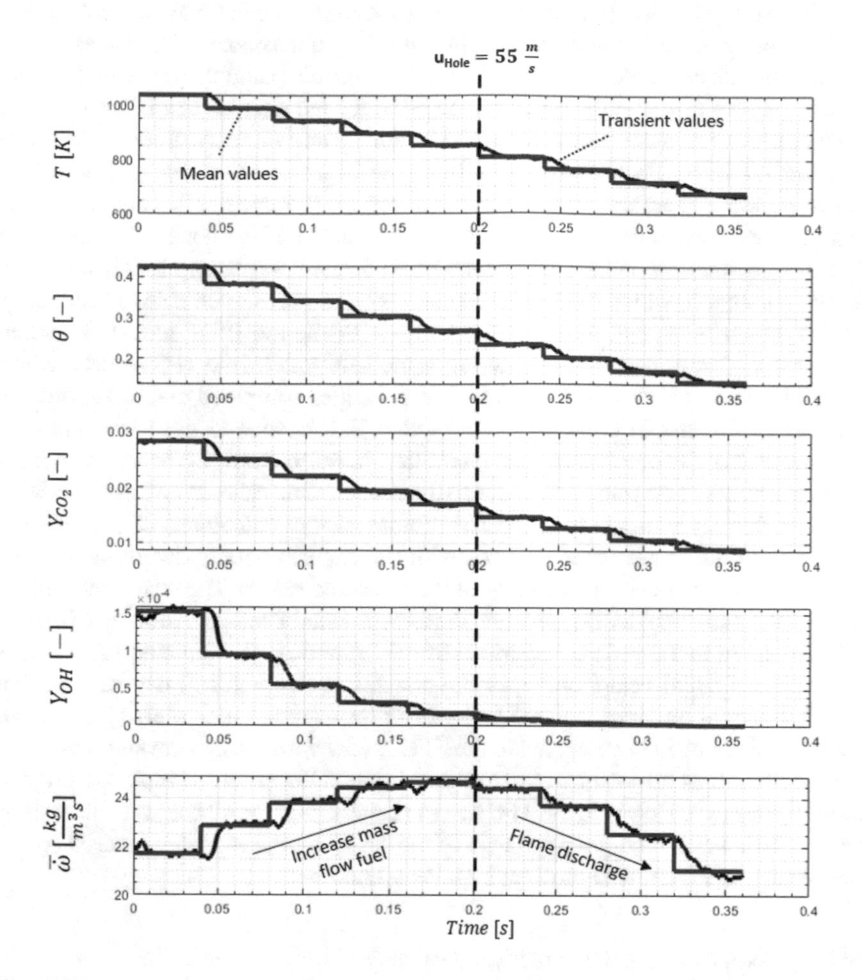

Fig. 7: Progression of globally volume averaged quantities over time in the LBO simulation with the TFC-model for the nozzle D_2 at T_0 = 100 °C and λ = 1,661

In the last diagram of Figure 7 the reaction rate is shown. Due to the increase of the incoming mass flow of fuel, the reaction rate initially increases. This is the previously described first stage in which the reaction zone expands, while maintaining a stationary flame root. By slowly discharging the flame in the second phase, the increased fuel flow can no longer be converted to the same extent in the combustion chamber, which results in a reduction of the reaction rate. It is reasonable to define the transition point between increase due to increasing fuel mass flow and the discharging of the flame as LBO. This characteristic point of consumption behavior can be determined very precisely by the beginning of the decline of the reaction rate and takes place at the velocity increase from 50 m/s to 55 m/s, which corresponds exactly to the experimentally determined value of 53 m/s. The resulting LBO points of all calculated conditions and used nozzles are plotted in a Peclet diagram in Figure 8. The diagram presents, in addition to experimentally determined LBO values (stars), the LBO points calculated with the JPDF model (circles) and those calculated with the TFC model (diamonds). It can be seen that for the nozzles at a DR = 2.8 there is very good agreement between both TCI models and the experimental data over the entire measured Peclet range. The JPDF model predicts the LBO values with an almost perfect agreement with the Peclet curve for the nozzles D_2 and D_3, while smaller deviations occurred at the highest Pe_u values of the nozzle D_1. Here, the highest velocities are present and the observed discrepancy may possibly be due to the cell resolution, which has been kept constant and may not be able to sufficiently resolve the turbulence at these large velocities. While the JPDF model in a small range both underestimates and overestimates the LBO values, the TFC model predicts the LBO values at generally higher blowout velocities compared to the experiments. The distance of the TFC points for the DR = 2.8 nozzles to the Peclet curve is almost constant. The deviation was quantified in more detail in the following: As measured values show some scatter due to the probabilistic character of the underlying turbulent flow, it is useful to compare the numerically calculated LBO limits not with the individual measured data but with the experimentally determined Peclet correlation, which provides an average of the measured values. The relative deviations between the numerical LBO points and the experimental Peclet curve were calculated and are shown in Table 1: Comparison of the numerically calculated LBO values to the experimental Peclet curve for the nozzles at DR = 2.8. It can be seen that the JPDF model on average predicts the LBO limits more accurately with a mean deviation of about 11% compared to the TFC model with a mean deviation of about 15%. Both values are comparable to the 6% mean deviation of the experimental values themselves. However, the TFC model scatters much less here, as evidenced by the low variance. The higher variance of the JPDF model is mainly caused by the overestimation of the LBO limits at high Peclet numbers.

Table 1: Comparison of the numerically calculated LBO values to the experimental Peclet curve for the nozzles at DR = 2.8

	JPDF-Model	TFC-Model
Mean relative deviation	11.3 %	14.8 %
Variance relative deviation	1.0 %	0.3 %

Two additional calculations to check the prediction of the dump ratio influence were performed with both models for nozzle D_5, which has a higher DR of 6 (red dots). Here, the JPDF model can predict the drop to lower Pe_u values, while the TFC model calculates the blowout limits at significantly higher values. Thus, the influence of the DR on the LBO limits cannot be predicted by the TFC model. It could be shown that a determination of the blowout limits and thus of the Peclet parameters is possible by CFD simulations with only moderate grid resolution. Both TCI models allow to predict the LBO limits for the nozzles at DR = 2.8 and this despite the fact that both models represent the flame shape as well as the blowout process itself very differently. The mean deviation between the resulting LBO is quite small for both models. However, the influence of a variation of the dump ratio on the LBO limits can only be described using the JPDF model.

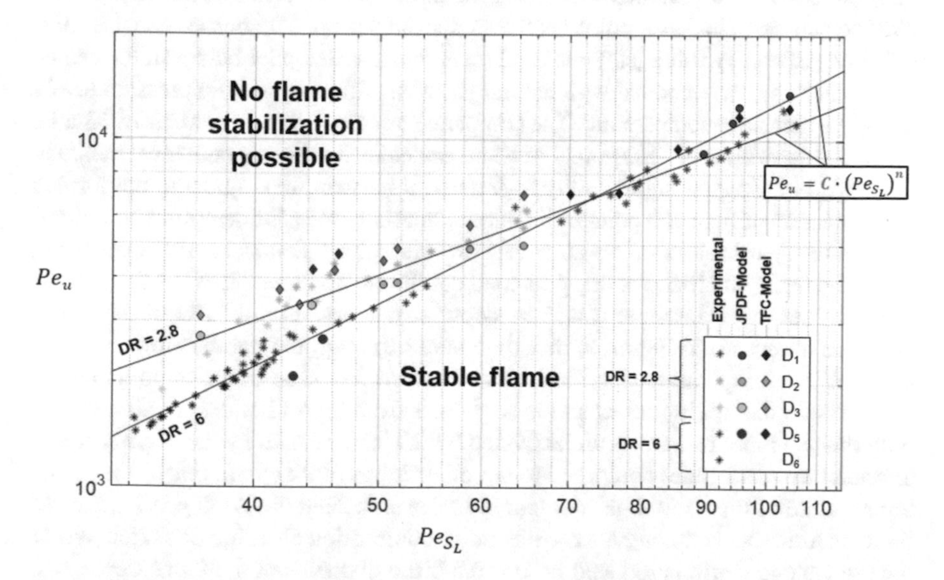

Fig. 8: Comparison of the numerically calculated LBO limits to the experiments in a Peclet diagram

5 Calculation of an industrial burner

The calculation of the model burner already showed that the investigated models are able to calculate the blowout of the flame with good accuracy. Anyhow, the calculation of industrial gas turbine combustors imposes a much bigger challenge, since the geometries are larger and considerably more complex in their overall geometry. As an additional difficulty, in these cases, additional air is usually introduced into the combustion chamber to cool down the walls and regulate the flame temperature and the combustors are operated with strongly preheated gases and at elevated pressure conditions. The goal was therefore to investigate whether the developed numerical setup can also be used to calculate LBO for a complex multi-jet flame burner under industrially more realistic operating conditions. To this end, the blowout point of an industrial prototype burner developed and experimentally investigated by Siemens was calculated numerically using the developed models. Since this burner has additional air inlets into the combustion chamber, the additional impact of the spatially and temporally varying mixture field on the reaction is required. The JPDF model already inherently includes the effect of mixing by the aid of the first and second statistical moment (the mean and the variance) of the mixture fraction. Because of this feature, and as, additionally, it was shown that this model was the only one capable of correctly predicting the effect of DR variation, this model was used to simulate the LBO point for the industrial set-up. The burner consists of a multitude of non-swirling jet flames and includes a swirled pilot burner in the center. The combustor was operated with natural gas at significantly elevate pressure levels and high preheating temperature. The complete numerical domain consists of 33 mio. cells. The experimental LBO point has been recalculated using the numerical methods developed to show its applicability to industrial burners. In contrast to the former determination of the LBO point of the model matrix burner, for determination of the LBO in the industrial set-up the air mass flow rate was set identical to the experiment, while the flame stabilized at this part-load condition. To induce LBO as the second step the fuel mass flow rate was then stepwise decreased until LBO of the flame could be observed. In Figure 9, meridian slices through the combustion chamber, showing the temperature distributions are shown for three different equivalence ratios inside the mixing tubes and is set leaner from top to bottom. In all pictures only the two inner fuel inlets are activated, which can be seen by the slightly lower temperature (dark blue) compared to the other inlets (dark cyan), due to the lower temperature of the natural gas compared to the preheated air. At $\Phi = 0.77$, a wide flame is formed, occupying almost the entire combustion chamber diameter. When the equivalence ratio is reduced to $\Phi = 0.57$, the diameter of the flame decreases, while still maintaining a stable flame, which is mostly located in the central inner recirculation zone. At these conditions, LBO was observed in the experiment, but as can be clearly seen, the simulation still predicts a stable flame. Only when the equivalence ratio is further reduced to the next equivalence ratio $\Phi = 0.48$, as can be seen in the lowest picture, the flame can no longer stabilize, and the reaction zone disappears. At these conditions, there is still a slightly elevated temperature in the recirculation zone (around 100 °C above the inlet temperature), but can be explained by the long residence time in this zone and the associated slow response. It can be assumed that if for this operation point the calculation is continued for a longer time a uniform temperature field would form. Nevertheless, the blowout is clearly noticeable visually. Summarizing the observations, it could be shown that even for high pressure level and complex geometry, which are important characteristics for industrial applications, the developed numerical setup using the JPDF model can be used to predict LBO. Flame blowout was calculated exactly in the experimentally observed interval.

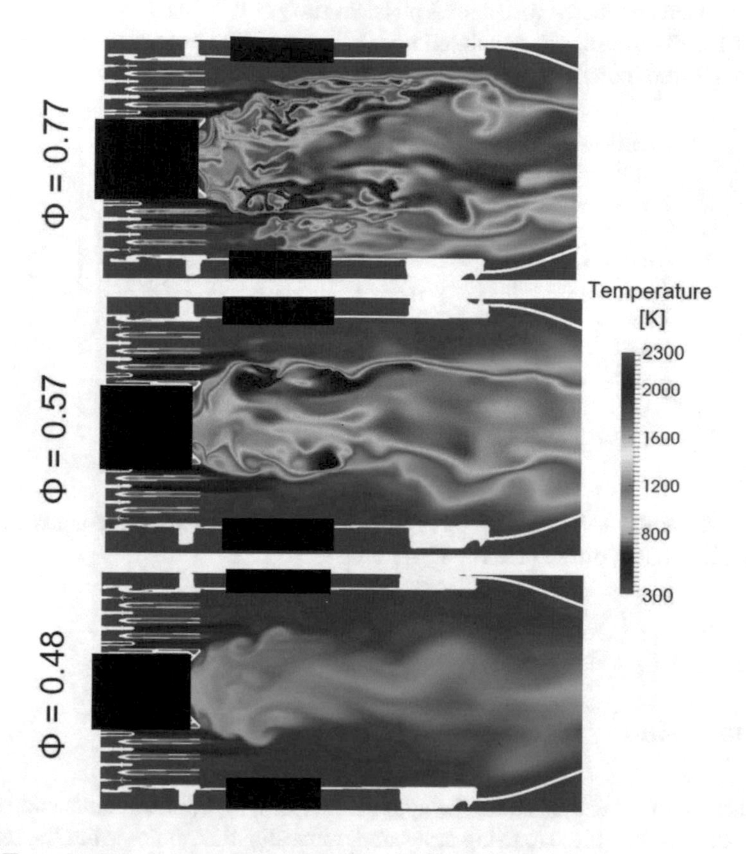

Fig. 9: Temperature distribution inside the combustion chamber at different equivalence ratios

6 Performance and required resources

Figure 10 presents the results of a performance and scalability test carried out on the ForHLR II. For this test, a configuration with 28 million cells was used. It can be seen that the code scales well up to 1000 cores and reasonably up to 2000 cores which corresponds to a minimum required number of cells per core of 14 000. Because the grid size of the model combustor (the matrix burner) with a total cell count of 1.3 mio. was rather small typically only 200 processors have been used. Due to the number of LBO points calculated and the needed long physical time to reach LBO around 1 mio core-h have been used to achieve the LBO calculation of the matrix burner. Lastly, the simulation of the industrial burner with a cell count of 33 mio. cells has been conducted with a total physical time of 0.7s that has been simulated to reach LBO. The case was calculated with typically 1000 processors and resulted in around 4.2 mio. core-h used.

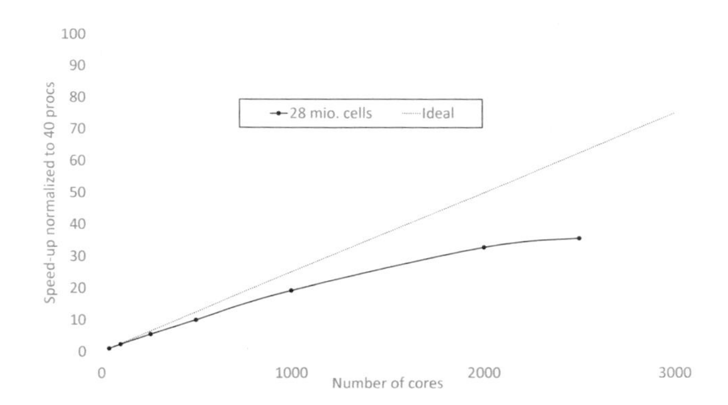

Fig. 10: Scaling of the used reactive Solver in OpenFOAM on the supercomputer FORHLR II using different number of cores

7 Conclusion

The current report demonstrates the applicability of two different combustion models for the calculation of the reacting cases under investigation. It describes the results of the reactive flow simulations that were applied for LBO calculations of a wide range of operating conditions. Lastly, an LBO simulation for a highly complex industrial burner has been successfully conducted. The numerically calculated LBO limit is in good agreement to the experimental value.

References

1. Robbin Bhagwan, Alexander Schwagerus, Christof Weis, Peter Habisreuther, Nikolaos Zarzalis, Michael Huth, Berthold Koestlin, and Stefan Dederichs. Combustion characteristics of natural gas fueled, premixed turbulent jet flame arrays confined in a hexagonal combustor. In *Turbo Expo: Power for Land, Sea, and Air*, volume 58615, page V04AT04A018. American Society of Mechanical Engineers, 2019.
2. Fabian Eiberger. Wechselwirkung zwischen turbulenz und wärmestrahlung, 2018.
3. Matthias Kern. *Modellierung kinetisch kontrollierter, turbulenter Flammen für Magerbrennkonzepte*. KIT Scientific Publishing, 2013.

4. Markus Klein, Amsini Sadiki, and Johannes Janicka. A digital filter based generation of inflow data for spatially developing direct numerical or large eddy simulations. *Journal of computational Physics*, 186(2):652–665, 2003.
5. Andrea Montorfano, Federico Piscaglia, and Giancarlo Ferrari. Inlet boundary conditions for incompressible les: A comparative study. *Mathematical and Computer Modelling*, 57(7-8):1640–1647, 2013.
6. C. Weis, A. Schwagerus, S. Faller, R. Bhagwan, P Habisreuther, and N. Zarzalis. Determination of a correlation for predicting lean blow off limits of gaseous fueled, premixed turbulent jet flame arrays enclosed in a hexagonal dump combustor. page S5_AIII_48, 2019.

Data-driven multiscale modeling of self-assembly and hierarchical structural formation in biological macro-molecular systems

P.N. Depta, M. Dosta and S. Heinrich

Abstract Self-assembly and hierarchical structural formation are essential in many systems of both nature and technology. Examples are the dynamic self-assembly of multi-protein clusters and the interdependency with catalytic activity, as well as the self-assembly of virus capsids critical for virus function. In an attempt to better understand and model these systems, we develop a multiscale modeling methodology to capture macro-molecular self-assembly on the micro-meter and milli-second scale. Focus of this report is the derivation and implementation of data-driven interaction potentials based on molecular dynamics simulations using Universal Kriging on the example of the hepatitis B core antigen (HBcAg).

1 Introduction

The dynamic self-assembly of multi-protein structures has received increasing interest in recent years [1]. This is especially true for the inter-dependency of structure and function. Understanding these processes remains a challenge and is critical in enabling the targeted modification and optimization of e.g. efficient bioreaction cascades [2].

Additionally, such hierarchical structural formation processes are at the basis of most macromolecular systems [3]. Namely, this includes material science and also structures of physiological systems in the human body, e.g. viruses. In the context of this work, viruses are investigated in the scope of virus-like-particles (VLP), which are formed out of monomers (primary proteins) of their structural capsid proteins providing critical properties for their function [4]. Examples are the hepatitis B viral protein (HBcAg) [5], which we are working on as an application for this framework, but potentially also, in light of current events, the SARS-CoV-2 virus.

Philipp Nicolas Depta, Maksym Dosta and Stefan Heinrich
Institute of Solids Process Engineering and Particle Technology, Hamburg University of Technology,
e-mail: nicolas.depta@tuhh.de, dosta@tuhh.de, stefan.heinrich@tuhh.de

W. E. Nagel et al. (eds.), *High Performance Computing in Science and Engineering '21*,
https://doi.org/10.1007/978-3-031-17937-2_32

The main challenge of investigating macro-molecular structural formation both experimentally and by means of simulation is that the physical phenomena involved therein typically spread over vast scales in length and time [6–8]. These scales are typically too fast and small for experimental methods to capture the dynamic structural formation / self-assembly in detail and are therefore limited to higher scales and rather steady-state investigations when structures have already formed.

On the simulation side, the main challenge is that methods that provide the appropriate molecular resolution (typically in the framework of molecular dynamics (MD) both atomistic and coarse-grained) are limited to system sizes below 100 nm due to high computational complexity. While many larger-scale methods have been proposed in the context of e.g. coarse-grained molecular dynamics (CG MD), Brownian dynamics (BD), dissipative particle dynamics (DPD), ballistic cluster-cluster aggregation (BLCA), diffusion-limited cluster(-cluster) aggregation (DLCA), reaction-limited cluster aggregation (RLCA), and others - methods capturing the μm and ms scale, while maintaining sufficient detail and being parameterized primarily from lower scales and first principles, are rather scarce. In this work, we investigate a modeling approach called by us the "Molecular Discrete Element Method" (MDEM) for abstracting entire macro-molecules as an object with a position and orientation and deriving their interaction in a data-driven fashion from molecular dynamics. In this context, we have already successfully established the means to describe and parameterize anisotropic diffusion and enforce the correct thermodynamics (i.e. canonical ensemble) [9]. This project and contribution focuses on the derivation of interaction potentials from MD for the hepatitis B core antigen (HBcAg).

2 Methods

2.1 Modeling framework

In the proposed modeling approach, each macro-molecule is abstracted as an object with a position and orientation. Newton's equation of motion is solved using explicit time-stepping and interaction between objects as well as the environment (thermodynamics) is modeled. An overview of the framework is shown in Fig. 1.

At the basis of the framework is the protein / molecular reference structure, which is assumed stable / stiff for the model, but is flexible during parameterization. For the purpose of this work, the reference structure and MD model are assumed given (see acknowledgments). Diffusion of the objects is modeled for both translation and rotation anisotropically using the previously published diffusion model [9] based on Langevin dynamics and implicit water modeling. In order to describe the interaction of objects (i.e. macro-molecules), we propose a data-driven potential field approach in which the interaction potential for relative positions and orientations (6-D) of pairwise contacts is saved in a field. Hierarchical fields (grids) are possible and implemented including automatic refinement. A numerical gradient operation can then be used to derive forces and torques acting between molecules during the simulation.

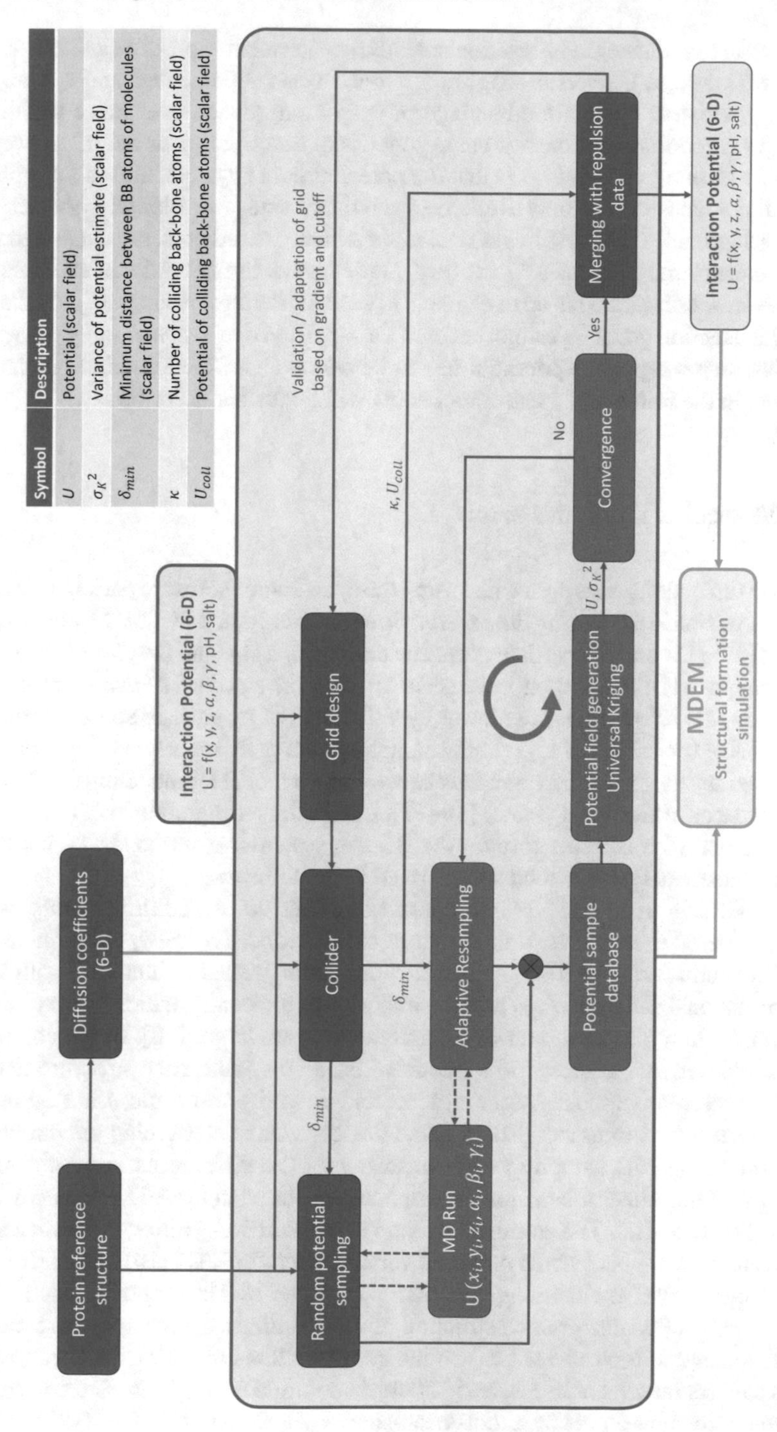

Fig. 1: Overview of the MDEM framework.

In order to estimate the interaction potential at each relative position and orientation from MD data, a Universal Kriging approach was implemented and a sampling strategy derived. The main advantage of such an approach over more traditional surrogate modeling approaches using functional descriptions or neural networks / machine-learning descriptions is that the potential field approach allows for arbitrary potential shapes and is only limited by memory size. Additionally, the Kriging approach provides the 'best linear unbiased estimate' based on certain mathematical requirements and presents a 'glass box' model providing not only an estimate, but also an error estimate. This error estimate is used for iterative resampling. As the MD based interaction potential cannot account for repulsion due to overlapping molecules, in a last step a repulsive potential has to be added as a function of the molecular overlap. In the following, these components will be further described.

2.2 Molecular dynamics setup

All MD simulations were performed using the open-source software package *Gromacs* [10, 11] version 2020.1 using the *Martini* force-field version 2.2P [12, 13]. Polarizable water (PW) [13] and the particle mesh Ewald (PME) technique [14] for electrostatics were employed in order to improve accuracy over the standard *Martini* water. Since atomistic MD simulations are slower by 1-2 orders of magnitude and consequently not possible for the iterative potential sampling approach, we rely on this previously employed and atomistically validated coarse-grained model for modeling the Pyruvate Dehydrogenase Complex (PDC) [15–17]. Credit for setting up the MD model and deriving the reference structure for the HBcAg system is given to Dr. U. Jandt (see acknowledgments) and will be summarized in the following.

The HBcAg virus capsid is composed of either 90 or 120 dimer units, which are considered the smallest unit structure in this work. Typically, two dimers (AB and CD) build from the HBcAg monomer are distinguished in literature with slight conformational differences. While the majority of the conformation is very similar, differences in regions for inter-dimer interactions are larger [18]. In the context of this work, only one dimer kind is modeled based on a reference structure derived from representative clustering. As done for the entire molecule, the regions for inter-dimer interactions are also modeled flexibly when determining the interaction potential using MD sampling and consequently the differences in conformation during the inter-dimer interaction are implicitly captured in the MD data and derived interaction potential. The atomistic structure for the HBcAg dimer was provided by Dr. M. Kozlowska based on a modified version of PDB 6HTX [19] and PDB 1QGT [18]. Representative clustering was then performed by Dr. U. Jandt based on the *martinized* [12] coarse-grained structure of the AB dimer using the linkage method as implemented in *Gromacs* [20] on the conformations of a 10 ns MD run with a 10 ps savings interval at 293 K and 150 mM sodium chloride ions. The determined reference structure differs by a root-mean-square deviation (RMSD) of 0.39 nm from the original conformation.

The MD setup is based on the 'new' parameter set for the *Martini* force-field with PW and PME unless otherwise stated. A time step of 20 fs was employed for all simulations unless otherwise stated, temperature maintained at 293 K, the Parrinello-Rahman barostat with a compressibility of 3×10^{-4} bar^{-1} and coupling constant of 12 ps used, and all systems charge neutralized with an additional 150 mM of sodium chloride ions added. Systems contained two dimers A and B at a specific relative position and orientation centered in a triclinic box with a minimum of 5.5 nm to any periodic boundary condition (PBC). A convergence study with a large distance of 8 nm to the PBC showed no notable differences.

The simulation procedure consisted of two energy minimizations using the steepest descent algorithm with tolerance of 10'000 kJ/mol/nm (first with normal *Martini* water and no PME, second with PW and PME); an equilibration for 50 ps using a time step of 5 fs with position restraints on the back-bone atoms and Berendsen barostat with a coupling constant of 4 ps to avoid oscillations; and lastly a production MD run for 0.6 ns. Energies between all groups (A, B, PW, ions) were calculated every 20 steps and saved along with trajectories every 500 steps.

Postprocessing was performed using *Gromacs* utilities. All energy components were extracted and reference structures fitted to determine relative positions and orientation. Energies, positions, and orientations were then averaged between 0.5 and 0.6 ns. Overall, the following potentials between respective groups were investigated for determining the overall interaction potential and Lennard-Jones and Coulomb potentials added when applicable: A-B, A-A + B-B, A-PW + B-PW, PW-PW, A-ions + B-ions, PW-ions, ions-ions, bonds, G96-angles, improper dihedral angles, Coulomb reciprocal. As it can be seen, not only the interaction between the molecules themselves, but also effects on the water, ions, bonds, and long range electrostatics in the reciprocal Coulomb term were evaluated.

2.3 Iterative multivariant interpolation using Universal Kriging

In order to estimate the interaction potential at each relative position and orientation on the grid, a Universal Kriging approach was implemented. Kriging is most frequently applied in the field of geostatistics. Due to the scope of this report, only a brief overview and not all details can be given. For more detail on the mathematical background the interested reader is referred to literature [21–23].

The goal of Kriging is the determination of optimal weights for the estimation of a spatially distributed (random) variable based on a linear combination of observations. Optimality refers to the minimum estimation variance. In the context of this work, the interaction potential $U_{krig,p}(\mathbf{x}, \mathbf{q})$ of a potential component p as a function of relative position $\mathbf{x}$ and orientation $\mathbf{q}$ has to be estimated based on N_{krig} observations as

$$U_{krig,p}(\mathbf{x}, \mathbf{q}) = \sum_{i=1}^{N_{krig}} w_i U_{p,i}(\mathbf{x}_i, \mathbf{q}_i) \,. \tag{1}$$

Typical Kriging is performed over a subset of observations ($N_{krig} \subseteq N_{tot}$) in the local neighborhood. Universal Kriging assumes that the underlying process, in this case potential U (for simplicity index p is dropped), can be decomposed into a systematic trend $\mu(\mathbf{x}, \mathbf{q})$ and random component $Y(\mathbf{x}, \mathbf{q})$ as

$$U(\mathbf{x}, \mathbf{q}) = \mu(\mathbf{x}, \mathbf{q}) + Y(\mathbf{x}, \mathbf{q}) , \tag{2}$$

where the systematic trend can be described by the linear combination of deterministic basic functions and is in this work determined in the lower-dimensional space of the minimum distance δ_{min} between back-bone atoms of molecule A and B referenced to zero by their collision distance 0.3 nm. Universal Kriging requires the remaining random component Y to be intrinsically stationary with zero mean. While the zero mean requirement is directly fulfilled in minimum distance space for all tested systems, intrinsic stationarity is strictly speaking not fulfilled in the case of molecular interaction, as the typical Gaussian distribution at small minimum distances tends towards a (degenerate) delta distribution of zero at large minimum distances. In order to resolve this, spatial continuity is investigated in sections for which intrinsic stationarity is approximately fulfilled. Spatial continuity of Y is described by the (residual) variogram γ_Y using the root-mean-square distance δ_r of the back-bone atoms as a distance measure. The optimal weights providing the unbiased and minimum prediction variance can then be determined by solving a linear system of equations. This is done separately for all potential components and the overall interaction potential can then be calculated as a superposition of all components. The overall procedure for iterative multivariant interpolation and resampling using Universal Kriging then becomes

1. **Trend fitting** using weighted-least-squares for all potential components;
2. **Sectional Variogram** determination and fitting of trend-compensated residual Y for all potential components using weighted-least-squares;
3. **Universal Kriging** for qualifying potential components. Convergence studies showed that a minimum of $N_{krig} = 500$ is necessary for potential estimation, while $N_{krig} = 100$ is sufficient for estimation of the variance;
4. **Summation** of all potential components (trend only for those not qualifying for Kriging) for determination of overall potential estimate U_{krig};
5. **Accounting of molecular collisions** by increasing potential as a function of atom collisions and proximity to MD data, as these conformations cannot be sampled using MD;
6. **Resampling** based on variance reduction and extrema (potential minima/maxima, gradient maxima) localization and specification. For variance reduction, virtual points are iteratively placed at the maximum variance location and the variance of the field recalculated (actual value at location not needed).

The algorithm was implemented in a custom C++ code with hybrid MPI+OpenMP parallelization using the *Eigen* library for solving the linear system of equations and employing *Matplotlib* in *Python* for fitting.

2-D Example A 2-D example of the algorithm can be found in the appendix.

Insertion of Experimental Findings and Expectations In order to incorporate experimental knowledge on e.g. binding locations, virtual data points can be inserted into the data set. Similar approaches are common in the generation of force-fields in MD, as a precise estimation of parameters based on lower-scale models is often not sufficient. These points do not influence the trend and variogram derived from the MD data set, but only the neighborhood where they are located. Multiple approaches were investigated and the final approach will be presented in the results section.

2.4 MDEM: Molecular Discrete Element Method

The modeling of structural formation based on Langevin dynamics was implemented in the open-source DEM simulation framework MUSEN [24]. The code is implemented in C++ and CUDA for the simulation on graphics processing units (GPU) using the CUDA Toolkit v11.2 by NVIDIA [25]. The leap-frog algorithm was used for time integration and contact detection performed using a Verlet list implementation [24]. The previously published diffusion model [9] was employed and the numerical gradient operation for deriving forces and torques of contacts from interaction potentials implemented. A temperature of 293 K, a dynamic viscosity of 0.001 Pa s, a time step of 10^{-13} s, and periodic boundary conditions were used unless otherwise indicated. As the assembly process is diffusion limited, a two step procedure was employed: In a first step, an accelerated simulation was performed at a lower dynamic viscosity of 0.0001 Pa s and lower temperature of 100 K. This enabled a time step of 10^{-12} s and reduced simulation time for 1 ms to approximately 3 days on an Nvidia Titan RTX. In a second step, equilibration was performed using the previously stated settings for 0.01 ms.

2.5 Parallelism and scaling

The framework for iterative resampling and generation of the interaction potential consists of five steps with varying degree of parallelism, which will be presented in the following. At multiple points within the jobs consistency and error checks were implemented to catch rare issues.

1. **Pre-processing (1 node):** First, the next resampling points are determined based on either (normalized) variance minimization, potential minima/maxima, or gradient maxima resampling. Variance minimization employs a hybrid parallelized (MPI+OpenMP) and extrema identification is largely single-threaded with partial OpenMP parallelized sections. Second, MD input files are generated in parallel.
2. **MD (16 - 64 nodes):** A custom implemented MPI-based scheduler runs the individual MD simulations within an overall job. The sub-jobs are run in descending system volume order to improve synchronization at the end (typically <5 min run-time differences at the end). For the HBcAg system typically either

16 or 32 cores per MD simulation were used. Each sub-job reads a 60 KB input file (generated in last step) from the parallel file system, performs all temporary operations in a temporary RAM disk to reduce file-server load, and saves one compressed results file of approximately 9-10 MB.

3. **Post-processing (1 node):** Post-processing was performed on one node using the *Gromacs* utilities in parallel for multiple jobs.
4. **Post-processing - Variogram (1 node):** Analysis of interaction data was performed to determine potential trends and spatial correlation to derive Variogram models (custom OpenMP parallelized code). For small data-sets, which did not require the large SMP nodes, this step was carried out within the previous post-processing.
5. **Kriging (10 - 16 nodes):** For interaction potential field generation, the custom Universal Kriging code was used. The code employs a hybrid MPI+OpenMP parallelization and MPI-IO for reading/saving of potential fields.

In order to efficiently use the resources granted to us, in a first step a parallel scaling analysis of all codes employed on more than one node was performed, specifically the MD code using *Gromacs* and the custom Kriging code. For the MD scaling analysis, 15 random HBcAg pairwise MD simulations at minimum atom distances of 0.5 - 2.5 nm between interaction partners using *Gromacs* 2020.1 on *AMC EPYC* 7742 CPUs at HLRS were performed. As individual MD runs are only performed on the node-level, scaling with an increasing number of cores was investigated. Results of the parallel efficiency can be seen in Fig. 2. Both full MD points (including single-core pre-processing), as well production MD only are compared. As it can be seen, parallel efficiency is good until 32 cores after which, the single-core pre-processing severely degrades performed. *Consequently, core counts of 16 or 32 were employed to ensure sufficient efficiency.*

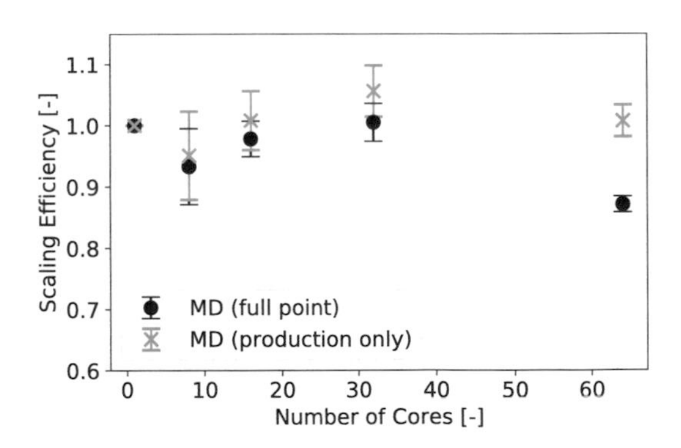

Fig. 2 Scaling efficiency of 15 random pairwise HBcAg MD simulations at minimum atom distances of 0.5 - 2.5 nm between interaction partners using *Gromacs* 2020.1 on *AMC EPYC* 7742 CPUs. Note that error bars indicate standard deviations and all cores were fully utilized during test. Efficiencies over 1 (super-linear speedup) are attributed to cache increase.

Furthermore, the scaling of the Kriging code was investigated based on the random VLP dataset and 0.63 nm grid. One MPI process was used per socket with 64 threads. Primarily, Kriging sizes of 100 (used during variance minimization) were investigated and differentiated between the full job (overall Kriging code) and only the Kriging portion (without parsing of input data). Reason for this is that parsing of input data takes place currently from a text-file format, as during method development changes are still frequent. Once the format is final, a shift towards a binary format will improve this efficiency. Additionally, Kriging sizes of 500 (convergent size for potential estimate, used for extrema resampling) were estimated. As these runs are computationally expensive (quadratic scaling leads to an increase by a factor of 25, leading to approx. 14'000 core-h per run), one test on 4 nodes was performed to determine the cost-increase of the Kriging portion to be 23.5 and then used to estimate scaling.

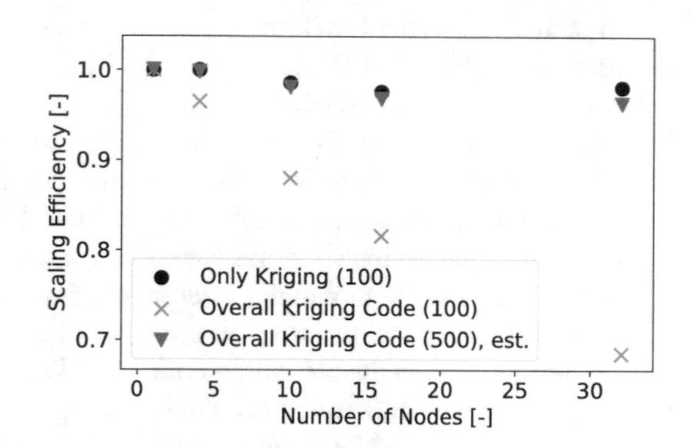

Fig. 3 Scaling efficiency of the Kriging code for two Kriging sizes (100 and 500) based on the random VLP dataset before iterative resampling and 0.63 nm grid. Note that only Kriging (without reading of input data and writing of results), as well as overall Kriging are shown.

As the results in Fig. 3 show, the Kriging portion ('only') of the code scales with a parallel efficiency of at least 98 % at 32 nodes (4096 cores). However, the overall code scales with a parallel efficiency of 96 % at 32 nodes for a Kriging size of 500 and only 69 % at 32 nodes for a Kriging size of 100. As discussed earlier, this is caused by the inefficient parsing of the input data from a text-file and is to be improved by binary parsing once the format is final. *Consequently, in order to use resources efficiently, during variance minimization with a Kriging size of 100 only 4-10 nodes (512-1280 cores) were used and for Kriging sizes of 500 up to 16 nodes (2048 cores).* Additionally, note that an analysis into using less than four cores per CCX showed no improvements in overall performance.

3 Results

In the following, the results of the interaction potential for HBcAg and virus-like-particle assembly will be presented. Beforehand, validation and convergence studies of components were carried out. This included a 2-D validation test (see appendix Fig. 7), a MD box size convergence analysis, and a convergence analysis of the number of Kriging points in a neighborhood based on the random HBcAg data set, which will be discussed in the following.

3.1 HBcAg interaction potential

In a first step, a random interaction data set was sampled from MD. For this, MD simulations were performed at random relative positions and orientations for molecules A and B at difference distance classes (minimum distance between atom centers). 20'000 simulations were performed between 0.4 - 0.5 nm with a focus of sampling binding locations, 5'000 simulations were performed in each 0.2 nm interval between 0.5 - 2.5 nm, and 5'000 simulations were performed in each 0.5 nm interval between 2.5 - 5.0 nm. This lead to a total of 95'000 random data points, which is doubled due to symmetry.

Upon analysis of the (random) interaction data it was found that potentials A-B (attractive), A-PB + B-PW (repulsive), PW-PW (attractive), and A-ion + B-ion (repulsive) posses a significant trend in δ_{min} space, while potentials A-A + B-B, bond/G96-angle/improper dihedral, PW-ion, and ion-ion posses no significant trend. This shows that no trends in intra-dimer conformation could be detected (including bonded) and of ion related potentials only the interaction with the molecules, but not the water and between ions is significant. Consequently, the dominating components to the potential are the molecular interaction, interplay with the water, as well as ion mediation effects. During resampling, the decision on significance of trends and incorporation into the overall potential was left flexibly to the algorithm. Of all residuals Y, only the A-B potential contained a significant spatial correlation and was further evaluated using Universal Kriging.

In the following, the interaction potential was iteratively refined and results can be seen in Fig. 4 and Fig. 5. Resampling was performed as following: ten iterations of variance minimization of each 5'000 samples, ten iterations of normalized variance minimization (focus at larger distances) of each 5'000 samples, followed by potential minima, potential maxima, and gradient extrema in consecutive order for three times with each 20'000 samples (15'000 at main extrema locations, 5'000 at random neighboring grid points). For variance related Kriging, $N_{krig} = 100$ was used and for all other $N_{krig} = 500$. A total of 375'000 data points were sampled.

As it can be seen in Fig. 4, during variance resampling the average changes in potential between iterations decrease continuously while the estimation variance remains essentially unchanged. Furthermore, the maximum changes in potential between iterations remained essentially unchanged for all iterations between 100 - 200 kJ/mol. This is attributed to the high dimensionality of the interaction space and large residual noise. Additionally, extrema resampling (iteration 21-29) lead to an increase in average potential change as well as estimation variance. This is attributed to the inclusion of comparatively larger potential difference from the trend, which also increases the overall variance of the variogram, leading to an increase in estimation variance. This attribution is also consistent as no further increase in maximum potential change is observed, indicating that these samples merely posses a larger variance due to their proximity to extrema locations (e.g. binding locations).

The overall potential can be seen in Fig. 5 as a function of δ_{min} (average and standard-deviation over all grid locations) as well as the minimum cross-section in X-Y. As it can be seen, the interaction potential possesses a slight potential barrier

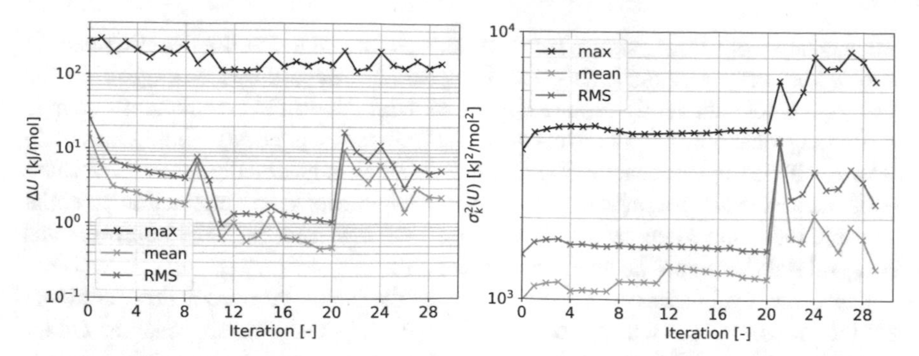

Fig. 4: Convergence of the iterative resampling procedure for potential changes (left, $U_i - U_{i-1}$) and variance development (right). Note that iterations 1-10 are variance resampling (5'000 samples each), iterations 11-20 are normalized variance resampling (5'000 samples each), and iterations 21-29 are extrema resampling (20'000 samples each).

at $\delta_{min} \approx 1.5$ nm, an intermediate potential well around $\delta_{min} \approx 0.5$ nm, and three regions of potential minima at the top left/right (aside the dimer spike) and underneath the dimer. As it can be seen in Fig. 5 (right) in the visualization of the binding locations, they are notably different and were found to be not sufficient for a stable capsid.

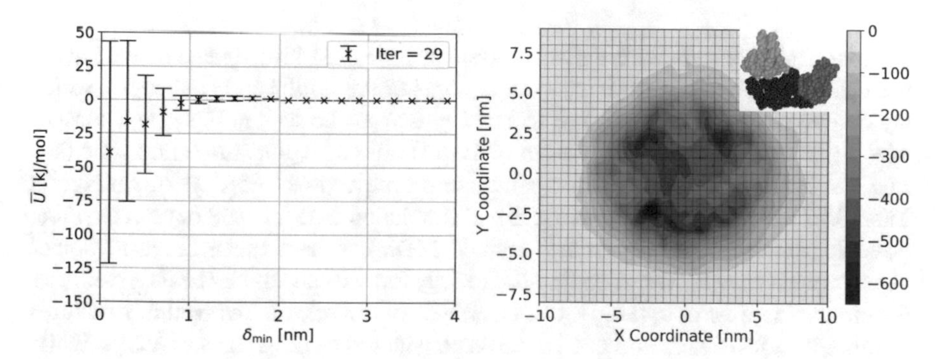

Fig. 5: **HBcAg interaction potential after resampling** over the minimum distance (left) and in X-Y cross-section (right, color scale in kJ/mol, minimum of all remaining degrees of freedom) with overlaid visualization of the binding locations on a trimer. Note that interaction potential and binding locations vary depending on molecular collision model.

Based on inspection of the data and binding modes, we attribute this to the main challenge that specific conformations are necessary for inter-molecular binding during capsid formation. The reference structure based on the *Martini* force-field and representative clustering shows some deviations, especially at the flexible residues involved in inter-molecular interaction. It appears that the differences might either be too large or not sufficiently captured by the *Martini* force-field to sufficiently sample the binding locations. In order to evaluate this further, extended MD simulations of 10 ns with 1'016 samples at each of the four main binding conformations were performed. While trimers exhibited improved stability and slight changes in interaction potential occurred including a decrease at the binding locations, no significant improvements in capsid stability could be achieved at this point.

Consequently, in a next step we investigated the impact of introducing additional artificial points at the binding locations into the data set. This approach is similar to the development of force-fields in MD, where experimental data is incorporated. Various approaches were tested and the following found to provide good results. Two sets of virtual points for potential AB were introduced into the data set (no influence on trends and variogram): A first set of constant potential $U_{\text{bind,center}} = -1400$ kJ/mol is located on a regular grid with -0.1, 0.0, and 0.1 nm distance in all directions (including rotation equivalent). A second set with Gaussian shape and a range of 1 nm and asymptotic value of -1000 kJ/mol is added on a regular grid of -0.4, -0.2, 0.0, 0.2, 0.4 nm up to a distance of 1 nm (δ_r) from the binding location and an increase of 10 back-bone atom collisions. VLPs based on this potential were found to be stable and assembly results will be discussed in the following.

3.2 VLP assembly

In order to study assembly, a variety of systems have been investigated thus far and a selected system will be presented to show some of the results and challenges, which are currently being addressed. The selected system can be seen in Fig. 6 and consists of a 1 μm^3 box with a concentration of 10 μM after 1.01 ms simulation time (see Sec. 2.4 for simulation procedure, simulation time approx. 3.5 days on an Nvidia Titan RTX). Note that both system size and simulation time are well beyond anything possible with traditional, also coarse-grained, MD. As it can be seen, the formation of capsid components as well as some close to fully formed capsids can be observed after the simulation time of approx. 1 ms. Over 20% of dimers are involved in structures of more than 85 dimers and can hence be considered as pre-stages of VLPs, while the remainder forms capsid components which can be clearly identified as fractions of spheres. Less than 0.3 % of dimers participate in structures of less than 5 dimers indicating along with the time-dependent data that structural formation to capsid components occurs very quickly. The largest fraction of structures after 1 ms consists of approx. 70 dimers.

During stability analysis of a fully formed capsid, it was found that such a capsid is very stable. During assembly, two main challenges have been identified: Firstly, the process is largely diffusion limited. Initial capsid components form at very small time scales and with growing size their diffusion kinetics slow down, leading to a reduction in assembly kinetics. In order to address this, the employed simulation

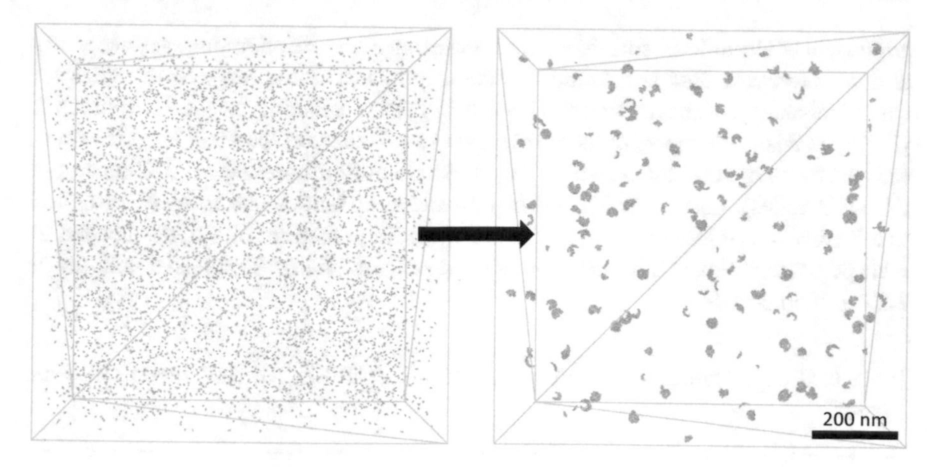

Fig. 6: Assembly of virus like particles from random state (left) after 1.01 ms (right).

approach with reduced water viscosity was developed. Furthermore, it was found that overgrowing of capsids occurs up to sizes of approx. 160 dimers. These capsids are not stable and change shape and size over time. Such phenomena are experimentally known, especially for larger concentrations, and are considered pre-stages of fully formed stable capids. In order to address this, we are working on further increasing the simulation times for improved equilibration, investigating lower concentrations to avoid kinetic traps during capsid formation, increasing system sizes for improved statistics, and exploring additional avenues (e.g. Monte-Carlo related).

3.3 Computational resources

In order to generate the presented results and to perform method development, a total of 5.85 million core-h on the Hawk system at HLRS were used, which were graciously provided in the context of federal project Acid 44178. GPU computations for the agglomeration studies were performed in-house. An overview of all computational resources can be found in Tab. 1 and will be discussed in the following. Overall, more than 383,000 molecular dynamics simulations of the HBcAg virus protein system in pairwise interaction were carried out providing more than 0.385 ms of overall simulation time and a wealth of information on the interaction of HBcAg necessary for virus capsid formation.

As it can be seen in Tab. 1, random sampling and iterative refinement accounted for approximately 2.0 million core-h and is consequently slightly over the anticipated 1.75 million core-h. However, three additional aspects were not accurately estimated during the planning of the project, which is attributed to the method development nature of the project: First, in the beginning, additional testing and validation including a MD box size and Kriging size convergence study was performed, leading to an additional 0.714 million core-h. Second, significantly more sampling was necessary and the increased data size made it necessary to utilize the large SMP nodes for statistical analysis. The authors are thankful for the availability of such resources at HLRS, as this made the analysis of such data sets feasible in the first place. However, the cost factor of 100 lead to an unanticipated 1.06 million core-h (0.01 million core-h on SMP nodes). Lastly, limitations of molecular dynamics had to be investigated in a limited set of longer simulations at the binding locations and it was needed to incorporate experimental information into the potential fields. This lead to additional 2.37 million core-h.

Table 1: Computational resources used on Hawk at HLRS. All **times are in million core-h**.

Context	MD	Kriging	Pre/Post
Testing, MD box size convergence, number of Kriging points	0.573	0.128	0.013
Random sampling	0.371	0.001	0.001
Iteration 1–20	0.384	0.017	0.020
Iteration 20-29	0.659	0.252	0.296
Full potential component spatial statistical analysis, biased sampling, Kriging with additional binding data	0.438	1.931	0.768
Sum	2.425	2.329	1.097
		Total	5.85

4 Conclusion

In conclusion, we have gained valuable insight into the proposed data-driven methodology for deriving macro-molecular interaction potentials from MD using Universal Kriging on the example of HBcAg for VLP assembly. The main challenges identified were found to be MD related, as force-fields are (currently) not specifically parameterized for potential sampling and capturing inter-molecular binding remains challenging, especially with CG-MD, which is required for sufficient sampling. We have proposed ways to overcome these limitations by biased MD sampling and inclusion of additional data similarly to traditional force-fields and achieved good results concerning VLP assembly. Overall, the proposed method showed merit in capturing self-assembly and post-processing MD data. Currently, we are exploring further approaches for improvement and are testing the method on the PDC system.

Appendix

In order to perform validation and visualize the algorithm, a random scalar 2-D example field (no units) between two spherical objects of radius 0.15 was generated using sequential Gaussian simulation and can be seen in Fig. 7. For this, a random truth field with similar statistical properties as typical MD data was generated (see Fig. 7 for details), overlaid with a Gaussian trend of -400 at contact and zero at range one, and a scaling to zero performed between a minimum distance of 0.4 and 1.2 using a Gaussian function. As it can be seen, the overall trends and binding locations (minima) are identified and the estimation error consists of small-scale discontinuities and noise.

Acknowledgements The authors would like to acknowledge the German Research Foundation (DFG) for funding within the focus program SPP 1934 (HE 4526/19-2), as well as the High-Performance Computing Center Stuttgart (HLRS, Acid 44178) for providing the computational resources. Furthermore, the authors would like to acknowledge Dr. Uwe Jandt for setting up the molecular dynamics system, and various MD related scripts in the context of the framework, as well as Dr. Mariana Kozlowska for providing the atomistic reference structures for the HBcAg system and many discussions in understanding the VLP assembly. Lastly, the authors thank the Institute of Bioprocess and Biosystems Engineering at TUHH for collaboration in the context of further application of the developed framework to the Pyruvate Dehydrogenase Complex, as well as critical reading of manuscripts.

References

1. M. Castellana, M.Z. Wilson, Y. Xu, P. Joshi, I.M. Cristea, J.D. Rabinowitz, Z. Gitai, N.S. Wingreen, Nat. Biotechnol. **32**(10), 1011 (2014). DOI 10.1038/nbt.3018
2. Y.H.P. Zhang, Biotechnol. Adv. **29**(6), 715 (2011). DOI 10.1016/j.biotechadv.2011.05.020
3. L.J. Sweetlove, A.R. Fernie, Nat. Commun. **9**(1), 2136 (2018). DOI 10.1038/s41467-018-04543-8
4. E.V. Grgacic, D.A. Anderson, Methods **40**(1), 60 (2006). DOI 10.1016/j.ymeth.2006.07.018
5. W. Fiers, M. De Filette, K.E. Bakkouri, B. Schepens, K. Roose, M. Schotsaert, A. Birkett, X. Saelens, Vaccine **27**(45), 6280 (2009). DOI 10.1016/j.vaccine.2009.07.007
6. K.A. Henzler-Wildman, M. Lei, V. Thai, S.J. Kerns, M. Karplus, D. Kern, Nature **450**(7171), 913 (2007). DOI 10.1038/nature06407
7. J.Z. Ruscio, J.E. Kohn, K.A. Ball, T. Head-Gordon, J. Am. Chem. Soc. **131**(39), 14111 (2009). DOI 10.1021/ja905396s
8. K. Steiner, H. Schwab, Comput. Struct. Biotechnol. J. **2**(3), e201209010 (2012). DOI 10.5936/csbj.201209010
9. P.N. Depta, U. Jandt, M. Dosta, A.P. Zeng, S. Heinrich, J. Chem. Inf. Model. **59**(1), 386 (2019). DOI 10.1021/acs.jcim.8b00613

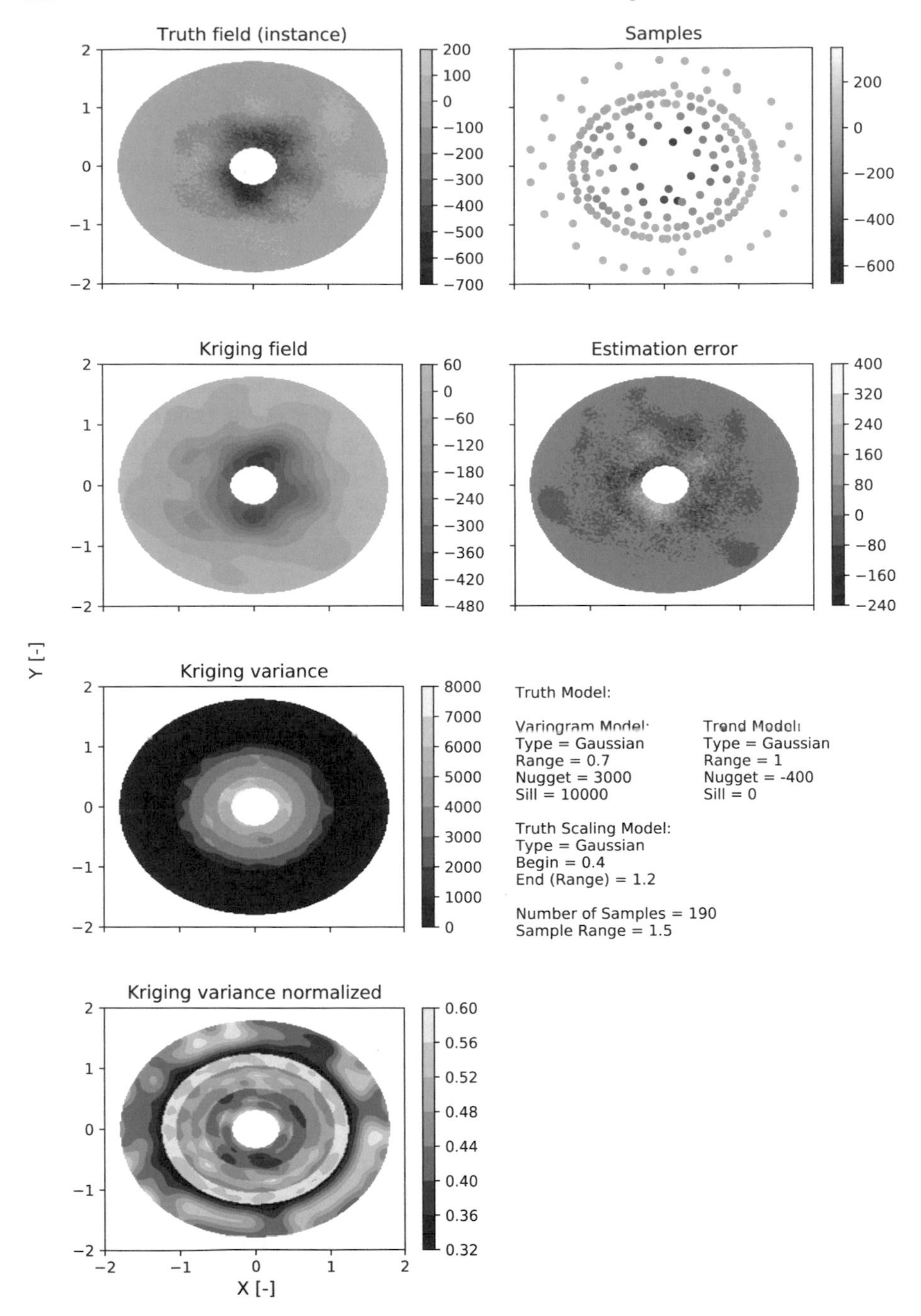

Fig. 7: 2-D Universal Kriging example after 17 iterations with 10 samples per iteration and 20 initial samples. For variogram determination the entire truth field was provided to ensure sufficient statistics.

10. M.J. Abraham, T. Murtola, R. Schulz, S. Páll, J.C. Smith, B. Hess, E. Lindahl, SoftwareX **1–2**, 19 (2015). DOI 10.1016/j.softx.2015.06.001
11. H. Berendsen, D. van der Spoel, R. van Drunen, Comput. Phys. Commun. **91**(1), 43 (1995). DOI 10.1016/0010-4655(95)00042-E
12. D.H. de Jong, G. Singh, W.F.D. Bennett, C. Arnarez, T.A. Wassenaar, L.V. Schäfer, X. Periole, D.P. Tieleman, S.J. Marrink, J. Chem. Theory Comput. **9**(1), 687 (2013). DOI 10.1021/ct3006 46g
13. S.O. Yesylevskyy, L.V. Schäfer, D. Sengupta, S.J. Marrink, PLoS Comput. Biol. **6**(6), e1000810 (2010). DOI 10.1371/journal.pcbi.1000810
14. T. Darden, D. York, L. Pedersen, J. Chem. Phys. **98**(12), 10089 (1993). DOI 10.1063/1.464397
15. S. Hezaveh, A.P. Zeng, U. Jandt, J. Phys. Chem. B **120**(19), 4399 (2016). DOI 10.1021/acs.jp cb.6b02698
16. S. Hezaveh, A.P. Zeng, U. Jandt, ACS Omega **2**(3), 1134 (2017). DOI 10.1021/acsomega.6b0 0386
17. S. Hezaveh, A.P. Zeng, U. Jandt, J. Chem. Inf. Model. **58**(2), 362 (2018). DOI 10.1021/acs.jc im.7b00557
18. S. Wynne, R. Crowther, A. Leslie, Molec. Cell **3**(6), 771 (1999). DOI 10.1016/S1097-2765(01)80009-5
19. B. Böttcher, M. Nassal, J. Mol. Biol. **430**(24), 4941 (2018). DOI 10.1016/j.jmb.2018.10.018
20. E. Lindahl, M.J. Abraham, B. Hess, D. Van Der Spoel, Zendo (2020). DOI 10.5281/ZENODO .3685920
21. N.A.C. Cressie, *Statistics for Spatial Data*, revised edition edn. (John Wiley & Sons, Inc, Hoboken, NJ, 2015)
22. R. Webster, M.A. Oliver, *Geostatistics for Environmental Scientists* (Wiley, 2007)
23. A. Lichtenstern, Kriging Methods in Spatial Statistics. Bachelorarbeit, Technische Universität München (2013)
24. M. Dosta, V. Skorych, SoftwareX **12**, 100618 (2020). DOI 10.1016/j.softx.2020.100618
25. NVIDIA Corporation, *CUDA Toolkit V11.2 Programming Guide* (NVIDIA Corporation, 2021)

Printed in the United States
by Baker & Taylor Publisher Services